Higgs Particle(s)

Physics Issues and
Experimental Searches in
High-Energy Collisions

ETTORE MAJORANA
INTERNATIONAL SCIENCE SERIES
Series Editor:
Antonino Zichichi
European Physical Society
Geneva, Switzerland

(PHYSICAL SCIENCES)

Recent volumes in the series:

A Continuation Order Plan is available for this series. A continuation order will bring delivery of
each new volume immediately upon publication. Volumes are billed only upon actual shipment.
For further information please contact the publisher.

Higgs Particle(s)

Physics Issues and Experimental Searches in High-Energy Collisions

Edited by

A. Ali
DESY
Hamburg, Federal Republic of Germany

Springer Science+Business Media, LLC

Library of Congress Cataloging-in-Publication Data

INFN Eloisatron Project Workshop on Higgs Particle(s): Physics Issues
  and Searches in High-Energy Collisions (1989 : Erice, Italy)
    Higgs particle(s) : physics issues and experimental searches in
  high-energy collisions / edited by A. Ali.
        p.   cm. -- (Ettore Majorana international science series.
  Physical sciences ; v. 50)
    "Proceedings of the Eighth INFN Eloisatron Project Workshop on
  Higgs Particle(s): Physics Issues and Searches in High-Energy
  Collisions, held July 15-26, 1989, in Erice, Italy"--T.p. verso.
    Includes bibliographical references and index.
    ISBN 978-1-4757-0910-0   ISBN 978-1-4757-0908-7 (eBook)
    DOI 10.1007/978-1-4757-0908-7
    1. Higgs bosons--Congresses.   I. Ali, A. (Ahmed) II. Title.
  III. Series.
  QC793.5.B62I54   1989
  539.7'21--dc20   90-7502
                                                        CIP

Proceedings of the Eighth INFN Eloisatron Project Workshop on
Higgs Particle(s): Physics Issues and Searches in High-Energy
Collisions, held July 15–26, 1989, in Erice, Italy

© 1990 Springer Science+Business Media New York
Originally published by Plenum Press, New York in 1990
Softcover reprint of the hardcover 1st edition 1990

PREFACE

The present volume is based on the proceedings of the 9th INFN
ELOISATRON Project workshop held at the Ettore Majorana Centre for Scientific Culture, Erice-Trapani, Italy, in the period July 15-26, 1989. The
topic of this workshop was:

Higgs Particles - Physics Issues and Experimental Searches in High
Energy Collisions

and it was attended by over forty participants.

The concept of Higgs mechanism, like most other scientific ideas,
evolved over several years, with many independent and important contributions in the sixtees to which Peter Higgs has referred in his historical
account in these proceedings. However, it was not until 1967, paraphra-
sing Weinberg, that "the right ideas were applied to the right problem"!
Higgs mechanism, i.e. spontaneous breaking of local gauge symmetries with
scalar particles becoming the longitudinal degrees of freedom of (otherwise) massless vector bosons and hence making them massive, is now an
integral part of the standard (Glashow-Salam Weinberg) theory of electroweak interactions. Fortunately for us, the structure of the standard
theory prevents at least one scalar field from being transformed away.
This is the Higgs particle in the standard theory - whose physics and
search strategies are the subject of this workshop.

The success of the standard model in explaining all electroweak
phenomena measured so far has moved the high energy physics community to
focus primarily on the missing pieces in this model - namely the Higgs
boson and the top quark. Very definitely, the top quark is heavier than
the $W^{\pm}$ boson, with direct searches at the Fermilab Tevatron setting a
lower bound of 78 GeV on its mass and indirect hints from electroweak
data suggesting its mass to lie well beyond 100 GeV. The situation with
the Higgs boson mass is less clear though recent Higgs searches at LEP
experiments have set a lower bound of 24 GeV on its mass.

Concerning the upper bounds, the top quark mass is bounded by the
consistency arguments of the electroweak data within the standard model
to lie below O(200 GeV). There is, however, no piece of data (or theoretical argument) which could persuasively be used to set an equivalent
upper bound on m_H. Arguments based on perturbative unitarity put m_H well
beyond 1 TeV though there are suggestions based on lattice studies of the
Higgs model that m_H may lie well below the unitarity limit. These arguments are discussed at length in these proceedings. With these premises,
it is not difficult to conclude that in order to get the Higgs boson,
experimental physicists may have to throw their net very widely! How
widely and in which direction must we move - these are interesting
issues. The aim of these proceedings is to contribute constructively to
the ongoing scientific debate on the best strategies of discovering the
Higgs boson (and top quark).

This book is organized in five parts. In part I, mostly theoretical issues concerning estimates of the Higgs mass and couplings (in particular self coupling) are discussed. Part II describes searches for light scalars, which include light Higgs bosons ($m_H < 10$ GeV) but also axions and dilatons. Particularly interesting are the decays of a Higgs boson having a mass in the range $2m_\pi < m_H < 2$ GeV, since the branching ratios in this mass range have considerable theoretical uncertainty. These issues are discussed in several papers here. Higgs searches in hadron collisions are discussed in part III, where search strategies for both the standard model Higgs boson as well as its counterparts in supersymmetric extensions are discussed. Because of the large lever arm in energy that future hadron colliders like the SSC and LHC (as well as the futuristic ELOISATRON machine) provide, searches for very massive Higgs ($m_H \sim$ O(1 TeV)) have received special attention here. Higgs searches in e^+e^- annihilation are summarized by Patricia Burchat in Part IV. Present experimental results, in particular those aspects related to the top quark and/or Higgs boson searches in e^+e^- and $\bar{p}p$ colliders, are summarized in part V. Attention is drawn in this part to the results of the first searches of a light Higgs boson in $\bar{p}p$ collisions at Fermilab by the CDF Collaboration This search initiates a long march on this front.

It should be pointed out here that by end of July 1989 (when this workshop ended) there were no results available from the LEP collider. However, as these proceedings are going to press, very significant bounds (especially on m_t and m_H) have become available from the SLC and LEP e^+e^- experiments. An update on some selective results has, therefore, been added to these proceedings.

The present workshop was supported by the Italian Ministry of Science and Technology, the Italian Ministry of Public Instructions, INFN ELOISATRON Project, and the Ettore Majorana Centre for Scientific Culture, Erice. The financial support of these organizations is gratefully acknowledged. I would also like to thank the warm hospitality of Prof. Antonino Zichichi at Erice. The help of Pinola Savali, Alberto Gabriele and Jerry Pilarsky made the organization and running of this (and other workshops at Erice with which I have been involved) a pleasure and lot of fun. I would like to thank them for their support. Finally, I would like to thank Frau Monika Kürzinger, of the University of Munich, for her help in the preparation of the manuscripts and Frau Helga Laudien and Frau Marianne Hausser of DESY for their secretarial help.

Ahmed Ali
Editor

Munich, Feb. 22, 1990

CONTENTS

INVENTING AN ELEMENTARY PARTICLE

Peter Higgs

Department of Physics
University of Edinburgh
Scotland

My contribution to this workshop is history not physics. I shall give a personal account of the "invention" of what a British Sunday paper recently described as "the mythical Higgs' particle". I hope to have convinced you by my presence at this meeting that I am not a myth, even if the particle named after me is.

The invention of elementary particles has usually been linked to some symmetry principle. The earliest example, Dirac's reluctant invention of the positron, involved TCP (implicitly). Others, such as the prediction of the Ω^-, have depended on approximate symmetries like flavour SU(3). The Higgs boson was a by-product of the search for viable models of elementary particles based on spontaneously broken symmetries, initiated by Nambu[1] in 1960.

I found the idea that mass splittings within particle multiplets (or even the masses themselves) might be generated by spontaneous breaking of the appropriate symmetries a very appealing one; it seemed to me that the internal symmetries of particles would be a little less mysterious if they were genuine (i.e. unbroken) symmetries at the level of Lagrangian field theory. However, as Steve Weinberg has told you, there was a serious obstacle to the realisation of Nambu's programme; this was the Goldstone theorem[2], which states that if a manifestly Lorentz invariant local field theory exhibits spontaneous symmetry breaking, it will contain massless spin-zero bosons, the non-existence of which is rather easy to establish experimentally! Thus it seemed as if spontaneous symmetry breaking was not enough; there would have to be explicit breaking as well to give mass to would-be Goldstone bosons, such as the pion in Nambu's original model.

Over the next few years a debate developed about whether the Goldstone theorem could be evaded. Anderson[3] pointed out that in a superconductor, where the broken symmetry is a local "gauge" symmetry, the Goldstone (plasmon) mode becomes massive due to the gauge field interaction, whereas the electromagnetic modes are massive (Meissner effect) despite the gauge invariance. He concluded that "the Goldstone zero-mass difficulty is not a serious one, because we can probably cancel it off against an equal Yang-Mills zero-mass problem". However, he did not discuss explicitly any relativistic model: since Lorentz

Higgs Particle(s)
Edited by A. Ali
Plenum Press, New York, 1990

invariance was a crucial ingredient of the Goldstone theorem, he had not provided a convincing demonstration that it could be evaded. Meanwhile, the Goldstone theorem had fallen into the hands of axiomatic field theorists, who proceeded to prove it with impeccable rigour.

In March 1964 Klein and Lee published a note[4] which provided the first clue to how the theorem could be evaded. They analysed the structure of the ground state expectation value of the commutator of a symmetry current with one of a multiplet of scalar fields - the object which was central to the proof of the theorem - in a "relativistic" description of a condensed matter system such as a superconductor, where the rest frame of the medium was identified by a time-like unit four-vector $\underline{n}$. They had no trouble in showing that the occurrence of $\underline{n}$ as well as the four-momentum $\underline{k}$ in the Fourier transform of this function allowed the theorem to be evaded. They speculated that it might be possible to find a truly relativistic model in which the same happened. I read their note while I was convalescing after an illness and it cheered me up considerably, but I could not see how to construct such a model.

Three months later, at the end of June, a reply from Gilbert appeared[5]. He pointed out that in a relativistic field theory with a Lorentz invariant vacuum as its ground state there is no medium whose rest frame could provide a special four-vector $\underline{n}$. Therefore the Goldstone theorem could not be evaded in this way. I got angry when I read Gilbert's paper, because I didn't want to believe it but I saw no way to bring back the banished vector $\underline{n}$.

But a few days later, early in July, it suddenly struck me that I had known for years a relativistic field theory involving just such a vector; it is called quantum electrodynamics! And the feature of this theory which permits the appearance of such a vector in the formalism, without loss of Lorentz invariance of the physical content, is gauge invariance. Thus I found a loophole in Gilbert's argument; gauge theories could evade the Goldstone theorem.

Now most people would probably not have thought of quantum electrodynamics as a theory involving such a vector. But I had long been a follower of Schwinger in preferring the Coulomb gauge formalism, which involves a choice of inertial frame $\underline{n}$ for the gauge, to the more commonly used covariant formalism, in which the photon propagator contains unphysical polarisations. More specifically, in parallel with my interest in spontaneous symmetry breaking I had been interested in an apparently unrelated problem, the connection between gauge invariance and zero-mass quanta. Schwinger[6] had shown that gauge invariance alone does not prevent the photon from being massive. He had invented a model in 1+1 dimensions in which this occurs, but he had no example of it in 3+1 dimensions. For me at this time, the most important thing which I had learned from Schwinger's papers was the most general form of the photon propagator in Coulomb gauge; it suggested to me how the vector $\underline{n}$ would occur in other vacuum expectation values, such as the one considered by Klein and Lee, in a gauge theory.

So far, all I had established was that Goldstone bosons need not occur when a _local_ symmetry is spontaneously broken in a physically Lorentz invariant theory. Obviously, the next thing to do was to find out what did happen in such a theory. Since the Goldstone phenomenon was known to occur in _classical_ scalar field theories, the simplest model to study was clearly the locally symmetric version of Goldstone's classical model. Linearizing the field equations, I saw at once that the relativistic version of the Anderson mechanism did indeed occur; the

Goldstone mode provided the third polarisation of a massive vector field. The other mode of the original scalar doublet remained as a massive scalar.

I quickly wrote a short paper, "Broken Symmetries, Massless Particles and Gauge Fields", which described how gauge theories may evade the Goldstone theorem, and submitted it to Physics Letters. It was received on 27 July and published 15 September[7]. Before writing up the work on what is now known as the Higgs model I spent a few days searching the literature to see whether it had been done before. I thought that Schwinger, in particular, might well have done something of the kind years earlier and I might have overlooked it. When I had satisfied myself that he hadn't, I wrote a second short paper, "Broken Symmetries and the Masses of Gauge Bosons", and submitted it too to Physics Letters. It was rejected. In his letter the Geneva editor wrote that it was not the kind of result which called for rapid publication in Physics Letters but that a fuller account of the work might be suitable for Il Nuovo Cimento.

I was indignant - I thought my discovery was important! My colleague Euan Squires, who spent the month of August 1984 at CERN in the Theory Division, later told me people there just didn't see the point of what I had done. In retrospect this is not surprising. In 1964 particle theory in Europe was dominated by S-matrix theory and the doctrines of Geoffrey Chew. Quantum field theory was out of fashion and I had been rash enough to base my claims on the linearized version of a classical field theory (invoking implicitly the de Broglie relations). What relevance could this possibly have to particle physics?

Realising that I had failed to sell my work sufficiently, I revised the paper and submitted it to Physical Review Letters. This time it was accepted[8], but the referee (who, I learned a few years ago, was Nambu) asked me to comment on the relation of my work to that of Englert and Brout[9], which was published in Physical Review Letters on 31 August, the day my paper was received, but of which I had been unaware since in 1964 the Brussels group did not send preprints to Edinburgh. The revised version differed from the original mainly in the last two paragraphs, the first of which sketched a model in which the broken gauge symmetry was the currently fashionable flavour SU(3): the final paragraph drew attention to the mythical bosons which are the subject of this workshop. I will quote it in full:
    "It is worth noting that an essential feature of the type of
    theory which has been described in this note is the prediction
    of incomplete multiplets of scalar and vector bosons. It is to
    be expected that this feature will appear also in theories in
    which the symmetry-breaking scalar fields are not elementary
    dynamic variables but bilinear combinations of Fermi fields".

During the following October I had discussions with Guralnik, Hagen and Kibble, who had discovered[10] how the mass of noninteracting vector bosons can be generated by the Anderson mechanism, and with Streater, who was involved in the more rigorous proofs of the Goldstone theorem. But it was not until September 1965, when I arrived in Chapel Hill on sabbatical leave at the invitation of Bryce DeWitt, that I settled down to work out the details of my Abelian model. The result of this work was my Physical Review paper[11], which appeared as a preprint in December 1965. In the New Year 1966 I received an invitation from one of the recipients of that preprint, Freeman Dyson, to give a colloquium in March at the Institute for Advanced Study. The previous summer, at the General Relativity Conference in London, Stanley Deser had invited me to give a talk at the joint seminar at Harvard some time during my year in the U.S.A., so I took the opportunity to arrange this for the day following my Princeton talk.

At tea before my Princeton talk on 15 March 1966, Klaus Hepp told me that there must be an error in my work, since Kastler, Robinson and Swieca had just proved the Goldstone theorem by C*-algebraic methods - the ultimate in rigour! Nevertheless I survived the questions of the Princeton axiomatists. Encouraged by this experience, I was ready for a rather different style of discussion the next day at Harvard. Years later, when I met Sidney Coleman again, he told me that he and his colleagues "had been looking forward to some fun tearing to pieces this idiot who thought he could get round the Goldstone theorem". Well, they did have some fun, but I had fun too!

By then I had already spent some time fruitlessly trying to construct a realistic model. The trouble was that, like so many people at that time, I was too preoccupied with the breaking of hadronic flavour symmetries: I was aware that leptonic symmetries had been proposed by various people, but I had not appreciated their significance. Shelly Glashow, in his Nobel lecture[12], said of Goldstone, Kibble and myself: "These workers never thought to apply their work on formal field theory to a phenomenologically relevant model. I had had many conversations with Goldstone and Higgs in 1960. Did I neglect to tell them about my SU(2) × U(1) model, or did they simply forget?" I should explain that I first met Glashow in 1960 at the first Scottish Universities' Summer School in Physics, where he was a participant and I was a member of the executive committee with the duties of steward. I do not recall hearing about the SU(2) × U(1) model there: my duties as steward kept me from taking part in the discussions (which continued far into the night) between Glashow, Cabibbo, Veltman and others. Glashow, referring later in the Nobel lecture to the failure of Bjorken and himself in 1964 to solve the problem of strangeness-changing neutral currents, said "I had apparently quite forgotten my earlier ideas of electroweak synthesis". He must have forgotten them still in 1966, since he was at my Harvard seminar!

My own attempts to find a phenomenologically relevant model continued after my return to Edinburgh in August 1966. Contrary to Glashow's belief, I was so absorbed in model building that I neglected to follow up the more formal aspects of spontaneously broken gauge theories. Tom Kibble's 1967 paper[13] dealt with the problem of the Goldstone theorem in the Lorentz gauge and with the structure of non-Abelian theories.

In August 1967, on my way to the "Particles and Fields" conference at Rochester, I visited my friend Michael Fisher at Brookhaven. While I was there I joined in a discussion between Steve Weinberg, David Boulware and others on the problem of understanding the $1^{\pm}$ and $0^{\pm}$ meson masses within the framework of chiral flavour symmetries. My contribution to the discussion, as I recall it, was to describe how I had found it impossible to generate these masses by spontaneous breaking of a local chiral symmetry. It was shortly after this that Steve realised (as he later put it in his Nobel lecture[14]) that he "had been applying the right ideas to the wrong problem" and turned his attention to broken leptonic symmetries and electroweak interactions.

I will conclude this talk by telling you about another contribution which I failed to make to this kind of theory. In my second 1964 paper[8] I referred to Anderson's work on superconductivity theory in a way which implied that he had discussed, in the context of this nonrelativistic system with composite scalars (Cooper pairs), all the excitations which

4

the relativistic Higgs model has. In other words, I believed that he already knew about the nonrelativistic counterpart of the Higgs boson. This was careless of me - I had read his papers some time previously and did not check them before writing mine. In 1981 (as was pointed out to me by Nambu five years ago) Littlewood and Varma of ATT Bell Laboratories published a paper entitled "Gauge invariant theory of the dynamical interaction of charge density waves and superconductivity"[15]. They extended Anderson's work to show that a recently detected, unexpected feature of the Raman spectrum of $NbSe_2$ was to be understood as due to "a massive collective mode which exists in all superconductors - the oscillation of the amplitude of the superconducting gap". This, which in fact had not been noticed by Anderson, is the nonrelativistic counterpart of a (composite) Higgs boson. So far, it is the only one which has been detected experimentally. It, at least, is no myth!

REFERENCES

1. Y. Nambu and G. Jona-Lasinio, Phys. Rev. 122, 345 (1961); 124, 246 (1961).
2. J. Goldstone, A. Salam and S. Weinberg, Phys. Rev. 127, 965 (1962).
3. P.W. Anderson, Phys. Rev. 130, 439 (1963).
4. A. Klein and B.W. Lee, Phys. Rev. Letters 12, 266 (1964).
5. W. Gilbert, Phys. Rev. Letters 12, 713 (1964).
6. J. Schwinger, Phys. Rev. 125, 397 (1962); 128, 2425 (1962).
7. P.W. Higgs, Phys. Letters 12, 132 (1964).
8. P.W. Higgs, Phys. Rev. Letters 13, 508 (1964).
9. F. Englert and R. Brout, Phys. Rev. Letters 13, 321 (1964).
10. G.S. Guralnik, C.R. Hagen and T.W.B. Kibble, Phys. Rev. Letters 13, 585 (1964).
11. P.W. Higgs, Phys. Rev. 145, 1156 (1966).
12. S.L. Glashow, Rev. Mod. Phys. 52, 539 (1980).
13. T.W.B. Kibble, Phys. Rev. 155, 1554 (1967).
14. S. Weinberg, Rev. Mod. Phys. 52, 515 (1980)
15. P.B. Littlewood and C.M. Varma, Phys. Rev. Letters 47, 811 (1981).

Triviality and Higgs Mass Bounds: A Status Report

M.A.B. Bég**

The Rockefeller University, New York, New York 10021

Abstract

We present a survey of theoretical bounds on the masses of
the Higgs boson and the top quark, focussing on the results that
stem from triviality. Background material is supplied in the
form of a review of the hypercolor/technicolor alternative--with
its freedom from triviality--to the canonical methodology; the
clumsy theoretical structure of the scenario is underlined and
the temptation to wield Ockham's razor is noted; that the subject
is nonetheless of great experimental interest is also
emphasized. Included in the extended introductory remarks is a
brief exegesis of the subject of triviality; the purpose is to
shed light on some aspects that lend themselves to clarification,
and thereby eliminate misconceptions that have been the source of
much pointless disputation; while this still leaves the bulk of
the subject shrouded in darkness, it permits identification of a
rich assortment of challenges and opportunities.

** Work supported in part by the U.S. Department of Energy
under Contract No. DE-AC02-87ER40325-Task B_1

1. <u>Genesis of Mass in the Salam-Weinberg Theory:</u>
   <u>Recapitulation of Some Well-Known Facts</u>

   Problems afflicting elementary scalar fields underlie most
of the blemishes that mar the beauty of the phenomenologically
successful standard model of elementary particle interactions[1,2]. Whether these are just superficial shortcomings, as
the word "blemish" connotes, or indicative of deeper fault lines
in the structure of the theory is a question which can not be
answered to everyone's satisfaction at this time. Attempts to go
beyond the standard model synthesis into the domain of
SU(5)-based grandunification appear to have been ill-advised and
have forced a retreat back to the standard model. That the
battle line is to be held at this point[3] is one of the basic
premises of this paper.

   The quintessential feature of the canonical methodology is
the injection of a mass into the theory through a non-vanishing
vacuum expectation value for a CP-even neutral scalar field; from
this ur-mass all masses flow. For fermions this comes about
through dimension-4 Yukawa couplings that lead to dimension-3
mass terms; for gauge fields, the mass-generation process is the
Higgs-Kibble mechanism: Goldstone bosons emerging from the
scalar sector combine with transverse Yang-Mills quanta to yield
massive spin-1 fields with three degrees of polarization.
Important for the purposes of this paper is the observation that
only <u>interacting</u> systems produce non-vanishing vacuum expectation
values and concomitant Goldstone bosons. The menace of
triviality must therefore be confronted. Before we do so, let us
note that triviality is a potential problem only in theories that
are not asymptotically free. Can we bypass the problem by
invoking symmetry-breaking schemes based on asymptotically free
theories? A tentative answer to this question is to be found in
the following section.

2. <u>Existing Alternatives to the Canonical Methodology:</u>
   <u>Reminder of Proposals That Seem Theoretically</u>
   <u>Flawed Yet Experimentally Interesting</u>

   It is tempting indeed to try to avoid the bane of
triviality[4] by switching to alternate schemes in which the
Higgs-Kibble mechanism, for symmetry-breaking <u>and</u>
mass-generation, can be realized through interactions that are
asymptotically free. One such scheme, developed for reasons
having to do with the aesthetic shortcomings of elementary scalar
fields[1,2]--as opposed to the concrete, tangible, menace of
triviality--already exists; it goes under various names[2], to wit:
technicolor, hypercolor, heavy-color etc.

In these polychromous constructs one introduces a new breed of quark, the hyperquark or techniquark which derives most of its mass from the Nambu-Goldstone mechanism associated with the spontaneous breakdown of the chiral groups in hyperflavor space, in much the same way in which the ordinary u and d quarks acquire a dynamical or infrared mass of about 300 MeV while creating a Goldstone mode--strictly speaking a pseudo-Goldstone mode that reduces to a Goldstone mode in the exact $SU(2)_L$ x $SU(2)_R$ limit, in which the current or ultraviolet masses of u and d are exactly zero--called the pion. To initiate the Higgs process that leads to masses for W^+, W^- and Z, one hyperquark doublet must have precisely zero <u>current</u> mass. Hyperquark interactions derive from QC'D, a theory analogous to QCD but with a crucial difference: whereas the parameter that sets the mass-scale of QCD is about 300 MeV, the relevant parameter in QC'D is about 300 GeV. (There can of course be other differences; for example, N_C' and N_F', the color and flavor numbers respectively in QC'D, need not be the same as in QCD; these, however, are matters of detail.)

An immediate triumph of QC'D is an automatic $|\Delta \vec{I}wk|$ = 1/2 rule, the rule that inter alia relates the neutral current parameters to those of the charged currents. In the canonical theory, it follows from the choice of a complex Higgs-<u>doublet</u>; here it emerges from the isotopic transformation property of the hyperquark condensates:

$$V(q') = \langle 0 | \bar{q}_R' \, q_L' | 0 \rangle \qquad (2.1)$$

Since the right-handed and left-handed fermi fields are singlets and doublets respectively, symmetry-breaking proceeds exactly as in the canonical theory; in particular one retrieves the sacrosanct relation:

$$\rho = m_w^2 / m_z^2 \cos^2 \theta_w = 1 \qquad (2.2)$$

m_w and m_z being the gauge field masses, θ_w the Weinberg-Salam angle ($\sin^2 \theta_w \doteq 0.22$).

There is also a rather exciting prospect, stemming from a feature shared by QC'D and QCD, to wit: paucity of parameters. Thus, many of the 19 constants, whose very existence as adjustable parameters is a serious blemish on the canonical theory, are calculable in terms of a smaller set. (We hedge, for the moment, about the exact number of these more basic constants.)

But wait. How is one to give any <u>current mass</u> to the quarks, <u>any</u> mass to the leptons? There is no way to do so. The hypercolor/technicolor scenario must therefore be enriched by injection of new interactions[2], characterized by a new mass scale, a scale sufficiently large to permit generation of masses that can simulate ultraviolet masses at energies of the order of 100 GeV. The prototype interaction can be examined grosso modo by considering two doublets:

$$\Psi = \binom{u}{d \cos \theta + s \sin \theta}_L + \binom{u'}{d'}_L \qquad (2.3)$$

θ being Cabibbo's mixing angle.

The horizontal SU(2) group generated by

$$Q_+ = \int d^3x \; (u_L^\dagger u'_L + d_L^\dagger d'_L \cos \theta + s_L^\dagger d'_L \sin \theta) \qquad (2.4a)$$

$$Q_- = Q_+^\dagger \qquad (2.4b)$$

$$Q_3 = \tfrac{1}{2} [Q_+, Q_-] \qquad (2.4c)$$

commutes with the electroweak group; gauging it permits one to transmute the constituent mass of hyperquarks into an effective current mass for quarks via the graph of Fig. 1.

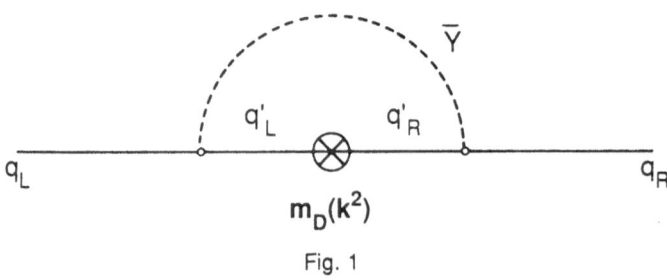

Fig. 1

Graph depicting mechanism whereby the dynamical hyperquark mass can manifest itself in the quark sector as an effective current mass at low energies. k, the internal fermion momentum, is the integration variable. Summations over all gauge fields, $\bar{Y}$--the overbar indicates that they are mass eigenfields (see second paper cited in ref. 5)--, and hyperquarks, contributing to any given element of the quark mass-matrix, are understood.

Since Q_3 contains strangeness-changing pieces, one is faced with an immediate problem: Can one generate current-like masses for all the quarks without generating unacceptably large couplings for flavor-changing neutral currents (FCNC's)? Early answers to this question were in the negative, making the FCNC problem in the enriched hypercolor/technicolor scenario one of the great problems of the day. Subsequent analyses[5] revealed that the problem was not quite so serious; one could find regions in the multi-dimensional parameter space, afforded by the plethora of Cabibbo-like mixing angles, in which there was no FCNC problem at all. Strictly speaking this is cheating; most of the parameters are calculable, and it is entirely possible that the calculated values will not fall in the trouble-free region of parameter space; in asserting that there need be no difficulty to resolve, one is hiding behind one's inability to calculate. (Incidentally, a tacit input in these analyses--that the coupling constants not vary rapidly with energy, in other words "walk" rather than "run"-- was explicitly identified as an important ingredient by several groups.[6])

Thus, while it is not possible to prove a firm no-go theorem that would enable us to dispense with the technicolor scenario--and move on--, it is a bitter pill to swallow because of its manifold theoretical shortcomings. To list but a few:

(i). The promise of reduction in the number of parameters is not fulfilled; indeed, the number actually goes up. Granted that many of these parameters are calculable, in principle; available calculational technology, however, is unequal to the task; in practice they remain incalculable--and we are faced with a proliferation of constants that can only be inferred from experiment. One is tempted to wield Ockham's razor.

(ii). Problems associated with the U(1) factor in the electroweak gauge group are in no way resolved by hypercolor/technicolor. The only way out here is to embed the electroweak group in a semi-simple gauge group; however, this generates a gauge hierarchy problem which defies solution in any dynamical framework. (Tumbling[7] is wishful thinking.)

(iii). At the electroweak level, a satisfactory hypercolor/technicolor-based working model that reproduces all features of the canonical methodology remains elusive.

Given the amount of effort expended, the last is perhaps the most telling argument against this mode of dynamical symmetry breaking. However, it is the kind of argument that _enhances_ the experimental appeal of the subject.

While searches for relatively light pseudo-Goldstone bosons, a characteristic feature of the scenario, have so far yielded only negative results, it is entirely possible that the masses of these bosons lie outside the mass-range scanned[8]. Incidentally, the neutral PGB's are CP-odd, unlike the left-over Higgs of the canonical theory which happens to be CP-even; this provides an interesting experimental handle[9]; however, to rule out the possibility that nature--in a deliberate attempt to be perverse--makes use of multi-Higgs systems with enough left-over (undevoured) Higgs-particles to mock this particular signature of hypercolor/technicolor, any PGB-sightings must be followed up with a search for hyper/techni-fermions. Such fermions are expected to have masses of the order of a TeV.

PGB-sightings plus evidence for a hyper/techni-hadron spectrum in the TeV range would add up to a rather compelling case for the hypercolor/technicolor scenario. The shortcomings listed earlier would then have to be viewed in an entirely new light! Until that happens, however, it seems prudent to stay on familiar terrain. To the canonical theory, with its elementary Higgses, we therefore return.

3. Intimations of Triviality

Forced by a singular lack of success in constructing satisfactory working models of dynamical symmetry breaking(DSB)--models that are intrinsically viable, adequate for generating the masses of _all_ massive particles without a heavy infusion of parameters that must be adjusted by hand--we retreat to elementary Higgs fields.

The situation is fraught with a measure of irony; the effect we now associate with the names of Higgs, Kibble and many other authors, conveniently abbreviating the credits to call it the Higgs mechanism, was first discovered by Anderson[10] as a _dynamical_ effect in many body theory--a context free of elementary scalars. A radical conceptual breakthrough in DSB--or, perhaps the emergence of a new idea in symmetry-breaking, of a nature as yet unknown--may break the impasse; until then, we have little choice but to accept elementary scalars as the least of many evils. However, the problem of triviality, vis-a-vis electroweak theory, then acquires a new dimension and must be confronted. To a brief sketch of the salient features of this theoretical-phenomenon[4] we now turn.

Triviality is a radical way out for a theory that can not muster a sufficiently benign asymptotic behavior to support a consistent theoretical structure; as stated earlier, it is a potential problem in theories that are not asymptotically free. (Asymptotic freedom is sufficient but not necessary to avert triviality.) Stopping short of going all the way to infinity, by arbitrarily cutting off all momenta at some finite value, is therefore always an option available for salvaging some of the structure; however, as noted elsewhere[10], one does not then have a consistent theory; at best one has a field theory based phenomenological construct that is not quite in accord with a basic principle of quantum mechanics, to wit unitarity. Cut-offs can be embarrassing indeed if they correspond to an energy accessible at accelerators.

Now the asymptotic limits of a theory, or idealized versions thereof, are determined by the properties of the β-functions that occur in the Callan-Symanzik equations[11]; this function always has a zero at the origin; in the theories under discussion, the origin is an infrared stable fixed point--relevant to the behavior in the low frequency limit. The high energy behavior is determined by the next zero of β, which would be an ultra violet stable fixed point; if no such zero exists, the theory avoids problems stemming from exploding effective couplings by choosing the path of triviality.

In the following, absence of an UV-stable fixed point will be taken to be operationally equivalent to triviality; theories than can make it to such a point of equilibrium shall be deemed to be consistent. More precisely, we are defining and restricting the discussion to a consistency class--hereinafter called the first consistency class--, the one containing theories that achieve non-triviality by having an UV-stable fixed point and attaining it. That other consistency classes may exist becomes obvious by considering theories with more than one coupling constant; limit-cycle[12] behavior and chaotic[13] behavior of the couplings, in bounded regions of coupling constant space, emerge as possible alternative paths to consistency in that neither entails infinite growth of any coupling, a necessary though not sufficient condition for the kind of inconsistency that forces triviality. (Note that an infinite value of a running coupling constant at infinite energy need not be a bar to consistency; indeed it is a feature of those theories of the first consistency class that have a fixed point at infinity.)

We proceed to consider the electrodynamic paradigm.

3.1 The U(1)-gauge theory

Triviality was first perceived as a problem in this theory, best exemplified by quantum electrodynamics, by Landau[14]. However, Landau did not have a proof; in modern language, what he had was a <u>conjecture</u> based on a renormalization group analysis in which only the one loop contribution to the Callan-Symanzik β functions had been retained. The endless, and pointless, disputations that have plagued physics in this matter appear to stem from the odd belief that a conjecture associated with the name of Landau somehow achieves the status of a proven result!

A precise statement of the conjecture is that β_{QED} has no zeros other than the one corresponding to zero coupling; if this be so, the theory can not go to an ultraviolet stable fixed point and must therefore be deemed to be mathematically inconsistent except for zero renormalized coupling--for <u>any</u> choice of renormalization point--when it reduces to a trivial theory.

The conjecture corresponds to one of the three possibilities that emerge from the analysis of Gell-Mann and Low[15]; to express them in our notation, consider the effective (or "renormalization group invariant" or "running") coupling constant $\bar{\alpha}(t)$ of the theory, at momentum p such that $t \equiv (1/2) \ln (p^2/\mu^2)$, μ being an arbitrary renormalization point; it satisfies the differential equation:

$$(d\bar{\alpha}/dt) = \beta(\bar{\alpha}); \quad \bar{\alpha}\,(0) = \alpha_R \qquad\qquad (3.1)$$

This equation is simply soluble:

$$t = \int_{\alpha_R}^{\bar{\alpha}(t)} d\alpha/\beta(\alpha) \qquad\qquad (3.2)$$

Thence the cases considered by Gell-Mann and Low [A is their case (b), B corresponds to (a), C is implicit in their paper but not explicitly stated.]:

(A) $\beta(x) = 0$ for some positive x $\left[\equiv \bar{\alpha}_{QED}\,(\infty)\right] < \infty$. $\bar{\alpha}\,(t)$ is then obviously positive and bounded and the theory is consistent. The bare charge is finite and independent of the renormalized charge; Z_γ, the photon wave function renormalization constant, is neither zero nor unity.

(B) $\bar{\alpha}\,(t)$ is positive but unbounded in the limit of infinite t; however, $\beta(x) <$ (non-zero constant). x, for large x. This corresponds to a fixed point at infinity; the bare charge is now infinite, $Z_\gamma = 0$.

(C) the integral in Eq. (3.2) converges and the theory is mathematically inconsistent

At the one loop level:

$$\beta(\bar{a}) = (2/3\pi)\ \bar{a}(t)^2 \qquad\qquad\qquad (3.3)$$

$$\bar{a}(t)^{-1} \stackrel{\circ}{=} \alpha_R^{-1} - (2/3\pi)\ t \qquad\qquad\qquad (3.4)$$

The integral in Eq. (3.2) indeed converges; positivity of $\bar{a}(t)$ can be guaranteed only if α_R^{-1} exceeds $2t/3\pi$ for <u>all</u> t; in other words $\alpha_R = 0$. Triviality thus cures inconsistency--lack of a fixed point--bearing out Landau's conjecture.

Most hard nosed physicists have stayed aloof from the entire controversy for the following reason: if one exercises the cut-off option and chooses Λ_{QED} so as to avoid reaching the ghost pole in Eq. (3.4), one finds, for $\mu \equiv m_e$ and $\alpha_R \simeq 1/137$:

$$\Lambda_{QED} \stackrel{<}{\sim} 10^{276}\ \ GeV \qquad\qquad\qquad (3.5)$$

A trans-astronomical energy indeed! The view that we are dealing here with a problem of academic philosophy not experimental science is hard to challenge. And many--Walter Thirring[16] was, I believe, the first--have observed that in a logical system devised for description of physical reality one should not insist on consistency unless the system is known to be complete--at least in some sense of the word; QED is not because at energies as low as 280 MeV a photon can dissociate into a pion pair, thereby bringing new physics in the form of strong interactions into the picture.

Recently there has been a flurry of speculation[17] concerning the possibility that QED does have a non-perturbative UV-stable fixed point and that in the strong coupling region beyond this point it leads to a Nambu-Goldstone realization of chiral symmetry in flavor space--in a manner somewhat similar to what happens in QCD--with concomitant pseudo-Goldstone bosons such as the electro-pion. It has also been suggested that confinement of charged particles occurs in the new phase of QED.

The theoretical basis for such speculation is to be found in continuum calculations[18] that retain only planar graphs and use the quenching approximation: no fermion-loop insertions in internal photon lines. Lattice simulations[19] presented in support of the fixed point proposal--the point is expected to manifest itself as a second order phase transition--also make use of the quenching approximation. Now quenching is exact only in the limit in which a function[20] $f(\alpha)$, the so-called Baker-Johnson function, vanishes; however, as Baker and Johnson themselves noted, there is then a contradiction[21]: the vanishing of $f(\alpha)$ is not compatible with the non-renormalization theorem for the

15

triangle anomaly--a circumstance that led them to state that "... [barring pathological possibilities] the finite theory does not seem to be a real possibility."

If one insists on the vanishing of $f(\alpha)$ at a real positive finite value of α--this is equivalent to the requirement that there be a finite fixed point--, one must face the possibility that a failure of the (perturbatively proven[22]) non-renormalization theorem occurs and leads to vanishing of the triangle anomaly--at least for the theory in the normal phase: $0 < \bar{\alpha} < \bar{\alpha}_{QED}(\infty)$. However, the vanishing of the anomaly in the conjectured strong coupling phase[17], $\bar{\alpha} > \bar{\alpha}_{QED}(\infty)$, may have potentially serious physical consequences. The mere existence of this phase would then pose a theoretical conundrum.

Is the new phase needed for phenomenological reasons? It has been suggested that the recently observed $e^+ e^-$ peaks may have something to do with this phase; the suggestion has been forcefully challenged[23]. The other possible application would lie in the resolution of certain problems that appear to afflict hypercolor/technicolor theories (vidé supra; Section 2) based on non-abelian gauge groups. It is not clear that switching to an abelian group is the remedy--even if one goes along with the view that the problem being tackled is in need of a solution.

To summarize: There is, at this time, no sound theoretical argument or any compelling experimental need for a finite fixed point in QED or for the existence of the much discussed new phase thereof. Landau lives! The only unassailable way out of triviality is the embedding of the U(1) gauge group in a semi simple gauge group, such as the grand unification group SU(5), thereby integrating electrodynamics into an asymptotically free theory--a procedure consistent with the early proposal of Thirring.

3.2 The $\lambda\phi^4$ theory

Triviality was recognized as a potential problem in the $\lambda\phi^4$ theory by Wilson[24]. Despite much work, since his pioneering analysis, a complete proof, rigorous at a level acceptable to all, is not yet in hand. However, given the overwhelming evidence in support of triviality, the word "rigorous" in the preceding sentence should be interpreted in the sense of Mark Kac. ["A proof," our late colleague used to say, "is sufficient for a reasonable person; a rigorous proof is needed for dealing with unreasonable individuals."]

We begin with a heuristic non-proof of the type that inspired Landau's conjecture in QED. At the one loop level,

$$\beta(\lambda) = 3\lambda^2/2\pi^2 \qquad (3.6)$$

$$\bar{\lambda}(t)^{-1} = \lambda_R^{-1} - (3/2\pi^2)\, t \qquad (3.7)$$

To maintain the positivity of $\bar{\lambda}(t)$, for all t, it becomes necessary to set:

$$\lambda_R = 0, \qquad (3.8)$$

as in QED.

More important than the similarity to the Landau "analysis" is a crucial difference: unlike the situation for QED, triviality of the $\lambda\phi^4$ theory is a problem of hard physics, not a matter of academic principle or a question of philosophy. This can be seen grosso modo through the following rough--admittedly imprecise--argument. In the Salam-Weinberg theory, the scalar self-coupling parameter is related to the mass of the left-over Higgs boson:

$$m_H = m_W \, (2\,\lambda\,\sin^2\theta_W/\pi\,\alpha)^{1/2} \qquad (3.9a)$$

$$\simeq \sqrt{\lambda}\ .\ 355\ \text{GeV} \qquad (3.9b)$$

Here α is the fine structure constant, θ_W is the Salam-Weinberg angle; the number in Eq. (3.9b) corresponds to $\sin^2\theta_W \simeq 0.22$.

If a cut-off Λ be introduced to avert triviality, then for $m_H \simeq 600$ MeV, a not-unreasonable a priori value of the Higgs mass, we find that

$$\Lambda \underset{\sim}{<} \quad 800\ \text{GeV} \qquad (3.10)$$

This energy is uncomfortably low, within the reach of existing accelerators! There is a problem to be solved, before physics can safely go forward.

Let us therefore return to the heuristic one-loop based argument leading to Eq. (3.8). Something strange happens at the two-loop level. [A similar phenomenon occurs in QED at the three loop level[21].] The β function, calculated by several authors[25], actually has a second zero:

$$\beta(\lambda) = (3/2\pi^2)\,\lambda^2 - (39/32\pi^4)\,\lambda^3 \qquad (3.11)$$

which indeed vanishes at

$$\lambda = 0 \qquad (3.12)$$

and

$$\lambda = 12.15 \qquad (3.13)$$

However, this second zero is spurious. For retention of just two loops in the loop expansion is justifiable only in the weak coupling limit; the magnitude of λ at which the second zero occurs hardly corresponds to weak coupling.

A brave attempt has been made to carry the calculation up to four loops[26], using the ε-expansion; however, the coefficient of the λ^4 term, <u>calculated in this way</u>, depends--and an answer to the question of the existence or non-existence of a fixed point hinges--on the choice of the renormalization point--not exactly a satisfactory state of affairs. As the authors of ref. 26 correctly conclude, "... the problem of a perturbative determination of a non-trivial solution of $\beta\,(\lambda) = 0$ is somewhat ambiguous."

What then do we know with certainty? A brief listing of judiciously selected results[27] that have been established with a measure of rigor follows.

It should be borne in mind though that to complete the argument leading to some of the following results, it is necessary to postulate that the solutions of the field equations are in accord with the Osterwalder-Schrader axioms[28]; this is to permit construction of the Minkowski space (real time) theory from the Euclidean formulation, the customary starting point in the functional approach. Also, unless otherwise stated, rigorous demonstrations are in hand, at this time, only for the case of a single real or complex field--that is to say, fields with no more than two components--with the theory in the symmetric phase: $\langle 0|\phi|0\rangle = 0$.

(i) Triviality cannot be proven in any finite order of perturbation theory.

This almost obvious result can be put on a firm basis, using the work of Glimm and Jaffe[29]. The failure of triviality to manifest itself in any <u>purely</u> perturbative calculational framework, of the type adhered to in canonical electroweak theory, is perhaps the principal reason for late emergence of interest in the subject. Incidentally, the situation here is not unprecedented; a perturbative expansion makes perfect sense in the Lee model[30], and leads to a non-trivial S-matrix; however, when one sums up the series, one finds that $S = 1$, and the theory is trivial.

(ii) The renormalized coupling constant lies in a bounded interval[31]:

$$0 \leq \lambda_{ren} \leq \lambda_{max} \quad \text{for } d \leq 4 \text{ (continuable to } d>4).$$

where d is the dimensionality of the space in which the theory is defined.

Thus the possibility of an upper bound on the Higgs mass--if the above inequality can be extended to the realistic case of four-component fields in the phase: $\langle 0|\phi_i|0\rangle \neq 0$ for one i.

(iii) For d>4, triviality has been established by Aizenman[32] and Frohlich[33].

(iv) The non-relativistic limit of the theory is trivial in 1+3 dimensions. The collision matrix vanishes and so does the renormalized coupling; the S-matrix thus reduces to the unit matrix.

While this result is fully consistent with the conjecture that the Lorentz-invariant $\lambda\phi^4$ theory is trivial, it does not by itself shed any light on the status of the relativistic theory. The reader who wishes to boost his way to a proof is cautioned that an essential feature of non-relativistic dynamics is the absence of production processes; in relativistic quantum field theory, however, the Aks theorem[34] tells us that scattering implies production; thus, the no-production constraint, unless it can somehow be relaxed, would--without benefit of any input about the nature of the field theory and the couplings therein--ensure a trivial theory in the relativistic domain, in a not very meaningful way.

(v) For d=4, Frohlich[33] has noted that triviality can be established if Z_3--the wave function renormalization constant--vanishes. If $Z_3 \neq 0$, the theory can be non-trivial only if it is asymptotically free[35]; in perturbation theory there is, of course, no evidence of asymptotic freedom.

(vi) The continuum limit of the lattice theory is trivial, or consistent with triviality, in <u>all</u> existing calculations[36]. When other non-perturbative calculational techniques are available, notably the 1/N expansion for the O(N)- symmetric theory, triviality again follows in the limit of infinite cut-off[37].

The preceding paragraph summarizes what appear to be, at this time, the most compelling reasons for believing that the theory is indeed trivial.

3.3 <u>Extended triviality</u>

By extended triviality[38] we shall mean triviality of the coupled $\lambda\phi^4$-theory--U(1)-gauge-field system, the theory commonly known as scalar electrodynamics. The most general formulation of this theory would, of course, include charged fermions with Yukawa couplings to the ϕ's; it may then be viewed as the high-frequency limit of the Salam-Weinberg theory--a circumstance that makes it worthy of detailed scrutiny. It is by no means obvious that nature opts for extended triviality, even if triviality be the fate of spinor electrodynamics or of the pure $\lambda\phi^4$ theory. Indeed, suggestions have been made in the literature[39,40] to the effect that it may be possible to achieve a measure of consistency, if not actually avoid extended triviality, by having the ratio $\bar{\lambda}(t)/\bar{a}(t)$ go to a fixed point in the limit $t \to \infty$. However, as was noted[10] shortly after these papers appeared, this is at best a provisional scenario albeit with a crucial feature that appears to redeem it: The cut-off required in the context of the Salam-Weinberg theory--the ratio considered there is $y(t) = (\bar{\lambda}/\bar{g}_1^2)$, g_1, being the U(1) gauge coupling, rather than $\bar{\lambda}/\bar{a}$--is fairly high[40]:

$$\Lambda_{S-W} \lesssim 10^{41} \quad \text{GeV} \tag{3.14}$$

This is to be compared with the alarmingly low value quoted earlier [Eq. (3.10)], 800 GeV, for the same parameter, and the estimate of Λ_{QED} in Eq. (3.5).

The cut-off actually used in ref. 40 was somewhat lower

$$\Lambda_{S-W} \lesssim 10^{37} \quad \text{GeV} \tag{3.15}$$

Interest in this analysis stemmed from the fact that if $y(t)$ is to attain a fixed point the parameter space of the theory can not be arbitrary; in particular, the Higgs mass is bounded from above:

$$m_H < M (\Lambda_{S-W}, m_t) \tag{3.16}$$

m_t being the top quark mass.

We defer to a later section, a discussion of these bounds; here we merely note that the cut-off dependence of parameters of <u>physical relevance</u> reopened the age-old question: What happens in the limit of infinite cut-off in a non-perturbative formulation of scalar electrodynamics?

The possibility that both couplings converge to a fixed point is discussed below, in the context of the Salam-Weinberg theory; only there can the physical implications of this alternative be fully appreciated. In so far as numerical simulations are concerned, the complete returns are not yet in; however, the evidence in hand is consistent with the conjecture that extended triviality prevails:

$$(\lambda_R, \alpha_R) \to (0,0) \text{ as } \Lambda \to \infty \tag{3.17}$$

Since renormalized quantities are, by definition, cut-off independent, there seems to be something prima facie absurd about the logical sequence in (3.17). Not so! The point is that in potentially trivial theories with positive semi-definite couplings the renormalized couplings are in general bounded from above, to ensure the positivity of the effective couplings; the

<u>bounds</u> can and do depend on the cut-off; triviality is the statement that they collapse to zero in the limit of infinite cut-off.

3.4 <u>Trivial theories as "effective low energy theories": questions of semantics(?)</u>

It has become customary to say that a theory that suffers from the affliction of triviality should be interpreted as an "effective low energy theory;" rarely is an effort made to define the terms; indeed few are aware that the words in quote marks are ambiguous.

To see this last point, consider non-relativistic potential theory. By any yardstick, it is an excellent description of physical reality at energies low enough to permit neglect of relativistic effects and phenomena such as production of new particles, at energies up to some value E_{max} beyond which new physics enters the picture--to switch to modern language. However, the theory <u>exists</u> at all energies; it simply does not correspond to what is observed beyond E_{max}; in so far as the structure of the theory is concerned, there is no need to introduce any cut-off parameter. In this sense we have an effective theory which happens to be mathematically consistent[38].

An example of a second quantized effective theory with the virtue of consistency is ordinary QCD without benefit of any top-quark field in the Lagrangian. It is well defined at all energies but can not be expected to conform to physics when top-quark effects become important. Furthermore, it can be derived from an ur-Lagrangian with top-quark fields by using the Appelquist-Carazzone decoupling theorem[41].

QED and the $\lambda\phi^4$ theory belong to a totally different genus of "effective theory." For if the cut-off is removed in these theories (that is to say, taken to infinity) their structure collapses at <u>all</u> energies.

To keep track of the distinction introduced above, we shall henceforth refer to type I and type II effective theories; non-relativistic potential theory and QCD with just light quarks are of type I; QED and the $\lambda\phi^4$-theory, with unremovable cut-offs understood, are of type II.

4. <u>The Salam-Weinberg Theory and the Trap of Triviality</u>

To study the stability of the canonical realization of the theory, inclusive of strong interactions [gauge group: $SU(3)_C$ x $SU(2)$ x $U(1)$], we seek an equilibrium configuration for the solution of the following system of equations[38], in the limit of large t:

$$d\bar{g}_i/dt = \beta_i(\bar{g}_j, \bar{\lambda}, \bar{G}) \qquad (4.1a)$$

$$d\bar{\lambda}/dt \;=\; \beta_\lambda(\bar{g}_j, \;\bar{\lambda}, \;\bar{G}) \qquad\qquad (4.1b)$$

$$d\bar{G}/dt \;=\; \beta_G(\bar{g}_j, \;\bar{\lambda}, \;\bar{G}) \qquad\qquad (4.1c)$$

Here g_j [j = 1 for U(1), 2 for SU(2), etc.] are the gauge couplings in the model, G is a generic symbol for fermion-Higgs Yukawa couplings, λ is of course the quartic Higgs self-coupling.

There are many ways in which so rich a system can go into an equilibrium configuration that would be physically acceptable if it is stable; we may have a fixed point or more than one such point, or limit cycles that might reveal their presence through a periodic fluctuation in the magnitude of effective couplings, or chaotic motion in a bounded region of parameter space with an experimental signature as yet uncharted. The absence of _any_ such configuration would imply that the theory must struggle under the cross of triviality. Quite obviously the problem of stability can be best tackled--in its full generality it may never be solved--via numerical simulations on supercomputers. Progress is impeded however by technical hurdles having little to do with the physics issues at hand, a prime example being the difficulty of putting fermions on a lattice in a satisfactory way. Nonetheless, Callaway and Petronzio[4] have taken up the challenge; when, in due course, they reach the mountain top--if given enough CRU's (computer resource units) on a CRAY, I have no doubt that they will--, they shall tell us how the cards fall. In the meantime, it is worthwhile to catalogue possible outcomes of the quest for stability and make use of available partial results to gain some insight into the nature of the solution that one might expect. To this end we therefore proceed.

4.1 Consistency through induction of asymptotic freedom

For a system of coupled fields, containing sub-systems that are by themselves asymptotically non-free, a path to consistency lies through the phenomenon of induced asymptotic freedom. The simplest illustrative example has been discussed earlier (Section 3.1): electrodynamics, through embedding in a larger gauge theory based on a semi-simple group such as SU(5), becomes asymptotically free. With enough non-abelian gauge fields, consistency can always be achieved in this way. Thus both coupling constants in the O(N)-symmetric $\lambda\phi^4$-theory coupled to an O(N)-gauge field theory vanish at infinite momentum in the large N limit. In the absence of an adequate number of non-abelian fields, one may try to achieve asymptotic freedom by imposing eigenvalue conditions[42]; however, this procedure yields consistency at the price of stability; for the domain of

attraction of the fixed point at the origin is a set of zero
measure, and the slightest perturbing force will render the
theory asymptotically non-free. (The origin, in such theories is
often referred to as an ultra-violet unstable fixed point.).

What about the Salam-Weinberg theory? The three non-abelian
gauge fields in the theory suffice to induce asymptotic freedom
in the $\bar{G}$ but not in $\bar{\lambda}$ or $\bar{g}_1^2$. While efforts to date have borne
no fruit, the possibility of embedding the theory in a larger
theory which achieves consistency for a substantial range of
input parameters remains open.

4.2 Ultraviolet limit of the canonical theory

We shall take $\bar{g}_2$ and $\bar{g}_3$ to be asymptotically free; that this
is indeed so in the full canonical theory is widely assumed but
has not been proven. To avoid complications stemming from
cross-coupling effects, all G's will be set equal to zero.
(Later on, we shall see that this approximation is invalid if the
mass of the top quark--or any other fermion--exceeds a critical
value $m^* \cong 80$ GeV.) The equations (4.1a-c) then reduce to but
two:

$$d\bar{\alpha}/dt = \beta_\alpha \ (\bar{\alpha}, \bar{\lambda}, t) \ ; \ \bar{\alpha}(0) = \alpha_R \qquad (4.2a)$$

$$d\bar{\lambda}/dt = \beta_\lambda \ (\bar{\alpha}, \bar{\lambda}, t) \ ; \ \bar{\lambda}(0) = \lambda_R \qquad (4.2b)$$

Here, we have chosen to work with $\bar{\alpha}$ rather than $\bar{g}_1^2$; the
t-dependence of the β's stems from $\bar{g}_2$ and $\bar{g}_3$.

Two dimensional systems of this type have been much studied
in the theory of non-linear oscillations[43]; we consider the
locus of the point $\{\bar{\alpha}\ (t),\ \bar{\lambda}\ (t)\}$ with increasing t
in the two-coupling plane.

4.2.1 Chaotic equilibrium

The point $(\bar{\alpha},\bar{\lambda})$ fills out a bounded region in the plane[13].

Without periodic driving terms in the β's the possibility of
realizing this mode is too remote to merit serious consideration.

4.2.2 Stable or semi-stable limit cycle

Non-linear systems have the remarkable capability of
self-exciting sustained oscillations with amplitude and period
determined by the system itself[43]. Hence the limit cycle
alternative.

A limit cycle is a closed orbit such that

$$\bar{\alpha}\ (t+t_o) = \bar{\alpha}\ (t) \qquad (4.3a)$$

$$\bar{\lambda}\ (t+t_o) = \bar{\lambda}\ (t). \qquad (4.3b)$$

t_0 being the period of the cycle. It is called semi-stable if a trajectory can fall into it either from without or from within, but not both; stable if the configuration is accessible from both sides.

This possibility warrants further scrutiny. I have nothing to say about it at this time.

4.2.3 Fixed point

The trajectory flows into a point and terminates:

$$\{\bar{a}(t), \bar{\lambda}(t)\} \rightarrow \{\bar{a}(\infty), \bar{\lambda}(\infty)\} \qquad (4.4)$$

This is the most desirable outcome in that it guarantees stability in everyone's sense of the word. We distinguish four cases[44]:

$$(i) \quad \bar{a}(\infty) < \infty, \; \bar{\lambda}(\infty) < \infty$$

We rule this out as physically unacceptable because it entails the vanishing of the $\pi^0 \rightarrow 2\gamma$ anomaly.

$$(ii) \quad \bar{a}(\infty) < \infty, \; \bar{\lambda}(\infty) = \infty$$

This is ruled out on two counts: the vanishing of the $\pi^0 \rightarrow 2\gamma$ anomaly and the known triviality of the pure $\lambda\phi^4$ theory. To see this last point, note that on the road to asymptopia the $\lambda\phi^4$ theory decouples out in this case and must make it to a fixed point on its own; since there is none for the isolated theory, it falls into the trap of triviality.

$$(iii) \quad \bar{a}(\infty) = \infty, \; \bar{\lambda}(\infty) < \infty$$

Through an argument similar to the one outlined in (ii), this case can be ruled out because of the triviality of pure QED.

$$(iv) \quad \bar{a}(\infty) = \infty, \; \bar{\lambda}(\infty) = \infty$$

If extended triviality be valid, this possibility is ruled out as well[38]. Note that extended triviality would suffice to establish the triviality of the Salam-Weinberg theory if one is allowed to ignore the non-abelian couplings $\bar{g}_2$ and $\bar{g}_3$ from the outset; the renormalization-group equations for the full theory would then be the same as for scalar electrodynamics.

4.3 Interpretation of the theory as a type -II effective theory

From the discussion in the preceding sections, we may conclude that, barring possibilities such as a limit-cycle rescue, the theory will fall into the trap of triviality unless an ultra-violet cut-off, Λ_S, be introduced[38]. This cut-off does not merely demarcate the region in which the theory is capable of reproducing physical reality, as is the case with type I theories; it is rather an unremovable cut-off; without it the entire edifice collapses. Thus a type-II effective theory--not exactly a consistent theory in the sense in which we use the words[10],--but a field theory based phenomenological construct.

4.3.1 Resolution of the anomaly problem

Contrary to naive expectation, a problem that afflicts the finite fixed point scenario (vidé supra) does not arise in the cut-off theory[44]; the $\pi^0 \to 2\gamma$ anomaly may be retrieved if $\Lambda_S \gg 250$ GeV.

4.3.2 Problems created and opportunities afforded by the cut-off

In writing down a renormalizable Lagrangian, corresponding to a formally consistent theory, one drops off terms of dimension greater than 4; there is no bar to inclusion of such terms, however, so long as they respect all the symmetries of the theory, before the cut-off needed to formulate the theory has been taken to infinity; the normal expectation is that if properly scaled they will go to zero in the limit of infinite cut-off and thus automatically guarantee the renormalizability of the primitive Lagrangian[45].

With a cut-off that stays finite, there is no compelling argument for removal of terms of dimensionality in excess of four[46]; their magnitude must be inferred, directly or indirectly, from experiment. The bright side of the coin is the opportunity afforded for exploration of interesting new physics without making any premature commitments about the logical completion of the standard model or any mechanism for the genesis of Λ_S. Thus the inclusion in the electroweak Lagrangian of a sequence of exotic gauge invariant terms:

$$\delta L_I = (g_6^2/8 \ \Lambda_s^2)(\Phi \cdot \Phi)^3$$

$$+ (g_p^2/2 \ \Lambda_s^2) \ \bar{e}^{\ C} \gamma^\lambda u_R \ \ \bar{d}^{\ C} \gamma_\lambda \ u_R$$

$$+ (g_{\mu e}^2 /2 \ \Lambda_s^2) \ \bar{e} \ \gamma^\lambda e_R \ \ \bar{e} \ \gamma_\lambda \mu_R + \ldots \quad (4.5)$$

Here Φ is the Higgs field expressed as an O(4) vector, u and d are quark fields of the indicated helicity, e the electron field and the superscript C means "charge conjugated". The second term in Eq. (4.5) will lead to proton decay via the B-L conserving process:

$$u + u \to e^+ + \bar{d} \quad (4.6)$$

The third term leads to the muon-number non-conserving process:

$$\mu^- \to e^+ e^- e^- \quad (4.7)$$

For a fuller discussion of unfamiliar--in the standard model context--of dimension six interactions, the reader is referred to ref. 46.

4.3.3 How big the stabilization cut-off?

Our discussion of anomaly retrieval combined with the obvious moral that may be drawn from the preceding paragraph leads to bounds on Λ_S:

$$250 \text{ GeV} \ll \Lambda_S < \Lambda_{NP} \qquad (4.8)$$

Λ_{NP} being the threshold for birth of new physics, physics that can not be accommodated in the canonical form of the standard model without any supplementary terms of the type introduced in Eq. (4.5). [Cf. Section 7 of ref. (38)].

5. Higgs Mass Bounds Before the Dawn of Triviality

Many attempts were made to bound the mass of the Higgs boson before the relevance of triviality to the subject was recognized; we list three that we deem significant.

5.1 The Weinberg-Linde lower bound

One loop radiative corrections lead to contributions to the effective Higgs potential, $\Gamma_H(\phi)$, that can destabilize the Higgs vacuum--that is to say, lead to $\Gamma_H(\langle 0|\phi|0\rangle) > \Gamma_H(0)$--if the self-coupling λ is too small. The requirement of vacuum-stability[47] leads to a lower bound on λ, which in turn implies a lower bound on m_H:

$$m_H{}^2 > G_F(3\sqrt{2}/16\pi^2)(2m_W{}^4 + m_Z{}^4 - 4m_t{}^4) \qquad (5.1)$$

Here, in the expectation that the top-quark may have a mass comparable to that of W and Z bosons, we have retained its contribution to the effective potential in deriving the bound; Weinberg's expression is recovered in the limit $m_t \to 0$. Eq. (5.1) may be rewritten as:

$$m_H > 6.8 \text{ GeV} \left[1-(m_t/m^*)^4\right]^{1/2} \qquad (5.2)$$

where

$$m^* \simeq 80 \text{ GeV} \qquad (5.3)$$

The bound thus vanishes at $m_t \simeq 80$ GeV, becoming imaginary for larger values of m_t. However, a very interesting effect then comes into operation: to assure positivity of the effective λ, the Higgs mass must satisfy a new lower bound[40,48] which increases rapidly with m_t. For example,

$$m_H > 140 \text{ GeV for } m_t = 150 \text{ GeV}. \qquad (5.4)$$

More about this positivity bound later.

5.2. The Lee-Quigg-Thacker upper bound

This stems from the requirement that the perturbative structure of electroweak theory be not jeopardized[49]; the requirement is given substance through the following condition: the Higgs mass is not to exceed a critical value, M_C, beyond which partial-wave unitarity is not respected by the tree diagrams for two-body scattering of gauge bosons. (Cf. the discussion in ref. 1 of the processes: $\nu_e + \bar{\nu}_e \rightarrow W^+ + W^-$, $e^+ + e^- \rightarrow W^+ + W^-$ and $Z + Z$). Thus these authors obtain:

$$m_H \leq M_C = (8\pi\sqrt{2}/3G_F)^{1/2} \simeq 1 \text{ TeV}. \qquad (5.5)$$

5.3 Upper and lower bounds from grand unificatiion scenarios

Cabibbo, Maiani, Parisi and Petronzio[48] devised a very interesting renormalization group-based technology for crossing the vast and barren wastes of Glashow's no-new-physics desert without sacrificing two cherished features of electroweak theory: its perturbative structure and the stability of its vacuum. Retention of the former feature, for example, throughout the energy range from 100 GeV to the grand unification mass is possible only if the initial value of $\bar{\lambda}$ is bounded from above; otherwise $\bar{\lambda}$ will grow to values that are unacceptably large, by the time one arrives at the end of the trail; thus the emergence of an upper bound on the Higgs mass.

Unfortunately the discussion of these authors is geared to a framework that is now deemed to be obsolete, in that nature does not seem to care for it: the (minimal) SU(5) based grand unification theory of Georgi and Glashow[1]. Terrestrial physics and astrophysics have combined forces to nail shut the SU(5) coffin[50].

Despite the collapse of the theoretical structure in which it was developed, the quintessence of the technology created by Cabibbo et al can be salvaged; this in effect is what was done by workers on triviality--even though some initially thought that they had found a brand new calculational tool. The bounds obtained by these authors are therefore best discussed in the context of triviality bounds.

6. Attempts to Obtain Non-perturbative Upper Bounds

These have not quite succeeded to date, at least not in a manner that would be deemed satisfactory by all. The reasons are manifold and may be gleaned from the following.

Numerical simulation is the principal non-perturbative method available at this time; this entails putting a discretized version of the theory on a lattice, solving for quantities of interest on a supercomputer, and recovering the continuum limit by allowing the lattice spacing to diminish and eventually extrapolating it to zero. In a purely continuum formulation, this last step corresponds to taking the ultra-violet cut-off to infinity. However, as we have noted earlier, putting fermions on a lattice leads to problems; the principal difficulty has to do with fermion doubling--often 2^n-pling, n>1--, the appearance of redundant poles in the lattice propagator that endow each fermion with a <u>Doppelganger</u> that does not go away in the continuum limit[51]; no gauge invariant procedure for removal of these unwanted fermions that does not lead to other problems, such as non local interactions, is at this time known.

To handle issues pertaining to the Higgs field, it has therefore become customary to decouple[52,53] the scalar sector from the rest of the standard model; problems pertaining to fermions are thereby neatly avoided. This, however, means solving a problem that can be solved rather than tackling the one that should be solved--somewhat in the manner of the savant who insisted on looking for his car keys under a street lamp even though he had lost them elsewhere, at a place where there was no light.

With heavy fermions (such as a top quark of mass $\geq$ 80 GeV) in the game, there is no way to justify decoupling; if there are none, one may do so for λ sufficiently large. In ref. 53 the large λ prerequisite is satisfied through imposition of a somewhat ad hoc <u>lower</u> bound on the Higgs mass:

$$m_H > G_F^{-1/2} \simeq 300 \text{ GeV} \qquad (6.1)$$

The next step is to specify the maximum momentum that the standard model is required to accommodate; the choice in ref. 53 is

$$\Lambda_s = 2\pi \, m_H \qquad (6.2)$$

which leads to

$$m_H < 640 \text{ GeV} \qquad (6.3)$$

To appraise the utility, and gauge the significance, of this bound, it is good to remember that a theory in which the cut-off is comparable to the Higgs mass does not really make much sense[52]. Improvement of the field-theoretic structure of the standard model would entail abandoning Eq. (6.2) and increasing Λ_s, to the extent that it can be done without running into new physics; since, in general, a larger Λ_s leads to a lower upper bound on m_H, this would lead to a better bound than the one quoted in Eq. (6.3). Finally, if the Higgs prefers to be weakly

28

coupled and chooses to be at a mass below that in Eq. (6.1)--as suggested in the following section, with the understanding that only experiment can settle the matter in a decisive way--the upper bound, Eq.(6.3), reduces to a tautology.

Additional objections to analyses advertised as non-perturbative have been voiced in the literature[54]; these have to do with the relation between the mass quantity studied and the physical Higgs mass--to make the identification, appeals are generally made to perturbation theory--and the use of infrared-geared technology to probe an effect stemming from the ultraviolet sector.

7. Bounds Predicated on the Loop Expansion

As we have emphasized in Section 3, triviality is a non-perturbative effect that can not be discerned in any finite order of perturbation theory. We may, however, get a handle on the problem through partial summations of the perturbation expansion; this is precisely what the Callan-Symanzik equations do for us at sufficiently high energies, where the seeds of triviality lie, if the β-functions therein are calculated through a loop expansion terminated at a finite number of loops.

To appreciate the potential of the methodology being used, consider the genesis of the pion decay parameter[2] f_π in massless QCD. Obviously f_π must vanish in every finite order of perturbation theory. However, we note that it satisfies a homogeneous Callan-Symanzik equation:

$$\mu \frac{\partial f_\pi}{\partial \mu} + \beta (g_3) \frac{\partial f_\pi}{\partial g_3} = 0 \qquad (7.1)$$

Here μ is the mass scale injected to define the Green's functions of the theory, and g_3 is the $SU(3)_C$ coupling in the notation of Section 4. If we make use of the one loop expression for β; we can solve Eq. (7.1):

$$f_\pi = C\mu \, \exp\left[-24\pi^2/(33 - 2N_F) \, g_3^2\right] \qquad (7.2)$$

N_F being the number of quark flavors, C a constant.

Observe that Eq. (7.2) corresponds to a highly non-perturbative situation, there being an essential singularity at $g_3=0$; every term in a formal Taylor expansion of f_π in powers of g_3 vanishes; yet we have managed to reproduce the expression through use of the one-loop expression for β.

With the understanding that the top quark is the only fermion with a mass non-negligible in the natural mass unit associated with weak interactions, $\langle 0|\phi|0\rangle \simeq 250$ GeV, the structure of the bounds obtained through use of one-loop expressions for the β-functions and/or the effective potential is as follows[40]:

(a) If m_t is less than a critical mass, $m^* \simeq 80$ GeV, the Higgs mass is bounded both above and below:

$$m_0 (m_t) \leq m_H \leq M_{max} (m_t, \Lambda_s), \quad (m_t < m^*) \tag{7.3}$$

where $m_0 (m_t)$ is the Weinberg-Linde bound[47],

$$m_0(m_t)^2 = (3\sqrt{2}.G_F/16\pi^2) (2m_W^4 + m_Z^4 - 4m_t^4) \tag{7.4}$$

(b) If m_t exceeds m^*, a new lower bound on m_H comes into being; it stems from the requirement that $\bar{\lambda}$ be positive, and rises rapidly with increasing m_t:

$$M_{min} (m_t, \Lambda_s) < m_H < M_{max} (m_t, \Lambda_s), \quad (m_t > m^*) \tag{7.5}$$

The Λ_s - dependence of the (upper bound)^2 is, approximately, inverse logarithmic; the lower bound is almost Λ_s independent.

(c) If m_t equals $M_F(\Lambda_s)$, the maximum fermion mass allowed in this formulation, the upper and lower bounds coalesce into one; the Higgs mass is then determined rather than bounded:

$$m_H = M_H(\Lambda_s) \quad \text{for } m_t = M_F(\Lambda_s) \tag{7.6}$$

These results are plotted in Figs. 2(a) and 2(b), for two choices of the cut-off:
$$\Lambda_s = 10^{15} \text{ GeV and } 10^{37} \text{ GeV.}$$

7.1 More on the cut-off dependence of the upper bound

One may make use of a serendipitous "compensation phenomenon", noted in ref. 38, to trace the cut-off dependence of the upper bound on the Higgs mass by analytical integration of the renormalization group equation for $\bar{\lambda}$. The region of stability of the standard model is defined by the requirement

$$\bar{\lambda} (t) < \infty. \tag{7.7}$$

This inequality constrains the maximum momentum the model can accommodate for any given $\bar{\lambda}(0)$ or, equivalently, m_H. Thus the plot of Fig. 3, in the limit $m_t = 0$; from it one may read off the Higgs-mass upper bound, $M_{max} (0, \Lambda_s)$ for any given Λ_s. [Actually, the $m_t = 0$ bound is good for a wide range of m_t; $M_{max}(m_t, \Lambda_s)$ is almost m_t independent until one gets to a mass of 80 GeV.]

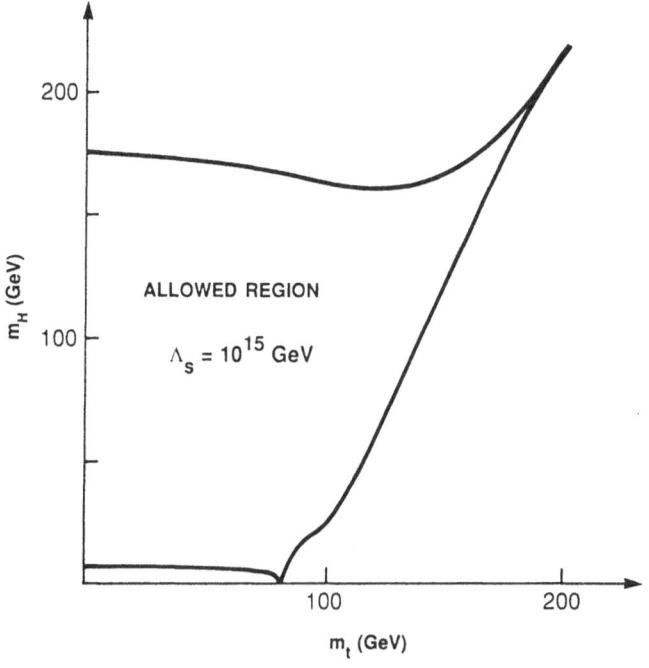

Fig. 2 (a)

Upper and lower bounds on m_H, the
Higgs boson mass, plotted as a function
of m_t, the top quark mass, under the
assumption that triviality is averted
by the emergence of new physics at
momenta $\gtrsim 10^{15}$ GeV. [Apart from minor
cosmetic changes, made to facilitate
comparison with Fig. 2(b), the plot is
that of ref. 48--where it was obtained
in the now-obsolete framework of
minimal SU(5)-based grand unification.]

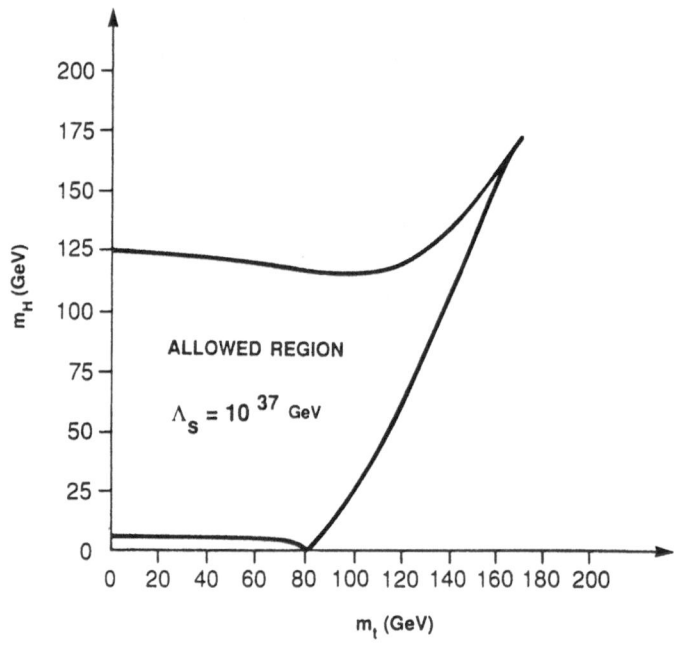

Fig. 2 (b)

The allowed values of m_H and m_t if the standard model, treated as an <u>isolated</u> system, is pushed close to the outer limits of its validity [see ref. 40]. Contrast with Fig. 2(a), and note the relatively small change in the bounds--less than sixty percent--as the cut-off is increased by twenty-two orders of magnitude!

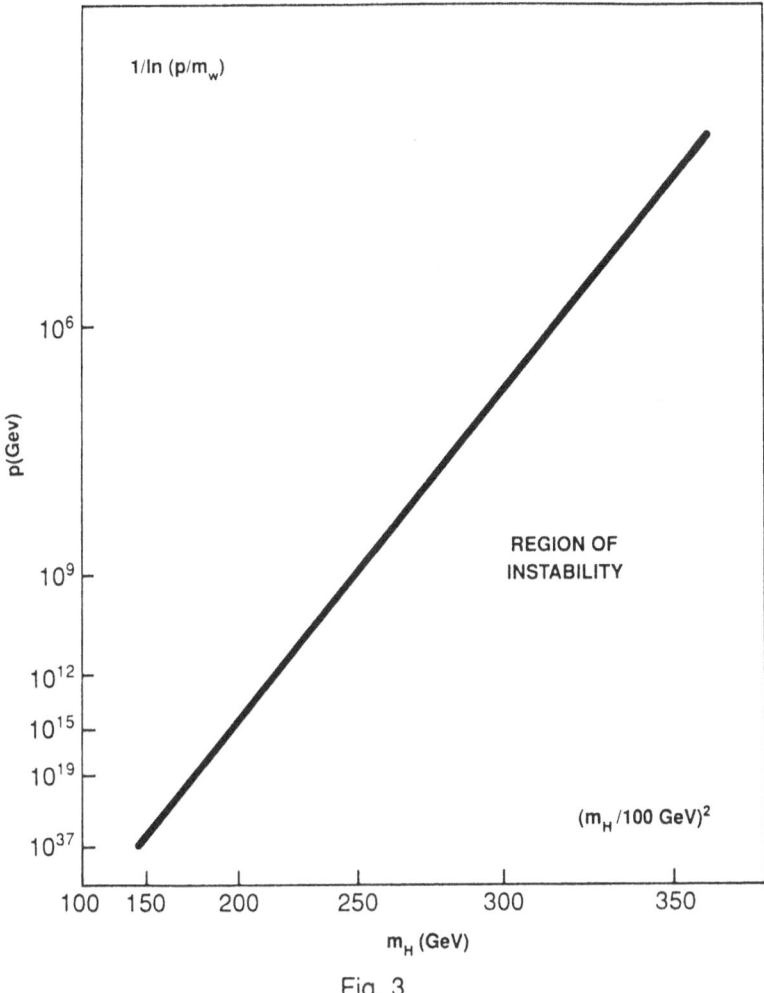

Fig. 3

The boundary of the "region of stability," approached from above, gives p_{max}, the upper bound on momenta in the standard model, for a given Higgs mass, m_H; Λ_S may thus be estimated via $\Lambda_S \lesssim p_{max}$. Note that the plot [from ref. 38] also gives upper bounds on m_H for various Λ_S; that it is a straight line is perhaps its most interesting feature. For $\Lambda_S > 10^{37}$ GeV, as well as $m_H > 350$ GeV, the calculational procedure becomes unreliable.

8. Concluding Remarks : Outlook

All Higgs mass bounds stem from the requirement that the Salam-Weinberg theory, and whatever features thereof that one's taste decrees to be compelling, be a viable theoretical construct. Triviality tells us that an effective type II theory, with a built-in stabilization cut-off Λ_S, is the best one can do--at this time.

Can one have $\underline{a}$ bound that does not require knowledge of Λ_S? Indeed yes, as has been known for at least six years[52]; the effective theory starts becoming meaningless if one reduces Λ_S to where it is comparable to m_H, becoming quite absurd at the cross-over point, $\Lambda_S = m_H$, beyond which one can not go; at this point one has the bound:

$$m_H \leq 1 \text{ TeV} \tag{8.1}$$

Three things about this bound are worth noting: (a) it is not an improvement over the earlier bounds of Lee et al[49]; (b) it can not be sensibly derived from a loop expansion for the β-function; since the expansion becomes invalid[38] for $m_H \gtrsim 355$ GeV; (c) it is in all likelihood tautological in that almost any plausible physical input leads to a vast improvement.

An extreme step, perhaps, was taken in ref. 40 where the standard model, viewed as an isolated system, was pushed to the outer limits of its validity by going to $\Lambda_S \simeq 10^{37}$ GeV. (Actually the breakdown occurs four decades later, at about 10^{41} GeV; however, the calculational procedure becomes unreliable--and, in addition, the computers start groaning--beyond the 37th power.) For not too high a top quark mass (< 80 GeV), this leads to the low upper bound:

$$m_H < 125 \text{ GeV} \tag{8.2}$$

Since Λ_S may not exceed Λ_{NP}, the threshold for the onset of new phenomena that can not be described by the standard model as it is presently conceived, one may derive upper bounds ranging from 125 GeV to 1 TeV. Thus if some of the ideas currently being discussed in the superstring context survive, new physics will emerge at the Planck mass, 10^{19} GeV, and not earlier; this corresponds to[38]:

$$m_H < 170 \text{ GeV} \tag{8.3}$$

Acceptance of grand unification at about 10^{15} GeV would lead to[48]:

$$m_H < 196 \text{ GeV} \tag{8.4}$$

If past experience in physics be any guide, it is likely that the desert will bloom earlier; in expectation thereof, results from the next generation of colliders [LEP (Large

Electron-Positron Accelerator), HERA (Hadron-Electron Ring Accelerator), SSC (Superconducting Super Collider) and the proposed TLC (Teraelectronvolt Linear Collider), CLIC (CERN Linear Collider) and the ELOISATRON (Eurasiatic Long Intersecting Storage Accelerator)] are eagerly awaited. However, I hope that I have underlined with sufficient emphasis the potential role of a low-mass Higgs as a stabilizer of the Salam-Weinberg theory, and given adequate justification for a dedicated search in the region $m_H <$ 300 GeV. Further subdivision of this region into domains corresponding to windows of visibility and invisibility at the various machines is a matter best left to the many experts assembled here. Nonetheless, I might say that mass values up to about 200 GeV are particularly interesting; to the theoretical reasons discussed at length above one should add the experimental/machine-design challenge they offer[55].

Footnotes and References

1. For an early review, with a voluminous listing of the primary sources, see: M.A.B. Bég and A. Sirlin, Annu. Rev. Nucl. Science 24, 379 (1974). Original contributions in this paper include a proposal to opt for the dynamical alternative to elementary Higgs.

2. M.A.B. Bég and A. Sirlin, Phys. Reps. 88, 1 (1982). This continuation of ref. 1 contains, inter alia, an extensive discussion of the hypercolor/technicolor scheme introduced by S. Weinberg [Phys. Rev. D13, 974 (1976) and L. Susskind [Phys. Rev. D20, 2619 (1979)]; for a dedicated review of technicolor, see: E. Farhi and L. Susskind, Phys. Reps. 74, 277 (1981).

3. Cf. H Georgi, E.E. Jenkins and E.H. Simmons, Phys. Rev. Lett. 62, 2789 (1989).

4. For a review and extensive references to the literature, see: D.J.E. Callaway, Phys. Reps. 167, 241 (1988).

5. M.A.B. Bég, Phys. Lett. B124, 403 (1983); ibid. B129, 113 (1983).

6. T. Appelquist, D. Karabali and L.C.R. Wijewardhana, Phys. Rev. Lett. 57, 957 (1986).

7. E. Farhi and L. Susskind, ref. 2.

8. S.-L. Wu, Phys. Reps. 107, 59 (1984)

9. M.A.B. Bég, H.D. Politzer and P. Ramond, Phys. Rev. Lett. 43, 1701 (1979).

10. M.A.B. Bég, in: Particles and the Universe--Proc. of the Int. Symp. held at Thessaloniki, Greece, June 24-29, 1985,

G. Lazarides and Q. Shafi eds. (North Holland, New York, 1986) p. 61.

11. C.G. Callan Jr., Phys. Rev. $\underline{D2}$, 1541 (1970); K. Symanzik, Comm. Math. Phys. $\underline{18}$, 227 (1970).

12. See, for example, N.V. Butenin, "Elements of the Theory of Nonlinear Oscillations," (Blaisdell, New York, 1965).

13. I am indebted to John Lowenstein and Mitchell Feigenbaum for enjoyable discussions on this topic.

14. L.D. Landau, in: "Niels Bohr and the Development of Physics," (McGraw Hill, New York, 1955).

15. M. Gell-Mann and F.E. Low, Phys. Rev. $\underline{95}$, 1300 (1954).

16. W. Thirring, "Principles of Quantum Electrodynamics," (Academic Press, New York, 1958). p.199.

17. See, for example, D.G. Caldi, Comments Nucl. and Part. Phys. (in press) and references cited therein; D.G. Caldi and A. Chodos, Phys. Rev. $\underline{D36}$, 2876 (1987).

18. C.N. Leung, S.T. Love and W.A Bardeen, Nucl. Phys. $\underline{B273}$, 649 (1986).

19. J. Kogut, E. Dagotto and a. Kociċ, Phys. Rev. Lett. $\underline{60}$, 772 (1988) and Urbana Preprint No.: NSF-ITP-88-139/ILL-(TH)-88-#32 (1988).

20. S.L. Adler, C.G. Callan, Jr., D.J. Gross and R. Jackiw, Phys. Rev. $\underline{D6}$, 2982 (1972).

21. M. Baker and K. Johnson, Physica (Utrecht) $\underline{96A}$, 120 (1979).

22. S.L. Adler and W.A. Bardeen, Phys. Rev. $\underline{182}$, 1517 (1969).

23. J.M. Cornwall and J. Tiktopoulos, Phys. Rev. $\underline{D39}$, 334 (1989).

24. For a review of this early work, see: K. Wilson and J. Kogut, Phys. Reps. $\underline{12C}$, 78 (1974).

25. E. Brezin, J.C. Le Guillou and J. Zinn-Justin, Phys. Rev. $\underline{D8}$, 434 (1973) [See, however, Section VI of this paper.]; I. Jack and H. Osborne, J. Phys. $\underline{A16}$, 1101 (1983); M.E. Machacek and M.T. Vaughn, Nucl. Phys. $\underline{B249}$, 70 (1985). The last two papers have caused some confusion in the literature; see, for example, K. Babu and E. Ma, Z. Phys. $\underline{C31}$, 451 (1986).

26. E. Brezin, J.C. Le Guillou and J. Zinn-Justin, Phys. Rev. $\underline{D9}$, 1121 (1974).

27. M.A.B. Bég and R.C. Furlong, Phys. Rev. $\underline{D31}$, 1370 (1985).

28. K. Osterwalder and R. Schrader, Comm. Math. Phys. $\underline{42}$, 281 (1975); V. Glaser, ibid. $\underline{37}$, 257 (1974).

29. J. Glimm and A. Jaffe, Phys. Rev. Lett. $\underline{33}$, 440 (1974).

30. T.-D. Lee, Phys. Rev. $\underline{95}$, 1329 (1954).

31. J. Glimm and A. Jaffe, Ann. Inst. Henri Poincaré $\underline{22}$, 97 (1975).

32. M. Aizenmann, Phys. Rev. Lett. <u>47</u>, 1 (1981).

33. J. Frohlich, Nucl. Phys. <u>B200</u> [FS4], 281 (1982).

34. S.Ø. Aks, J. Math. Phys. <u>6</u>, 516 (1965).

35. This follows from the fact that the dimension of the field is canonical; the theory thus asymptotically tends to one in which the two-point function is that of a free field--in other words, a free theory.

36. See, D.J.E. Callaway, ref. 4, for references up to early 1988. Recent papers upholding this result include the one cited in ref. 53 below.

37. W.A. Bardeen and M. Moshe, Phys. Rev. <u>D28</u>, 1372 (1983) and references cited therein.

38. M.A.B. Bég, Comments Nucl. Part. Phys. <u>17</u>, 119 (1987).

39. D.J.E. Callaway, Nucl. Phys. <u>B233</u>, 189 (1984).

40. M.A.B. Bég, C. Panagiotakopoulos and A. Sirlin, Phys. Rev. Lett. <u>52</u>, 883 (1984).

41. T. Appelquist and J. Carazzone, Phys. Rev. <u>D11</u>, 2856 (1975).

42. N.P. Chang, A. Das and J. Perez-Mercador, Phys. Rev. <u>D22</u>, 1429 (1980) and references therein; E.S. Fradkin and O.R. Kalashnikov, Nuovo Cim. Lett. <u>29</u>, 455 (1980) and references therein.

43. N.V. Butenin, ref. 12.

44. M.A.B. Bég, Phys. Rev. <u>D39</u>, 2373 (1989).

45. Cf. J. Polchinski, Nucl. Phys. <u>B231</u>, 269 (1984).

46. M.A.B. Bég, Comments Nucl. Part. Phys. <u>18</u>, 215 (1988).

47. S. Weinberg, Phys. Rev. Lett. <u>36</u>, 294 (1976); A.D. Linde, JETP Lett. <u>23</u>, 73 (1976).

48. N. Cabibbo, L. Maiani, G. Parisi and R. Petronzio, Nucl. Phys. <u>B158</u>, 295 (1979).

49. B.W. Lee, C. Quigg and H.B. Thacker, Phys. Rev. <u>D16</u>, 1519 (1977).

50. Absence of proton decay at the rate predicted is the principal argument against minimal SU(5). See, for example, W. Lucha, Comments Nucl. Part. Phys. <u>16</u>, 155 (1986). For cosmological objections, see: A.D. Linde, in Proc. of XXIV Int. Conf. on High Energy Physics, R. Kotthaus and J. Kuhn eds. (Springer Verlag, Berlin 1989) p. 357.

51. H.B. Nielsen and M. Ninomiya, Phys. Lett. <u>B105</u>, 219 (1981); Nucl. Phys. <u>B185</u>, 20 (1981) and <u>B193</u>, 173 (1981).

52. R. Dashen and H. Neuberger, Phys. Rev. Lett. <u>50</u>, 1897 (1983).

53. J. Kuti, L. Lin and Y. Shen, Phys. Rev. Lett. <u>61</u>, 678 (1988).

54. M.B. Einhorn and D.N. Williams, Phys. Lett. <u>B211</u>, 457 (1988).

55. R. Blankenbecler and S.D. Drell, Phys. Rev. Lett. <u>61</u>, 2324 (1988).

Lattice Studies of the Higgs System

J. Jersák

Institute of Theoretical Physics E, RWTH Aachen

D-5100 Aachen, W. Germany

and

HLRZ at KFA Jülich

D-5170 Jülich, W. Germany

Abstract

Investigation of the coupled Higgs and gauge fields on the lattice has elucidated the gauge invariant formulation and several non-perturbative aspects of the Higgs mechanism, in particular its properties for strong gauge coupling and its relationship to confinement. However, until now no indication has been found for the gauge field to inhibit the vanishing of the Φ^4 coupling in the limit of infinite cut-off. The scalar sector dominates the properties of the Higgs mechanism, and the cut-off cannot be removed.

With the gauge sector treated therefore only perturbatively, extensive analytic and numerical calculations in the pure scalar sector of the Standard Model have been performed recently on the lattice. The results indicate that the cut-off parameter in this regularization can be substantially greater than the Higgs boson mass only if this mass is not much bigger than 640 GeV, and the scalar sector is not strongly interacting.

Lattice studies of the Higgs system including fermions have been initiated. Modifications of the phase diagram due to the spontaneous chiral symmetry breaking for fermions with vectorial gauge coupling have been observed. In models with strong Yukawa coupling the fermion masses increase as the expectation value of the Higgs field decreases. Therefore a method of putting chiral fermions on the lattice in a gauge invariant way and removing the unwanted fermion doublers by means of a strong coupling of the Wilson-Yukawa type to the Higgs field is very promising and currently under investigation.

Higgs Particle(s)
Edited by A. Ali
Plenum Press, New York, 1990

1 Introduction

According to our contemporary knowledge, in the $SU(2) \otimes U(1)$ gauge theory of electroweak interactions the masses of gauge bosons and fermions are generated via the Higgs mechanism [1], which relies on the occurence of spontaneous symmetry breaking (SSB) in the 4-dimensional Φ^4 theory. However, the work of Wilson, Aizenman, Fröhlich, Lang [2] and of many others (for a review see [3]) strongly suggests (though a rigorous proof is still lacking) that this theory is non-interacting ("trivial") when a regularization parameter is removed. It seems, therefore, that we have to accept that the electroweak theory is an "effective theory". containing inherently some finite cut-off parameter Λ.

Thus it is quite natural to investigate the electroweak theory on the euclidean space-time lattice with a finite lattice spacing $a = \Lambda^{-1}$. In addition, the lattice regularization has two important technical advantages: it preserves gauge invariance [4] and allows the use of non-perturbative computational methods like the strong coupling expansion and, in particular, the numerical Monte Carlo simulation. These methods, as well as several very useful concepts (phase transition, critical behaviour, correlation, and, in particular, universality) originate in statistical mechanics, which contributes in a profound way to our modern understanding of quantum field theory.

The exciting possibility is that Λ is not far beyond the reach of the next generation accelerators, as we would then have to invent a new theory soon. An indication for such a low value of Λ would be a large mass of the Higgs boson, as the renormalized quartic self-interaction of the scalar field would have to be rather large. Similar reasoning might apply also to a large mass of the top quark or some further heavy fermion. Thus there is a strong motivation to investigate the Higgs systems non-perturbatively with the lattice methods and to determine how large the Higgs boson mass and the quartic coupling could actually be if the cut-off parameter should be substantially larger than the Higgs boson mass.

Another possibility is that non-perturbative effects of the gauge or fermion fields change the non-interacting character of the Φ^4 theory and that the cut-off can be removed. Also this requires a lattice investigation.

This lecture reviews the results and present activities in the lattice studies of the Higgs systems, both without and with fermion fields. It is intended for non-specialists in lattice gauge theories. There exist several reviews which elaborate on various aspects of the subject of this lecture in much more detail [5-13] and I shall refer to them in the corresponding parts of this text.

2 Non-perturbative properties of the lattice Higgs models

The lattice Higgs models (coupled scalar and gauge fields) have been the subject of interest since the beginning of the lattice gauge theory in 1974 [4], and the numerical studies started nearly 10 years ago. I can describe only briefly the most important results. Numerous further details can be found in the review articles [5-7,9,13].

2.1 Formulation of the Higgs systems on the lattice

The lattice Higgs models are defined by means of the path integral on the euclidean hyper-cubic lattice. Let us first summarize the notations we use for the scalar fields on such a lattice:

$x = (\vec{x}, x_4)$: sites on the euclidean space-time lattice
$V = L^4$: lattice volume or size
a : lattice constant
$\mu = 1, \ldots, 4$: directions on the lattice
$x + \mu$: site which is the nearest neighbour to x in direction μ
φ_x^α : dimensional real scalar field defined at sites
$\alpha = 0, \ldots, 3$: O(4) index (absent for the one-component field)
$\Phi_x^\alpha = a.\varphi_x^\alpha / \sqrt{2\kappa}$: dimensionless real scalar field defined at sites
$\hat{\Phi}_x$: 2×2 matrix composed of the components Φ_x^α
κ and λ : bare hopping parameter and quartic coupling
$(\Phi_{x+\mu} - \Phi_x)/a$: lattice derivative.

It is usual to interpret $\Lambda = 1/a$ or sometimes π/a (the maximal value of one component of the momentum in the Brillouin zone) as a cut-off parameter.

The Lagrangian density of a one-component scalar field on the lattice can have, up to a simple transcription of the derivative, the same form as in the euclidean continuum,

$$\sum_\mu \frac{1}{2} \left(\frac{\varphi_{x+\mu} - \varphi_x}{a} \right)^2 + \frac{1}{2} m_0^2 \varphi_x^2 + \frac{1}{4!} g_0 \varphi_x^4. \tag{2.1}$$

But on the lattice it is convenient to reparametrize the scalar field theory by rescaling the field and the quartic coupling and introducing the hopping parameter κ instead of m_0^2:

$$\varphi_x = \sqrt{2\kappa}\, \Phi_x / a. \tag{2.2}$$

$$g_0 = \frac{6\lambda}{\kappa^2}. \tag{2.3}$$

$$m_0^2 = \frac{1 - 2\lambda - 8\kappa}{a^2 \kappa}. \tag{2.4}$$

The action is then

$$S_\Phi = -2\kappa \sum_{x\mu} \Phi_x \Phi_{x+\mu} + \lambda \sum_x (\Phi_x^2 - 1)^2 + \sum_x \Phi_x^2. \tag{2.5}$$

As is usual on the lattice, we mostly set numerically $a = 1$, i.e. the dimensional quantities will be determined in lattice units.

To get some feeling for the meaning of the values of the dimensionless hopping parameter κ ($\kappa \geq 0$) we note that the case $g_0 = 0$, $m_0^2 = 0$ corresponds to $\kappa = 1/8$. The limits $m_0^2 \to +\infty$ and $m_0^2 \to -\infty$ at fixed g_0 correspond to $\kappa \to 0$ and $\kappa \to \infty$, respectively. This can be easily derived from the relations (2.3) and (2.4).

Thus we have obtained a system which is very similar to the statistical mechanics models. An approach to the continuum limit, i.e. the cut-off removal, requires that some correlation

41

length ξ grows much larger than a (scaling region). This is possible only in the vicinity of a critical point corresponding to a phase transition of second or higher order. As $m = 1/\xi$ has the physical interpretation of a mass, it means that some mass in lattice units has to approach zero, $am \to 0$. Dimensionless ratios of physical observables like masses should approach constants during the continuum limit.

The first term in the action (2.5) is the nearest neighbour (nn) coupling between the field variables. Let me remark that one can put the scalar field on a lattice in many different ways, varying e.g. the lattice geometry or including other coupling terms like a next-to-the-nearest-neighbour coupling. According to the universality hypothesis, these differences are unimportant in the scaling regions and the continuum limits should be the same.

It is instructive to consider the model (2.5) in the limit

$$\lambda \longrightarrow \infty, \quad \kappa \text{ fixed.} \tag{2.6}$$

The factor $exp\{-\lambda(\Phi_x^2 - 1)^2\}$ in each integral $\int d\Phi_x$ of the path integral vanishes unless $\Phi_x^2 = 1$. Thus in the sum over the field configurations only those with all

$$\Phi_x = \pm 1 \tag{2.7}$$

do really contribute. Therefore the path integral reduces to the partition function of the Ising model on the $d = 4$ lattice. The parameter κ corresponds to the inverse temperature in this model.

As we know from statistical physics that the Ising model has a second order phase transition, we conlude that the scalar field theory on the $d = 4$ lattice has for $\lambda = \infty$ a critical point

$$\kappa = \kappa_c = 0.0748. \tag{2.8}$$

For $\kappa < \kappa_c$ the system is disordered, whereas for $\kappa > \kappa_c$ the symmetry symmetry is broken spontaneously. $\langle \Phi \rangle \neq 0$.

It has been demonstrated non-perturbatively (cf.[14]) that a similar situation arises also for λ finite. Thus there is a line of critical points $\kappa = \kappa_c(\lambda)$ in the κ, λ plane, and equivalently in the m_0^2, g_0 plane. Therefore the lattice Φ^4 theory in $d = 4$ has two phases: the symmetric phase and the SSB phase. Warning: There is no SSB in the continuum limits of the lattice Φ^4 theory as in this limit the Φ^4 theory is presumably non-interacting in four-dimensional space-time.

A generalization to the 4-component scalar field model is straightforward. The action is in this case

$$S = -2\kappa \sum_{x\mu} \Phi_x^\alpha \Phi_{x+\mu}^\alpha + \lambda \sum_x \left[(\Phi_x^\alpha)^2 - 1 \right]^2 + \sum_x (\Phi_x^\alpha)^2, \tag{2.9}$$

where the summation over α is implied. In the $\lambda = \infty$ limit it reduces to the O(4) non-linear σ-model with $\kappa_c = 0.304$.

The gauge fields are introduced following the original proposal of Wilson [4]. The parallel transporters $U_{x,\mu}$, called link variables or simply gauge fields, are defined on the lattice links connecting the neighbour points x and $x + \mu$. Their values are from the fundamental representation G of a unitary gauge group and their relation to the gauge potentials $A_\mu(x)$ is, for small a,

$$U_{x,\mu} \simeq e^{-igaA_\mu(x)}, \tag{2.10}$$

g being the gauge coupling constant. Note that on the lattice *no gauge fixing is necessary* as the path integral is performed over the compact manifold of the gauge group. The local gauge transformations are, in the typical cases,

$$U_{x,\mu} \to G_x U_{x,\mu} G_{x+\mu}^{-1}, \tag{2.11}$$

$$\hat{\Phi}_x \to G_x \hat{\Phi}_x, \tag{2.12}$$

where $\hat{\Phi}_x$ is a suitable combination of the scalar field components transforming according to the fundamental representation G, too. There are numerous lattice Higgs models varying in the choice of the gauge group and also of the representation under which the scalar fields transform [5]. I shall mainly discuss the SU(2) Higgs model arising from the four-component scalar field model (2.9) when the SU(2) gauge field is introduced. The action is

$$\begin{aligned}
S_H = &-\frac{\beta}{4} \sum_P Tr(U_P + U_P^\dagger) \\
&-\kappa \sum_x \sum_{\mu=1}^4 \frac{1}{2}(Tr\hat{\Phi}_x^\dagger U_{x,\mu}\hat{\Phi}_{x+\mu} + c.c.) \\
&+\lambda \sum_x (Tr\hat{\Phi}_x^\dagger \hat{\Phi}_x - 1)^2 + \sum_x Tr\hat{\Phi}_x^\dagger \hat{\Phi}_x .
\end{aligned} \tag{2.13}$$

Here β is the gauge coupling parameter usually used on the lattice. It is related to the bare continuum coupling constant g,

$$\beta = \frac{4}{g^2}. \tag{2.14}$$

U_P is the product of $U_{x,\mu}$ along a lattice plaquette P and $\hat{\Phi}$ is the 2×2 matrix

$$\hat{\Phi} = \begin{pmatrix} \Phi^0 + i\Phi^1 & i\Phi^3 + \Phi^2 \\ i\Phi^3 - \Phi^2 & \Phi^0 - i\Phi^1 \end{pmatrix} \tag{2.15}$$

which is proportional to a SU(2) matrix. The O(4) symmetry of (2.9) is, due to the isomorphy O(4) $\simeq$ SU(2) $\otimes$ SU(2)/Z(2), extended to the symmetry

$$SU(2)^{(local)} \otimes SU(2)^{(global)} \tag{2.16}$$

of the action (2.13).

2.2 Phase diagram of the SU(2) Higgs model on the lattice

The first step in the investigation of a lattice model is the understanding of its phase structure. I shall only summarize what we know since several years about the model (2.13), and refer to the review articles [5, 7, 9] for details and earlier references. It is actually sufficient to understand the phase structure for a fixed λ when λ is large ($\lambda > 1$), as shown in Fig. 1. The limit cases are:
(i) $\beta = \infty$ (i.e. $g = 0$), where the O(4) Φ^4 model with its symmetric and SSB phases and the second order phase transition at $\kappa = \kappa_c(\infty)$ is found.

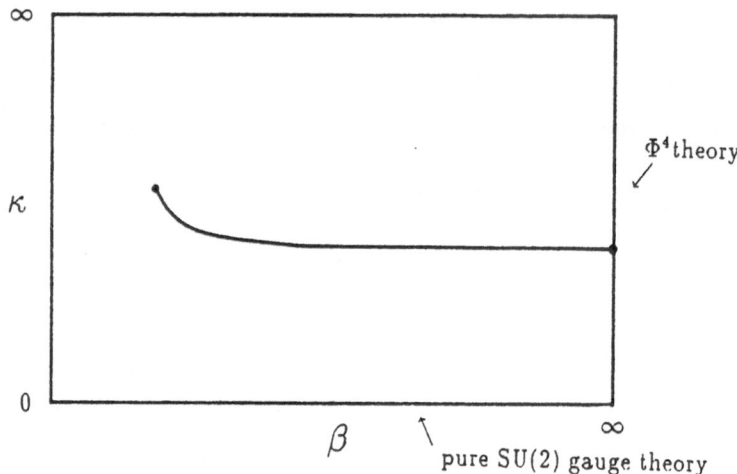

Figure 1. *Schematic phase diagram of the SU(2) Higgs model for fixed large* λ*. The line of Higgs phase transitions has critical points at both ends, otherwise it is most probably of first order.*

(ii) $\kappa = 0$ (infinitely heavy scalar field), where the model reduces to the pure SU(2) gauge field on the lattice that, as far as we know, has no phase transition and is confining for all β.
(iii) $\beta = 0$ (infinitely strong gauge coupling), where the gauge fields $U_{x,\mu}$ enter the action (2.13) only linearly and can be thus easily integrated out in the path integral. For $\lambda = \infty$ the model is thus easily solved with the result that no phase transition occurs. (This is not so for very small λ, but this fact is of no particular relevance for the continuum physics, so I will not consider that case).
(iv) $\kappa = \infty$. which corresponds to $m_0^2 = -\infty$. The scalar field is deeply in the SSB regime and, as seen in the unitary gauge, the gauge fields are frozen. Thus there is no phase transition there.

The inside of the phase diagram can now be nearly guessed: The SSB phase transition of the pure Φ^4 model at $\beta = \infty$ extends for $\beta < \infty$ as a line of Higgs phase transitions to low β-values, where it ends at certain $\beta > 0$. The region above this line, the *Higgs region*, is the place where the Higgs mechanism operates. We believe that we understand this region quite well perturbatively with gauge fixing, using e.g. the unitary gauge. It should contain the Higgs boson H and the massive gauge boson triplet W. In particular, $\langle \Phi \rangle \neq 0$.

The region below the line is the *confinement region*, called so because it has a strong resemblance with the lattice QCD. Think about QCD where the SU(3) color gauge group is replaced by SU(2) and the quarks, transforming as the fundamental representation of the gauge group, are replaced by a massive (as κ is small) scalar SU(2) doublet. Up to the spin (and baryons) the analogy is remarkable. As spin should not be relevant for the existence of confinement, this system should be confining. Let us call it "small QCD". From the physical point of view, the most interesting fact, which is established rigorously for local observables [15, 16], is the *analytic connection* between the confinement and Higgs regions.

2.3 Gauge invariant formulation of the Higgs mechanism and its relation to the confinement

If two regions of a phase diagram belong to the same phase, then they must have many similar properties, in particular the spectrum must contain the same states, though the energy differences can be quite different in various regions. Thus we should be able to understand the spectrum in the Higgs region by thinking in the QCD-like concepts. Is this possible?

The "small QCD" tells us that in the confinement region there are "mesons", consisting of scalar "quarks". Confinement implies that there is no free particle ("quark") created from the vacuum by the field Φ. The same must be true also in the Higgs region. Only states which can be produced from the vacuum by gauge invariant operators with compact support like

$$Tr \; \hat{\Phi}_x^\dagger \hat{\Phi}_x, \quad Tr \; \hat{\Phi}_x^\dagger U_{x,\mu} \hat{\Phi}_{x+\mu}, \quad \text{etc.,} \tag{2.17}$$

are expected in the asymptotic spectrum. Therefore both the Higgs particle and the intermediate W-boson have to be related to the operators (interpolating fields) like (2.17) [17-20]. This relationship has been discussed rigorously by Fröhlich, Morchio and Strocchi [21]. Note that the operators usually used in the unitary gauge are quite complicated (non-local) if written in a gauge invariant form. For example for the Higgs boson we would have to take the square root of the first expression in (2.17).

There is still another difficulty we have to clarify: The regions are connected analytically, and therefore there exists no local quantity (order parameter) which would vanish in one region and be non-zero in the other. As the model (2.13) is gauge invariant, and on the lattice no gauge fixing is needed, we can calculate $\langle \Phi \rangle$ in the Higgs region in a gauge invariant way and easily convince ourselves that

$$\langle \Phi \rangle \equiv 0 \tag{2.18}$$

identically. This is a consequence of the theorem that there is no spontaneous breakdown of a local gauge symmetry [22]. The usual assumption $\langle \Phi \rangle \neq 0$ is consistent only in some suitable, e.g. unitary gauge.

Thus the possibility of maintaining the gauge invariance forces us to rethink the perturbative picture of the Higgs mechanism, which usually uses the unitary gauge, and sharpen our understanding of the quantum field theory. For example, we are not claiming that the Higgs boson is a composite particle if it is created by a composite operator. A composite operator is a mathematically defined concept, whereas a composite particle in the quantum field theory is an intuitive concept, assuming that we can somehow observe the constituent particles and use them for explaining the properties of the composite particle. Still, the gauge non-invariant Higgs field is not an interpolating field of the Higgs boson (a gauge invariant state). They come together in the unitary gauge only.

Further observation is that we do not have a clear-cut criterion to decide whether the Higgs mechanism operates or not. The non-vanishing gauge invariant quantities like $\langle Tr \hat{\Phi}^\dagger \hat{\Phi} \rangle$, which could substitute for $\langle \Phi \rangle$ in the perturbation theory, do not vanish even below the Higgs phase transition line as they are positive definite. They are only small with respect to their values above this line.

Even more astonishing is the fact that also the W-mass am_W does not vanish below the line, though we do not expect the Higgs mechanism to operate there. Quite to the contrary, it shoots up when κ decreases [23, 24] (Fig. 2). This can be understood within the "small QCD": here W is a kind of "vector meson" which gets heavy as the "quark" mass increases with decresing κ. In the "small QCD" one can also think about "quark"-less states – the gauge invariant glueballs. A glueball with quantum numbers 1^{--} would be a sort of vector boson, too. However, its mass is non-zero, though it remains finite even in the limit $\kappa \to 0$.

One finds that there is no region in the phase diagram where some massless vector boson would exist. The perturbation analysis is, from this point of view, qualitatively wrong, as it ignores non-perturbative effects in the confinement region.

This observation led some time ago Abbott and Farhi [19] to the suggestion that the standard model could be formulated also in the confinement region, without assuming the Higgs mechanism. Indeed, such a "strongly coupled standard model" seems possible, though some problems with the spontaneous chiral symmetry breakdown might arise when fermions are included (see Subsec.4.1). What I want to stress here is that this model is not more confining than the usual Higgs mechanism.

By this statement I mean the concept of confinement in the sense of non-existence of charged states in the asymptotic spectrum. This criterion for confinement in the presence of matter fields has been formulated rigorously by Fredenhagen and Marcu [25, 9], and shown numerically to be valid in both regions of the SU(2) Higgs model [26, 27]. The difference between the confinement in both regions is, roughly speaking, that whereas in the confinemet region an introduced charge is screened by essentially only one heavy scalar "quark" carrying the anticharge as in QCD, in the Higgs region it is screened by the scalar field condensate, as in a plasma. Actually the Higgs mechanism has been suggested long ago by Mack to be a model for confinement in gauge theories [17].

2.4   Calculations of the ratio m_H/m_W in the full Higgs model

The results for m_H and m_W shown in Fig. 2 have been obtained at $\beta = 2.4$ which corresponds to a value of g which is quite far from the weak coupling strength, $g^2 \simeq 0.4$. We have to choose $\beta \simeq 8 - 10$ to approach the physically realistic region, which makes the numerical simulation difficult. Two attempts, by Anna Hasenfratz and Neuhaus [28], and by Langguth and Montvay [29], have been made to calculate the ratio m_H/m_W under these conditions. The idea was to estimate the upper bound for m_H/m_W by choosing $\lambda = \infty$, as here the renormalized quartic coupling is expected to be maximal, and then to determine for this value of λ the possible values of the mass ratio. The result of both works is consistent,

$$m_H/m_W \ \leq \ 9. \tag{2.19}$$

The problem with this result is that its reliability is not sufficiently under control. In particular, it has been pointed out [30] that the results even for smaller values of β can be sizeably influenced by the finite size of the lattice used in the numerical simulation. These finite size effects are caused by the small value of am_W for large β (am_W should vanish at $\beta = \infty$ as W then goes over into the massless Goldstone boson) which means that the correlation length is comparable with the size of the lattice. It is not yet clear how to eliminate such effects. As I will discuss in Sec.3., with the present techniques it is more realistic to study numerically the upper bound on m_H/m_W within the O(4) Φ^4 lattice model.

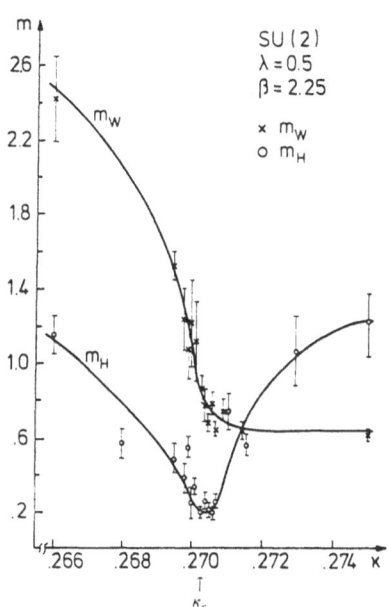

Figure 2. *The κ-dependence of the Higgs boson and W-boson masses in lattice units very close to the Higgs phase transition determined by means of the correlation functions of the operators (2.17). The W-mass is small in the Higgs region immediately above the phase transition, but it rises with increasing κ (not shown in the figure) as expected from perturbative calculations. Below κ$_c$, in the confinement region, this mass is non-vanishing too, and rises sharply if determined by means of the same correlation function as in the Higgs region. The figure is based on the data obtained in Ref. [24].*

But later one certainly should return to the investigation of the spectrum in the full Higgs model.

2.5 Search for new critical points caused by the presence of gauge fields

Lattice models differ very much from the field theories in continuum unless all correlation lengths (inverse particle masses) are large with respect to the lattice constant. In the language of statistical mechanics it means having the possibility of tuning the bare parameters to the vicinity of some critical point, i.e. of a phase transition of second (or higher) order. Such critical points have to be found. and it has to be investigated whether the behaviour of the system in their vicinity meets the requirements of the continuum physics. Putting the Higgs system on a lattice we create a complex four-dimensional statistical mechanics model (2.13) with 3 coupling parameters β, κ and λ. With the exception of D. P. Landau and J. L. Xu, who are statistical physicists [26, 27, 31], its investigation has remained in the hands of particle physicists, however.

One line of critical points is well known: it is the phase transition line $\kappa_c(\lambda)$ of the pure Φ^4 model at $\beta = \infty$. But most probably [2] the whole line is in the attraction domain of the Gaussian fixed point at $\lambda = 0$, thus yielding a trivial continuum limit. Let me mention that the inclusion of the gauge fields does not change the character of this perturbative fixed point [32]. We may hope that, due to the presence of gauge fields. at some finite β some another, non-perturbative critical point might exist which could lead to an interacting continuum theory [6, 13]. The only candidate is the manifold of the Higgs phase transition. Unfortunately, it has turned out that this transition is of first order nearly everywhere where it has been investigated more closely [5]. This is consistent with the famous calculation by Coleman and E. Weinberg [33]. Exceptions are the endpoints at small β and a region at large λ and β where the investigation is difficult.

The problem is very technical: how to distinguish a second-order phase transition from a weak first-order one on finite lattices? Statistical physicists have a lot of methods available for that, but they require a variation of the linear lattice size in large intervals. which is not possible in four dimensions.

In a recent investigation [31] a region has been found ($\lambda = 0.5$, $\beta > 2.6$) where no signal of a first order behaviour at the Higgs phase transition has been observed even in large calculations, so that some possibility that a critical point somewhere at $2.6 < \beta < \infty$ exists still remains. In an analogous SU(2) Higgs model. with the scalar field in the adjoint representation, some indication for the existence of such a point has been found by the Monte Carlo renormalization group studies [34]. But we shall not know for long.

The critical point at the lower end of the Higgs phase transition line in Fig. 1 has been recently localized rather precisely [31]. Its properties are not yet known. but it is improbable that it is of use for the electroweak theory as its position at $\beta \simeq 2$ presumably implies a too strong renormalized gauge coupling. It might be of a more general field-theoretical interest, however. though some preliminary investigation by means of the renormalization group method indicates the opposite [35].

It thus seems that the most probable region where the lattice Higgs models can contribute to the electroweak theory is that of large β. Therefore we now want to consider the Φ^4 theory on the lattice more closely.

3 Striving for precision in the Φ^4 theory on the lattice

The main conclusion from the previous section is that the gauge fields probably play only a passive role in the Higgs mechanism, and so it can be expected that this mechanism can be investigated also quantitatively in the $\beta = \infty$ limit of the Higgs models – in the Φ^4 models. Effects of the gauge fields can then be taken into account perturbatively. As suggested by Dashen and Neuberger [36], the scalar sector should be investigated non-perturbatively, however, in order to determine reliably the upper bound on the Higgs boson mass. The existence of such a bound follows from the triviality of the Φ^4 theory. But triviality à priori does not mean that the interaction is weak, it just cannot be arbitrarily strong. How strong it can be, i.e. how high the Higgs boson mass could be, is still a non-perturbative problem, as one has to investigate the extreme case of the largest possible coupling.

Some non-perturbative renormalization group calculations in continuum have been performed by Hasenfratz and Nager [37], but their reliability is not known. The problem seems to be clearly the case for numerical simulations on the lattice. However, it has been demonstrated by Lüscher and Weisz [38, 39] that in the Φ^4 models on the lattice there exists a non-perturbative *analytic* technique which provides reliable and remarkably accurate results even in this extreme situation. The results are quite competitive with high precision Monte Carlo calculations, so I will discuss both. For more details on this topics I refer to the review articles [10, 12].

Let me recall that in order to investigate the upper bound in the Dashen-Neuberger [36] approximation, we have to calculate the ratio

$$R_\Phi = \frac{m_\sigma}{F} , \tag{3.1}$$

where m_σ is the σ-boson (Higgs boson) mass and

$$F = \frac{\langle \Phi \rangle}{\sqrt{Z}} = \langle \Phi_R \rangle. \tag{3.2}$$

Z being the renormalization constant of the field Φ. As the vector boson mass is [10]

$$m_W = \frac{1}{2} g_R F, \tag{3.3}$$

where g_R denotes the renormalized SU(2) gauge coupling constant, $g_R^2 \simeq 0.4$, we have finally

$$\frac{m_H}{m_W} = \frac{2 R_\Phi}{g_R}. \tag{3.4}$$

Thus in calculations in the O(4) model we have to determine 3 quantities: $\langle \Phi \rangle$, Z and m_σ. The renormalized coupling in the broken phase can be most conveniently defined [38] as

$$\lambda_R = \frac{1}{2} R_\Phi^2. \tag{3.5}$$

which has the virtue that the relation between m_σ, F and λ_R remains the same as on the tree level.

3.1 Analytic results

In a series of papers, Lüscher and Weisz [38, 39] investigated the Φ^4 models with Z(2) and O(N) symmetries and nn coupling on an infinite hypercubic lattice, eqs. (2.5) and (2.9), both in the symmetric and broken phases. Let me describe briefly the crucial steps of their analysis:

i) The (non-perturbative) κ-expansion of several quantities like the renormalized mass m_R and the renormalized coupling λ_R is performed to the 14-th order and truncation errors estimated. It turns out that even if κ is so close below κ_c that the correlation length $\xi = m_R^{-1}$ is about $\xi \simeq 2a$, the expansion is still controlled and λ_R is already quite small.

ii) This allows to perform the second step perturbatively. They assume that the only fixed point of the Φ^4 theory is a Gaussian fixed point, i.e. that the continuum limit is non-interacting. Using the 3-loops β-function the quantities are continued by means of the renormalization group equations from the $\xi \simeq 2a$ region until $\kappa = \kappa_c$. This is consistent because the renormalization group drives λ_R to still smaller values as κ approaches κ_c. As ξ grows, the lattice effects get unimportant and the quantities show typical scaling behaviour, known from the perturbative RG analysis in the continuum theory, like

$$am_R = C \, (\beta_0 \lambda_R)^{-\beta_1/\beta_0^2} \exp(-1/\beta_0 \lambda_R). \qquad (3.6)$$

Only the coefficient C carries the non-perturbative information.

iii) Similar scaling behaviour is found also for $\kappa > \kappa_c$ close to κ_c. The corresponding coefficient C' is related to C by means of the massless Φ^4 theory valid in the critical regime at $\kappa \simeq \kappa_c$, in which the mass is introduced as a perturbation.

iv) Thus the scaling formulae carry the information further into the broken phase, until they become unreliable when the correlation length becomes again small, $\xi < 2a$. This is sufficient, as the rest of the SSB phase, which is without the scaling properties, is of no interest for the continuum physics anyhow.

The results are impressive tables where one can find the values of m_R, λ_R and the renormalization constant Z for arbitrary values of the parameters λ and κ within the scaling strip along the critical line. And all that with error estimates! Let me show for illustration the lines of constant λ_R, Fig. 3, in both phases. It shows quantitatively how lines with large values of λ_R leave the phase diagram at $\lambda = \infty$ at relatively large distances from κ_c. This is a nice illustration of the triviality of the continuum limit, as for a fixed non-vanishing λ_R the correlation length ξ remains limited and the cut-off cannot be completely removed. The theory can still be used for description of the continuum physics, provided that the cut-off is substantially larger than the physical masses and energies. It is then called "effective theory". Requiring that the cut-off is at least twice as high as m_R ($\xi \geq 2a$), Lüscher and Weisz [39] obtain in the O(4) case the upper bound

$$m_H < 630 \text{ GeV}. \qquad (3.7)$$

This result has been obtained for a particular regularization of the Φ^4 theory: with the nn coupling on a hypercubic lattice. The dependence on the regularization scheme is not known. Nevertheless, it is the first non-perturbative solution of the Φ^4 theories claiming reliability.

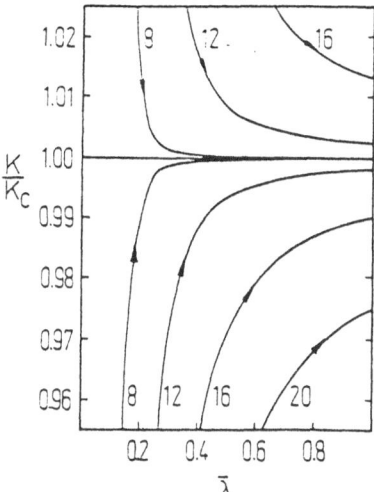

Figure 3. *Lines of constant λ_R in the Φ^4 theory obtained by Lüscher and Weisz [39]. (Interval $0 \leq \bar{\lambda} \leq 1$ corresponds to $0 \leq \lambda \leq \infty$.) As for any $\lambda_R > 0$ these lines do not meet the critical line $\kappa = \kappa_c(\lambda)$, it is impossible to construct an interacting continuum field theory.*

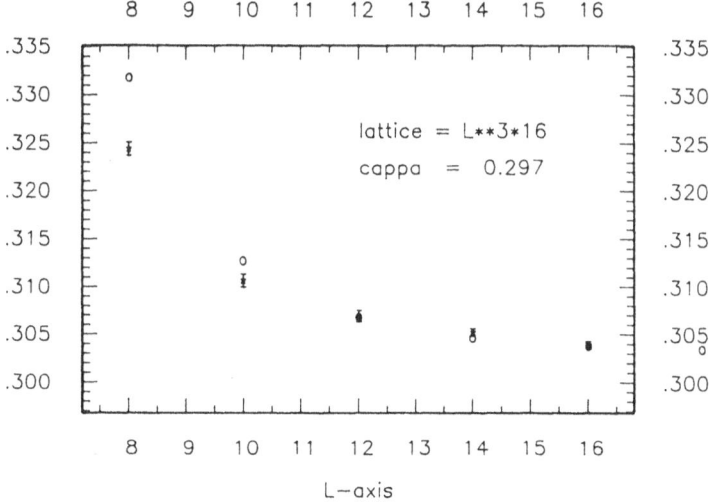

Figure 4. *The particle mass obtained in the $O(4)$ Φ^4 theory in the symmetric phase on lattices of different sizes L (asterisks). The circles indicate the results of perturbative calculations of the mass in finite volumes. The agreement for larger L allows us to interpolate the results to infinite lattices by means of the perturbative formula. The figure is from Ref. [43].*

3.2 High precision Monte Carlo calculations in some relatively simple cases

With all due respect for the analytic non-perturbative calculations, for $\xi \simeq 2a$ in the SSB phase, i.e. in the region where the upper bound is determined, the numerical simulations should be ultimately more accurate and reliable. But this requires good understanding of all phenomena which might distort the results of Monte Carlo calculations, in particular the control of the finite lattice size effects, as the calculations are performed on finite lattices, usually with periodic boundary conditions. These effects can be close to catastrophic: for example, in finite volumina there cannot be any spontaneous breakdown of symmetry, and without some precaution we shall always find $\langle \Phi \rangle = 0$ in numerical simulations even in the SSB phase.

Many finite size effects can be brought under control by analytic means, and can be even very useful. As an example, think about the mass $m(L)$ of a particle in a finite volume of linear size L with periodic boundary conditions. One vacuum polarization contribution to $m(L)$, which causes its L-dependence, consists in emitting and reabsorbing another particle which makes a trip "around the world". Such a contribution can be estimated e.g. by means of the perturbative quantum field theory in finite volumes [40]. If $m(L)$ is determined numerically in some range of L it provides an invaluable information about the scattering amplitude between the particles [40]. But most important, using the known L-dependence, one can then determine the actual mass m_R in the infinite volume without having to perform calculations on so huge lattices that the finite size effects are completely negligible.

The importance of these techniques is so high that I want at least to mention the exercises made with them in relatively simple cases of the one-component Φ^4 model and of the O(4) Φ^4 model in the symmetric phase, before turning to the real thing – to the O(4) Φ^4 theory in the SSB phase. Montvay and others [41, 42] performed a high statistics investigation of the one-component model (2.5) in the Ising limit ($\lambda = \infty$) in the symmetric phase for various L and found that using the formulae of Refs. [40] one can reliably extract am_R, λ_R and some other quantities on lattices of sizes a few times the correlation length only. This kind of analysis has been recently extended to the O(4) model in the symmetric phase (Fig. 4) [43].

The broken phase of the one-component model presents a further challenge. As a symmetry cannot be broken spontaneously for finite L, the magnetization of the configurations flips ("tunnels") during the Monte Carlo simulation between the two ground states with positive and negative magnetizations. This introduces a new kind of finite size effects which cannot be taken into account by perturbative methods, as those work always in the vicinity of one ground state only. Kuti and Shen [44] attacked the problem by performing simulations with a constraint which fixes the magnetization of all configurations to some predetermined value during the calculation. This is a method suitable for determining the effective potential [45], from which then m_R and λ_R can be derived. Another approach introduces an external field to stabilize the system [46]. In an elaborate investigation in the Ising limit [47, 48] an opposite attitude has been adopted and the flip rate has been wastly enhanced by means of a wonderful non-local cluster algorithm invented by Swendsen and Wang [49, 42], which can flip whole clusters of spins simultaneously. This allows one to investigate the flip phenomenon quantitatively, and, using its relation to the quantum mechanical tunnel effect, to get it under analytic control. Then, again, one can determine m_R and λ_R in the infinite volume limit [48].

The merit of these investigations is twofold: First, they represent a substantial methodological progress in numerical simulations of quantum field theories. Second, when the observables obtained in Monte Carlo calculations are correctly extrapolated to the infinite volume limit, they agree to an impressive degree with the analytic results of Lüscher and Weisz for the Φ^4 models. This is a demonstration of the reliability of both methods.

3.3 $\langle \Phi \rangle$ and Z in the broken phase of the O(4) model

A naive attempt to determine $\langle \Phi \rangle$ in the broken phase of the O(4) Φ^4 model by calculating $\langle \Phi^\alpha \rangle$ as an average over configurations produced in a Monte Carlo run ends up soon even on quite large lattices with a value consistent with zero. The direction of the magnetization of individual configurations,

$$M^\alpha = \frac{1}{V} \sum_x \Phi_x^\alpha, \tag{3.8}$$

drifts quickly in the O(4) space due to the presence of 3 light Goldstone bosons.

The most obvious remedy is to consider the positive definite quantity $|M^\alpha|$, and to assume that

$$\langle \Phi \rangle \simeq \langle |M^\alpha| \rangle. \tag{3.9}$$

This has been done in some simulations [50-52]. Alternatively, one can use also the effective potential method [51, 53, 54]. Unfortunately, it is not à priori clear how reliable these approaches are. One can imagine that on larger lattices domains of different magnetization could form which would distort the relation (3.9).

Neuberger [55, 56] was the first who noticed that for the extraction of F (eq. (3.2)) from the data obtained in the O(4) Φ^4 model on finite lattices one can exploit in finite volumes the non-linear σ-model usually applied for description of the low energy properties of Goldstone bosons ("pions"). Actully, several very useful analytic relations due to Gasser and Leutwyler [57, 58] and based on the low energy effective pion Lagrangian have been available in the context of the study of finite size effects in the lattices QCD, and more are to come [59-61]. It just took some time before it has been realized [52,62,10] that they apply to the O(4) Φ^4 model in the broken phase as well. I shall describe only the idea and the most important formula. More details can be found in Refs. [58, 10] and in a work in preparation [59].

The idea behind the method of Gasser and Leutwyler is that the finite size effects in systems with spontaneous breakdown of a continuous symmetry are mainly due to the Goldstone bosons of a very small mass m_G. This assumes that some other possibly present particles – in our case the σ-particle of mass m_σ – are much heavier than m_G, and that the lattice is sufficiently large so that σ itself causes no finite size effects. This amounts to the requirements

$$m_\sigma \gg m_G , \quad 1/m_\sigma \ll aL. \tag{3.10}$$

Then the system is well approximated by the low energy effective Lagrangian for the Goldstone bosons, in our case the non-linear O(4) σ-model. The model is stabilized against the drift of the system through the set of degenerate ground states by the introduction of a small constant external source j breaking the O(4) symmetry explicitly (inserting the term $j \sum_x \Phi_x^0$ into the action (2.9)). It gives the Goldstone bosons in the infinite volume a non-vanishing mass $m_G \propto j$. The two free parameters of the model are F^2, entering as the coupling constant, and $\langle \Phi \rangle$, which determines the effective strength of the interactions with the external source j. The Langrangian density is ($\varphi = \sqrt{2\kappa}\Phi/a$)

$$\mathcal{L}(x) = \frac{1}{2}F^2 \sum_{\alpha=0}^{3} (\partial_\mu U^\alpha(x))^2 \; - \; j\langle\varphi\rangle U^0(x), \qquad \sum_{\alpha=0}^{3} U^\alpha(x)U^\alpha(x) = 1. \qquad (3.11)$$

This model is then used in the continuum euclidean space-time of finite volume L^4. In the analytic calculations the quantities $1/L^4$ and j are treated as small expansion parameters of the same order, and suitable power expansions in these quantities are performed. In this way we get useful formulae which describe the properties of the model for $L < \infty$ and $j > 0$ in terms of F and $\langle\varphi\rangle$, the parameters of the infinite volume model with $j = 0$.

An example of such formulae is, in the lowest order of the power expansion, the relation

$$jL^4 \, \langle\varphi^0_x\rangle_{L,j} = u^2\eta(u), \qquad (3.12)$$

where

$$u = \langle\varphi\rangle \, jL^4 \qquad (3.13)$$

and

$$\eta(u) = \frac{1}{u}\,\frac{I_2(u)}{I_1(u)}. \qquad (3.14)$$

Here I_1 and I_2 are Bessel functions of imaginary argument and $\langle\varphi^0_x\rangle_{L,j}$ is the expectation value of the scalar field for finite L and non-vanishing j. Notice that these formulae include the standard expression for the magnetization in the SSB case, namely

$$\langle\varphi\rangle = \lim_{j\to 0} \lim_{L\to\infty} \langle\varphi^0_x\rangle_{L,j}. \qquad (3.15)$$

The immediate conclusion is that if such formulae are to be used, then one should simulate on the lattice the Φ^4 model with an external source (which does not need to be given any physical interpretation). The numerical determination of $\langle\varphi^0_x\rangle_{L,j}$ is straightforward and doing it only for one pair of values of the parameters L and non-vanishing j is, in principle, sufficient to determine $\langle\varphi\rangle$ in the infinite volume and $j = 0$! Let me add that numerical simulations for $j > 0$ are much easier than those for $j = 0$ as the ground state drift is suppressed.

Can such a miraculous method be correct? The obvious test is to fix the bare parameters κ and λ and to perform calculations in the O(4) Φ^4 theory with the external source j for different values of j and L comparing the results for $\langle\varphi\rangle$. This has been done in a recent study [62, 63] and a typical result is presented in Fig. 5. The figure shows very different values of $\langle\varphi^0_x\rangle_{L,j}$ obtained in simulations with various L and j for one κ-value (we have taken $\lambda = \infty$). The curves correspond to the expression for $\langle\varphi^0_x\rangle_{L,j}$ obtained from (3.12) with only one common value of $\langle\varphi\rangle$. As the agreement is excellent, we are confident that $\langle\varphi\rangle$ has been determined reliably. Its value is indicated by the bold dot in the figure. Please, notice how different from $\langle\varphi^0_x\rangle_{L,j}$ it is.

Expressions analogous to eq. (3.12) exist also for the propagators, and can be used for the determination of Z. Also these formulae have been tested successfully [62, 63]. So we conclude that the Gasser-Leutwyler formulae are applicable to the lattice O(4) Φ^4 theory in the SSB phase. The results for $\langle\varphi\rangle$ are also consistent with those obtained by means of $|M^\alpha|$, eq. (3.8), which increases the confidence in earlier results obtained by methods based on the relation (3.9) [50-52]. It should be pointed out, however, that the κ-range already investigated is still quite distant from the critical point $\kappa_c = 0.304$. We have kept $am_\sigma > 0.8$ in order to satisfy safely the constraints (3.10). Further studies closer to κ_c are necessary.

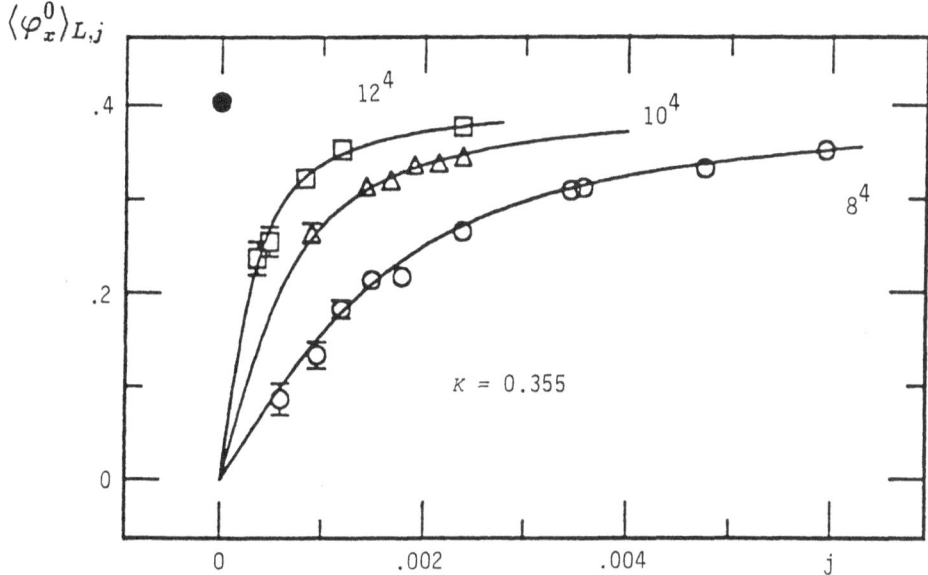

Figure 5. *All data for* $\langle \varphi_x^0 \rangle_{L,j}$ *obtained for various lattice sizes* L *and external sources* j *agree with values obtained from* $\langle \varphi \rangle$ *(the dot in the upper left corner) by means of the Gasser-Leutwyler formula (3.12) (including some higher order corrections). The figure is from Ref.* [62].

3.4   Higgs boson mass

The two groups which have recently determined the upper bound on the Higgs boson mass in large scale numerical simulations of the O(4) Φ^4 theory on the lattice use different techniques for determining the σ-particle mass m_σ (which is then identified with the Higgs boson mass m_H). One way is to look at the decrease of a two-point function in the configuration space with the distance τ and fit it to the form $A + Be^{-m_\sigma \tau}$ corrected for the lattice periodicity [50, 52, 63]. Another method is to determine the Fourier transformation of the two-point function and compare it with the free particle propagator on the lattice [51, 53, 54]. The second group determines m_σ also by means of the effective potential. The results are consistent on the precision level which presumably suffices for the determination of the upper bound. They also agree with the results of Lüscher and Weisz for the O(4) model [39]. But some details remain to be clarified.

One of them is the dependence of m_σ on the lattice size L. The currently available analytic results within the Gasser-Leutwyler method ignore the σ-particle completely. Kuti, Lin and Shen [64] investigated the L-dependence of m_σ by means of the renormalized perturbation theory for the effective potential obtaining

$$m_\sigma(L) \simeq m_\sigma(\infty) + \frac{A}{L^2}. \qquad (3.16)$$

This formula gives sizeable (about 5 per cent) correction to the values of $m_\sigma(L)$ obtained on lattices of various sizes up to 16^4 [63]. But the formula has not yet been sufficiently verified by means of a systematic investigation of the L-dependence of $m_\sigma(L)$ in analogy to the simpler cases of the symmetric phase [43] or of the Z(2) model [41, 47, 48]. So we have an uncertainty of possibly a few percent in m_σ.

This is related to a methodological problem: how to determine $m_\sigma(L)$ on relatively small lattices with a high accuracy needed for a study of the validity e.g. of the formula (3.16) and for a reliable extrapolation of $m_\sigma(L)$ to $L = \infty$. Due to the presence of the light Goldstone bosons the spectrum of the model in finite volumes is very complex. As has been demonstrated within the framework of the non-linear σ-model, the system has various excited states with energies which can be close to $m_\sigma(L)$ [65]. (The spectrum is actually described by the spectrum of a quantum-mechanical particle moving on the manifold of the SU(2) group and is quite different from a multi-Goldstone spectrum of the infinite volume theory even if the Goldstone particles are assumed to have a finite mass.) Any correlation function suitable for a determination of $m_\sigma(L)$ will get contributions also from some of these states. We do not yet know how to separate these contributions which might distort the results for $m_\sigma(L)$.

The third problem concerns the instability of the Higgs particle. On the lattice the σ-particle does not have any decay channel open until L is so large that the energy of two Goldstone bosons with the lowest non-zero momentum $2\pi/aL$ is approximately equal to m_σ.

$$am_\sigma \simeq 2 \cdot \frac{2\pi}{L}. \qquad (3.17)$$

For $am_\sigma \simeq 0.5$ this amounts to $L \simeq 25$, much too large sizes with respect to what has been possible until now. So we cannot yet investigate the effects of instability of the Higgs boson on the lattice, though the theoretical analytic framework for such studies is ready [66, 67]. Fortunately, it seems that for the Higgs boson masses allowed by the upper bound of not much more than 600 GeV, the decay width will be no more than about 20 per cent of the mass [38, 10] and its effects thus should not be very important.

3.5 Upper bound on the Higgs mass and tests of the Standard Model

The results on the upper bound obtained by Monte Carlo calculations of the group around Kuti [51, 53, 54], by our collaboration [50, 52, 63] and by analytic methods [39] agree very well if L-dependence of m_σ of the form (3.16) in the numerical calculations on finite lattices is assumed. For definiteness I quote the result of Kuti et al.:

$$m_H \leq (640 \pm 40) GeV, \qquad (3.18)$$

if the ratio of the cut-off to the Higgs mass, Λ/m_H, should be greater than 2. The results of all 3 groups have been summarized by Kuti at the Munich conference [12] (note that Kuti formulated them in terms of the cut-off parameter π/a). The most important aspect of this value for the upper bound is that the scalar sector is actually never strongly interacting, even if one starts with an infinite bare coupling λ. The maximal value of λ_R is about 2/3 of the tree level unitary bound only [39].

I stress that, strictly speaking, this result applies only to the O(4) Φ^4 model regularized on the hypercubic lattice with nn coupling, eq. (2.9). However, some calculations on lattices with other geometries or couplings give similar, though not equal results [68]. Of course, it would be very helpful to have non-perturbative results also for some other regularizations than on the lattice to compare with. Unfortunately, they are not available except the calculation by P. Hasenfratz and Nager [37] whose reliability is not known but whose results are also not far from (3.18). One should also take into consideration that the lattice regularization is fully acceptable from the physical point of view. The deviations from the continuous rotational invariance are not a serious drawback as they can be well controlled and suppressed by a suitabble choice of the lattice action [69]. From all that one has the impression that the quantitative results obtained until now by means of the lattice regularization are physically relevant and that the triviality of the Φ^4 theory has really the phenomenological consequence of impossibility of a Higgs boson mass much larger than (3.18) and of the strongly interacting scalar sector in the Standard Model. However, some caution [70] is certainly still in place.

It should be clear that the bound obtained in a given regularization is still slightly dependent on the degree of scaling violations one is ready to allow. Should the ratio of the cut-off to the Higgs mass, Λ/m_H, be greater than 2, or, say, 5? The latter requirement can lower the bound by 100 GeV [53]. It is actually more sensible to look at the "envelope", showing what would be the maximal value of R_Φ at a chosen value of the cut-off $\Lambda/m_\sigma = 1/am_\sigma$. In Fig. 6 the analytic results for the envelope [39] (diamonds), the finite-volume corrected (eq. (3.16)) results of Kuti et al. [51, 53, 54] (curve 2) and the similarly corrected results of our group [50, 52, 63] (curve 1) are shown. The shapes of the curves are determined by the assumed scaling laws

$$
\begin{aligned}
\langle \Phi_R \rangle &\propto \sqrt{t} \ (\ln t)^{1/4} \\
m_\sigma &\propto \sqrt{t} \ (\ln t)^{-1/4} \\
t &= \frac{\kappa - \kappa_c}{\kappa_c}.
\end{aligned}
\qquad (3.19)
$$

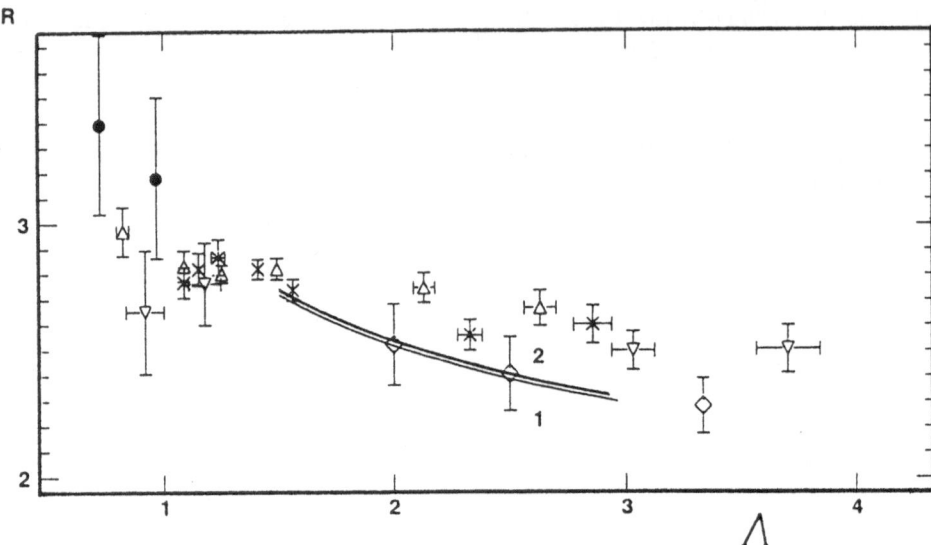

Figure 6. *The envelope of the possible values of the ratio R_* (eq. (3.1)) for different values of the cut-off parameter Λ as obtained by Lüscher and Weisz (diamonds), our group (curve 1) and Kuti et al. (curve 2). Curves are drawn according to the scaling laws (3.19) after interpolation to the infinite lattice. Some raw data on finite lattices are shown too. The figure is from Ref. [63].*

Raw data of our group on $12^4 - 16^4$ lattices, without a finite size correction for m_σ, are shown in this figure, too. We can see that the shift of R_* due to finite size effects might be about 5 per cent, but we do not know yet for sure.

We have achieved remarkable reliability and accuracy, which can still be improved in the future. What should be our goals? The question of a "precise" value of the upper bound is a question with intrinsic unprecision and contains several uncertainties [70, 71]. The actual question, which has a more precise meaning, is: How accurately can we control the electroweak theory in the case when the renormalized quartic coupling is nearly as large as allowed by the bound? In this case the cut-off is low and it might be possible to detect experimentally some, at first only minor departures from the present day effective theory. Thus the emphasis is on precision of the numerical determination of various quantities in some range of the cut-off values.

In particular, we should study the possibilities of detecting scaling violations, like the dependence on the regularization scheme. We should investigate the differences between the positions and shapes of the envelopes for various lattice formulations [68]. These envelopes have to coincide for large cut-off Λ/m_σ as they all have to obey the same scaling laws (3.19) for small t. But they will differ at lower values of Λ/m_σ [10] indicating that the Φ^4 theory is too much regularization dependent there to be useful as an effective theory for physical phenomena. Therefore I find it more appropriate to quote the upper bounds like (3.18) with error bars (instead of saying that $m_H < 680\,\text{GeV}$), in order to be able to compare the results for various regularizations.

I conclude that in the future we should try to determine the onset of the common scaling behaviour (3.19) for various regularizations which might require numerical calculations at relatively large $\Lambda/m_\sigma = \xi$ and thus a very good control of the finite size effects for large correlation lengths. Much remains to be done in this respect.

4 Introducing fermions into the Higgs models

Until now we have completely ignored fermions in the electroweak theory. This might be correct quantitatively if no heavy fermions exist, but it is not satisfactory for general theoretical reasons. And if a heavy fermion with strong Yukawa coupling exists, then we must include fermions into the non-perturbative lattice study of the Higgs system to maintain the reliability of our calculations.

Of course, the trouble is that it is so difficult to include fermions on the lattice. Fermion doubling, wrong chirality of some doublers, and difficult computational methods are the obstacles. But things got moving and I shall describe the (mostly preliminary) progress in this field. Useful review articles on this topics are Refs. [7, 8, 11].

Let me just mention that there are presently two methods of lattice formulation of fermionic fields. One is the "staggered fermion" or Kogut-Susskind method, where different Dirac components of a fermion field live at different neighbouring lattice sites. This formulation preserves a continuous remnant of the chiral symmetry of the continuum theory and reduces the number of fermions from 16 to 4. For its description I have to refer to other works, e.g. [72]. The other method is due to Wilson [4, 73], and I will describe it later in some detail, as this method is more suitable for the lattice study of the electroweak interactions. In numerical simulations the fermionic degrees of freedom are often treated in the "quenched" approximation which neglects effects of the virtual fermion loops, i.e. the dynamics of fermions.

4.1 Chiral symmetry breaking in the Higgs models with fermions

It is a well established result of numerical investigation of lattice QCD that at low temperatures the chiral symmetry of QCD is spontaneously broken and the fermion condensate is non-vanishing, $\langle \bar{\Psi}\Psi \rangle \neq 0$, in the limit of massless quarks. Thus, when the Higgs models are enriched by fermions with vector-like coupling to the gauge field,

$$\frac{1}{2} \sum_{x\mu} \eta_{x,\mu} (\bar{\chi}_x U_{x,\mu} \chi_{x+\mu} - \bar{\chi}_{x+\mu} U_{x,\mu}^\dagger \chi_x),\tag{4.1}$$

χ_x being the "staggered" fermions and $\eta_{x,\mu}$ certain phase factors [72], we expect also in these systems that the chiral symmetry is spontaneously broken as long as the effects of the scalar field are small, i.e. for small κ. Then the question arises, what is the influence of the scalar field on $\langle \bar{\Psi}\Psi \rangle$ for large values of κ and what is the interplay between the spontaneous chiral symmetry breaking and the Higgs mechanism related to the SSB in the Φ^4 theory.

In the case of the SU(2) Higgs model this question is of direct relevance to the electroweak theory as the SU(2) sector of this theory with chiral fermions can be rewritten in a vector-like

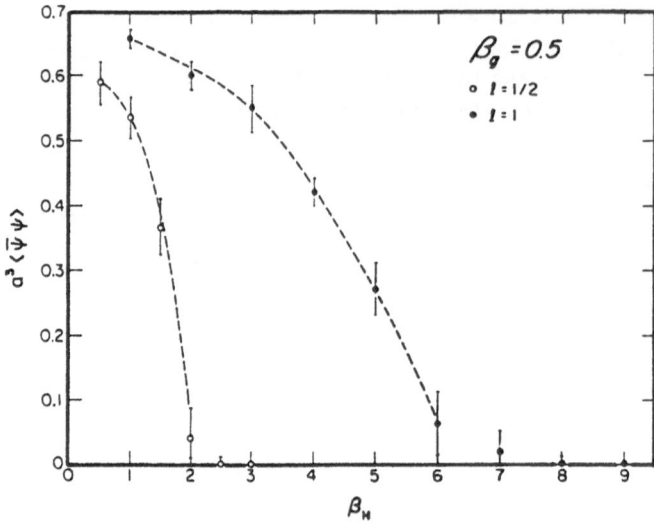

Figure 7. *Chiral condensate* $\langle \bar{\Psi}\Psi \rangle$ *as a function of* $\beta_H = 2\kappa$ *in the* $SU(2)$ *Higgs model at* $\beta = 0.5$ *in the quenched approximation. The figure is from Ref.* [75].

form. The argument based on the reality of the group $SU(2)$ is due to Georgi (cited in [20]) and explained in detail in Ref. [74].

The pioneering work has been done by Lee and Shigemitsu [75] who found numerically for staggered fermions in the quenched approximation that for small β the chiral condensate $\langle \bar{\Psi}\Psi \rangle$ decreases with increasing κ and vanishes at some finite value of κ Fig. 7. This is not in contradiction with the theorem on the analytic connection between the confinement and Higgs regions (Subsec. 2.2) because $\langle \bar{\Psi}\Psi \rangle$ is, expressed in terms of the scalar and gauge fields when the fermionic integration in the path integral has been carried out, a highly non-local quantity. For fermions with weak isospin $I = 1/2$ this κ value is lower than for those with $I = 1$. Lee and Shrock [76] have shown later analytically, using at $\beta = 0$ the mean field method for the order parameter $\langle \bar{\Psi}\Psi \rangle$, that this is a genuine phase transition and the single phase of the $SU(2)$ Higgs model thus splits into 2 phases, with $\langle \bar{\Psi}\Psi \rangle > 0$ for low κ and $\langle \bar{\Psi}\Psi \rangle = 0$ for large κ. This result, and a systematic analytic and numerical investigation of the chiral phase transition line for $\beta \geq 0$ in the $SU(2)$ model by several authors [76-81] lead to the following picture (see Fig.8):

As κ increases and the system approaches the Higgs region, the configurations of the gauge field are increasingly restricted by the interaction with the scalar field in such a way that the effective fermion-fermion interaction gets weaker. This causes the restoration of the chiral symmetry, which takes place at higher κ if the fermion field has larger effective gauge coupling for higher I.

The very interesting fact demonstrated by De and Shigemitsu [78] is that the chiral phase transition for $I = 1/2$ seems to coincide with the Higgs phase transition at $\beta = 2.3$ (the Higgs phase transition line extends from $\beta = \infty$ until $\beta \simeq 2$). The chiral transition

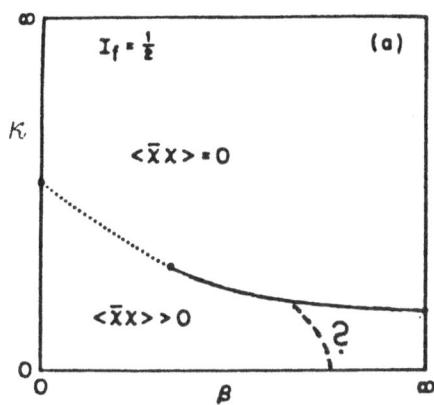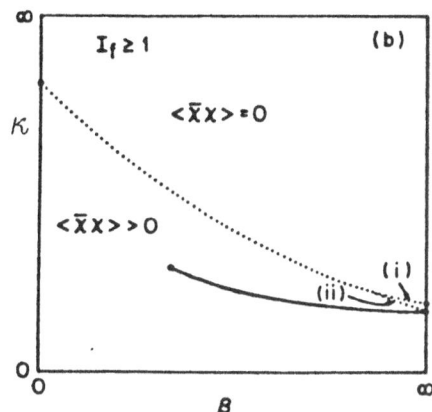

Figure 8. *Schematic phase diagrams for the $SU(2)$ Higgs model with (a) $I = 1/2$ and (b) $I = 1$ fermions, as they are expected on the basis of calculations performed for $\beta \leq 2.3$. For larger β, i.e. in the physically relevant region, the positions of the chiral phase transition (doted and dashed lines) is not really known, however. The figure is adapted from Ref. [82].*

line at low β also looks like an analytic continuation of the Higgs transition line to lower β until $\beta = 0$ (Fig. 8). Thus for $I = 1/2$ the Higgs region is apparently free of the chiral condensate and fermion masses can arise only through the Yukawa coupling – as expected in the continuum electroweak theory. On the contrary, the Abbott-Farhi [19, 20] scenario (the "strongly coupled standard model") does not seem to be realized in this theory as fermions would acquire large mass of the order of the weak interaction scale, $O(10^2 \text{GeV})$, below the Higgs transition due to the chiral symmetry breaking. It should be stressed, however, that the phase diagram at large β is not yet accessible to numerical simulations, and that there is no real evidence that the chiral phase transition continues along the Higgs phase transition for large β [81]. The phase diagram could also be as indicated in Fig. 8 by the dashed line [11]. In this case the Abbott-Farhi scenario would not be excluded.

Similar argument can be also used to speculate about absence of the $I = 1$ fermions in the physical spectrum [82]. If the phase diagram for $I = 1$ looks really to be as shown in Fig. 8, and for $\beta = 2.3$ this is the case [78], then an approach to the continuum limit in the vicinity of the Higgs phase transition would have to be performed with $I = 1$ fermions in the broken chiral symmetry phase with the result that such fermions would end up with too large mass to have already been observed.

The effect of the Yukawa coupling y on the chiral phase transition in the $SU(2)$ Higgs model is to extend the region without the chiral symmetry breaking. According to some strong β coupling calculations [74], for sufficiently high y the chiral symmetry breaking completely vanishes.

The $U(1)$ Higgs model with vector-like coupling of fermions to the gauge field is perhaps less interesting from the point of view of the electroweak theory, but it is an excellent laboratory for the study of the above effects as this model on the lattice also has the Higgs-confinement phase with SSB of chiral symmetry in the confinement region. Much understanding of the chiral symmetry breaking in the lattice Higgs models has been achieved also within this model [83, 84]. Good reviews of the whole topic are in Refs. [7, 11].

4.2   Yukawa coupling in simple scalar field theories with fermions

Non-perturbative investigations of the fermion mass generation through the Yukawa coupling to scalar fields, and of some other effects of this coupling, has become recently one of the focuses of research in the lattice quantum field theories. The most important long term reasons for this interest are:

(i) Investigation of the question whether there is an upper bound on the Yukawa generated fermion masses [85], in analogy to the upper bound on the Higgs boson mass, and possibly its determination.

(ii) Study of the influence of the strong Yukawa coupling on the scalar sector, in particular on the upper bound on the Higgs boson mass [11, 54] and on the envelope.

(iii) Search for new critical points suitable for the construction of a – possibly non-trivial – continuum limit of some lattice formulation of electroweak theory [13].

Most of the work until now has been done with toy models with Z(2) and U(1) symmetry only, disregarding the problems with fermion doubling and other phenomenological short-comings of these models and leaving out gauge fields. The methodological contributions of these studies are invaluable, however, and some qualitative aspects of the results might well be of some relevance to physics, too.

As an example, let us take the Z(2) lattice Yukawa model with one naive lattice fermion field [86], which in the continuum limit describes 16 degenerate fermions:

$$
\begin{aligned}
S &= -2\kappa \sum_{x\mu} \Phi_x \Phi_{x+\mu} + \lambda \sum_x (\Phi_x^2 - 1)^2 + \sum_x \Phi_x^2 \\
&\quad + \frac{1}{2} \sum_{x\mu} \bar{\Psi}_x \gamma_\mu (\Psi_{n+\mu} - \Psi_{n-\mu}) + y \sum_x \Phi_x \bar{\Psi}_x \Psi_x.
\end{aligned}
\tag{4.2}
$$

The model is symmetric with respect to the global "chiral" $Z(2)$ transformation

$$
\Phi_x \to -\Phi_x, \quad \Psi_x \to i\gamma_5 \Psi_x, \quad \bar{\Psi}_x \to \bar{\Psi}_x i\gamma_5,
\tag{4.3}
$$

which is a generalization of the Z(2) symmetry of the one-component scalar field model. Alternatively, one can use staggered fermions [87-90] instead of the naive ones, reducing the number of degenerate fermions to four. One cannot add an explicite fermion mass term without violating the Z(2) symmetry.

Let us consider the fermion mass m_F. For small y we expect in the SSB phase

$$
am_F \simeq y \langle \Phi \rangle
\tag{4.4}
$$

and

$$
am_F \to 0 \quad \text{as} \quad \kappa \searrow \kappa_c.
\tag{4.5}
$$

For large y the situation is different. Rescaling the fermion field, $\sqrt{y} \cdot \Psi \to \Psi$, the fermionic kinetic term in eq. (4.2) gets the factor $1/y$. Performing then the strong coupling expansion in the powers of $1/y$, which is a standard tool in the lattice theories, one finds that m_F *increases* as $\kappa \searrow \kappa_c$ [86]. Also a mean field calculation (within a more complex model discussed in Subsec. 4.4) at strong y gives [91]

$$m_F \propto \frac{1}{a^2 \langle \Phi \rangle}.$$ (4.6)

For some recent analytic arguments see also Ref. [92].

Thus there are two qualitatively different regions of Yukawa coupling. Firstly, the weak y (perturbative) region, where the relation (4.4) holds. Secondly, the strong y region, where m_F increases as κ approaches the critical point in the SSB phase. In the limit of the infinite cut-off the fermions get presumably infinitely heavy there. Numerical calculations should clarify what separates these regions – some cross-over or even a new phase transition with some interesting critical points?

The present situation of the numerical study of this and some related questions is as follows. The weak coupling region of the $Z(2)$ model with staggerd fermions has been investigated quite early by Shigemitsu [87] in the quenched approximation. She found that am_F and also the renormalized Yukawa coupling,

$$y_R = \frac{am_F}{\langle \Phi \rangle_R},$$ (4.7)

decrease for $\kappa \searrow \kappa_c$. The situation resembles that in the pure Φ^4 theory ($m_F \leftrightarrow m_\sigma$, $y_R \leftrightarrow \lambda_R$). Later Anna Hasenfratz and Neuhaus [86] found a distinctly different behavior for large y, illustrated in Fig. 9. Here it is demonstrated that for $y > 2$ the fermion mass am_F increases as $\langle \Phi \rangle$ decreases. This calculation also has been performed in quenched approximation, but that approximation should be reasonable good both for small y (i.e. the weakly coupled fermions) and large y (fermions so heavy that their loops are strongly suppressed). Recently similar phenomenon has been observed also in the $SU(2)$ model [93, 94]. Thus the existence of two different regions seems to be established.

The question what separates these regions must be answered in calculations with dynamical fermions. Kuti and his co-workers observed in such calculations with staggered fermions [90], when increasing y, a smooth behaviour of several observables and no signal for a phase transition even for quite large y, the results staying consistent with the perturbation theory.

However, we know that the phase diagram of the $Z(2)$ model with dynamical fermions is quite different from that in the quenched case [88,89,92,95,96] and, in addition, depends on the way how the Yukawa coupling is transcribed on the lattice. The difference between the phase diagram in the quenched and dynamical cases is indicated schematically in Fig. 10. The phase transition line separating the broken and symmetric phases is found in the dynamical fermion case at κ values much lower than κ_c of the pure Φ^4 model, as the Yukawa coupling tends to order the scalar field and thus enlarges the SSB phase. Similar change of the phase diagram has been found also in the Yukawa model with $U(1)$ symmetry and dynamical fermions [97]. These facts suggest that the separation between the weak and strong y regions can lie at very different places for quenched and dynamical fermions. Further study of this question is thus required. First calculation of the masses in the dynamical fermion case for an $U(1)$ model by Thornton indicate the precence of the strong coupling region indeed [98].

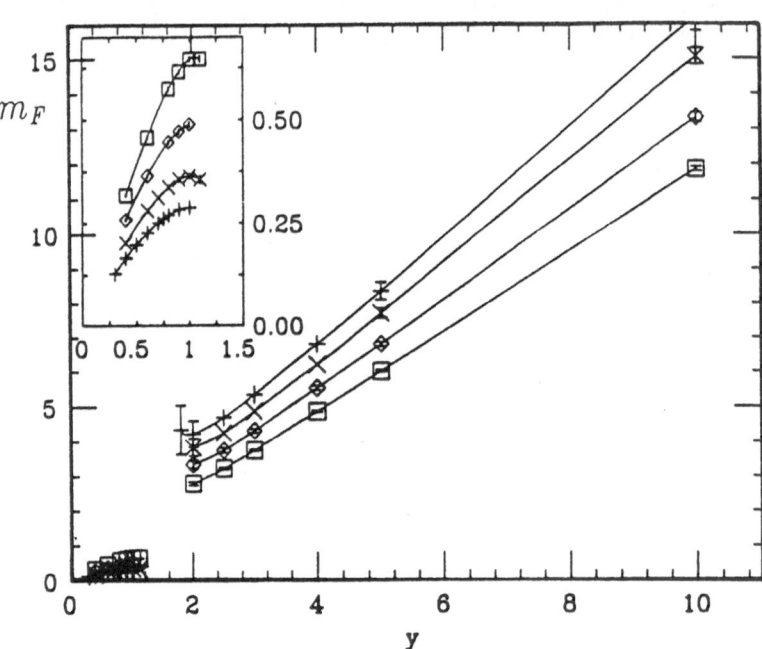

Figure 9. *Fermion mass generated by the Yukawa coupling in the SSB phase of the model (4.2) for 4 κ-values as a function of y in the quenched approximation. For y > 2 the fermion mass increases as κ decreases from larger values (squares) to smaller ones (crosses). The figure is from Ref. [86].*

4.3 Spectrum of chiral fermions on the lattice

Now we come to the most difficult problem of the formulation of weak interactions on the lattice – to the problem of the fermion spectrum. With the lattice regularization comes the problem of fermion doubling. Under quite general assumptions Nielsen and Ninomiya [99] have shown that lattice transcriptions of a continuum left-handed Weyl field contain an equal number of left- and right-handed fermions. Karsten and Smit [100] pointed out that this is related to the manifest gauge invariance of the lattice regularization even in the case of anomalous chiral gauge theories, as the doubling results in cancellation of anomalies. These facts do not exclude, however, that the unwanted additional fermions – "doublers" – get very heavy or do not interact with that part of the spectrum which would be phenomenologically acceptable. So there is some hope left, and several interesting suggestions how to circumvent the problem have been made.

I start by recalling how the fermion doubling problem is solved by the Wilson method [4, 73] in the case of vector-like gauge theories like QCD. For clarity I shall reconstitute in this subsection the dimensions of the fields and write explicitly the lattice constant a. The

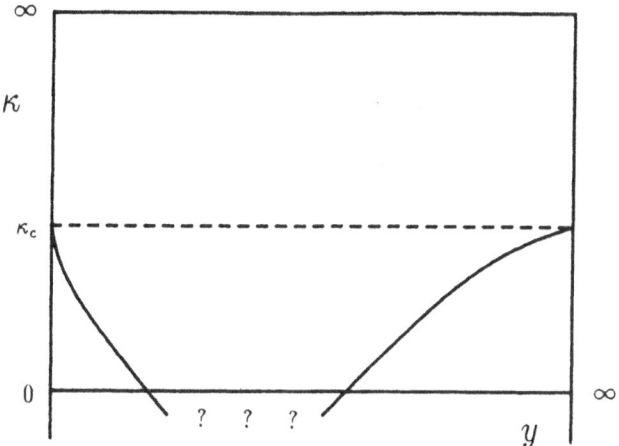

Figure 10. *Schematic phase diagram of the Φ^4 theory with a Yukawa coupling y to fermions in the quenched approximation (dashed line) and for dynamical fermions (full line).*

action S^Ψ of a free Dirac field Ψ is written on the lattice in the form

$$S^\Psi = S_n^\Psi + S_W^\Psi, \tag{4.8}$$

where the first term is called the "naive" action

$$S_n^\Psi a^{-4} = \frac{1}{2a} \sum_{x\mu} \bar{\Psi}_x \gamma_\mu (\Psi_{x+\mu} - \Psi_{x-\mu}) + \sum_x m_0 \bar{\Psi}_x \Psi_x, \tag{4.9}$$

and the second one the Wilson term with the dimensionless Wilson parameter r.

$$S_W^\Psi a^{-4} = \frac{r}{a} \sum_{x\mu} (\bar{\Psi}_x \Psi_x - \frac{1}{2}\bar{\Psi}_{x+\mu}\Psi_x - \frac{1}{2}\bar{\Psi}_x\Psi_{x+\mu}). \tag{4.10}$$

The inverse propagator in the momentum space is then

$$S_F^{-1}(k) = \sum_\mu i\frac{1}{a}\gamma_\mu \sin ak_\mu + m_0 + \frac{r}{a}\sum_\mu (1 - \cos ak_\mu). \tag{4.11}$$

The sixteen fermions are seen when the momentum components k_μ are either near zero or close to their maximal value within the Brillouin zone,

$$k_\mu = p_\mu + P_\mu^D \quad . \quad D = 0, \ldots 15, \tag{4.12}$$

$$P_\mu^D = (0,0,0,0) , \ (\frac{\pi}{a}, 0, 0, 0), \ \ldots, \ (\frac{\pi}{a}, \frac{\pi}{a}, \frac{\pi}{a}, \frac{\pi}{a}). \tag{4.13}$$

For each D one recovers for small ap_μ one continuum Dirac field propagator from S_F. If the Wilson term (4.10) is absent ($r = 0$ in (4.11)). all fermions have the same mass $m_F = m_0$. For $r > 0$ the doublers have higher mass m_D.

$$m_F = m_0. \quad m_D = m_0 + 2\frac{r}{a}n_D . \quad n_D = 1, \ldots, 4 \ \text{ for } D \geq 1. \tag{4.14}$$

65

This is fine, as in the continuum limit, i.e. for $a \rightarrow 0$, the doublers get an infinite mass. (Numerically it is of course quite difficult to achieve sufficiently small a).

In the case of chiral gauge theories the doublers do not only duplicate the spectrum, but in addition some of them transform in a "mirror" way with respect to the chiral transformations (i.e. the transformation properties of the left- and right-handed components are interchanged) so that we must remove them at least from the low mass spectrum. The problem is that the Wilson term (4.10) cannot be used for the removal of the doublers as it is not manifestly invariant under the chiral gauge transformation for the same reason for which also the mass term in (4.9) is non-invariant, namely it contains products of the type $\bar{\Psi}_L \Psi_R$ and $\bar{\Psi}_R \Psi_L$ (the subscripts L and R denote the left- and right-handed fields, respectively).

One way how to circumvent the problem has been proposed by Montvay [101]. He pointed out that, if the doublers with mirror properties have to arise in the spectrum [102, 13], it is better to include them from the very beginning as fundamental fields in order to better control the possibilities of their removal. Thus he deliberately doubles the fermion spectrum by introducing the "mirror" Dirac fermion χ such that χ_R transforms with respect to the chiral transformation in the same way as Ψ_L, and analogously χ_L transforms as Ψ_R. Then one can write invariant terms $\bar{\Psi}_L \chi_R$ and $\bar{\Psi}_R \chi_L$, both local and of the Wilson type (4.10), which removes the doublers of both the usual fermions and the mirrors. Choosing suitably some of the parameters, we can hope to make the mirror fermions sufficiently heavy so that they are consistent with the present day phenomenological knowledge. One should also keep in mind the possibility that heavy mirror fermions do exist in the nature. Montvay discussed some aspects of his approach in his Erice lecture last year [13].

Another idea is to compensate for the gauge non-invariance of the Wilson term by introducing some counter-terms whose coefficients are chosen in such a way that the Ward identities are satisfied in the perturbation expansion up to positive orders of the lattice constant a [103].

It seems to me that the most natural idea is to modify the Wilson term in the case of a chiral gauge theory on the lattice in the same way as the fermion mass term in the continuum theory had to be modified: insert a suitably transforming scalar field φ so that for a Dirac field Ψ the products of the type $\bar{\Psi}_L \varphi \Psi_R$ etc. are manifestly gauge invariant. Then the Wilson term gets the form of a new, Wilson-Yukawa (or "point-split Yukawa") coupling term. Aoki and others [104] suggested that φ can be an auxiliary field without a kinetic part. But the interaction with fermions presumably generates the kinetic term so that one can as well include it from the very beginning.

The most natural possibility is to use the Higgs field itself. This has been suggested in 1980 by Smit [105, 8] and later by Swift [106]. In the simple model (4.2) with the global $Z(2)$ symmetry such a Wilson-Yukawa term can be written (using now, according to (2.2), the dimensional scalar field $\varphi = \sqrt{2\kappa}\Phi/a$) in the form

$$w \sum_{x\mu} (\bar{\Psi}_x \varphi_x \Psi_x - \frac{1}{2}\bar{\Psi}_{x+\mu}\varphi_x \Psi_x - \frac{1}{2}\bar{\Psi}_x \varphi_x \Psi_{x+\mu}). \tag{4.15}$$

Here w is a new Wilson-Yukawa coupling constant. Local gauge invariance of this term can easily be achieved by inserting suitable gauge field variables on the links in (4.15). Perturbatively, we expect from eq. (4.14) that in the SSB phase

$$m_F = y\langle\varphi\rangle$$

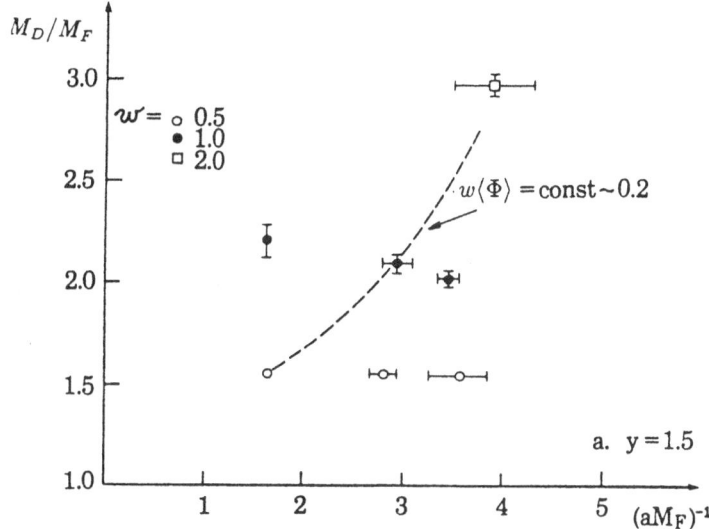

Figure 11. *Ratio of the doubler to fermion masses in the quenched approximation in the one-component Φ^4 theory with the Wilson-Yukawa coupling term (4.15). Different symbols denote different values of w. The dashed line indicates the possibility to increase the ratio by increasing w and keeping $w\langle\Phi\rangle$ constant. The figure is from Ref. [107].*

$$m_D = y\langle\varphi\rangle + 2w\langle\varphi\rangle n_D , \quad D = 1, ..., 15. \tag{4.16}$$

Notice that for dimensional reasons the factor $1/a$ present in eqs. (4.10) and (4.14) is absent in (4.15) and (4.16). As we have to perform the continuum limit in such a way that the expectation value of the scalar field with physical dimension, $\langle\varphi\rangle \simeq 250$ GeV, remains constant, the perturbative formula (4.16) suggests that the doublers can be made heavy only by increasing w. But it could also be, and recent results indicate that this is the case [94], that for large w the relationship between m_D and $\langle\varphi\rangle$ is very different from (4.16), rather analogous to (4.6), and values of m_D of the order of the cut-off Λ can be even achieved for large but fixed w. This is similar to the non-perturbative behaviour of the fermion mass m_F for strong Yukawa coupling described in Subsec. 4.2. In any case this means that the question of giving the doublers a high mass is a non-perturbative dynamical problem in the coupling w. Fortunately, the lattice regularization is suitable for dealing with such problems, so this is not prohibitive. Thus, at least in principle, there *is* a possibility to use the lattice regularization in the chiral gauge models [105, 8, 106], it has just been overlooked because of its relative complexity.

Shigemitsu [107] pioneered the numerical (quenched) investigation of this mechanism in the Z(2) model and found that to some extent it is possible to increase the ratio m_D/m_F simultaneously with decreasing am_F, the fermion mass in lattice units, i.e. when the continuum limit is approached (Fig. 11). Some similar results have been obtained recently also in the 2-component model when the dynamics of the fermions have been taken into account [98]. A study of the physically more realistic chiral SU(2)⊗SU(2) model will be discussed in the next subsection.

4.4 Smit-Swift formulation of the standard model on the lattice

The parameters of the model are: the gauge field coupling parameter β, the hopping parameter κ, the quartic coupling λ, the usual local Yukawa coupling y and the Wilson-Yukawa coupling w. The model has the symmetry

$$SU(2)_L^{(local)} \otimes SU(2)_R^{(global)} \tag{4.17}$$

with respect to the chiral transformations

$$
\begin{aligned}
\Psi &\rightarrow (V_L P_L + V_R P_R)\Psi, \quad \overline{\Psi} \rightarrow \overline{\Psi}(V_L^\dagger P_R + V_R^\dagger P_L) \\
\hat{\Phi} &\rightarrow V_L \hat{\Phi} V_R^\dagger, \\
V_L &\in SU(2)_L^{(local)}, \quad V_R \in SU(2)_R^{(global)}
\end{aligned}
\tag{4.18}
$$

with $\hat{\Phi}$ defined in eq. (2.15). The action is

$$S = S_H + S_F + S_Y + S_W. \tag{4.19}$$

The meaning of the individual terms in S is the following: S_H is the action of the SU(2) Higgs model, eq. (2.13). S_F describes the "naive" Dirac fermions

$$S_F = \frac{1}{2} \sum_x \sum_{\mu=1}^{4} \{\overline{\Psi}_x \gamma_\mu P_L (U_{x,\mu} \Psi_{x+\mu} - U_{x-\mu,\mu}^\dagger \Psi_{x-\mu}) + \overline{\Psi}_x \gamma_\mu P_R (\Psi_{x+\mu} - \Psi_{x-\mu})\}. \tag{4.20}$$

S_Y is the Yukawa coupling term

$$S_Y = y \sum_x \overline{\Psi}_x (\hat{\Phi}_x P_R + \hat{\Phi}_x^\dagger P_L)\Psi_x, \tag{4.21}$$

and S_W is the Wilson-Yukawa coupling term

$$
\begin{aligned}
S_W = \; & w \sum_x \sum_{\mu=1}^{4} \{\overline{\Psi}_x (\hat{\Phi}_x P_R + \hat{\Phi}_x^\dagger P_L)\Psi_x \\
& -\frac{1}{2}\overline{\Psi}_x (\hat{\Phi}_x P_R + \hat{\Phi}_{x+\mu}^\dagger P_L)\Psi_{x+\mu} - \frac{1}{2}\overline{\Psi}_{x+\mu}(\hat{\Phi}_{x+\mu} P_R + \hat{\Phi}_x^\dagger P_L)\Psi_x\}.
\end{aligned}
\tag{4.22}
$$

As usual,

$$P_L = \frac{1}{2}(1 - \gamma_5), \quad P_R = \frac{1}{2}(1 + \gamma_5). \tag{4.23}$$

This lattice model, and its extension to the SU(2)$\otimes$U(1) case, has been suggested by Smit [105, 8] and Swift [106]. Recently, it has been shown that the right-handed massless neutrino present in this model decouples from the physical spectrum [108].

For $w = 0$ this model describes 16 SU(2)-doublets of fermions of the same mass. Some of the 15 doublers have the "mirror" coupling to the gauge field. The hope is that due to the presence of the Wilson-Yukawa coupling w the 15 unwanted doublers get sufficiently massive to be in agreement with the phenomenology. This is an intrinsically non-perturbative problem as the coupling w has to be made large.

Until now the model has been investigated numerically in the limit $\beta = \infty$, i.e. without the gauge fields in the quenched simulation [93, 94]. (For other special cases see Subsec. 3.1.) I just want to mention two results: It has been demonstrated that in the quenched

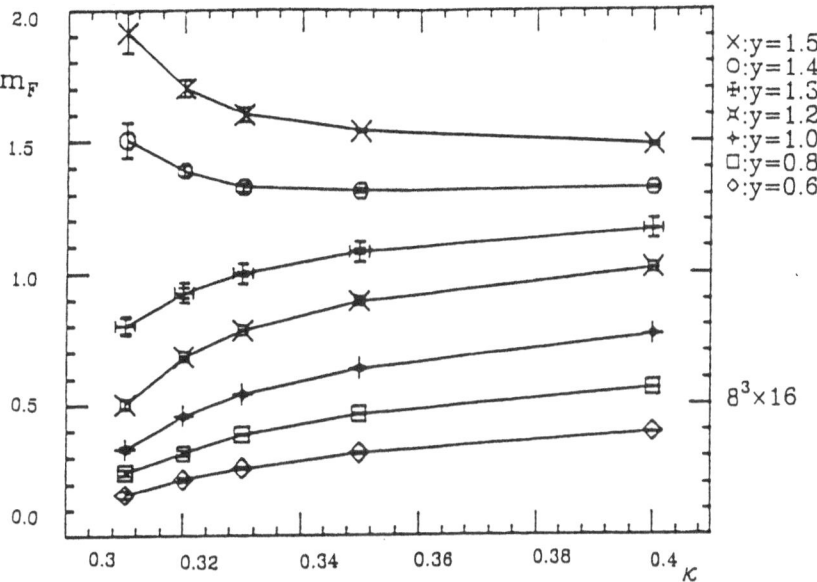

Figure 12. *Fermion mass in the quenched approximation for various fixed values of y as a function of κ in the SSB phase of the SU(2) Smit-Swift model (4.19) without the gauge fields and with $w = 0$. For $y \geq 1.4$ the fermion mass increases as κ approaches the critical point at $\kappa_c \simeq 0.304$. The figure is from Ref. [93].*

approximation the model in the SSB phase has both the weak and strong y coupling regions. In the strong coupling region the fermion mass (and also the doubler masses) increase with decreasing distance from the critical line (Fig. 12). First results [94] for various values of w indicate that the doubler masses can be eleminated from the physical part of the spectrum. They can be namely made as large as the cut-off ($am_D \approx 1$) by a proper choice of w even for fixed w while am_F can be made small by a suitable choise of y. These results are very encouraging. (Fig. 13).

5 Outlook

Further work investigating the Smit-Swift model and other lattice models for weak interactions is in progress and the results are promising. Furthermore, a fast "hybrid Monte Carlo" algorithm for fermions has been recently introduced [109]. So I think that in a few years we shall have the non-perturbative properties of the Standard Model on the lattice numerically under control also in the case that some heavy fermions exist. We shall know the phase diagram and whether there exist some new critical points where one might obtain a – possibly non-trivial – continuum limit. An upper bound on the fermion mass – if it exists – will be determined. But the precision of the calculations will remain probably limited for a long time.

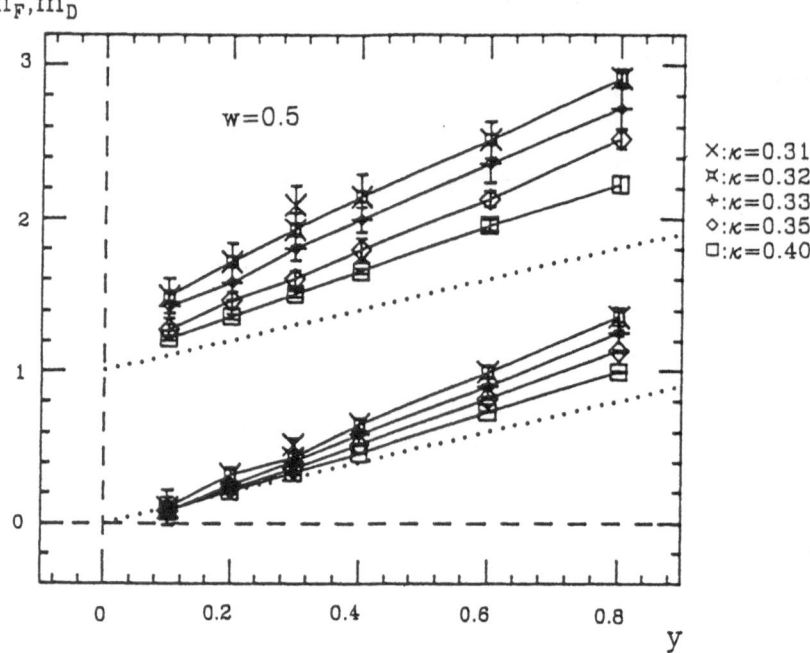

Figure 13. *Fermion and doubler mass as a function of y for various κ in the same model as in Fig. 12 but for w = 0.5. It demonstrates that in the quenched approximation it is possible to vary κ and y in such a way that the fermion mass (lower set of lines) approaches zero in lattice units as κ → κ_c but the doubler mass (upper set of lines) stays larger than the cut-off (am_D > 1). Thus the doublers do not appear in the physical spectrum. The figure is from Ref. [94].*

Concerning the scalar field calculations, we shall be able to achieve a high precision indeed. My optimism is due to the recent development of a new cluster algorithms for the O(4) Φ^4 model [110], based on an idea of Wolff [111], allowing its investigation on substantially larger lattices and with much increased statistical effectiveness. Thus fine tests of the validity of the Standard Model will be possible in the near future in the approximation treating the gauge fields and fermions only perturbatively

Concerning the lattice Higgs models, we do not have substantially improved algorithms for the gauge field yet, so that these models will have to wait for some time for a renewed interest.

Acknowledgement. I thank A. Ali and A. Zichichi for the opportunity to deliver these lectures and for the hospitality in Erice. I am grateful to A.K. De, C.B. Lang and J. Smit for many helpful suggestions. I have also benefited from discussions with A. Ali, M.A.B. Bég, W. Bock, K. Jansen, H.A. Kastrup, H. Leutwyler, I. Montvay and T. Neuhaus. I further thank Justina Smižanská-Jersák for typing these lecture notes and T. Trappenberg for some suggestions and help with the manuscript. A support of the Deutsches Bundesministerium für Forschung und Technologie is acknowledged.

References

[1] P.W. Anderson, Phys. Rev. 130 (1963) 439;
P.W. Higgs, Phys. Lett. 12 (1964) 132; Phys. Rev. Lett. 13 (1964) 508; Phys. Rev. 145 (1966) 1156;
F. Englert and R. Brout, Phys. Rev. Lett. 13 (1964) 321;
G.S. Guralnik, C.R. Hagen and T.W.B. Kibble, Phys. Rev. Lett. 13 (1964) 585;
T.W. Kibble, Phys. Rev. 155 (1967) 1554.

[2] K.G. Wilson, Phys. Rev. B4 (1971) 3184;
K.G. Wilson and J. Kogut, Phys. Rep. 12C (1974) 76;
M. Aizenman, Phys. Rev. Lett. 47 (1981) 1; Commun. Math. Phys. 86 (1982) 1;
J. Fröhlich, Nucl. Phys.B200 (1982) 281;
C.B. Lang, Phys. Lett. 155B (1985) 399; Nucl. Phys. B265 [FS15] (1986) 630.

[3] D.J.E. Callaway, Phys. Rep. 167 (1988) 241.

[4] K.G. Wilson, Phys. Rev. D10 (1974) 2445.

[5] J. Jersák, in *Lattice Gauge Theory - A Challenge in Large-Scale Computing*, Ed. B. Bunk, K.H. Mütter and K. Schilling (Plenum Press, 1986).

[6] I. Montvay, in *Proceedings of the International Europhysics Conference on High Energy Physics*, Uppsala 1987, ed. O. Botner, p.298.

[7] R.E. Shrock, Nucl. Phys. B (Proc. Suppl.) 4 (1988) 373.

[8] J. Smit, Nucl. Phys. B (Proc. Suppl.) 4 (1988) 451.

[9] H.G. Evertz and M. Marcu, in *Proceedings of the 12th John Hopkins Workshop on Current Problems in Particle Theory, TeV Physics*, Baltimore 1988, ed. G. Domokos and S. Kovesi-Domokos (World Scientific 1988).

[10] P. Hasenfratz, Nucl. Phys. B (Proc. Suppl.) 9 (1989) 3.

[11] J. Kuti, Nucl. Phys. B (Proc. Suppl.) 9 (1989) 55.

[12] J. Kuti, in *Proceedings of the XXIV Int. Conf. on HEP*, Munich 1988, ed. R. Kotthaus and J. H. Kühn, Springer 1989.

[13] I. Montvay, DESY 88-158 (Eloisatron Project Workshop, Erice 1988).

[14] J. Glimm, A. Jaffe, Quantum Physics (Springer 1987).

[15] K. Osterwalder and E. Seiler, Ann. Phys. 110 (1978) 440.

[16] E. Fradkin and S. Shenker, Phys. Rev. D19 (1979) 3682.

[17] G. Mack, Lectures presented at the *Topical Conference on Quantum Chromodynamics* at Chania (Crete) 1977 (DESY report 77/58 (1977), unpublished).

[18] S. Dimopoulos, S. Raby and L. Susskind, Nucl. Phys. B173 (1980) 208.

[19] L.F. Abbott and E. Farhi, Phys. Lett. 101B (1981) 69; Nucl. Phys. B189 (1981) 547.

[20] M. Claudson, E. Farhi and M. Jaffe, Phys. Rev. D34 (1986) 873.

[21] J. Fröhlich, G. Morchio and F. Strocchi, Phys. Lett. 97B (1980) 249; Nucl. Phys. B190 [FS3] (1981) 553.

[22] S. Elitzur, Phys. Rev. D12 (1975) 3978.

[23] I. Montvay, Phys. Lett. 150B (1985) 441.

[24] H.G. Evertz, J. Jersák, C.B. Lang and T. Neuhaus, Phys. Lett. 171B (1986) 271.

[25] K. Fredenhagen and M. Marcu, Commun. Math. Phys. 92 (1983) 81; Phys. Rev. Lett. 56 (1986) 223.

[26] H.G. Evertz, V. Grösch, J. Jersák, H.A. Kastrup, D.P. Landau, T. Neuhaus and J.L. Xu, Phys. Lett. 175B (1986) 335.

[27] W. Bock, H.G. Evertz, K. Jansen, J. Jersák, K. Kanaya, H.A. Kastrup, D.P. Landau, T. Neuhaus and J.L. Xu, PITHA 88/14 (submitted to Z. Phys. C).

[28] A. Hasenfratz and T. Neuhaus, Nucl. Phys. B297 (1988) 205.

[29] W. Langguth and I. Montvay, Z. Phys. C36 (1987) 725.

[30] H.G. Evertz, E. Katznelson, P.G. Lauwers and M. Marcu, Phys. Lett. B221 (1989) 143.

[31] W. Bock, H.G. Evertz, J. Jersák, D.P. Landau, T. Neuhaus and J. L. Xu, Bielefeld BI-TP 89/17 (submitted to Phys. Rev. D).

[32] A. Hasenfratz and P. Hasenfratz, Phys. Rev. D34 (1986) 3160.

[33] S. Coleman and E. Weinberg, Phys. Rev. D7 (1973) 1888.

[34] R. Baier, C.B. Lang and H.-J. Reusch, Nucl. Phys. B (Proc. Suppl.) 4 (1988) 407; Nucl. Phys. B305 [FS23] (1988) 396.

[35] D.J.E. Callaway and R. Petronzio, Nucl. Phys. B267 (1986) 253.

[36] R. Dashen and H. Neuberger, Phys. Rev. Lett. 50 (1983) 1897.

[37] P. Hasenfratz and J. Nager, Z. Phys. C37 (1988) 477.

[38] M. Lüscher and P. Weisz, Nucl. Phys. B290 [FS20] (1987) 25; Nucl. Phys. B295 [FS21] (1988) 65; M. Lüscher, in *Nonperturbative Quantum Field Theory*, ed. G. 't Hooft et al. (Plenum 1988).

[39] M. Lüscher and P. Weisz, Phys. Lett. 212B (1988) 472; Nucl. Phys. B318 (1989) 705.

[40] M. Lüscher, in *Progress in Gauge Field Theory*, ed. G. 't Hooft et al. (Plenum 1984); Commun. Math. Phys. 104 (1986) 177; Commun. Math. Phys. 105 (1986) 153.

[41] I. Montvay and P. Weisz, Nucl. Phys. B290 [FS20] (1987) 327.

[42] I. Montvay, G. Münster and U. Wolff, Nucl. Phys. B305 [FS23] (1988) 143.

[43] C. Frick, K. Jansen, J. Jersák, I. Montvay, G. Münster and P. Seuferling, DESY 89-090 (to be published in Nucl. Phys. B).

[44] J. Kuti and Y. Shen, Phys. Rev. Lett. 60 (1988) 85.

[45] L. O'Raifeartaigh, A. Wipf and H. Yoneyama, Nucl. Phys. B271 (1986) 653.

[46] T. Munehisa and Y. Munehisa, Yamanashi-89-1.

[47] K. Jansen, J. Jersák, I. Montvay, G. Münster, T. Trappenberg and U. Wolff, Phys. Lett. 213B (1988) 203.

[48] K. Jansen, I. Montvay, G. Münster, T. Trappenberg and U. Wolff, Nucl. Phys. B322 (1989) 698.

[49] R.H. Swendsen and J.-S. Wang, Phys. Rev. Lett. 58 (1987) 86.

[50] A. Hasenfratz, K. Jansen, C.B. Lang, T. Neuhaus and H. Yoneyama, Phys. Lett. B199 (1987) 531.

[51] J. Kuti, L. Lin and Y. Shen, Phys. Rev. Lett. 61 (1988) 678.

[52] A. Hasenfratz, K. Jansen, J. Jersák, C.B. Lang, T. Neuhaus and H. Yoneyama, Nucl. Phys. B317 (1989) 81.

[53] J. Kuti, L. Lin and Y. Shen, Nucl. Phys. B (Proc. Suppl.) 4 (1988) 397.

[54] J. Kuti, L. Lin and Y. Shen, in *Lattice Higgs Workshop*, Tallahassee, ed. B. Berg et al. (World Scientific 1988) p.140.

[55] H. Neuberger, Nucl. Phys. B (Proc. Suppl.) 4 (1988) 501; Phys. Rev. Lett. 60 (1988) 889; Nucl. Phys. B300 [FS22] (1988) 180.

[56] U.M. Heller and H. Neuberger, Phys. Lett. B207 (1988) 189.

[57] J. Gasser and H. Leutwyler, Ann. Phys. 158 (1984) 142; Phys. Lett. B184 (1987) 83; Phys. Lett. B188 (1987) 477; Phys. Lett. B189 (1987) 197; Nucl. Phys. B307 (1988) 763.

[58] H. Leutwyler, Nucl. Phys. B (Proc. Suppl.) 4 (1988) 248.

[59] P. Hasenfratz and H. Leutwyler, to be published.

[60] M. Göckeler and H. Leutwyler, to be published.

[61] H. Leutwyler, private communication.

[62] A. Hasenfratz, K. Jansen, J. Jersák, C.B. Lang, H. Leutwyler and T. Neuhaus, FSU-SCRI-89-42 (submitted to Phys. Rev. D).

[63] T. Neuhaus, Nucl. Phys. B (Proc. Suppl.) 9 (1989) 21.

[64] J. Kuti, L. Lin and Y. Shen, Nucl. Phys. B (Proc. Suppl.) 9 (1989) 26.

[65] H. Leutwyler, Phys. Lett. B189 (1987) 197.

[66] U.-J. Wiese, Nucl. Phys. B (Proc. Suppl.) 9 (1989) 609.

[67] M. Lüscher, DESY 88-156 (Les Houches 1988).

[68] G. Bhanot and K. Bitar, Phys. Rev. Lett. 61 (1988) 798.

[69] C.B. Lang, UNI GRAZ-UTP-05-07-89 (to be published in Phys. Lett. B).

[70] M.B. Einhorn in *Lattice Higgs Workshop*, Tallahassee, ed. B. Berg et al. (World Scientific 1988) p. 69.

[71] H. Neuberger, in *Lattice Higgs Workshop*, Tallahassee, ed. B. Berg et al. (World Scientific 1988) p.197.

[72] J. Kogut, Rev. Mod. Phys 55 (1983) 775.

[73] K.G. Wilson, in *New Phenomena in Subnuclear Physics*, ed. A. Zichichi (Plenum 1977).

[74] I-H. Lee and R.E. Shrock, Nucl. Phys. B305 [FS23] (1988) 305.

[75] I-H. Lee and J. Shigemitsu, Phys. Lett. B178 (1986) 93.

[76] I-H. Lee and R.E. Shrock, Phys. Lett. B201 (1988) 497.

[77] S. Aoki, I-H. Lee and R.E. Shrock, Phys. Lett. B207 (1988) 471; Phys. Lett. 219B (1989) 355.

[78] A.K. De and J. Shigemitsu, Nucl. Phys. B307 (1988) 376.

[79] R.E. Shrock, Nucl. Phys. B (Proc. Suppl.) 4 (1988) 373.

[80] B. Pendleton, Nucl. Phys. B (Proc. Suppl.) 9 (1989) 82.

[81] J. Kuti, L. Lin, P. Rossi and Y. Shen, Nucl. Phys. B (Proc. Suppl.) 9 (1989) 87.

[82] I-H. Lee and R. Shrock, Phys. Lett. B196 (1987) 82.

[83] I-H. Lee and R. Shrock, Phys. Rev. Lett. 59 (1987) 14; Nucl. Phys. B290 [FS20] (1987) 275; Phys. Lett. B199 (1987) 541; Nucl. Phys. B305 [FS23] (1988) 286.

[84] E. Dagotto and J. Kogut, Phys. Lett. B208 (1988) 475.

[85] M.B. Einhorn and G.J. Goldberg, Phys. Rev. Lett. 57 (1986) 2115.

[86] A. Hasenfratz and T. Neuhaus, Phys. Lett. B220 (1989) 435.

[87] J. Shigemitsu, Phys. Lett. B189 (1987) 164.

[88] J. Polonyi and J. Shigemitsu, Phys. Rev D38 (1988) 3231.

[89] J. Shigemitsu, in *Lattice Higgs Workshop*, ed. B. Berg et al. (World Scientific 1988) p.229; Phys. Lett. B226 (1989) 364.

[90] J. Kuti, L. Lin, P. Rossi and Y. Shen, Nucl. Phys. B (Proc. Suppl.) 9 (1989) 99.

[91] J. Smit, Nucl. Phys. B (Proc. Suppl.) 9 (1989) 579.

[92] I-H. Lee, J. Shigemitsu and R.E. Shrock, BNL-43211 (submitted to Nucl. Phys. B).

[93] W. Bock, A.K. De, K. Jansen, J. Jersák and T. Neuhaus, HLRZ Jülich 89-41 (to be published in Phys. Lett. B).

[94] W. Bock, A.K. De, K. Jansen, J. Jersák, T. Neuhaus and J. Smit, HLRZ Jülich 89-57 (submitted to Phys. Lett. B).

[95] A. Hasenfratz, Nucl. Phys. B (Proc. Suppl.) 9 (1989) 92.

[96] I-H. Lee, J. Shigemitsu and R.E. Shrock, BNL-43098 (submitted to Nucl. Phys. B).

[97] D. Stephenson and A.M. Thornton, Phys. Lett. B212 (1988) 479;
A.M. Thornton, Phys. Lett. B214 (1988) 577; Phys. Lett. B221 (1989) 151;
Edinburgh 89/454.

[98] A.M. Thornton, Edingburgh preprint 89/467.

[99] H. Nielsen and M. Ninomiya, Nucl. Phys. B185 (1981) 20.

[100] L.H. Karsten and J. Smit, Nucl. Phys. B183 (1981) 103.

[101] I. Montvay, Phys. Lett. B199 (1987) 89; Phys. Lett. B205 (1988) 315; Nucl. Phys. B (Proc. Suppl.) 4 (1988) 443.

[102] I. Montvay, Nucl. Phys. B307 (1988) 389.

[103] A. Borrelli, L. Maiani, G.C. Rossi, R. Sisto and M. Testa, Phys. Lett. B221 (1989) 360.

[104] S. Aoki, in *Lattice Higgs Workshop*, ed. B. Berg et al. (World Scientific 1988), p.13; Nucl. Phys. B (Proc. Suppl.) 4 (1988) 479; Phys. Rev. Lett. 60 (1988) 2109; Phys. Rev. D38 (1988) 618;
K. Funakubo and T. Kashiwa, Phys. Rev. Lett. 60 (1988) 2113;
T.D. Kieu, D. Sen and S.-S. Xue, Phys. Rev. Lett. 61 (1988) 282.

[105] J. Smit, Nucl. Phys. B175 (1980) 307; Acta Physica Polonica B17 (1986) 531.

[106] P.V.D. Swift, Phys. Lett. B145 (1984) 256.

[107] J. Shigemitsu, OSU preprint DOE/ER/01545-397 (unpublished).

[108] M.F.L. Golterman and D.N. Petcher, Phys. Lett. 225B (1989) 159.

[109] S. Duane, A.D. Kennedy, B.J. Pendelton and D. Roweth, Phys. Lett. B195 (1987) 216.

[110] C. Frick, K. Jansen and P. Seuferling, HLRZ Jülich 89-38.

[111] U. Wolff, Phys. Rev. Lett. 62 (1989) 361.

Addendum

Since this text has been finished I have realized that I was not aware of or forgot to mention in the review several works which certainly merit attention. With an apology to their authors I want to include some of them at least in this addendum.

Subsec. 2.5: Nill [112] investigated ingeniously the U(1) lattice Higgs model by means of the perturbation expansion and found that a part of the Higgs phase transition manifold for small coupling constants λ and $e^2 = 1/\beta$ is of second order. This does not necessarily contradict the old Coleman-Weinberg observation [33] of a first order transition as the ratios of λ and e^2 are different in both approaches. The result of Nill raises the hope that the gauge fields could change the non-interacting character of the pure Φ^4 theory in continuum.

Subsecs. 3.2-3.4: I should have pointed out that the finite size effects are being studied predominantly by statistical physicists who accumulated a lot of experience and knowledge I cannot refer here to. The particle physicists concentrate mostly on special questions and of course benefit very much from the statistical physics also in this respect.

Subsec. 4.1: The existence of the chiral phase transition in the Higgs models with fermions raises the question whether this transition could not be used for a construction of the continuum limit. Horowitz et al. [113] find, however, that the continuum field theory obtained in this way is most probably non-interacting, similar to the pure Φ^4 theory.

[112] F. Nill, 2 preprints of Freie Universität Berlin (1988);
H. Meyer-Ortmanns and F. Nill, to appear.

[113] A.M. Horowitz, Phys. Lett. B219 (1989) 329;
A.M. Horowitz, S. Meyer and B. Pendleton, Kaiserslautern preprint (to appear in Phys. Lett. B).

SCALAR PARTICLES IN SUPERSTRING MODELS

Pierre Binétruy

LAPP, Chemin de Bellevue, BP 110
F-74941 Annecy-Le-Vieux Cedex, FRANCE

1   INTRODUCTION

Although the superstring tidal wave seems to have somewhat receded, important progress has been made recently in the detailed knowledge of these theories. After the days of the "superstring revolution" where the possible relevance of strings to the fundamental structure was recognized, came the time when the basic problems of this approach were identified: in particular the painfully large number of possible candidates and the issue of supersymmetry breaking. However, since one has started to address these problems, some definite progress has been made.

There are many different ways to present these advances but a particularly convenient one is to consider the status of scalar particles in these models. This is due to a striking property of strings: *in string theory, all relevant dimensionless parameters are vacuum expectation values of some scalar fields.* There is indeed only one fundamental string scale M_S: M_S is given in terms of the string coupling α' (which sets the normalisation of the 2-dimensional metric) by

$$M_S = \alpha'^{-1/2}. \tag{1.1}$$

All other relevant scales in the theory (Planck scale M_{Pl}, compactification scale M_{comp}, grand unification scale M_{GUT}, scale of supersymmetry breaking...) are expressed in terms of M_S and the vacuum expectation values (*vevs*) of associated scalar fields.

In the following Section, I will review some aspects of superstring models with an emphasis on the progress achieved in the last years. This is of particular relevance for phenomenologically-oriented studies since most of the so-called superstring-inspired phenomenology has dealt with the first models proposed and ignored later developments.[*] Section 3 reviews the structure of the scalar sector of string models and describes how the dilaton and the so-called moduli fields are invoved in the basic

---

[*] One should stress however that some of these developments have in fact strengthened basic hypotheses made in the early days, which were based at that time on little more than prejudice.

Higgs Particle(s)
Edited by A. Ali
Plenum Press. New York, 1990

issues of string theories. Finally Section 4 deals with the subject of this Meeting: the Higgs field. This together with the spectrum of supersymmetric particles, is certainly the place where most of the superstring-related phenomenology will take place in the future. There is no definite prediction yet from superstring but, as we will see, rather stringent bounds can be derived in most models, which put the Higgs sector within the reach of the next generation of accelerators. If nothing was found, this would point towards some fairly pathological models among the many present candidates. By the time these experimental searches are completed, one may hope that theoretical studies would have themselves pointed these models out. Otherwise string models would be in serious trouble.

2 SUPERSTRINGS MODELS

2.1 The original E_6 model

It is important to recall what is the content of the original model and how it was derived because it still is the unavowed basis of many superstring-related phenomenology discussions. It was obtained by Candelas, Horowitz, Strominger and Witten[1] by compactification of the 10-dimensional heterotic string theory[2] on a 6-dimensional compact Calabi-Yau manifold (the Calabi-Yau property ensuring that the 4-dimensional theory incorporates one supersymmetry charge).

The spectrum of this class of models is rather simple. It consists of:

(i) a supergravity multiplet which includes the graviton, its supersymmetric partner the gravitino, an antisymmetric tensor field, the dilaton and its fermionic partner.

(ii) chiral supermultiplets describing the "shape" of the compact manifold, of which more later.

(iii) a gauge non-singlet sector associated with gauge group $E_6 \otimes E_8'$ or one of its subgroups $\mathcal{H} \otimes \mathcal{H}'$ ($\mathcal{H} \subset E_6$ and $\mathcal{H}' \subset E_8'$). This sector divides itself into two:

– an observable sector where one finds the usual quarks, leptons, gauge fields and Higgs; it consists indeed of gauge supermultiplets (spin 1, spin 1/2) associated with $\mathcal{H}$ and matter chiral supermultiplets (spin 0, spin 1/2) singlet under $\mathcal{H}'$ and charged under $\mathcal{H}$: in the case of E_6, which we will take for illustration purpose, they are in full representations 27 and $\overline{27}$; otherwise, they are in representations of $\mathcal{H}$ contained in 27 and $\overline{27}$.

– a sector of gauge supermultiplets associated with $\mathcal{H}'$; these fields are neutral under $\mathcal{H}$ and thus interact with the observable sector only through gravitational interactions. They are said to form a *hidden sector*.

It proved to be a more difficult task to derive the couplings of these fields. The first attempt was an educated guess by Witten[3], who found his inspiration in a toy model of torus compactification supposed to reproduce the basic features of Calabi-Yau compactification. Witten started with the Lagrangian of 10-dimensional supergravity which, to a first approximation, is the field theory limit of the heterotic string and truncated it so as to obtain a 4-dimensional model with a $E_6 \otimes E_8'$ gauge supermultiplet and *one* matter supermultiplet in a 27 of E_6, coupled to supergravity.

Quite generally, in a supergravity model, all the couplings are determined by only three field-dependent functions[4]:

(a) the normalisation of the Yang-Mills kinetic term f_{ab},

$$\mathcal{L}_{YM} = -\frac{1}{4} f_{ab} F^{a\mu\nu} F^b_{\mu\nu} + \cdots \qquad (2.1)$$

where the indices a and b run over all the gauge group generators,

(b) the Kähler potential K which describes the normalisation of the kinetic term for the scalar fields:

$$\mathcal{L} = \frac{\partial^2 K}{\partial \Phi^A \partial \bar{\Phi}^{\bar{B}}} \partial^\mu \Phi^A \partial_\mu \bar{\Phi}^{\bar{B}} + \cdots \qquad (2.2)$$

(c) the superpotential W, an analytic function, which yields the scalar potential (F-term):

$$V = e^K \left[K_{A\bar{B}} (\frac{\partial W}{\partial \Phi^A} + W \frac{\partial K}{\partial \Phi^A})(\frac{\partial \bar{W}}{\partial \bar{\Phi}^{\bar{B}}} + \bar{W} \frac{\partial K}{\partial \bar{\Phi}^{\bar{B}}}) - 3|W|^2 \right] \qquad (2.3)$$

Witten's result is expressed in terms of the multiplets introduced earlier: the matter fields $\Phi^i, i = 1, \ldots, 27$, the dilaton φ and the squared radius T of the compact manifold: [†]

$$f_{ab} = S \, \delta_{ab}, \qquad (2.4)$$
$$K = -\ln(S + \bar{S}) - 3\ln(T + \bar{T} - \sum_i \Phi^i \Phi^{i*}), \qquad (2.5)$$
$$W = d_{ijk} \Phi^i \Phi^j \Phi^k, \qquad (2.6)$$

where

$$S + \bar{S} = (T + \bar{T})^3 \varphi^{-3}. \qquad (2.7)$$

Eqs.(2.4-2.7) fully determine the couplings in the model.

It is useful to see more closely how the field T appears because it provides a first example of the fields described in (ii) above, the so-called moduli fields. The radius R of the compact manifold defined for instance from the metric g_{IJ} (I, J are 6-dimensional real indices describing the compact manifold) by[3]

$$g_{IJ} = R^2 \delta_{IJ} \qquad (2.8)$$

is precisely given in string units in terms of the real part of the T field:

$$R = (ReT)^{1/2} M_S^{-1} \equiv M_{comp}^{-1}. \qquad (2.9)$$

The imaginary part of T, on the other hand, is deduced from the antisymmetric tensor contribution ($\epsilon_{12} = -\epsilon_{21} = \epsilon_{34} = -\epsilon_{43} = \epsilon_{56} = -\epsilon_{65} = 1$):

$$ImT = B, \quad B_{IJ} = B \, \epsilon_{IJ}. \qquad (2.10)$$

These two equations provide the full complex scalar component of the unique moduli superfield present in this toy model (in general, there are as many such scalar fields as there are 27 representations in the matter sector[5]).

Eq.(2.9) is the first example that we encounter of the property stated in the Introduction: the compactification scale, which turns out in these models to be also

---

[†] All scalar fields will be taken to be dimensionless.

the grand unification scale, is given in string units by the *vev* of the T field. Similarly[6], the Planck scale is expressed in terms of the *vev* of the S field introduced in (2.7): [‡]

$$M_{Pl} = (ReS)^{1/2} M_S. \tag{2.11}$$

Another important parameter is similarly obtained by considering Eqs.(2.1,2.4); the gauge coupling is obtained from the *vev* of the S field:

$$1/g^2 = ReS. \tag{2.12}$$

2.2 Problems with the original model

The model described in the last subsection covers more or less what was known in 1985 and served as a starting point for many superstring-inspired phenomenology papers. It has however some severe drawbacks which had to be addressed.

First of all the number of generations (only one 27) is not realistic and nothing is known of the couplings of matter fields in $\overline{27}$.

Secondly, the truncation procedure used in Ref.3, which consists in setting the heavy fields to zero, is a very dangerous and misleading procedure. The standard procedure would consist in integrating the heavy modes out but this is very difficult to implement in the case of Calabi-Yau compactification. Indeed, in order to do so, one should know the metric of the compact manifold; the choice (2.8) is obviously an oversimplification. Actually, there is no Calabi-Yau manifold for which the metric is known explicitly (this is in fact a long-standing mathematical problem). This seemed to lead to a dead end and the possibility of reliably computing the matter couplings in a realistic case seemed infinitely remote.

Finally, the compactification procedure used above, which starts from the field theory limit of the string, makes sense only if the heavy string modes lie at a scale much larger than the scale of compactification, i.e. $M_S \gg M_{comp}$. From (2.9), this means that ReT which measures the radius in string units has to be large in order for the whole approach to be consistent. But the grand unification scale given by (2.9) and (2.11)

$$M_{GUT} = M_{comp} = \frac{M_{Pl}}{(ReS)^{1/2}(ReT)^{1/2}} \tag{2.13}$$

together with the value (2.12) of the gauge coupling at grand unification are constrained by present data. This in turn sets the *vev* of ReT to be of order 1, in contradiction with our consistency condition[6,7].

2.3 The modern formulation

In the modern formulation, the point of view changes completely. Instead of considering the problem at the level of a many-dimensional spacetime and having to deal with the intricacies of compactification, one looks at it from the 2-dimensional point of view of *Conformal Field Theory*[8].

[‡]The *vev* symbol $< \cdots >$ which should be present whenever we discuss mass scales is suppressed throughout.

A closed string which propagates in a D-dimensional space describes a 2-dimensional surface known as the *world-sheet*. Because the string is itself a one-dimensional object, the oscillations propagating along the string in one direction (say to the left) do so independently of the oscillations propagating in the other direction (to the right). Conformal transformations are precisely the reparametrisations of the world-sheet which leave untouched this distinction between left-movers and right-movers.

In the heterotic string case, which was our starting point, space is 10-dimensional; out of the 26 left-moving degrees of freedom, 16 are internal (i.e. quantum numbers[§]) and the remaining 10 describe, together with the 10 right-moving modes, the ten spacetime dimensions, six of which are compact.

It was soon realized that a compact dimension is in fact not so different from an internal degree of freedom: the associated momentum is quantized and may be interpreted as a quantum number. This led to the concept of 4-dimensional string models where all but 4 degrees of freedom are interpreted as internal.

From the world-sheet point of view, it is even more similar since all these degrees of freedom, whether continuous, compact or internal are interpreted as 2-dimensional scalar (or fermion) fields.

Similarly, supersymmetry can be described on the world-sheet. A simple count of the number of degrees of freedom of a supersymmetric charge shows that at least two 2-dimensional supersymmetric charges are necessary to provide $N = 1$ supersymmetry in 4 dimensions[9,10].

The models compactified on Calabi-Yau manifolds actually correspond to 4 supersymmetry charges (2 left-moving and 2 right-moving, hence their name of (2,2) models) and the corresponding 2-dimensional theories have been identified[11]. Methods using (super)conformal symmetry then allow to compute the couplings between physical states. Each of the physical states is represented by an operator (*vertex operator*) whose explicit form is known; computing the couplings between given states amounts to computing the correlation functions of the corresponding operators on the 2-dimensional surface –a task familiar in statistical physics–. Thus within a given model it is in principle possible to compute all the couplings of interest (one is of course limited by the huge number of possible models).

The next Section describes the main results that were obtained using these methods.

3 THE STRUCTURE SCALARS

By "structure scalars" we mean (i) the dilaton and (ii) the moduli fields which were introduced in the compactification models of subsection 2.1 as the fields describing the shape of the compact manifold. These fields interact only gravitationally with the matter scalars to be described in the next Section and are therefore *hidden* from them. They nevertheless play an important role in issues of key importance for low energy predictions, such as supersymmetry breaking and the discrimination between the large number of models.

3.1 The dilaton field

We have seen in subsection 2.1 that the dilaton appears in the 4-dimensional

---

[§]In the Calabi-Yau case, two of these numbers are identified with the local properties of the curved compact manifold –remember that general relativity is a gauge theory– and the remaining 14 are the $6 + 8$ quantum numbers of $E_6 \otimes E'_8$.

supermultiplet spectrum as the scalar field S of Eq.(2.7). It is this field that we call here dilaton and it is as such that it provides the ratio M_{Pl}/M_S as given in Eq.(2.11).

Indeed one finds that the properties of S obtained in the toy model are quite general. In particular, it remains true at the tree level of string theory that the Yang-Mills normalisation function f_{ab} is still simply given by S as in (2.4): the value of the gauge coupling at grand unification is obtained from the vev of S

$$1/g^2 = ReS. \tag{3.1}$$

Also as before S does not appear in the superpotential, which means that the corresponding term in the scalar potential is zero (cf. (2.3)):

$$V(S) = 0. \tag{3.2}$$

Thus the S field corresponds to a flat direction of the potential and its vev remains undetermined at this stage. This is certainly not a desirable feature of the model since it prevents us from determining M_{Pl} as well as the value of the gauge coupling.

There is however an important property which relates the presence of flat directions of the potential with unbroken supersymmetry. One can prove the following theorem[12]: *as long as supersymmetry is not broken, flat directions are not lifted to all orders of perturbation theory.* Thus, the lifting of the degeneracy associated wth a flat direction is connected with supersymmetry breaking and it must occur through non-perturbative effects. We will come back to this in a moment but let us first introduce other flat directions of the scalar potential: the moduli fields.

3.2 The moduli fields.

We have already encountered an example of such a field in our example of subsection 2.1: the T field whose vev was connected with the radius of the compact manifold through (2.9). This is the representative of one class of moduli which we may call of the radius type and generically represent by T. They somehow determine the topological class of the manifold[9]. There is however another class of moduli which we will denote by the letter C: these determine the complex structure of the manifold.

Let us illustrate the difference on the simplest of the compact manifolds: the torus. As is well-known, the torus can be represented as a rectangle whose opposite sides are identified. We choose to represent it in the complex plane by a square of unit length (Fig.1(a)). Changing the radius of this torus amounts to multiply all its dimensions by a factor λ (Fig.1(b)). On the other hand, the case of Fig.1(c) where only the "imaginary" direction is dilated cannot be described in the complex plane by a holomorphic transformation $z \to \lambda z$; it corresponds to a change in the complex structure of the torus.

Using (2,2) superconformal invariance, one can prove the following theorem[5]: in a given model, there are as many T fields as there are 27 representations in the matter sector and there are as many C fields as there are $\overline{27}$. One can check this result on our toy model (respectively one and zero).

Some of the T fields play an important role. We have already encountered the overall radius of the manifold and will see its role in connection with duality in the next subsection. In the case of orbifold models, one also finds among the T fields the *blowing modes.*

Crudely speaking, orbifolds are obtained by symmetrizing tori (or more generally manifolds)[13]. If we take for example the torus of Fig.1(a) and symmetrize it with respect to its center, we need to fold it along the dotted lines in Fig.1(d) in order to make the correct identifications (we can also disregard half of its surface, say the half above the diagonal). We thus obtain a tetrahedron whose vertices correspond to the fixed points of the symmetrization procedure. It is possible to obtain a Calabi-Yau manifold from an orbifold by blowing up the singularities of curvature (vertices): one

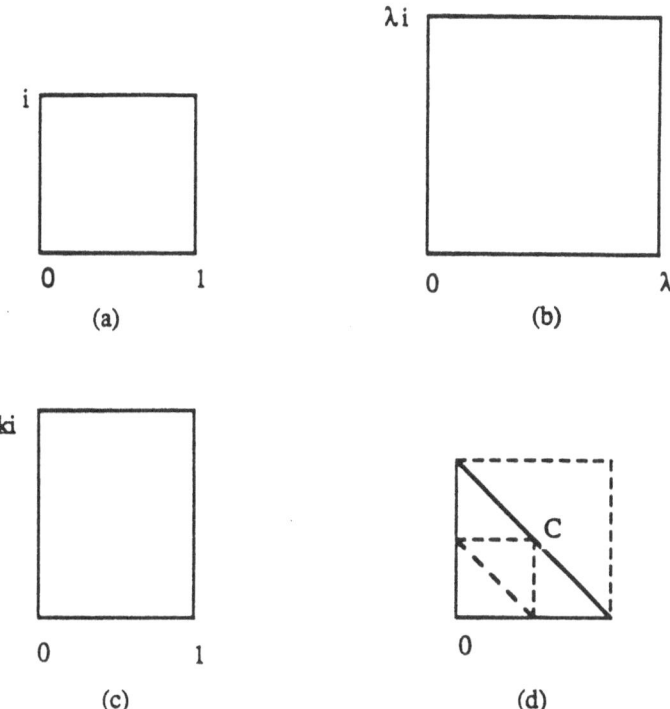

Figure 1. (a) Torus represented by a square of unit length in the complex plane: opposite sides are identified. (b) Another torus with the same complex structure. (c) Another torus with a different complex structure. (d) Orbifold obtained by symmetrizing torus (a) with respect to its center C.

replaces each vertex by a Calabi-Yau manifold. It turns out that, for each fixed point, there is a T field whose vev is the radius of the Calabi-Yau manifold attached to it. Hence $<T>=0$ corresponds to the orbifold limit whereas $<T>\neq 0$ corresponds to the Calabi-Yau case. This provides a simple way for blowing up the singularities of orbifolds and studying the Calabi-Yau models in their vicinity[5,13,14].

Most of the progress made in the last year has been in the determination of the properties of moduli. Remember that it is sufficient to know the dependence of the Kähler potential and of the superpotential (cf. (2.2) and (2.3)). One finds[15] that in the case of (2,2) models, the same holomorphic function determines the superpotential of the matter fields and the Kähler potential of the moduli (until now, the only connection we had made between the two was that they appear in equal numbers). More explicitly, both the Kähler potential $K_1(T)$ for the T fields and the superpotential W for the matter fields in the 27 are fixed by the same function $\mathcal{F}_1(T)$:

$$K_1(T) = -\ln Y_1, \tag{3.3}$$

$$Y_1 = \sum_{a=1}^{N_1} (\partial_a \mathcal{F}_1 + \partial_{\bar{a}} \overline{\mathcal{F}}_1)(T^a + \bar{T}^a) - 2(\mathcal{F}_1 + \overline{\mathcal{F}}_1), \tag{3.4}$$

$$W = \frac{1}{3} \partial_a \partial_b \partial_c \mathcal{F}_1 \ \Phi^a \Phi^b \Phi^c, \tag{3.5}$$

where N_1 is the number of T fields (i.e. the number of matter fields Φ in a 27).

Note that for $N_1 = 1$ and $\mathcal{F}_1 = T^3$ one recovers the couplings originally found by Witten (Eqs.(2.5,2.6)) [¶]. Also, setting the matter fields Φ to zero, one automatically gets a vanishing potential. Thus, as promised, the moduli fields correspond to flat directions of the scalar potential.

A similar relation exists between the Kähler potential $K_2(C)$ for the C fields and the superpotential for the matter fields in $\overline{27}$. Finally, the normalisation of the kinetic term (2.2) for the matter fields can be expressed in terms of $K_1(T)$ and $K_2(C)$.[15]

All this might seem rather academic from a phenomenological point of view since the moduli remain *hidden* from the observable sector. Nevertheless, we will see in the next subsection that these fields play an important role in issues of key importance for superstring phenomenology. Moreover, even though their couplings are of a gravitational type, they could be of direct relevance for present day experiments. Take for example $N_1 = 1$ and

$$\mathcal{F}_1 = \frac{1}{8} \lambda T^4 / M_{Pl}, \tag{3.6}$$

which yields a superpotential for the matter fields:

$$W = \frac{\lambda}{M_{Pl}} T \Phi^3. \tag{3.7}$$

Such a term typically gives a coupling in the Lagrangian of the form

$$\frac{\lambda}{M_{Pl}} < H > \overline{\Psi} \Psi \ T \tag{3.8}$$

where H is a Higgs field which we set to its *vev* (say 300 Gev) and the Ψ are quarks or leptons.

The coupling (3.8) induces a force between the constituents of ordinary matter which is mediated by the T field. Its range depends on the mass of this modulus. Quite generally, the C fields acquire a rather large mass through supersymmetry breaking because this breaking determines a particular choice for the complex structure (hence it induces a structure in the otherwise flat direction C). The case of the T fields is somewhat different: some acquire a mass (for example[16], in some cases, the T field corresponding to the overall radius acquires a mass $m_{3/2} M_{GUT}/M_{Pl}$) but other directions remain flat. If the T field considered in (3.6) is one of them, it will eventually obtain a non-zero mass at one loop through the couplings (3.8). Cvetič[17] evaluated it

---

[¶] A note on notations: in Eq.(2.6), there is only one 27 and the index i runs over the components of this 27. In (3.3-3.5), there are N_1 T fields, hence N_1 27 and $a = 1 \ldots N_1$; the index i is suppressed.

to be possibly as small as 10^{-18}Gev. Then the above force mediated by T would be long-ranged (< 1000m) and could be tested in fifth force experiments.

3.3 The importance of structure scalars

If superstring theory has not come yet with definite predictions, it is due mainly to two obstacles: the huge number of possible models and the uncertainties associated with the supersymmetry-breaking mechanism. In both instances, moduli play a very special role.

In principle, a criterion that would discriminate between the different string models and determine the ground state among them would require to fully develop a quantum field theory of strings. We are right now quite far from having such a criterion at our disposal. On the other hand, one has gradually become aware of the important role played by the symmetry known as *duality* in comparing different string theories.

It is beyond the scope of this presentation to give a detailed description of duality and its many uses but one can give an idea of how it works. Consider the compactification of one string coordinate on a circle of radius R. Remember that one can define independently left-movers and right-movers i.e.

$$
\begin{aligned}
X(\tau, \sigma) &= X_L(\tau + \sigma) + X_R(\tau - \sigma), \\
X_L(\tau + \sigma) &= x_L + 2\alpha' p_L(\tau + \sigma) + \text{oscillating modes}, \\
X_R(\tau - \sigma) &= x_R + 2\alpha' p_R(\tau - \sigma) + \text{oscillating modes}, \quad (3.9)
\end{aligned}
$$

where τ and σ are coordinates on the world-sheet, $x = x_L + x_R$ is the centre of mass of the string and we have written explicitly only the zero (non-oscillating) modes. Because the coordinate X describes a circle, the associated total momentum $p = p_L + p_R$ is quantized: $p = m/R$. On the other hand, since the string can wind around the circle n times (i.e. $X(\sigma + \pi) = X(\sigma) + 2n\pi R$), we have

$$
\begin{aligned}
p_L &= \frac{1}{2\alpha'} nR + \frac{m}{2R} + \text{oscillating modes}, \\
p_R &= -\frac{1}{2\alpha'} nR + \frac{m}{2R} + \text{oscillating modes}. \quad (3.10)
\end{aligned}
$$

As is well-known, the eigenstates of the string mass-squared operator

$$
\frac{1}{2} M^2 = p_L^2 + p_R^2 + \text{oscillating mode contribution} \quad (3.11)
$$

constitute the spectrum of particles of the theory. The key observation is that this spectrum is invariant under the transformation

$$
\begin{aligned}
n &\longleftrightarrow m \\
R &\longleftrightarrow \alpha'/R. \quad (3.12)
\end{aligned}
$$

This is the *duality transformation*[18]. Although its derivation is straightforward, it is rather surprising to uncover a symmetry which relates compact manifolds with very large radius to ones with very small radius. One can immediately note the role it could play in our case since, as was stressed at the end of subsection 2.2, the consistency of the Candelas *et al.*[1] approach required $R \gg 1$ whereas grand unification constraints seem to indicate that R is of order 1.

We have interpreted the radius as a modulus –the unique one in the case of compactification on a circle–. In more complicated models, duality generalizes to a discrete transformation on the whole space of moduli. To get an idea of the type of structure that is involved, we will describe briefly the case of 2-dimensional torus compactification which is very similar to the compactification described in subsection 2.1. Indeed Eqs.(2.8-2.10) can be translated *verbatim* (with now I,J 2-dimensional real indices describing the torus) and we see that torus compactification is described by a single complex modulus $T = R^2 + iB$ and the modulus space is half of the complex plane, which we write in coset space notation $\frac{SL(2,\mathbf{R})}{U(1)}$. In this case, the duality transformations are simply $SL(2,\mathbf{Z})$ transformations (also called modular transformations):

$$T \rightarrow \frac{aT - iB}{icT + d}, \quad ad - bc = 1, \quad a, b, c, d \in \mathbf{Z}. \tag{3.13}$$

This case is fairly representative of the general structure of the moduli space. Just for the sake of completeness, let us indicate that in the case of the heterotic string compactified on a D-dimensional torus, the moduli space has locally the structure of the coset space[19] $\frac{SO(16+D,D)}{SO(16+D)\otimes SO(D)}$. The general duality transformations form the discrete group[20] $SO(16 + D, D, \mathbf{Z})$. The uses of duality transformations are many: they play an important role in comparing different string theories[21], in finding models with enhanced gauge symmetry (self-dual models)[22], in discussions of strings at high temperature[23], in statistical models[24]...

Another topic where the structure scalars play an important role is supersymmetry breaking. Both the dilaton S and the moduli fields correspond to flat directions of the potential and we have pointed out earlier that the lifting of the corresponding degeneracy is a sign of supersymmetry breaking through nonperturbative effects. It is clear that supersymmetry breaking is an important issue but all the more in superstring models since we need to determine the *vevs* of the structure scalars in order to fix the value of such important quantities as M_{Pl}, M_{GUT} or the value of the gauge coupling g (cf Eqs.(2.11)-(2.13)).

Another problem specific to string is the fact that all scales turn out to be of the same order. We mentionned in subsection 2.2 that constraints on M_{GUT} and g impose to consider *vevs* for ReS and ReT of order 1. Thus a realistic model would necessarily have[6,7]

$$M_S \simeq M_{Pl} \simeq M_{GUT} \simeq M_{comp}. \tag{3.14}$$

In these conditions, it remains to explain how a scale as small as M_W can ever be generated. We know that M_W is related to the mass term for the Higgs field of $SU(2) \otimes U(1)$, itself determined by supersymmetry breaking (all the fields present at low energy are found among the massless modes of the superstring; hence any mass in this sector has a supersymmetry-breaking origin). Thus supersymmetry breaking is the place where the string theory has to address the problem of hierarchy, in the hard way.

There are several ways of dealing with supersymmetry breaking in the models that we consider (for a recent review see for example Ref.25). To emphasize the issues at stake, I will concentrate on a specific mechanism based on gaugino condensation[26]. In models compactified on a Calabi-Yau manifold, we found a hidden sector of gauge supermultiplets corresponding to the gauge symmetry $\mathcal{H}' \subset E_8'$. One expects that, when the $\mathcal{H}'$ gauge coupling becomes strong, the corresponding gauginos form condensates which break local supersymmetry in this *hidden* sector: the gravitino acquires a non-zero mass $m_{3/2}$.

In order not to generate a huge cosmological constant, one usually accompany this with a second supersymmetry-breaking mechanism whose role is to cancel the vacuum energy. A favourite way is to generate a *vev* for the field strength of the antisymmetric tensor:

$$H_{IJK} = \partial_I B_{JK} + \partial_K B_{IJ} + \partial_J B_{KI},$$
$$< H_{IJK} > = \text{c}\, M_{Pl}^3\, \epsilon_{IJK} \tag{3.15}$$

(this time I, J, K are 3-dimensional complex coordinates describing the 6-dimensional compact manifold; ϵ_{IJK} is the completely antisymmetric tensor). The end result is that a superpotential is generated

$$W = \text{c} + h e^{-\frac{3S}{2b_0}} \tag{3.16}$$

where the second term takes its origin from gaugino condensation. This induces a non-trivial potential in the S direction which fixes $< S >$; $< T >$ is then determined at one loop. One finds as expected that both of these *vevs* are of order 1; more precisely one generates at most two orders of magnitude between $m_{3/2} = e^{K/2}|W|$ and M_{Pl}.

A detailed analysis of radiative corrections[16] in the model of subsection 2.1 shows that no soft supersymmetry breaking terms are generated either at tree level or at one loop in the observable sector (this includes gaugino and scalar masses). The reason for this has recently been identified[27]: it is the presence of a $SL(2, \mathbf{R})$ symmetry for the full Lagrangian[28]. More precisely,

$$T \rightarrow \frac{aT - iB}{icT + d}, \quad ad - bc = 1,$$
$$\Phi^i \rightarrow \frac{\Phi^i}{icT + d}, \quad a, b, c, d \in \mathbf{R}. \tag{3.17}$$

This symmetry forbids to send the information of supersymmetry breaking from the *hidden* sector (where $m_{3/2} \neq 0$) to the observable sector. Luckily, this symmetry has an anomaly which breaks nonperturbatively the invariance. An analysis[27] based on effective Lagrangian methods[29] shows that a scale $\tilde{m}$ of global supersymmetry breaking is thus generated in the observable sector which can easily account for the 16 orders of magnitude between M_W and M_{Pl} ($\tilde{m} \approx M_W$).

Of course, the similarity between (3.13) and (3.17) suggests that the whole scheme might be general. One should note however that the symmetry that we have just discussed is continuous whereas duality is discrete. It would be very interesting to see in the light of all this how the discrete character of duality appears at the level of field theory[30]. Of possible relevance are the quantization conditions that appear in discussing supersymmetry breaking (for example, topological arguments indicate that c is quantized, which in turn forces $m_{3/2}$ to take only discrete values).

4 THE MATTER SCALARS. THE HIGGS

We now consider the scalars of the observable sector.

4.1 The low energy scalar potential

In the type of model of subsection 2.1, there are cases where the gauge group $\mathcal{H}$

is smaller than E_6: they correspond to models where part of the gauge symmetry is broken by the compactification process (the so-called breaking by Wilson loops)[31]. In such instances, the scalars that we consider are in representations of the gauge group $\mathcal{H} \subset E_6$ which are contained in 27 and $\overline{27}$ of E_6. To be explicit, we give in Table 1 the field content of a complete representation 27. Fields in $\overline{27}$ have opposite quantum numbers. As we will stress later, one could expect in principle fields which do not belong to these representations.

Table 1 . Value of the charges under the $SU(3)_C \otimes SU(3)_L \otimes SU(3)_R$ and $SU(3)_C \otimes SU(2)_L \otimes U(1)_{Y_L} \otimes U(1)_{Y_R} \otimes U(1)_{T_{3R}}$ subgroups of E_6 for the components of the 27 of $E_6 (Y = Y_L + Y_R)$.

$SU(3)_c \otimes SU(3)_L \otimes SU(3)_R$		$SU(3)_c \otimes SU(2)_L$	Y_L	Y_R	T_{3R}
$(3,3,1)$ $Q = \begin{pmatrix} u \\ d \end{pmatrix}$		$(3,2)$	$1/3$	0	0
D		$(3,1)$	$-2/3$	0	0
$(\bar{3},1,\bar{3})$ u^c		$(\bar{3},1)$	0	$-4/3$	$-1/2$
d^c		$(\bar{3},1)$	0	$2/3$	$1/2$
D^c		$(\bar{3},1)$	0	$2/3$	0
$(1,\bar{3},3)$ $H_1 = \begin{pmatrix} H_1^+ \\ H_1^0 \end{pmatrix}$ or $E^c = \begin{pmatrix} E^+ \\ N_E^c \end{pmatrix}$		$(1,2)$	$-1/3$	$4/3$	$1/2$
$H_2 = \begin{pmatrix} H_2^0 \\ H_2^- \end{pmatrix}$ or $E = \begin{pmatrix} \nu_E \\ E^- \end{pmatrix}$		$(1,2)$	$-1/3$	$-2/3$	$-1/2$
$L = \begin{pmatrix} \nu \\ e^- \end{pmatrix}$.		$(1,2)$	$-1/3$	$-2/3$	0
e^c		$(1,1)$	$2/3$	$4/3$	$1/2$
N'		$(1,1)$	$2/3$	$-2/3$	$-1/2$
N		$(1,1)$	$2/3$	$-2/3$	0

Looking at Table 1, one realizes that there are two types of fields which can be used to break $\mathcal{H}$ down to $SU(3) \otimes U(1)$:

– fields with the quantum numbers of a Higgs field under $SU(3) \otimes SU(2) \otimes U(1)$; we will refer to them generically as H; as in any supersymmetric model, there are two types of such fields, H_1 and H_2, coupling respectively to the u-type and d-type (upper and lower component of a weak doublet) quarks;

–fields singlet under $SU(3) \otimes SU(2) \otimes U(1)$ referred to as N. In Table 1, there are actually two types of such fields: one has allowed trilinear couplings to the Higgs, i.e.

$$W = \lambda H_1 H_2 N, \tag{4.1}$$

whereas the other one, which we will call N', has couplings to the Higgs forbidden by E_6 gauge symmetry.

There are of course other trilinear couplings involving the other matter fields as well as possibly non-renormalisable terms (quartic terms in the superpotential...).

Disregarding these for a moment, we obtain the following supersymmetric potential (cf Eq.(2.3) where we ignore all non-renormalisable terms):

$$V = \lambda^2[|N|^2(|H_1|^2 + |H_2|^2) + |H_1|^2|H_2|^2] + \mathrm{D-terms}. \tag{4.2}$$

Of course, supersymmetry breaking is needed in order to generate non-zero masses for the scalars. As explained in the last section, supersymmetry breaking mechanisms should yield a scale $\tilde{m}$ of global supersymmetry breaking in the observable sector which is typically of the order of M_W. When local supersymmetry is broken in a hidden sector, the most general supersymmetry breaking terms which appear in the Lagrangian of the observable sector are of only three different types: gaugino mass, scalar mass and A-term (term in the potential which is proportional to the superpotential)[32]

$$\mathcal{L}' = \frac{1}{2}m_\lambda\bar{\lambda}\lambda - m_s^2|\Phi|^2 + A[W(\Phi) + W(\Phi)^*]. \tag{4.3}$$

Hence supersymmetry breaking generates new terms in the scalar potential:

$$V' = m_1^2|H_1|^2 + m_2^2|H_2|^2 + m_N^2|N|^2 + m_{N'}^2|N'|^2 - A_\lambda[\lambda H_1 H_2 N + h.c.]. \tag{4.4}$$

In principle, the complete scalar potential for the H and N fields alone,[||] is given by Eqs.(4.2),(4.4). In these expressions however, one should replace all couplings by running ones $m(\mu), A(\mu), \lambda(\mu)\dots$ i.e. include large logarithmic [**] radiative corrections of order $\ln(M_{GUT}/\mu)$. The equations of evolution of these running couplings are the well-known Renormalisation Group Equations (RGE) whose explicit form can be computed once the spectrum and the couplings of the model are determined. The supersymmetry breaking contributions that we discussed in Section 4 appear actually at compactification and provide boundary conditions for these RGE:

$$m_1(M_{GUT}) = m_1^{(0)}, m_2(M_{GUT}) = m_2^{(0)}, m_N(M_{GUT}) = m_N^{(0)}, \cdots \tag{4.5}$$

where $m_1^{(0)}, m_2^{(0)}, \dots$ are all of order $\tilde{m}$.

As we go down in scale, we reach a region where some of the mass-squared in (4.4) become negative; some of the scalar fields acquire a non-zero *vev*, part of the gauge symmetry is broken: one refers to this as a *radiative breaking of gauge symmetry*[34]. Solving for the ground state of the μ-dependent scalar potential $V_0 = V + V'$ yields vacuum expectation values

$$< H_1 > = v_1(\mu), < H_2 > = v_2(\mu), < N > = n(\mu), < N' > = n'(\mu). \tag{4.6}$$

4.2 The different types of models

Until now, we have neglected the nonrenormalisable terms that may arise as in any effective theory obtained by decoupling heavy modes. Let us consider the N field alone for a while.

[||]Including only renormalisable terms; see below.

[**]In principle, quadratic divergences appear in the hidden sector where one breaks supersymmetry but these divergences are softened by the gravitational coupling between hidden and observable sector (which contributes a factor $1/M_{Pl}^2$) and only logarithmic divergences are generated in this sector. This is precisely the reason of breaking supersymmetry in a hidden sector.[33]

Suppose that there exists a term of the type $(27 \times \overline{27})^n$ in the superpotential

$$W = \frac{\tilde{\lambda}}{M_{Pl}^{2n-3}} \, N^n \overline{N}^n, \quad n \geq 2, \tag{4.7}$$

where $\overline{N}$ is the field in $\overline{27}$ with opposite quantum numbers as N. Then the supersymmetric part of the potential reads

$$V = \left(n \frac{\tilde{\lambda}}{M_{Pl}^{2n-3}} \right)^2 (|N|^{2(n-1)}|\overline{N}|^{2n} + |N|^{2n}|\overline{N}|^{2(n-1)}) + D-terms. \tag{4.8}$$

The D-terms vanish for $N = \overline{N}$ since these fields have opposite quantum numbers. In this direction, after we have included the supersymmetry breaking part V' in (4.4),

$$V = m_N^2 N^2 + 2 \left(n \frac{\tilde{\lambda}}{M_{Pl}^{2n-3}} \right)^2 |N|^{2(2n-1)}. \tag{4.9}$$

When m_N^2 becomes negative, N acquires a non-zero vev:

$$< N > = (\sqrt{-m_N^2} M_{Pl}^{2n-3})^{\frac{1}{2(n-1)}}. \tag{4.10}$$

Since $\sqrt{-m_N^2} \approx \tilde{m} \approx M_W$, one notes that $< N >$ is larger than 10^{10}GeV and the gauge symmetry under which N is charged is broken at a scale intermediate between M_W and M_{Pl}. This is referred to as "*intermediate scale*" breaking[35]. If all N-type fields present in the model obtain such an intermediate vev, then we are left only with H fields at low energy.

Let us note that, so far, we have suppressed a generation index. In fact, in realistic models, we might have 3 generations for each type of fields: $H_1^i, i \leq 3, H_2^j, j \leq 3, \cdots$. It is however desirable that only one H_1 and one H_2 get a non-zero vev.Indeed, if there are too many H_1^i, H_2^j with non-zero $vevs$, one runs into problems with the evolution of the gauge coupling constant: M_{GUT} becomes too small. This problem is actually generic to all models with a minimal gauge and field content[36]. One sometimes call "unHiggs" the H fields which do not acquire a vev.[37]

The prototype of the models with only H fields has therefore only 2 "real" Higgs H_1 and H_2 which survive at low energy. It is usually referred to as the Minimal Supersymmetric Model (MSM). Out of the eight scalar degrees of freedom, two are eaten in the Higgs mechanism and the physical scalar fields are:
  - 2 neutral Higgs h^0, H^0,
  - 1 pseudoscalar A^0,
  - 1 charged Higgs $H^{\pm}$.

The phenomenological aspects of this model will be treated by H.Haber[38]. Let me just say here that the low energy mass spectrum in the scalar sector depends on only 2 parameters which we can choose to be m_{H+} and β, $tg\beta = v_2/v_1$. One thus obtains relations between the different masses which can be turned into mass limits:

$$m_{H+} \geq M_W, \; m_{H_0} \geq M_Z, \; m_{A_0} \geq m_{h^0},$$
$$m_{h^0} \leq M_Z |\cos 2\beta|. \tag{4.11}$$

For a large class of orbifolds however, one can show[39] that there are no such non-renormalisable terms as in 4.7. In this case, if there are N or N' fields in the massless spectrum, they will only acquire a nonzero vev in the Tev region. The minimal model of this type has 3 scalars: H_1, H_2 and N (models with N' have phenomenological problems of their own[40]). The spectrum after gauge symmetry breaking includes 3 neutral Higgs, 1 pseudoscalar and 1 charged Higgs.

Another possibility arises with the (numerous) models which have a gauge symmetry smaller than E_6. It may happen in this case that N or N' are present in the massless spectrum but are gauge singlets. An example of this is provided by the so-called flipped $SU(5) \otimes U(1)$ model[41]. This model was obtained in an effort to reproduce the grand unified $SU(5)$ model. It turns out that in string models it is very difficult to have the same representation in the gauge sector and in the scalar sector (as for the 24 of grand unified $SU(5)$). The string model which comes closest symmetry is precisely the flipped $SU(5) \otimes U(1)$ model: one is obliged to enlarge the gauge symmetry and to "flip" the $SU(5)$ assignments of quarks and leptons. In the TeV region, one is left with H_1, H_2 and N which is precisely a gauge singlet in this model. A further difference with the previous cases is that now a renormalisable N^3 term is allowed in the superpotential. The phenomenology of all these models will be discussed in J. Gunion and H. Haber's talks.

References

1. P. Candelas, G. Horowitz, A. Strominger and E. Witten, *Nucl. Phys.* **B258**, 46 (1985).

2. D. Gross, J. Harvey, E. Martinec and R. Rohm, *Nucl. Phys.* **B256**, 253 (1985), *Nucl. Phys.* **B267**, 75 (1986).

3. E. Witten, *Phys. Lett.* **155B**: 151 (1985).

4. E. Cremmer, S. Ferrara, L. Girardello, and A. Van Proeyen, *Nucl. Phys.* **B212**: 413 (1983).

5. L. Dixon, *in* Proceedings of the Summer Workshop in High Energy Physics and Cosmology, Trieste 1987, ed. G.Furlan, J.C.Pati, D.W.Sciama, E.Sezgin and Q.Shafi, *ICTP Series in Theoretical Physics*, Vol.4, p.67.

6. M. Dine and N. Seiberg, *Phys. Rev. Lett.* **55**, 366 (1985).

7. V. Kaplunovsky, *Phys. Rev. Lett.* **55**, 1036 (1985).

8. See for example M. Peskin, Proc. of the 1986 Theoretical Advanced Study Institute in Particle Physics, Santa Cruz.

9. E. Witten, *Nucl. Phys.* **B268**: 79 (1986).

10. A. Sen, *Nucl. Phys.* **B278**: 289 (1986); J. Distler, *Phys. Lett.* **188B**: 431 (1987).

11. D. Gepner, *Nucl. Phys.* **B296**, 757 (1988), *Phys. Lett.* **199B**: 380 (1987).

12. E.Witten, *Nucl. Phys.* **B188**: 513 (1981).

13. L.Dixon, J.Harvey, C.Vafa and E.Witten, *Nucl. Phys.* **B261**, 678 (1985), **B274**, 285 (1986).

14. M. Cvetič, in *Proc. of the International Workshop on Superstrings, Composite Structures and Cosmology*, College Park, 1987 (World Scientific). S. Hamidi and C. Vafa, *Nucl. Phys.* **B279**, 465 (1987).

15. S. Cecotti, S. Ferrara and L. Girardello, *Nucl. Phys.* **B308**: 436 (1988),*Phys. Lett* **213B**: 443 (1988) and CERN preprint TH-5080; M. Cvetič, J. Louis and B. Ovrut, *Phys. Lett* **206B**: 227 (1988) and University of Pennsylvania preprint UPR-0380T; M. Cvetič, J. Molera and B. Ovrut, University of Pennsylvania preprint UPR-0376T; D. Kutazov, Weizmann Institute preprint WIS-88/55/Oct-PH; L.J. Dixon, V.S. Kaplunovsky and J. Louis, SLAC–PUB–4959, UTTG–19–89 (1989)

16. P. Binétruy, S.Dawson, M.K. Gaillard and I. Hinchliffe, *Phys. Lett.* **192B**: 377 (1987), and Phys. Rev. **D37**: 2633 (1988).

17. M. Cvetič, *Phys. Lett.* **229B**: 41 (1989).

18. K. Kikkawa and M. Yamasaki, *Phys. Lett.* **149B**: 357 (1984); N. Sakai and I. Senda, *Prog. Theor. Phys.* **75**: 692 (1984)

19. K. Narain, *Phys. Lett.* **169B**: 41 (1986); K. Narain, M. Sarmadi and E. Witten, *Nucl. Phys.* **B279**: 369 (1987).

20. V.P. Nair, A. Shapere, A. Strominger and F. Wilczek, *Nucl. Phys.* **B287**: 402 (1987); A. Giveon, E. Rabinovici and G. Veneziano, *Nucl. Phys.* **B322**: 167 (1989).

21. J. Dai, R.G. Leigh and J. Polchinski, preprint UTTG-12-89; M. Dine, P. Huet and N. Seiberg, *Nucl. Phys.* **B322**: 301 (1989).

22. P. Ginsparg and C. Vafa, *Nucl. Phys.* **B289**: 414. (1987)

23. J. Atick and E. Witten, *Nucl. Phys.* **B310**: 291 (1988).

24. R. Savit, *Rev. Mod. Phys.* **52**: 453 (1980).

25. P. Binétruy, "The issue of supersymmetry breaking in strings," to be published in *Proc. of the XII Symposium on Elementary Particle Physics*, Kazimierz, 1989; preprint LAPP-TH-271/89.

26. M. Dine, R. Rohm, N. Seiberg and E. Witten, *Phys. Lett.* **156B**: 55 (1985).

27. P. Binétruy and M.K. Gaillard, *Phys. Lett.* **232B**: 83 (1989); preprint LAPP-TH-273/89.

28. S.P. Li, R. Peschanski and C.A. Savoy, *Phys. Lett.* **178B**: 193 (1986)

29. G. Veneziano and S. Yankielowicz, *Phys. Lett.* **113B**: 231 (1982).

30. S. Ferrara, D. L üst, A. Shapere and S. Theisen, *Phys. Lett.* **225B**: 363 (1989),*Phys. Lett* **233B**: 147 (1989).

31. Y. Hosotani, *Phys. Lett.* **126B**: 309 (1983).

32. L. Girardello and M. Grisaru, *Nucl. Phys.* **B194**, 65 (1982).

33. R. Barbieri, S. Ferrare and C. Savoy, *Phys. Lett.* **119B:** 343 (1989); A.H. Chamseddine, R. Arnowitt and P. Nath, *Phys. Rev. Lett.* **49**, 970 (1982); L. Hall, J. Lykken and S. Weinberg, *Phys. Rev.* **D27:** 2359 (1983).

34. L. Alvarez-Gaumé, M. Claudson and M.B. Wise, *Nucl. Phys.* **B207**, 96 (1982); L. Alvarez-Gaumé, J. Polchinski and M.B. Wise, *Nucl. Phys.* **B221**, 495 (1983); J. Ellis, J.S. Hagelin ,D.V. Nanopoulos and K. Tamvakis, *Phys. Lett.* **125B:** 275 (1983); L.E. Ibáñez and C. Lopez, *Nucl. Phys.* **B233**, 511 (1984).

35. M.Dine, V. Kaplunovsky, M.Mangano, C. Nappi and N. Seiberg, *Nucl. Phys.* **B259**, 549 (1985).

36. A. Font, L.E. Ibáñez, H.-P. Nilles and F. Quevedo, *Phys. Lett.* **210B:** 101 (1988).

37. J.Ellis, D.V. Nanopoulos, S.T. Petcov and F. Zwirner, *Nucl. Phys.* **B283**, 93 (1987).

38. H. Haber, these Proceedings.

39. M. Cvetič, *Phys. Rev.* **D37**, 2366 (1988).

40. B. Campbell, J. Ellis, M.K. Gaillard, D.V. Nanopoulos and K.A. Olive, *Phys. Lett.* **180B:** 77 (1986).

41. I. Antoniadis, J. Ellis, J.S. Hagelin and D.V. Nanopulos, *Phys. Lett.* **194B:** 231 (1987).

A RELATION BETWEEN THE SCALES OF

WEAK AND STRONG INTERACTIONS?

Christof Wetterich

Deutsches Elektronen Synchrotron

Notkestr. 85

2000 Hamburg 52

FRG

ABSTRACT

I explore the possibility that long distance physics determines the Fermi scale independently of the precise parameters of the short distance scalar potential. Then the ratio between the scales of weak and strong interactions would become predictable.

1. The connection problem

The standard model of electroweak and strong interactions exhibits two basic mass scales: Strong interactions are characterized by the scale Λ_{QCD} which determines the masses of baryons. The Fermi scale φ_0 measures the magnitude of electroweak spontaneous symmetry breaking. The masses of leptons and W^+, Z-bosons are proportional φ_0. Also the quark masses are proportional φ_0 and the light pseudo Goldstone mesons like pions and kaons have a mass $m \sim (\varphi_0 \Lambda_{QCD})^{1/2}$. The scale ratio $\gamma = \Lambda_{QCD}/\varphi_0 \approx 10^{-3}$ is one of the most basic quantities for our explanation of nature.

As it stands, the standard model is completely consistent up to a very high energy scale M (acting as a physical momentum cutoff). At this scale it should emerge as an effective theory from a more fundamental theory which is hoped to unify all forces and explain the many free parameters present in the standard model. The characteristic scale for unification with gravity is the Planck mass $M_p \approx 10^{19}$ GeV, implying $M \lesssim M_p$. So far nothing indicates that M must be much smaller than M_p and I assume here that the unification scale M is indeed in the vicinity of M_p (say $M \approx 10^{17}$ GeV). Then a simple question arises: Why is φ_0 so "near" Λ_{QCD}? Looked upon from the short distance scale M the two scales Λ_{QCD} and φ_0 are almost degenerate, with y appearing as a "fine structure", similar in size to typical fermion mass ratios. Is this a pure accident, or is there a deeper reason for a connection between the two long distance scales? This is the connection problem [1].

In the tree approximation the Fermi scale φ_0 is determined by the minimum of the scalar potential

$$ V_M(\varphi) = -\mu^2 \varphi^\dagger \varphi + \frac{\lambda}{2}(\varphi^\dagger \varphi)^2 $$

$$ \varphi_0^2 = \mu^2 / \lambda \qquad\qquad (1) $$

For small coupling constants S. Coleman and E. Weinberg [2] computed the one loop corrections to the effective protential. Within the standard model, "naive perturbation theory"[1] implies that the Fermi scale remains determined by the (renormalized) mass parameter μ^2 similar to (1). Only for very small μ^2 and λ (and not too large quark masses) the running quartic scalar coupling vanishes at some scale $\mu_0^2 > \mu^2$. In this case (Coleman-Weinberg spontaneous symmetry breaking [2]) one obtains $\varphi_0^2 \approx \mu_0^2$. Both μ^2 and μ_0^2 are essentially free parameters of the electroweak sector of the standard model, and so is the scale φ_0. (The most natural scale for μ^2 is of the order M^2. A solution of the connection problem would simultaneously answer the question why $\varphi_0^2 \ll M^2$ - the gauge hierarchy problem [3]). On the other hand, the scale Λ_{QCD} is determined by the strong coupling constant at the scale M, $\alpha_s(M)$, which is just another free parameter of the standard model. The scale ratio y cannot be predicted. One concludes that the connection problem has no solution within the standard model if naive perturbation theory reliably describes the physics below the scale M. In this talk I will question the validity of naive perturbation theory. (This also questions the validity of the theoretical

---

1) The reason for this name for the Coleman-Weinberg perturbative expansion will become clear in a moment.

lower bounds on the mass of the Higgs scalar which are derived from naive perturbation theory.)

2. Long distance and short distance effects and the need for a scale dependent effective action

For the purpose of this discussion we may neglect the electroweak gauge bosons and leptons and reduce the standard model to the coupled system of scalar doublet, quarks and gluons with local SU(3) and global SU(2)xU(1) symmetry. The scalar φ is coupled to the strong interaction degrees of freedom (quarks q and gluons G) only through the Yukawa couplings of the quarks, h_q. Imagine that a brilliant calculator (comparable to Laplace's demon) integrates out the QCD degrees of freedom and computes the effective action for the scalar field, $\hat{S}[\varphi]$

$$exp - \hat{S}[\varphi] = \int \mathcal{D}G \, \mathcal{D}q \, exp - S[G, q, \varphi] \qquad (2)$$

In addition to the original potential $V_M(\varphi)$ the effective action $\hat{S}[\varphi]$ will contain many new scalar interactions which vanish only for zero Yukawa couplings. These interactions are nonlocal at all length scales smaller than $\approx \Lambda_{QCD}^{-1}$ (more precisely m_π^{-1}). For example, spontaneous chiral symmetry breaking in QCD through (light) quark condensates[2] induces interactions of the type [1]

$$V_{\langle \bar{q}q \rangle} = h_q \langle \bar{q}q \rangle_0 \, \phi_{k_0} + \sum_{n>2} \beta_n \left(h_q \phi_{k_0} \right)^n \langle \bar{q}q \rangle_0^{\frac{4-n}{3}} \qquad (3)$$

Here ϕ_k is the average of φ over a volume V_k with typical size k^{-4} and k_0 is some scale near Λ_{QCD}. For n > 2 the interactions (3) are nonlocal and involve products $\varphi(x_1)\varphi(x_2)...\varphi(x_n)$ with x_i separated by a typical distance k_0^{-1}.[3] These are, of course, by far not the only nonlocal interactions. Nonlocalities arise on all scales k $> \Lambda_{QCD}$. As another example, the fluctuations of the top quark give typical interactions (one loop)

_____

2) I denote by $\langle \bar{q}q \rangle_0$ the quark condensate in the zero quark mass limit.

3) The interactions (3) should be considered as a very crude approximation, demonstrating the qualitative coupling of φ to the quark condensate.

$$V_t = -\frac{3}{16\pi^2} h_t^4 \phi_{k_t}^4 \ln\left(h_t^2 \phi_{k_t}^2 / k_t^2\right) \qquad (4)$$

which are nonlocal on scales $k_t \approx m_t$.

How important are the induced nonlocal interactions (3)(4) for the determination of the Fermi scale? Let me assume that the mass term of the short distance scalar potential (1) is of the order of the unification scale, $\mu^2 \approx M^2$. For φ of the order 10^2 GeV a simple order of magnitude estimate gives $V_M \sim 10^4$ GeV^2 M^2 whereas $V_{\langle \bar{q}q \rangle}$ is of order Λ_{QCD}^4 and $V_t \lesssim m_t^4$. Naively, one could immediately conclude that $V_{\langle \bar{q}q \rangle}$ and V_t are completely negligible. There is, however, an important qualitative difference between $V_{\langle \bar{q}q \rangle}$, V_t and V_M: Whereas V_M is local at the scale M, the two other contributions are strongly nonlocal. There is no way for a direct comparison of such effects. It is indeed possible that tiny nonlocal effects dominate over the local interactions at large enough distances - a typical example being the very weak nonlocal magnetic interactions responsible for the formation of Weiss domains in ferromagnets. Any comparison of the different contributions needs first an extrapolation of the short distance action to an effective action at longer distances. In particular, one has to compute the effective potential for averages of the scalar field over volumes with size k^{-4}. Renormalization group equations will be an important tool to relate the effective potential at different scales k. Knowledge of the effective potential for averages of the scalar field over a volume with size Λ_{QCD}^{-4} is required before making a direct estimate about the relative importance of the nonperturbative contribution $V_{\langle \bar{q}q \rangle}$. As I will show below, naive perturbation theory breaks down for a computation of the effective potential at long distances. An investigation of the connection problem within the standard model has to go beyond the Coleman-Weinberg loop expansion and provide a reliable estimate for the long distance effective potential.

3. The average action: "block spin" concepts in continuous spacetime

A computation of the effective action for averages of fields in the standard model is not a simple task. As a first step I restrict myself to the pure scalar sector and discuss the average action for the N-component φ^4 theory. For even N I use a complex formulation with Euclidean action

$$S[\chi] = \int d^4x \left\{ -\mu^2 \chi^\dagger \chi + \frac{\lambda}{2} (\chi^\dagger \chi)^2 + \partial^\mu \chi^\dagger \partial_\mu \chi \right\} \qquad (5)$$

Formally, the average action Γ_k can be defined by functional integration with a constraint which keeps the average ϕ_k of the field χ at a fixed value φ:

$$exp - \Gamma_k[\varphi] = \int \mathcal{D}\chi \prod_x \delta(\phi_k(x) - \varphi(x)) exp - S[\chi] \qquad (6)$$

In the lattice formulation the average action involves a product over δ-distributions at block lattice sites x and coincides with the standard block spin action [4]. I will present here a formulation [5] where both $S[\chi]$ and $\Gamma_k[\varphi]$ are integrals over a continuous spacetime, preserving the symmetries of rotations and translations. (I use an Euclidean formulation and regularization by a momentum cutoff.)

The average of χ over a volume with typical scale k is defined by

$$\phi_k(x) = \int d^4y \, f_k(y-x) \chi(y) \qquad (7)$$

with f_k decreasing fast for $|x-y| \gg k^{-1}$ (for example $f_k(x) = \pi^{-2}k^4 exp(-k^2 x^\mu x_\mu)$) and fullfilling[4)]

$$\int d^4x \, f_k(x) = 1$$
$$\int d^4z \, f_{k_2}(x-z) f_{k_1}(z-y) = f_k(x-y) \qquad (8)$$

for suitable $k(k_1, k_2)$. In momentum space the Fourier components are related by[5)]

$$\phi_k(q) = f_k(q)\chi(q) \qquad (9)$$

It is convenient to define the constraint in momentum space [6]

$$\prod_x \delta(\phi_k(x) - \varphi(x)) \rightarrow C \, exp\{-\int d^4q \, \nu(q) |f_k(q)\chi(q) - \varphi(q)|^2\} \qquad (10)$$

---

4) This implies that the average in a volume V_{k_2} over average fields ϕ_{k_1} is again an average field in V_k, with $k > \max(k_1, k_2)$. For the Gaussian $f_k(x)$ above one has $k^{-2} = k_1^{-2} + k_2^{-2}$.

5) For convenience I use different normalizations for $f_k(q)$ and $\chi(q)$, $\phi_k(q)$: $f_k(x) = (2\pi)^{-4}\int d^4q f_k(q) exp(-iq^\mu x_\mu)$, $\chi(x) = (2\pi)^{-4}\int d^4q \chi(q) exp(-iq^\mu x_\mu)$.

by a Gaussian with large $\mathcal{V}(q)$. I will concentrate on the choice

$$\mathcal{V}(q) = \mathcal{V}\left(1 - f_k(q)^2\right)^{-1} \tag{11}$$

with $\mathcal{V} \gg \mu^2$.

The average action can be written as the logarithm of the partition function from a "constrained action" $S_k[\varphi, \chi]$

$$\exp - \Gamma_k[\varphi] = \int \mathcal{D}\chi \, \exp - S_k[\varphi, \chi] \tag{12}$$

$$S_k[\varphi,\chi] = \int d^4q \left\{ \chi^+(q)(P(q) - \mu^2) \chi(q) \right.$$

$$- \mathcal{V}(q) f_k(q)\left(\varphi^+(q)\chi(q) + \chi^+(q)\varphi(q)\right)$$

$$\left. + \mathcal{V}(q)\varphi^+(q)\varphi(q) \right\}$$

$$+ \frac{\lambda}{32\pi^4} \int d^4q_1 \, d^4q_2 \, d^4q_3 \left(\chi^+(q_1)\chi(q_2)\right)\left(\chi^+(q_3)\chi(q_1 - q_2 + q_3)\right) \tag{13}$$

The constrained action contains a linear source term for χ $(\sim \varphi^+(q)\chi(q))$. The nonlocality of the constraint finds its expression in the modified propagator ("average propagator")

$$P(q) = q^2 + f_k^2(q)\mathcal{V}(q) \tag{14}$$

which cuts off all small momenta $q^2 \ll k^2$. Functional integration over χ is therefore essentially an integration over the high momentum modes with $q^2 \gtrsim k^2$. In turn, the average action Γ_k is the effective action for the low momentum modes, where all momenta $q^2 \gtrsim k^2$ have been integrated out. It can be shown [6] that there is a one to one correspondence between the expectation values of operators involving the low momentum modes $\varphi(q)$, evaluated with the effective action $\Gamma_k[\varphi]$, and corresponding operators with the same functional dependence on the average fields $\phi_k(q)$, evaluated with the original action $S[\chi]$. Thus Γ_k is the effective action for averages of fields over volumes with size k^{-4}.

The definition of the average action Γ_k preserves the full symmetries of spacetime rotations and translations. We can therefore expand as usual in

100

the number of derivatives. The average potential $U_k(\varphi)$ is the nonderivative part of the Lagrangian, obtained for $\varphi(x) = $ const

$$\Gamma_k = \int d^4x \; U_k(\varphi) + \text{derivative terms} \qquad (15)$$

In the infinite volume limit $k \to 0$ it approaches the usual effective potential as defined by a Legendre transform of the logarithm of the partition function in presence of sources. The average potential must therefore become convex for $k \to 0$ [7][8]. For finite k, however, U_k is not necessarily convex since it relates to averages of fields over finite volumes. The Fermi scale φ_0 corresponds to the absolute minimum of the average potential of the full standard model[6] for finite, small k. This includes the QCD contributions discussed above. The average potential is the tool we need for an estimate of the importance of the nonlocal interactions which involve averages over large volumes.

4. The average potential in the inner region: breakdown of naive perturbation theory

It is intuitively clear that for large enough length scales k^{-1} the average potential will behave qualitatively different for φ^2 larger or smaller $\varphi_M^2 = \mu^2/\lambda$. In the outer region of the potential (large φ^2) the constrained action (with φ constant) is minimized by the constant solution $\chi = \varphi$. The steepest descent expansion around this minimum leads to a valid perturbative series for small λ. The loop expansion with a constant background field $\chi = \varphi$ (naive perturbation theory) should be reliable in the outer region. Using renormalization group improved perturbation theory it can indeed be shown [5][6] that U_k approaches the perturbative Coleman-Weinberg potential in the outer region as $k \to 0$. (Corrections to the Coleman-Weinberg potential from the finiteness of the average volume are of the order $k^2/\lambda\varphi^2$.) In the pure φ^4 theory the minimum of the perturbative Coleman-Weinberg potential is always within the outer region [6].

In the inner region of the potential (small φ^2) the constant solution

---

6) Inclusion of the SU(2)xU(1) gauge bosons requires a refinement of the definitions, since the average field is not a locally gauge invariant quantity.

$\chi = \varphi$ does no longer minimize the constrained action. For example, an average value $\phi_k = 0$ can be obtained by a configuration with $|\chi|^2 = \varphi_M^2$, where the phase of χ changes smoothly in space such that the average of χ vanishes in a volume V_k. For a model with continuous symmetries (N>2) only additional kinetic energy is needed compared to the solution with average $|\phi_k| = \varphi_M$. A simple order of magnitude estimate gives $U_k \sim -k^2 \varphi^2 + const$ for the inner region - in sharp contrast to naive perturbation theory where the potential energy difference between configurations with constant $\chi = \varphi_M$ or $\chi = 0$ is of order $\mu^2 \varphi_M^2$. It is indeed well known that naive perturbation theory breaks down for the inner region of the effective potential (the $k \to 0$ limit of U_k): First, the effective potential must be convex [7], whereas naive perturbation theory gives only small corrections to the nonconvex tree potential (1). Second, naive perturbation theory leads to an imaginary part in the one loop effective potential. This is an indication that the constant background field $\chi = \varphi$ corresponds to an unstable saddlepoint rather than a minimum of the constrained action [9]. This instability of fluctuations around a constant background field already appears for small enough finite k, as can be seen easily by a quadratic approximation of S_k for small fluctuations around $\chi = \varphi$ [5]. One concludes that naive perturbation theory is unreliable for the inner region of the average potential. One needs a new steepest descent expansion around the true minimum of S_k, accounting for the qualitative behaviour $U_k \sim -k^2 \varphi^2$ in the inner region. In particular, one cannot trust the results of naive perturbation theory in the context of the connection problem: For φ_M^2 of the order of the unification scale M^2 the Fermi scale φ_0 is far inside the inner region of the average potential. We need to compute U_k in the inner region before being able to assess the relative importance of nonlocal contributions like $V_{<\bar{q}q>}$ or V_t (3, 4).

5. Spin wave solutions and the loop expansion for the average potential in the inner region

It is fortunately possible to find for the inner region a new exact solution [5] of the nonlocal field equations derived from the constrained action S_k (for N > 3). This spin wave solution corresponds to the absolute minimum of S_k and reads (for φ in the one direction, $\varphi_1 = \varphi_1^* = \varphi = const$)

$$\chi_1(x) = \varphi = const.$$

$$\chi_2(x) = h\sqrt{1 - \varphi^2/h^2}\ exp\left(-i\, p_0^\mu x_\mu\right)$$

$$\chi_a(x) = 0 \quad , \quad a > 2 \tag{16}$$

The modulus of χ is constant

$$|\chi(x)|^2 = h^2 = \frac{1}{\lambda}\left(\mu^2 - \bar{k}^2\right) \tag{17}$$

whereas the average of χ is given by φ, up to small corrections of order k^2/ν.

$$\phi_{k_1}(x) = \varphi$$

$$\phi_{k_2}(x) = f_k(p_0)\chi_2(x) = O(\bar{k}^2/\nu)\chi_2(x) \tag{18}$$

The inverse wavelenght p_0 corresponds to the minimum of the average propagator $P(q)(14)$, with

$$P(p_0) = \bar{k}^2 = min\ P(q) \tag{19}$$

It follows from the general properties of $P(q)$ that both p_0^2 and $\bar{k}^2$ are of the order k^2. (The precise relation depends on the choice of $f_k(q)$.) The inner region only appears for $\bar{k}^2 < \mu^2$ and covers all $\varphi^2 < h^2$. For small $\bar{k}^2$ the inner region extends almost to the minimum of the tree potential, $h^2 \approx \varphi_M^2$.[7] As argued above, only gradient energy instead of potential energy is needed to have $|\phi_k|^2 \ll \varphi_M^2$.

The tree approximation to the average potential in the inner region obtains by inserting the spinwave solution (16) into the constrained action (13)

$$U_k^{(0)} = -\bar{k}^2 \varphi^\dagger \varphi - \frac{\lambda}{2} h^4 \tag{20}$$

---

7) In a renormalization group improved treatment [6] the inner region extends to the minimum of the perturbative potential, instead of φ_M^2 (minimum of the tree potential).

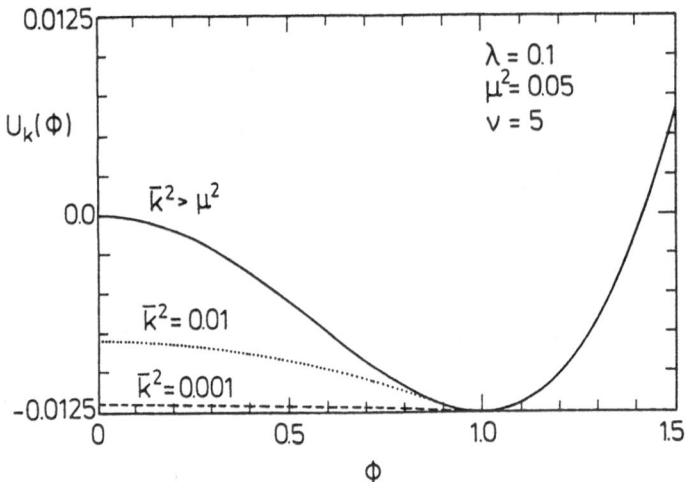

Fig. 1. The average potential in the tree approximation.

It becomes flat for k→0 and the average potential becomes convex, as it should be (compare fig.1). For the one loop contribution, one has to compute the determinant for the quadratic fluctuations around the spin wave solution [5].

$$U_k^{(1)} = \frac{1}{32\pi^4} \int_{q^2 < \Lambda^2} d^4q \; \ln\left\{ (P(q)-\bar{k}^2)\left[(P_+(q)-\bar{k}^2)(P_-(q)-\bar{k}^2) \right.\right.$$

$$+ (\mu^2-\bar{k}^2)(P_+(q)+P_-(q)-2\bar{k}^2) \right]$$

$$+ \lambda\varphi^\dagger\varphi \left[2(P_+(q)-\bar{k}^2)(P_-(q)-\bar{k}^2) - \right.$$

$$\left.\left. - (P(q)-\bar{k}^2)(P_+(q)+P_-(q)-2\bar{k}^2) \right]\right\} + const. \tag{21}$$

with

$$P_\pm(q) = P(q \pm p_0) \tag{22}$$

It is easily checked that the action for small fluctuations is always positive semidefinite and $U_k^{(1)}$ has no imaginary part (as it should be for an expansion around the absolute minimum). Evaluation of the integral (21) gives [5], for $\bar{k}^2 \ll \varphi^2 \ll h^2$,

$$U_k^{(1)} = -\frac{\lambda}{16\pi^2} c_2 \frac{\bar{k}^4}{\mu^2} \varphi^\dagger\varphi - \frac{\lambda^2}{16\pi^2} c_4 \frac{\bar{k}^4}{\mu^4} (\varphi^\dagger\varphi)^2 \tag{23}$$

with c_2, c_4 of order one. The one loop corrections are small in this region, suppressed compared to the tree approximation by additional powers of $\bar{k}^2/\mu^2$. Again, the one loop average potential becomes convex for k→0. The steepest descent approximation around the spin wave solution does not suffer from the problems of naive perturbation theory. Extending this approach to the full standard model should give a reliable answer to the connection problem.

6. The critical index α and the relative importance of nonperturbative QCD effects

I hope I have convinced the reader that the description of long distance physics needs an effective action characteristic for this distance - the average action - and that the average potential becomes flat in the inner region for k→0. The crucial question, however, is more quantitative: "How fast the average potential becomes flat for small k?" Let me concentrate on the region of interest for the connection problem,

105

namely very small $\varphi^2 \ll M^2$, and define the index α as a measure how fast the average potential becomes flat for small fields

$$\alpha = \Delta_{\bar{k}}^{-1}(\varphi_0) \frac{\partial}{\partial \ln \bar{k}} \Delta_{\bar{k}}(\varphi_0)$$

$$\Delta_{\bar{k}}(\varphi_0) = U_{\bar{k}}(\varphi_0) - U_{\bar{k}}(0) \tag{24}$$

(Here φ_0 is a small field which I choose to coincide with the Fermi scale $\sim$175 GeV.) For α independent of $\bar{k}$ one solves for Δ_k with the initial condition $\Delta_M(\varphi_0) = -\mu^2 \varphi_0^2$ and obtains for the potential difference

$$\Delta_{\bar{k}}(\varphi_0) = -\mu^2 M^{-2} (\bar{k}/M)^{\alpha-2} \bar{k}^2 \varphi_0^2 \tag{25}$$

(For α depending on $\bar{k}$ one may use in (25) a suitably defined average of α between the scales $\bar{k}$ and M). In the spontaneously broken phase the critical index α is a universal quantity independent of the precise definition of the short distance physics. (More precisely, this holds for $\alpha(\bar{k})$ for $\bar{k} \ll M$. Its value in the standard model decides on the relative importance of the nonlocal interactions discussed in section 2.

Let me describe three possible scenarios: Consider first the case $\alpha >$ 2. Even for $\mu^2 = M^2$ the average potential is already very flat at the Fermi scale $\bar{k} = \varphi_0$, $\Delta_{\varphi_0} \approx - (\varphi_0/M)^{\alpha-2} \varphi_0^4$, since φ_0/M is a very small quantity ($\approx 10^{-15}$). At the strong interaction scale $\bar{k} = \Lambda_{QCD}$ the average potential is even flatter

$$\Delta_{\Lambda_{QCD}}(\varphi_0) \approx - 10^{-18(\alpha-2)+6} \Lambda_{QCD}^4 \tag{26}$$

and we conclude that the nonperturbative QCD contribution $V_{\langle \bar{q}q \rangle}(3)$ may become important for α-2 > 1/3. (This value decreases for $\mu/M \ll 1$). It is perhaps interesting to note what would happen if $U_{\Lambda_{QCD}}$ is dominated by $V_{\langle \bar{q}q \rangle}$: The dominant contribution to the φ dependence of $V_{\langle \bar{q}q \rangle}$ comes from the s quark condensate, and one estimates [1] a minimum of $V_{\langle \bar{q}q \rangle}$ at

$$\varphi_0 \approx \langle \bar{s}s \rangle_0^{1/3} / h_s \tag{27}$$

with h_s the Yukawa coupling of the strange quark. The Fermi scale comes out in the right order of magnitude. This scenario predicts [1] a very small mass for the physical scalar

$$m^2 \approx m_s \langle \bar{s}s \rangle / \varphi_0^2 \approx (200 \, keV)^2 \qquad (28)$$

This may already be excluded by experiment, but some caution is required to evaluate the effective coupling of the physical scalar to nucleons and mesons.

My other two scenarios are more conservative and assume $\alpha = 2$ (or α in the vicinity of two). In this case the nonperturbative QCD contribution $V_{\langle \bar{q}q \rangle}$ is completely negligible. At the Fermi scale $\Delta\varphi_0$ is of the order $-\varphi_0^4$. The fluctuations of the top quark, however, give still a contribution V_t of the order φ_0^4, if the top quark mass is large. Our second scenario assumes that the long range fluctuations of the top quark (and W, Z bosons) lead to a local minimum of the average potential at φ_0. This could arise from combining a quadratic term $-\tilde{c}\, \bar{k}_t^2 \phi^2_{kt}$ with a quartic term like V_t for suitable $\tilde{c}$. It also assumes that this minimum is lower than the minimum of the naive perturbative potential. In this case the long range fluctuations would determine φ_0, essentially independent of $\mu^2(M)$, $\lambda(M)$. If the long distance physics "forgets" the precise values of the parameters of the scalar potential at short distances, the Fermi scale and the physical scalar mass m must be predictable in terms of the other couplings of the standard model. There is simply no free parameter (in addition to the dimensionless gauge and Yukawa couplings) whose variation leads to a variation of φ_0 and m. This reduction of parameters occurs in general whenever the long distance physics determines the average potential for small φ independently on the precise form of the short distance potential and the true minimum lies in this region. In my second scenario a solution of the connection problem may be related to the observation that the relatively large Yukawa coupling h_t of the top quark increases sizably towards smaller scales once the strong gauge coupling becomes sufficiently large at scales near Λ_{QCD}. The (perturbative) β-function for the quartic scalar coupling becomes negative for large h_t, and one expects that a positive φ^4 term (for small φ) is generated once $m_t \gtrsim 80$ GeV. (This fits well with the present experimental bounds on m_t).

In my third scenario, finally, the minimum of the naive perturbative potential is the true minimum of the average potential even for small k. In this case the flattening of the average potential in the inner region is irrelevant. Naive perturbation theory can be justified a posteriori and there is no solution to the connection problem within the standard model.

What do we know about the index α ? Simple energy considerations indicate that α should be at least two (one) for models with spontaneously broken continuous (discrete) symmetries. (One expects a nonvanishing surface energy for the case of discrete symmetry, N=1). The one loop calculation seems to indicate α =2 (20) for the pure scalar theory including entropy effects. Since the entropy effects may be underestimated in this simple approach I consider α =2 rather as a lower bound. A renormalization group improved computation of α is in progress. (Also wave function renormalization has to be included.) The effect of Yukawa couplings etc. for the average potential and α has still to be worked out. Already at this stage, however, it is obvious that the naive perturbative estimate $\Delta_k \approx -\mu^2 \varphi^2_0$ ($\alpha \approx 0$) is completely misleading. It is a very interesting problem in field theory to work out the average potential in the inner region for the standard model. One has to find out if the last scenario is realized or if the arguments discussed in this talk open the door to a new understanding of the physics of electroweak spontaneous symmetry breaking.

References

1) C. Wetterich, Phys. Lett. 209B (1988) 59.

2) S. Coleman and E. Weinberg, Phys. Rev. D7 (1973) 1888.

3) E. Gildener and S. Weinberg, Phys. Rev. D13 (1976) 3333.

4) K.G. Wilson, Phys. Rev. B4 (1971) 3174, 3184.
   K.G. Wilson and I.G. Kogut, Phys. Rep. 12 (1974) 75.
   L.P. Kadanoff, Physics 2 (1966) 263.

5) A. Ringwald and C. Wetterich, DESY preprint 89-068.

6) C. Wetterich, to appear.

7) K. Symanzik, Commun. Math. Phys. 16 (1970) 48.
   I. Iliopoulos, C. Itzykson and A. Martin,
   Rev. Mod. Phys. 47 (1975) 165.

8) For perturbative approximations to the convex effective potential, see:
   R. Fukuda and E. Kyriakopoulos, Nucl. Phys. B85 (1975) 354.
   L. O'Raifeartaigh and G. Parravicini, Nucl. Phys. B111 (1976) 501.
   Y. Fujimoto, L. O'Raifeartaigh and G. Parravicini,
   Nucl. Phys. B212 (1983) 268.
   C.M. Bender and F. Cooper, Nucl. Phys. B224 (1983) 403.
   D.J.E. Callaway and D.J. Maloof, Phys. Rev. D27 (1983) 406.
   R.M. Haymaker and J. Perez-Mercader, Phys. Rev. D27 (1983) 1948.
   D.J.E. Callaway, Phys. Rev. D27 (1983) 2974.

F. Cooper and B. Freedman, Nucl. Phys. B239 (1984) 459.

R.J. Rivers, Z. Phys. C22 (1984) 137.

L.O'Raifeartaigh, A. Wipf and Y. Yoneyama, Nucl. Phys. B271 (1986) 653.

9) S. Coleman, "Aspects of symmetry", Cambridge University press, 1985.

E. Weinberg and A. Wu, Phys. Rev. D36 (1987) 2474.

NON-MINIMAL HIGGS BOSONS: THEORY AND PHENOMENOLOGY

Howard E. Haber

Santa Cruz Institute for Particle Physics
University of California, Santa Cruz, CA 95064

ABSTRACT

I discuss the properties of elementary scalar Higgs bosons which appear in multi-Higgs extensions of the Standard Model of electroweak interactions. The phenomenology of the two-Higgs-doublet model (including the minimal supersymmetric extension of the Standard Model) is examined in detail. Finally, I consider the circumstances under which the most likely Higgs structure of a given theory is minimal.

1. INTRODUCTION

The Standard Model of electroweak interactions has been very successful in describing all experimentally observed electroweak phenomena. Nevertheless, the mechanism of spontaneous symmetry breaking has yet to be ascertained. The existence of a massive W and Z is proof that there is an underlying $SU(2) \times U(1)$ gauge symmetry which is spontaneously broken to $U(1)_{EM}$. The longitudinal components of the W and Z are precisely the Goldstone bosons associated with the spontaneous breakdown of the electroweak symmetry group. This observation is completely general, independent of a particular mechanism for spontaneous symmetry breaking. In the minimal version of the Standard Model, the neutral component of a complex doublet of elementary scalar fields acquires a vacuum expectation value. The W and Z acquire mass via the Higgs mechanism, and one physical neutral scalar, the Higgs boson (ϕ^0) remains in the scalar spectrum. More complicated models with non-minimal Higgs content are easily constructed. Such models can possess numerous charged and neutral Higgs scalars.[1] Despite the simplicity of this approach, the existence of fundamental scalars in field theory is problematical. If the electroweak model is embedded in a more fundamental structure characterized by a much larger energy scale (e.g., the Planck scale, which must appear in any theory including gravity), the Higgs boson mass would tend to acquire mass of order the large scale due to radiative corrections. Only by adjusting (i.e. "fine-tuning") the parameters of the Higgs potential "unnaturally" can one in-

Higgs Particle(s)
Edited by A. Ali
Plenum Press, New York, 1990

sure a large hierarchy between the Planck scale and the scale of electroweak symmetry breaking. (These problems have been reviewed in many places; see *e.g.* ref. 2.) Two classes of solutions of this "hierarchy" problem have been advanced. In one class of models, supersymmetry is invoked to solve the hierarchy problem.[3] In these models, the Higgs bosons are elementary scalars with properties similar to non-minimal Higgs bosons of the Standard Model. In a second class of models, the Higgs bosons are replaced by composite bound states of fundamental fermions. Examples of models of this type include technicolor models[4] and composite Higgs models.[5] The experimental verification of the elementary Higgs model or one of its alternatives is one of the fundamental programs of particle physics today. It is expected that with the advent of new colliders during the coming decade (1990s), it will be possible to answer decisively the question of the existence of the Higgs boson.

In this paper, I will focus only on those models in which the Higgs bosons are elementary scalar fields. By omitting a discussion of models in which the Higgs boson is not elementary, I am not implying that such models are inferior to models with elementary Higgs bosons. Models of elementary Higgs bosons have received a lion's share of the attention in the literature primarily because it is much easier to calculate in models with weakly interacting elementary scalars. This probably represents a shortcoming of the theorist rather than a shortcoming of the other approaches. Nevertheless, the advantage of models with elementary Higgs bosons is that its phenomenology is well predicted and in principle can be verified or falsified by the next generation of colliders. The great challenge of the coming decade will be to isolate the experimental signals associated with the mechanism of spontaneous electroweak symmetry breaking and to deduce its theoretical origin.

There is a large literature devoted to the search for the Higgs boson of the Standard Model.[1] As remarked above, most theorists believe that the minimal model with one physical Higgs scalar is not the whole story, since such a model cannot account for the hierarchy between the Planck and electroweak scales. Nevertheless, because of its simplicity, the minimal model is useful for setting benchmarks for the design of future experiments which hope to probe the origins of electroweak symmetry breaking. Among the electroweak models with extended Higgs sectors, the non-minimal model which has attracted the most attention is the two-Higgs-doublet model. One reason for its popularity is that the minimal supersymmetric extension of the Standard Model contains precisely two Higgs doublets. Such a model could then be consistent with the a solution to the hierarchy problem, while possessing a Higgs sector whose phenomenology is in many ways similar to the minimal Higgs model. Thus, I will first summarize the main search techniques which will be employed in hunting the Standard Model Higgs boson. I will then discuss theoretical aspects of the two-Higgs-doublet model. The phenomenological implications of this model will be examined, with particular attention paid to new features not present in the phenomenology of the minimal Higgs boson. The minimal supersymmetric extension of the Standard Model is an important example of a two-Higgs-doublet model; the constraints of supersymmetry are both significant and interesting. Non-minimal Higgs models with richer Higgs structures are certainly possible. However, I will argue that in multi-Higgs models which possess a high energy scale (beyond the scale of electroweak symmetry breaking), the effective low energy theory possesses a minimal Higgs sector. Thus, the phenomenology of minimal Higgs models is applicable to a much larger class of models than previously imagined.

2. SEARCH STRATEGIES FOR THE STANDARD MODEL HIGGS BOSON

The mass of the Higgs boson is not predicted in the Standard Model. It is generally accepted that the Standard Model Higgs boson must have mass less than of order 1 TeV.[6] A theoretical lower limit for the Higgs mass (the Linde-Weinberg bound[7]) can be derived, but for $m_t \approx m_W$, this lower limit vanishes. Thus, experimentalists must be prepared to conduct their Higgs boson searches to cover all masses between 0 and 1 TeV. At present, there is limited experimental information which may be relevant for ruling out a Standard Model Higgs boson with mass below 5 GeV. For higher masses, one must turn to the Z-factories and to future accelerators. The appropriate search strategies vary dramatically depending on which accelerator is used and which Higgs mass range is being probed. A detailed review of the available techniques which are presently known is given in ref. 1. Thus, I shall simply provide a very cursory summary here. Some of these techniques will also be described in more detail in other reports in these Proceedings.

Current experimental data (rare K decay and B decay searches) probably rule out all Higgs masses with $m_\phi < 2m_\tau$. (For a summary of the relevant experiments which leads to this conclusion, see ref 8.) It is important to emphasize that such a conclusion is only valid for the Higgs boson of the minimal electroweak model. In non-minimal Higgs models, it may be possible to have a lighter neutral Higgs boson without contradiction from current experimental data. To make further progress, new accelerators will be required, from SLC and LEP which have just recently begun to take data, to the hadron supercolliders (LHC, SSC, and the ELOISATRON) and possible e^+e^- supercolliders of the future. Some of the various mechanisms for Higgs production and detection are summarized below.

At e^+e^- colliders, the following considerations are relevant. Due to the $ZZ\phi^0$ coupling, a Higgs boson (if kinematically allowed) will be produced in Z decay: $Z \to \phi^0 Z^*$ where the virtual Z^* can couple to lepton or quark pairs. Probably, the most fruitful final state to examine is $Z \to \phi^0 \ell^+ \ell^-$ (ℓ = electron or muon) or $Z \to \phi^0 \nu\bar{\nu}$. With a data sample of 10^6–10^7 Z's, it should be possible to observe Higgs masses up to 30–50 GeV.[9] (The rare decay $Z \to \phi^0 \gamma$ is competitive only in the case of maximal luminosity and Higgs bosons near the upper end of the indicated mass range.) At LEP-II, with $\sqrt{s} = 200$ GeV, the $ZZ\phi^0$ coupling controls the process $e^+e^- \to Z\phi^0$ via s-channel Z-exchange. It should be possible to reach Higgs masses around 80–85 GeV with an integrated luminosity of 1000 pb^{-1}.[10] Toponium (Θ), if it exists in a mass range accessible to LEP-II, will be a good source of Higgs bosons via the decay $\Theta \to \phi^0 \gamma$. Assuming an integrated luminosity of 100 pb^{-1}, there should be enough events to detect a Higgs boson if $m_t \lesssim 80$ GeV.[11] However, this possibility is becoming more and more remote as the experimental limits on the top quark mass continue to increase. (The most recent measurement at the Tevatron by the CDF Collaboration finds that $m_t > 77$ GeV at 95% confidence level.[12]) At a future e^+e^- supercollider, the process $e^+e^- \to \phi^0 \nu\bar{\nu}$ via WW-fusion dominates $e^+e^- \to Z\phi^0$ as the primary source of Higgs bosons. Higgs masses of $(0.4$–$0.5)\sqrt{s}$ are accessible[13] given sufficient integrated luminosity (10 to 30 fb^{-1}).

A Standard Model Higgs boson cannot be detected at present hadron colliders (ACOL and Tevatron). Thus, I shall focus the discussion on future supercolliders. In discussing Higgs searches at hadron supercolliders, three mass regions have been distinguished: "intermediate" ($m_Z/2 \lesssim m_\phi \lesssim 2m_Z$), "heavy" ($2m_Z \lesssim m_\phi \lesssim 800$

GeV) and "obese" ($m_\phi \gtrsim 800$ GeV). The boundaries between these regions are not sharp. The dominant Higgs production mechanisms are gluon-gluon fusion and WW fusion. (Gluon-gluon fusion dominates for the lighter Higgs masses; WW fusion becomes dominant for heavier Higgs masses.) For example, if the design luminosity of the SSC is achieved, one should expect an integrated luminosity of 10^4 pb^{-1} in a one year run. This implies that between 10^6 and 10^4 Higgs bosons will be produced per year at the SSC, if 100 GeV $\lesssim m_\phi \lesssim$ 1 TeV.[14] The challenge of Higgs hunting at a hadron collider is to extract a Higgs boson signal out of a potentially large Standard Model background.

The intermediate mass regime is the most problematical. The dominant decay mode of the Higgs boson in this regime is into $t\bar{t}$ pairs (or $b\bar{b}$ pairs if $m_\phi < 2m_t$), and such a final state is extremely difficult to observe due to the presence of large Standard Model jet backgrounds. One will have to resort to rarer production or decay modes with more distinguishing characteristics. Three possible mechanisms have been studied: (i) $gg \to \phi^0 \to ZZ^*$, where both on-shell and off-shell Z's decay to e^+e^- or $\mu^+\mu^-$; (ii) $gg \to \phi^0 \to \gamma\gamma$, and (iii) $q\bar{q} \to W^* \to W\phi^0$, where $\phi^0 \to b\bar{b}$. Each of these mechanisms may provide signatures which allow experimenters to probe part of the intermediate mass regime. Nevertheless, it is clear that it will be a difficult task to completely rule out an intermediate mass Higgs at a hadron collider.[*]

In the heavy Higgs mass regime, there is one "gold-plated" signature: $\phi^0 \to ZZ$, where both Z's decay into electron or muon pairs. This should allow for the discovery of a Higgs boson with mass up to 600 GeV with the canonical 10^4 pb^{-1} integrated luminosity.[15] If the luminosity can be increased by a factor of 10 in a high luminosity running mode of the SSC, it may be possible to detect Higgs bosons with masses as high as 800 GeV.[17] The LHC, running at a $\sqrt{s}$ which is a factor of 3 to 4 lower that of the SSC will have a somewhat smaller Higgs discovery reach.[18] The ELOISATRON, with $\sqrt{s} = 200$ TeV, at the canonical luminosity of 10^4 pb^{-1} per year, will have more than enough data to cover the entire Higgs mass range (above $2m_W$) up to and beyond 1 TeV, using the gold-plated mode.[19] A second "silver-plated" mode is: $\phi^0 \to ZZ$ where one Z decays into e^+e^- or $\mu^+\mu^-$ and the other Z decays into neutrinos.[20] Although this signature has larger backgrounds to contend with, it should be possible to reject the background with an appropriate set of cuts. Because of the larger branching ratio of Z into neutrino pairs, this signature has the potential of reaching higher Higgs masses than the gold-plated signature. In principle, one would also like to have access to the W^+W^- modes of the Higgs boson, which are (about) twice as prolific as ZZ and have larger leptonic branching fractions. However, the possible signatures (e.g., $\phi^0 \to W^+W^-$, where one W decays leptonically and the second W decays either leptonically or hadronically) have very large Standard Model backgrounds which are generally difficult to deal with. Nevertheless, interesting techniques have been suggested, and the study of such signatures may eventually be quite fruitful.[21] Finally, the "obese" region presumably corresponds to that regime where the WW and ZZ scattering is becoming strong (at least in the scalar channel).[6,22] Experimental signatures of a strongly interacting WW and ZZ sector at the SSC have been considered in refs. 22 and 23. Thus, experimentalists should be prepared to search for evidence of such strong interaction phenomena if no evidence for a Higgs boson is found.

---

[*] The ability to probe the intermediate mass Higgs is one of the strengths of an Intermediate e^+e^- Linear Collider with $\sqrt{s} \simeq$ 300–500 GeV. Unfortunately, there are many engineering and accelerator physics problems to overcome before such a machine could be built.

3. CONSTRAINTS ON EXTENDED HIGGS SECTORS

Despite the lack of direct experimental information on the Higgs sector, there are in fact some constraints on extended Higgs models. First, it is an experimental fact that $\rho = m_W^2/(m_Z^2 \cos^2 \theta_W)$ is very close to 1.[24] In the Standard Model, the ρ–parameter is determined by the Higgs structure of the theory. It is well known[25] that in a model with only Higgs doublets (and singlets), the tree-level value of $\rho = 1$ is automatic, without adjustment of any parameters in the model. Although the minimal Higgs satisfies this property, so does any version of the Standard Model with any number of Higgs doublets (and singlets). There are also other ways to satisfy the $\rho \approx 1$ constraint. First, there are an infinite number of more complicated Higgs representations which also satisfy $\rho = 1$ at tree level.[26] The general formula is

$$\rho \equiv \frac{m_W^2}{m_Z^2 \cos^2 \theta_W} = \frac{\sum_{T,Y} [4T(T+1) - Y^2] \mid V_{T,Y} \mid^2 c_{T,Y}}{\sum_{T,Y} 2Y^2 \mid V_{T,Y} \mid^2}, \tag{1}$$

where $\langle \phi(T,Y) \rangle = V_{T,Y}$ defines the vacuum expectation values of each neutral Higgs field, and T and Y specify the total $SU(2)_L$ isopsin and the hypercharge of the Higgs representation to which it belongs, and

$$c_{T,Y} = \begin{cases} 1, & (T,Y) \in \text{complex representation,} \\ \frac{1}{2}, & (T, Y = 0) \in \text{real representation.} \end{cases} \tag{2}$$

Here, I have used a rather narrow definition of a real representation as consisting of a real multiplet of fields with integer weak isopsin and $Y = 0$. The requirement that $\rho = 1$ for arbitrary $V_{T,Y}$ values is

$$(2T + 1)^2 - 3Y^2 = 1. \tag{3}$$

The possibilities beyond $T = 1/2$, $Y = \pm 1$ are usually discarded since the representations are rather complicated (the simplest example is a Higgs representation with weak isospin 3 and $Y = 4$). Second, one can take a model with multiple copies of "bad" Higgs representations, and arrange a "custodial" $SU(2)$ symmetry among the copies, which then naturally imposes $\rho = 1$ at tree level. An example of this type of model can be found in ref. 27, and will be discussed further elsewhere in these Proceedings.[19] Finally, one can always choose arbitrary Higgs representations and fine tune the parameters of the Higgs potential to arrange $\rho \approx 1$. This latter "unnatural" possibility will be discarded from further consideration.

The second major theoretical constraint on the Higgs sector comes from the severe limits on the existence of flavor-changing neutral currents (FCNC's). In the minimal Higgs model, tree-level flavor changing neutral currents are automatically absent, because the same operations that diagonalize the mass matrix automatically diagonalize the Higgs-fermion couplings. In general, this ceases to be true in non-minimal Higgs models. One then has two choices. First, by arranging the parameters of the model so that the Higgs masses are large (typically of order 1 TeV), tree-level FCNC's mediated by Higgs exchange can be suppressed sufficiently so as not to be in conflict with known experimental limits. The second choice is more elegant, and is based on a theorem of Glashow and Weinberg[28] concerning FCNC's in models with more than one Higgs doublet. The theorem states that tree-level FCNC's mediated by Higgs bosons will be absent if all fermions of a given electric charge couple to no more than one Higgs doublet. If we require this theorem to be satisfied, the Higgs couplings to fermions is constrained, but not unique. One example of a model satisfying this requirement is the minimal supersymmetric extension of the Standard Model. This model (discussed in section 6 in more detail) possesses two Higgs doublets of opposite hypercharge; the $Y = -1$ doublet couples only to down-type quarks and charged leptons, and the $Y = 1$ doublet couples only to up-type quarks and neutrinos. In supersymmetric models, this choice is not arbitrary, but is in fact required by the supersymmetry in order to give masses to both up- and down-type quarks and leptons. A second example of a two-Higgs-doublet model which avoids FCNC's is a model in which one Higgs doublet does not couple to fermions at all (due to a discrete symmetry) and the other Higgs doublet couples to fermions in the same way as in the minimal Higgs model.[29] The resulting phenomenology of Higgs-fermion interactions is quite different from that in the first example given above.

A final set of conditions that must be satisfied by any model of electroweak symmetry breaking arises from the requirement that the $V_L V_L \to V_L V_L$ and $f_+ \bar{f}_+ \to V_L V_L$ amplitudes not violate unitarity bounds. (V can be either W or Z, and the $+$ subscript on the f indicates positive helicity.) There are two aspects to this requirement. First, as an automatic consequence of the gauge structure and renormalizability of any gauge theory, no partial wave amplitude can grow with energy. This condition requires non-trivial cancellation among Feynman diagrams which contribute to a given process. For example, in $WW \to WW$ scattering, the cancellation of the growing energy terms is guaranteed in the Standard Model because of the tree-level relation $g_{\phi^0 WW} = g m_W$, where g is the gauge coupling. In models with more complicated Higgs sectors, it is no longer necessary that a single scalar boson alone cure these unitarity problems. For example, in models containing only doublet and singlet Higgs fields, the following sum rules for the scalar boson VV and $f\bar{f}$ couplings must be obeyed:[30,31]

$$\sum_i g^2_{h^0_i VV} = g^2_{\phi^0 VV} , \tag{4}$$

and

$$\sum_i g_{h^0_i VV} g_{h^0_i f\bar{f}} = g_{\phi^0 VV} g_{\phi^0 f\bar{f}} . \tag{5}$$

where the sums are taken over all CP-even Higgs scalars (h^0_i) of the model, and the couplings on the right hand side are those of the minimal Higgs boson (ϕ^0) of the Standard Model.

The second aspect of the unitarity bound requirement deals with the size of the constant (energy independent) term of the partial wave amplitude. Tree-level unitarity would limit the size of this constant, which would then impose an upper limit on the Higgs boson mass(es). In models with a non-minimal Higgs sector, such an analysis requires that at least one neutral scalar must have mass below about 1 TeV.[30] Furthermore, those Higgs bosons with mass of order 1 TeV or below must approximately saturate the coupling constant sum rules given in eqs. (4) and (5). This is an interesting result, since it allows the existence of grand unified models which possess superheavy Higgs bosons, without violating tree-level unitarity. By the observation above, such superheavy bosons must be extremely weakly coupled to $W^+ W^-$ and ZZ. Finally, it should be remarked that the requirement of tree-level unitarity (the second aspect of the unitarity bound) can only be taken as a general guide, since if it is violated then the Higgs sector is presumably strongly interacting. In this case, perturbative arguments become extremely suspect.

The main conclusion to draw from the above discussion is that there is still plenty of freedom for the Higgs sector in the Standard Model, although the choice is not totally arbitrary. It is also clear that models with multiple doublets are the preferred models for non-minimal Higgs structures, although there is still room for investigation outside this framework. Finally, the two-Higgs-doublet version of the Standard Model is particularly attractive because:

1. It is an extension of the minimal model which adds new phenomena (*e.g.* observable charged Higgs bosons).

2. It is a minimal extension in that it adds the fewest new arbitrary parameters.

3. It satisfies theoretical constraints of $\rho \approx 1$ and the absence of tree-level FCNC's (if the Higgs-fermion couplings are appropriately chosen).

4. Such a Higgs structure is required in "low energy" supersymmetric models.

4. TWO HIGGS DOUBLET MODELS—THEORY

Let us investigate the minimal extension of the Higgs sector—the Standard Model with two Higgs doublets.[29,32-36] Let ϕ_1 and ϕ_2 denote two complex $Y = 1$, $SU(2)_L$ doublet scalar fields. The Higgs potential which spontaneously breaks $SU(2)_L \times U(1)_Y$ down to $U(1)_{EM}$ is[33]

$$
\begin{aligned}
V(\phi_1, \phi_2) = & \lambda_1 (\phi_1^\dagger \phi_1 - v_1^2)^2 + \lambda_2 (\phi_2^\dagger \phi_2 - v_2^2)^2 \\
& + \lambda_3 \left[(\phi_1^\dagger \phi_1 - v_1^2) + (\phi_2^\dagger \phi_2 - v_2^2) \right]^2 \\
& + \lambda_4 \left[(\phi_1^\dagger \phi_1)(\phi_2^\dagger \phi_2) - (\phi_1^\dagger \phi_2)(\phi_2^\dagger \phi_1) \right] \\
& + \lambda_5 \left[\mathrm{Re} \, (\phi_1^\dagger \phi_2) - v_1 v_2 \right]^2 \\
& + \lambda_6 \left[\mathrm{Im} \, (\phi_1^\dagger \phi_2) \right]^2 ,
\end{aligned}
\tag{6}
$$

where the λ_i are all real parameters. This potential is the most general one subject to gauge invariance and a discrete symmetry, $\phi_1 \rightarrow -\phi_1$, which is only softly violated (by dimension-two terms). The latter constraint is a technical one which is related to insuring that flavor changing neutral currents are not too large. For simplicity, I have also assumed that the Higgs sector is CP-invariant. The above potential guarantees the correct pattern of electroweak symmetry breaking over a large range of parameters. That is, if all the λ_i are non-negative, then the minimum of the potential is manifestly:

$$
\langle \phi_1 \rangle = \begin{pmatrix} 0 \\ v_1 \end{pmatrix}, \qquad \langle \phi_2 \rangle = \begin{pmatrix} 0 \\ v_2 \end{pmatrix},
\tag{7}
$$

which breaks the $SU(2)_L \times U(1)_Y$ down to $U(1)_{EM}$, as desired. In fact, the allowed

117

range of the λ_i corresponding to this desired minimum is somewhat larger, (and can be deduced by requiring that all squared Higgs masses are positive).

A key parameter of the model is the ratio of the vacuum expectation values

$$\tan \beta = v_2/v_1. \tag{8}$$

It is straightforward to remove the Goldstone bosons and determine the physical Higgs states. In the charged sector, the charged Goldstone boson is

$$G^\pm = \phi_1^\pm \cos \beta + \phi_2^\pm \sin \beta, \tag{9}$$

and the physical charged Higgs state is orthogonal to $G^\pm$:

$$H^\pm = -\phi_1^\pm \sin \beta + \phi_2^\pm \cos \beta, \tag{10}$$

with mass $m_{H^\pm}^2 = \lambda_4(v_1^2 + v_2^2)$. Due to the CP-invariance assumed above, the imaginary parts and the real parts of the neutral scalar fields decouple. In the imaginary (CP-odd) sector, the neutral Goldstone boson is:

$$G^0 = \sqrt{2} \left(\text{Im} \, \phi_1^0 \cos \beta + \text{Im} \, \phi_2^0 \sin \beta \right), \tag{11}$$

and the orthogonal neutral physical state is:

$$A^0 = \sqrt{2} \left(-\text{Im} \, \phi_1^0 \sin \beta + \text{Im} \, \phi_2^0 \cos \beta \right), \tag{12}$$

with mass $m_{A^0}^2 = \lambda_6(v_1^2 + v_2^2)$. The real ($CP$-even) sector contains two physical Higgs scalars which mix through the following mass-squared matrix:

$$\mathcal{M} = \begin{pmatrix} 4v_1^2(\lambda_1 + \lambda_3) + v_2^2\lambda_5 & (4\lambda_3 + \lambda_5)v_1v_2 \\ (4\lambda_3 + \lambda_5)v_1v_2 & 4v_2^2(\lambda_2 + \lambda_3) + v_1^2\lambda_5 \end{pmatrix}. \tag{13}$$

The physical mass eigenstates are:

$$\begin{aligned} H^0 &= \sqrt{2} \left[(\text{Re} \, \phi_1^0 - v_1) \cos \alpha + (\text{Re} \, \phi_2^0 - v_2) \sin \alpha \right], \\ h^0 &= \sqrt{2} \left[-(\text{Re} \, \phi_1^0 - v_1) \sin \alpha + (\text{Re} \, \phi_2^0 - v_2) \cos \alpha \right]. \end{aligned} \tag{14}$$

The corresponding masses are:

$$m_{H^0, h^0}^2 = \frac{1}{2} \left[\mathcal{M}_{11} + \mathcal{M}_{22} \pm \sqrt{(\mathcal{M}_{11} - \mathcal{M}_{22})^2 + 4\mathcal{M}_{12}^2} \right], \tag{15}$$

and the mixing angle α is obtained from:

$$\begin{aligned} \sin 2\alpha &= \frac{2\mathcal{M}_{12}}{\sqrt{(\mathcal{M}_{11} - \mathcal{M}_{22})^2 + 4\mathcal{M}_{12}^2}}, \\ \cos 2\alpha &= \frac{\mathcal{M}_{11} - \mathcal{M}_{22}}{\sqrt{(\mathcal{M}_{11} - \mathcal{M}_{22})^2 + 4\mathcal{M}_{12}^2}}. \end{aligned} \tag{16}$$

Note that according to eq. (15), $m_{H^0} \geq m_{h^0}$ as suggested by the notation.

To summarize, this model possesses five physical Higgs bosons: a charged pair $(H^{\pm})$; two neutral CP-even scalars (H^0 and h^0, where, by convention, $m_{H^0} > m_{h^0}$); and a neutral CP-odd scalar (A^0), often called a pseudoscalar. Instead of the one free parameter of the minimal model, this model has six free parameters: four Higgs masses, the ratio of vacuum expectation values, $\tan\beta$, and a Higgs mixing angle, α. Note that $v_1^2 + v_2^2$ is fixed by the W mass: $m_W^2 = g^2(v_1^2 + v_2^2)/2$.

It is worthwhile to examine the couplings of the physical Higgs bosons to vector bosons and fermion pairs, since these couplings control the production and decay of the Higgs bosons. First, consider the couplings to vector bosons. To understand the pattern of couplings, consider the fact that the Standard Model, *in the absence of the quarks and leptons*, separately conserves C and P. Thus I can assign unique J^{PC} quantum numbers to all the bosons of the theory, if the fermions are ignored. The quantum number assignments are displayed in Table 1.

Table 1

Quantum numbers of Higgs and Gauge Bosons

When C and P are separately conserved			
	J^{PC}		J^P
γ	1^{--}	$W^{\pm}$	1^-
Z	1^{--}	$H^{\pm}$	0^+
H^0	0^{++}		
h^0	0^{++}		
A^0	0^{+-}		
When C and P are violated but CP is conserved			
	J^{PC}		J^P
γ	1^{--}	$W^{\pm}$	$1^-, 1^+$
Z	$1^{--}, 1^{++}$	$H^{\pm}$	$0^+, 0^-$
H^0	$0^{++}, 0^{--}$		
h^0	$0^{++}, 0^{--}$		
A^0	$0^{+-}, 0^{-+}$		

At first glance, the C and P assignments for A^0 are surprising.[*] Formally, one could look at the bosonic sector of the Lagrangian, observe that certain terms are missing, and note that the C, P choices of Table 1 are sufficient to explain the absence of $A^0 W^+ W^-$ and $A^0 ZZ$ couplings. Physically, one can understand the results as

---

[*] The existence of the ZH^+H^- vertex implies that Z is a 1^{--} vector boson, and the $H^0 h^0 h^0$ vertex implies that H^0 is a 0^{++} scalar. It then follows from the existence of the $ZH^0 A^0$ vertex that A^0 is both C-odd and CP-odd as indicated in Table 1.

follows. In a one doublet model, the imaginary part of the neutral Higgs field is the Goldstone boson which is "eaten" and becomes the longitudinal component of the Z. This field, like all Goldstone boson fields, is derivatively coupled and is, therefore, CP-odd. Since the bosonic sector conserves C and P separately, the Goldstone boson must in fact have the $C = -1$ quantum number of the Z. Its $P = +1$ quantum number is the same as that of the other scalar components, and is opposite in sign from the parity of the vector bosons due to the one unit difference in spin. In a two-doublet model, there are two neutral Higgs fields with imaginary components. One linear combination of the imaginary components is the Goldstone boson, and the other linear combination is A^0. Both these fields must have the same C and P quantum numbers; hence, the 0^{+-} assignment for A^0 given in Table 1.

A second argument can be given for the absence of a tree-level coupling of the A^0 to vector boson pairs. First, let us recall that the coupling of the CP-even scalar Higgs boson(s) to a pair of massive vector bosons arises from the covariant derivative $(D_\mu \phi)(D^\mu \phi)$ terms in the Lagrangian after replacing one of the ϕ's by its vacuum expectation value. However, in a CP conserving theory this mechanism does not generate a coupling for the CP-odd A^0. (This is because in the convention adopted here, where the vacuum expectation value of ϕ is taken to be real, the A^0 originates from the imaginary component of ϕ.) Nor does this tree-level mechanism generate a coupling of a CP-even Higgs to a massless vector boson pair ($\gamma\gamma$ or gg). Both types of coupling occur only at the one-loop level. More formally, since the A^0 is CP-odd, a gauge invariant interaction must take the form: $\epsilon^{\mu\nu\alpha\beta} F_{\mu\nu} F_{\alpha\beta} A^0$. However, this is a dimension-five term which cannot appear in the fundamental Lagrangian, and is only generated by loop graphs. Similarly, at one-loop, dimension-five couplings of vector boson pairs to a CP-even Higgs, of the form $F_{\mu\nu} F^{\mu\nu} h$, are also generated; indeed, the coupling of two photons or two gluons to a CP-even Higgs arises in exactly this way.

In common parlance, the A^0 is usually referred to as a *pseudoscalar*. This is technically incorrect, since we have seen above that in the absence of fermions, the A^0 has $P = +1$ (and $C = -1$). Incorporating the fermions into the theory, C and P are no longer separately conserved, although CP remains a good quantum number (to a very good approximation). Thus, it is more precise to refer to A^0 as being CP-odd. The reason that A^0 is sometimes referred to as being a pseudoscalar will be revealed shortly. In any case, with the assignments of Table 1 in hand, it is easy to see which boson couplings are forbidden. The coupling of the Z to a pair of identical Higgs bosons is forbidden by Bose symmetry. For a pair of non-identical Higgs bosons, the coupling is only present when the two Higgs bosons have opposite CP quantum numbers; *i.e.* $ZA^0 H^0$ and $ZA^0 h^0$ are allowed. The ZZA^0 and $W^+ W^- A^0$ couplings are absent for the reasons discussed above. In terms of the quantum number assignments in Table 1, we would say that these couplings are forbidden by C-invariance! Of course, since C-invariance is broken when we introduce the quarks and leptons, these couplings are generated radiatively through fermion loops. There are a few other vertices forbidden at tree level for other reasons. Vertices involving neutral particles only and one or two photons clearly vanish at tree level, although they are generated at one-loop. The same is true for the coupling of all neutral Higgs bosons to a pair of gluons. The radiatively generated $A^0 gg$, $H^0 gg$, and $h^0 gg$ vertices are important since two-gluon fusion is one of the major production mechanisms for neutral Higgs bosons at a hadron collider. Two other vertices, $H^+ W^- \gamma$ and $H^+ W^- Z$, also vanish at tree level. The $H^+ W^- \gamma$ tree-level vertex is zero as a consequence of the conservation of the electromagnetic current. The vanishing of the H^+W^-Z vertex is more model dependent; it turns out to be a general feature of models with only Higgs doublets and singlets.[37–39] Again, these vertices are radiatively generated at one-loop, and lead to interesting rare decays of the charged Higgs.[40,41] All other three-point tree-level vertices involving gauge and Higgs bosons are allowed.

Probably the most important vertices for phenomenology are the couplings of H^0 and h^0 to W^+W^- and ZZ. These couplings tend to be somewhat suppressed compared to their values in the minimal-Higgs model, as is apparent from the sum rule (4) which reduces in the present case to

$$g^2_{H^0VV} + g^2_{h^0VV} = \left[g^{minimal}_{\phi^0VV} \right]^2 \tag{17}$$

and holds separately for $V = W$ or Z. In terms of the angles α and β defined earlier, we have[43]

$$\frac{g_{h^0VV}}{g_{\phi^0VV}} = \sin(\beta - \alpha)$$

$$\frac{g_{H^0VV}}{g_{\phi^0VV}} = \cos(\beta - \alpha), \tag{18}$$

and the sum rule of eq. (17) is obviously satisfied. Without specific predictions for α and β, one might be tempted to say that the scalar Higgs coupling to vector boson pairs should be down by roughly a factor of $\sqrt{2}$ compared with the minimal Higgs model. However, we will see later that in the supersymmetric model this expectation is generally false; in particular $\cos(\beta - \alpha)$ tends to be quite small, and $\sin(\beta - \alpha)$ is near 1.

The couplings of W (or Z) to a pair of Higgs bosons are also phenomenologically important. For example, such couplings introduce the following new Higgs boson decay channels: $A^0 \rightarrow Zh^0$, $A^0 \rightarrow W^{\pm}H^{\mp}$, $H^0 \rightarrow W^{\pm}H^{\mp}$ and $H^{\pm} \rightarrow W^{\pm}h^0$. Such decay modes can be very important when m_{A^0} and $m_{H^{\pm}}$ are large compared to m_{h^0}, due to. the longitudinal modes of the final state vector boson. However, the size of the required couplings is very model dependent. Once again there are sum rules which somewhat constrain the relevant couplings. Defining the Feynman coupling for $H^+W^-h_i$ to be $-ig_{H^+W^-h_i}(p+p')\cdot\epsilon_W$ (where p and p' are the four-momenta of H^+ and h_i, respectively) and using an analogous definition for the A^0ZH coupling, we have

$$g^2_{H^+W^-h^0} + g^2_{H^+W^-H^0} = \left[g^2_{A^0Zh^0} + g^2_{A^0ZH^0} \right] \cos^2\theta_W = g^2/4. \tag{19}$$

This sum rule is required in order to satisfy unitarity constraints in $H^+W^- \rightarrow H^+W^-$ and $A^0Z \rightarrow A^0Z$.[31] In order to avoid violating unitarity in $A^0Z \rightarrow W^+W^-$, the ZA^0h^0 and ZA^0H^0 coupling strengths must have exactly the correct ratio: one finds[42,31] that the h^0, H^0 coupling strengths to ZA^0 are proportional to $\cos(\beta - \alpha)$, $\sin(\beta - \alpha)$, respectively. Again, a specific model is required to determine both the division of the coupling strengths and the relation between the H^+, H^0 and h^0 masses.

Let us now consider the Higgs-fermion couplings. As discussed above, in the two-Higgs doublet model, the Higgs-fermion coupling is model dependent. Even if one imposes the theorem of Glashow and Weinberg[28] to forbid tree-level FCNC's induced by Higgs exchange, one still has numerous choices for how to couple the quarks and leptons to the two Higgs doublets. A set of discrete symmetries can always be concocted to make a particular choice natural (in the technical sense). For example, one can couple one Higgs doublet to all quarks and leptons and decouple the second Higgs doublet from the fermions.[29] A second choice is one where one Higgs doublet couples only to up-type quarks and charged leptons and the second Higgs doublet couples only to down-type quarks and neutrinos. Axion models and supersymmetric models are examples of this choice. Other choices can be made where the quark and lepton couplings are treated asymmetrically. Although the Feynman rules for the Higgs-fermion interaction differ depending on the choice made, there are a number of common features among all such models. First, the Higgs-fermion couplings can be either enhanced or suppressed compared to the minimal-Higgs model, depending on the parameters of the model. Second, introducing the fermions means that C and P are no longer separately conserved. However, in the approximation where CP is conserved, the Higgs and vector bosons can be thought of as admixtures of two eigenstates of definite C and P as indicated in Table 1. Consider then the coupling of the neutral Higgs boson to a fermion-antifermion pair. It is well known that an $f\bar{f}$ pair has $P = (-1)^{L+1}$ and $C = (-1)^{L+S}$ for total spin S and orbital angular momentum L, and thus cannot couple to 0^{--} and 0^{+-}. Therefore, for the neutral Higgs couplings to $f\bar{f}$, the H^0 and h^0 behave as pure 0^{++} scalars, whereas A^0 behaves as a pure 0^{-+} pseudoscalar. As a result, it is common to refer to A^0 as a pseudoscalar, even though this is only a correct statement in the context of its interaction with an $f\bar{f}$ pair.

The two choices for the Higgs-quark interactions which were mentioned above will now be explicitly displayed. The case in which the quarks and leptons do not couple to the first Higgs doublet (ϕ_1), but couple to the second Higgs doublet (ϕ_2) in a manner analogous to the minimal Higgs model will be called Model I.[29] The case in which ϕ_1 couples only to down-type quarks and charged leptons and ϕ_2 couples only to up-type quarks and neutrinos will be called Model II. Consider a three-generation model with diagonal (positive) quark matrices M_U and M_D (for the charge 2/3 and $-1/3$ quarks respectively) and Kobayashi-Maskawa (KM) mixing matrix K. Then, in Model I, the Higgs-fermion interaction takes the following form:[29,44]

$$
\begin{aligned}
\mathcal{L}_{Hf\bar{f}} = &- \frac{g}{2m_W \sin\beta}\, \overline{D} M_D D (H^0 \sin\alpha + h^0 \cos\alpha) - \frac{ig\cot\beta}{2m_W}\, \overline{D} M_D \gamma_5 D A^0 \\
&- \frac{g}{2m_W \sin\beta}\, \overline{U} M_U U (H^0 \sin\alpha + h^0 \cos\alpha) + \frac{ig\cot\beta}{2m_W}\, \overline{U} M_U \gamma_5 U A^0 \qquad (20) \\
&+ \frac{g\cot\beta}{2\sqrt{2}m_W} \left(H^+ \overline{U} \left[M_U K (1-\gamma_5) - K M_D (1+\gamma_5) \right] D + \text{h.c.} \right).
\end{aligned}
$$

In this case, $\tan\beta \equiv v_2/v_1$, where v_2 is the vacuum expectation value of the Higgs field which couples to *both* up and down-type quarks (whereas the other Higgs field is decoupled from the quarks). In contrast, the Model II interaction is:[34,44]

$$\mathcal{L}_{Hf\bar{f}} = -\frac{g}{2m_W \cos\beta}\,\overline{D}M_D D(H^0 \cos\alpha - h^0 \sin\alpha) + \frac{ig\tan\beta}{2m_W}\,\overline{D}M_D\gamma_5 D A^0$$

$$-\frac{g}{2m_W \sin\beta}\,\overline{U}M_U U(H^0 \sin\alpha + h^0 \cos\alpha) + \frac{ig\cot\beta}{2m_W}\,\overline{U}M_U\gamma_5 U A^0$$

$$+\frac{g}{2\sqrt{2}m_W}\left(H^+\overline{U}\left[\cot\beta M_U K(1-\gamma_5) + \tan\beta K M_D(1+\gamma_5)\right]D + \text{h.c.}\right).$$

(21)

This time, we define $\tan\beta \equiv v_2/v_1$, where v_1 (v_2) is the vacuum expectation value of the Higgs field which couples only to down-type (up-type) quarks.

In the two equations above, U and D are column matrices consisting of three generations of quark fields. (In both Models I and II, the Higgs-lepton couplings can be read off from the expressions above by replacing (U, D) with the corresponding lepton fields, replacing quark mass matrices with the corresponding *diagonal* lepton mass matrices, and setting $K = 1$.) Note that as previously advertised, the neutral Higgs interactions are flavor diagonal. In addition, the structure of the charged Higgs interactions involving the Kobayashi-Maskawa matrix is analogous to that of the ordinary charged current mediated by the W.

Since the Model II choice for the Higgs-fermion couplings is the required structure for the minimal supersymmetric model (to be discussed in detail in section 6), the results of eq. (21) are summarized for the couplings of the neutral Higgs bosons relative to the canonical Standard Model values below (using 3rd family notation):

$$H^0 t\bar{t}: \quad \frac{\sin\alpha}{\sin\beta} \qquad H^0 b\bar{b}: \quad \frac{\cos\alpha}{\cos\beta}$$

$$h^0 t\bar{t}: \quad \frac{\cos\alpha}{\sin\beta} \qquad h^0 b\bar{b}: \quad \frac{-\sin\alpha}{\cos\beta} \qquad (22)$$

$$A^0 t\bar{t}: \quad \cot\beta \qquad A^0 b\bar{b}: \quad \tan\beta,$$

where we must keep in mind that A^0 is coupled via a γ_5 to a $q\bar{q}$ pair. One may check that the unitarity sum rule, eq. (5), is satisfied by the h^0 and H^0 fermion-antifermion couplings given in eq. (22) and VV couplings given in eq. (18). We note that the $A^0 u\bar{u}$ ($A^0 d\bar{d}$) coupling is suppressed (enhanced) if $\tan\beta > 1$, and vice versa if $\tan\beta < 1$. Similar results hold for H^0 and h^0, although these couplings also involve the mixing angle α which can reduce the size of the couplings somewhat. The charged Higgs boson of Model II has a coupling to the $t\bar{b}$ channel given by:

$$g_{H-t\bar{b}} = \frac{g}{2\sqrt{2}m_W}\left[m_t \cot\beta(1+\gamma_5) + m_b \tan\beta(1-\gamma_5)\right]. \qquad (23)$$

Note that the t-quark-mass piece is suppressed for $\tan\beta > 1$. Finally, we note that the pattern of suppressed and enhanced couplings of Model II is quite distinct from that of Model I. In the latter case, the couplings of the pseudoscalar and charged Higgs bosons to *all* fermion types are uniformly suppressed (enhanced) if $\tan\beta > 1$ ($\tan\beta < 1$). A similar remark can be made concerning the neutral scalar Higgs-fermion couplings, although one must also take the dependence on the mixing angle α into account [see eq. (20)].

5. TWO HIGGS DOUBLET MODELS—PHENOMENOLOGY

The phenomenology of two Higgs doublets can now be discussed. Let us first consider Higgs searches at a hadron supercollider. Because of the fact that we now have six free parameters in the Higgs sector, there are only a few general statements one can make concerning the search for Higgs bosons at a hadron supercollider of the future. First, let us consider whether search techniques that worked for the Standard Model Higgs will also be appropriate in a two-doublet model.

The scalar Higgs bosons (H^0 and h^0) can be detected in the same manner as the minimal Higgs boson of the Standard Model, so long as they share relatively equally the VV coupling strength. If a scalar Higgs has a mass between about $2m_Z$ and 800 GeV *and its couplings to WW and ZZ are similar to Standard Model strength*, then it should be possible to detect this Higgs boson at the SSC by observing its decay into a pair of vector bosons (followed by subsequent decay of the vector bosons into lepton pairs). On the other hand, for masses less than $2m_W$, we are in the regime of the "intermediate mass Higgs", in which the Higgs boson mainly decays into the heaviest quark pair which is kinematically allowed. The difficulty of the intermediate mass regime is still present for the reasons described in section 2.

Consider next the pseudoscalar Higgs (A^0). For reasons that were explained above, the pseudoscalar does not couple to vector boson pairs at tree level. The phenomenological implications of this fact are potentially devastating. First, the important vector boson fusion mechanism for production of a Higgs boson is absent. Thus, the primary production mechanism will be via gg fusion. Second, the dominant decay of the pseudoscalar Higgs will probably be into the heaviest quark pair available, independent of the Higgs mass. This decay is a poor signature for Higgs production due to large Standard Model backgrounds, so we must examine other possible decay modes. The decay branching ratio of $A^0 \to \gamma\gamma$ is smaller compared to the $\gamma\gamma$ decay of the Standard Model Higgs boson, due to the absence of W boson loop graphs for $A^0 \to \gamma\gamma$. The absence of VVA^0 couplings also implies that the ZZ^* mode is absent at tree-level in A^0 decays. Thus, these rare decays will not be useful in searching for the A^0. There are other possible decay modes such as $A^0 \to Zh^0$, $A^0 \to ZH^0$ and $A^0 \to W^\pm H^\mp$ which may be useful in identifying an A^0 signal. If a scalar with the above properties could be found, and were shown to have a mass larger than $2m_W$, then the absence of decays into vector boson pairs would be strong evidence for the CP-odd nature of this scalar. (An exception to this conclusion occurs in supersymmetric models, which predict that the heavy Higgs *scalar*, H^0, has suppressed couplings to the vector boson channels. Nevertheless, such an observation would be definitive evidence for a non-minimal Higgs sector.)

Finally, consider the charged Higgs boson. As remarked above, there is no coupling of the charged Higgs boson to vector boson pairs (WZ and $W\gamma$) at tree level, so that its decays are likely to be dominated by the heaviest allowed quark channel. In addition, the single particle inclusive cross section for production of the charged Higgs boson is smaller than that typical of a neutral Higgs boson. The gluon-gluon-fusion and vector-boson-fusion mechanisms are not available in this case, so that inclusive production of $H^\pm$ must occur by other mechanisms. If the top quark has a moderate mass, but $m_t > m_{H^\pm} + m_b$, then the rate for $gg \to t\bar{t}$ followed by $t, \bar{t}$ decay to H^+, H^- is very large. Relative to the t decay rate to charged W's we have:

$$\frac{\Gamma(t \to H^+ b)}{\Gamma(t \to W^+ b)} = \frac{p_{H^+}}{p_{W^+}} \frac{m_t^2(m_t^2 - m_{H^+}^2)}{(m_t^2 + 2m_W^2)(m_t^2 - m_W^2)} \cot^2 \beta, \qquad (24)$$

where p_{H^+} and p_{W^+} are the center-of-mass momenta of the H^+ and W^+ for the respective decays. Thus, the H^+ channel is fully competitive with the W^+ mode. As a result, it may be possible to discover the charged Higgs boson via $t\bar{t}$ production followed by $t \to bH^+$. In fact, in this circumstance, the main challenge would be to discover the top-quark itself. Once the t-quark is found, anomalous t decays might point in the direction of a charged Higgs boson. This may be the only case where one could find evidence for Higgs bosons at the Tevatron or ACOL!

If $m_t < m_{H^\pm} + m_b$, then the most important production mechanisms for the $H^\pm$ derive from the subprocesses $gb \to H^- t$ and $g\bar{b} \to H^+ \bar{t}$. These are computed in ref. 14 where it is also demonstrated that a naive computation of H^+ (H^-) production using $t\bar{b}$, ($b\bar{t}$) fusion would be very inaccurate in the mass region of interest.[45] The charged Higgs boson cross section is comparable in magnitude to the gg fusion cross section of the Standard Model Higgs at the same Higgs mass. This is illustrated in fig. 1, taken from ref. 14.

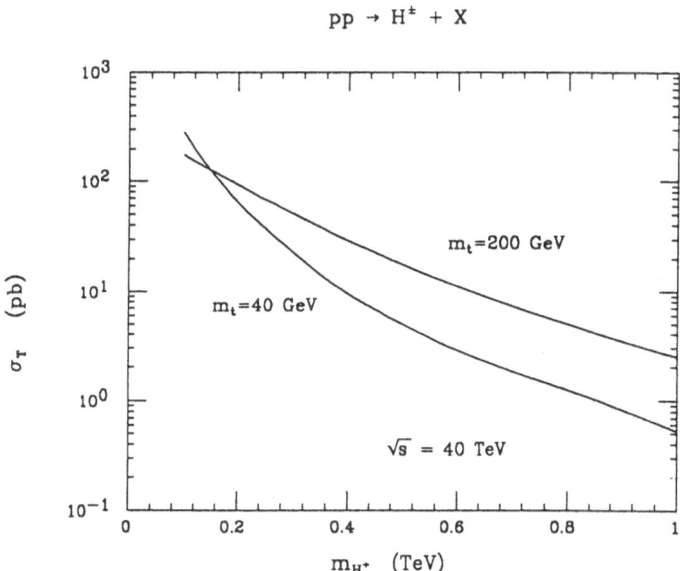

Figure 1. The cross section at the SSC for single charged Higgs production (summed over both charges) coming from $gb \to H^- t$ and $g\bar{b} \to H^+ \bar{t}$ as a function of $m_{H^\pm}$. Top quark masses of $m_t = 40$ GeV and $m_t = 200$ GeV are considered. This figure is taken from ref. 14.

The predicted cross section in the region $m_{H^\pm} > m_t$ is such that the raw number of charged Higgs events is substantial. However, for a given Higgs decay mode, the desired signal is generally swamped by huge backgrounds. To have any chance of seeing a signal, a trick must be employed. One trick that has been explored is that of a 'stiff lepton trigger', first proposed in ref. 14. In the production mechanisms, $\bar{b}g \to \bar{t}H^+$ and $bg \to tH^-$, one attempts to trigger on the t or $\bar{t}$ produced in association with the charged Higgs. One approach to doing this is to note that the final state t and $\bar{t}$ quarks are typically moving nearly parallel to the original beam. Ordinarily, they would just be lost inside the beam jets. However, if the t-quark decays semi-leptonically, the electron or muon will be kicked out with sufficiently large p_T (of order $m_t/2$), so that it can be used to trigger the desired event. Even the leptons coming from decays of the secondary b quarks that arise from t decay will contribute to this trigger, so that a trigger in which a stiff lepton with $p^l_T > 10$ GeV is required retains $\sim 45\%$ of the $H^{\pm}$ events, while rejecting all but 1% to 2% of most types of background processes.[14]

Even with the large sample of charged Higgs bosons described above, it will be extremely difficult to isolate a signal above Standard Model backgrounds (if $m_{H^{\pm}} > m_t + m_b$). It is clear that QCD backgrounds to observing the H^+ via its $t\bar{b}$ decay are very large. This has been studied and quantitative background levels are summarized in ref. 46. So far, no cuts have been found that make this mode worth pursuing. Thus, one must concentrate on the search for the charged Higgs boson via rarer decay modes. Among the various possibilities are: $H^{\pm} \to W^{\pm}\gamma$, $H^{\pm} \to W^{\pm}+$ quarkonium, $H^{\pm} \to W^{\pm}h^0$, and $H^{\pm} \to \tau\nu$. Unfortunately, one would have to have an anomalously large branching ratio for one of these rare decays in order to have a feasible method for detecting the charged Higgs boson. By considering realistic branching ratios and the signatures of the possible rare decays,[46] one comes to the conclusion that the charged Higgs boson cannot be detected at a hadron collider if $m_{H^{\pm}} > m_t + m_b$.

Higgs searches at e^+e^- colliders tend to be more straightforward. In particular, in contrast to the discussion above, it is rather simple to search for charged Higgs bosons at an e^+e^- collider. Light charged Higgs bosons are already ruled out by experiment. The cross-section for charged Higgs pair production (due to virtual γ-exchange), is equal to $0.25\beta^3$ units of R, where $\beta^3 \equiv (1 - 4m^2_{H^{\pm}}/s)^{3/2}$ is the p-wave threshold factor, and one unit of $R = \sigma(e^+e^- \to \gamma^* \to \mu^+\mu^-)$. Limits have been obtained at PEP and PETRA from the failure to observe the H^+ in $e^+e^- \to \gamma^*, Z^* \to H^+H^-$, excluding $2m_\tau \lesssim m_{H^+} \lesssim 19$ GeV.[47] I anticipate that by the end of 1989, the data from LEP will increase this limit to $m_{H^+} \gtrsim 40$ GeV, if a signal is not seen. The region of $m_{H^+} \lesssim 2m_\tau$ is apparently excluded by failure to observe the decay $B \to H^+X_c$.[48]

Let us now briefly consider Higgs searches at SLC, LEP, LEP-II and future e^+e^- supercolliders. I will focus here on the changes in Higgs phenomenology as compared with that of the minimal Higgs boson described in section 2. If one (or more) of the neutral Higgs bosons is lighter than some quarkonium state, then the decays of a $V(1^{--})$ quarkonium state to $h^0\gamma$, $H^0\gamma$, or $A^0\gamma$ (if allowed) can be either enhanced or suppressed by the square of the relevant coupling given in eq. (22). Next, let us reconsider the Higgs search at a Z factory. For the scalar bosons, the rate for $Z \to h\ell^+\ell^-$ ($h = h^0$ or H^0) will generally be somewhat suppressed. Nonetheless, both scalar Higgs bosons would probably be detectable in this mode (presuming both are light enough) unless one Higgs completely saturates the sum rule of eq. (17), in which case the other scalar will not be detectable in this mode. The neutral pseudoscalar has no tree-level couplings to VV. Hence, the rate for $Z \to A^0\ell^+\ell^-$ (which occurs only at one-loop) is extremely small. However, new decays of an on-shell Z are possible in the two doublet model:

$$Z \to A^0h^0, \quad Z \to A^0H^0, \quad Z \to H^+H^-, \tag{25}$$

which lead to simultaneous production of a scalar Higgs boson and the pseudoscalar Higgs boson or of a charged Higgs pair. The decay rates normalized to the partial width of Z into one generation of neutrinos are:

$$\frac{\Gamma(Z \to A^0 h^0)}{\Gamma(Z \to \nu\bar{\nu})} = \tfrac{1}{2}\cos^2(\beta - \alpha)B^3$$

$$\frac{\Gamma(Z \to H^+ H^-)}{\Gamma(Z \to \nu\bar{\nu})} = \tfrac{1}{2}\cos^2 2\theta_W B^3 \tag{26}$$

where $B = 2|\vec{p}|/m_Z$ with $|\vec{p}|$ being the magnitude of the three momentum of one of the final Higgs particles. For $\Gamma(Z \to A^0 H^0)$, replace $\cos(\beta - \alpha)$ with $\sin(\beta - \alpha)$ in the first expression above. As long as the angle factors are not particularly small, the branching fraction into Higgs pairs, if kinematically allowed, can be as large as a few percent, and would constitute the dominant source of Higgs production in Z decays!

At higher energy e^+e^- machines, appropriate for discovering more massive Higgs bosons, the major scalar Higgs production mechanisms: $e^+e^- \to ZH^0$ and $e^+e^- \to \nu\bar{\nu}H^0$ are dependent upon substantial VV couplings. Thus, as for the hadron collider case, so long as the two neutral scalars share fairly equally the allowed VV couplings [see eq. (17)], their detection should be quite straightforward at a machine with adequate energy. In contrast, the A^0 may be particularly difficult to find at an e^+e^- machine, since it has no VV couplings. The main pseudoscalar production mode that is available is $Z^* \to A^0 h^0$ or $A^0 H^0$. The cross sections for these processes are easily computed, and we find [49]

$$\sigma(e^+e^- \to A^0 h^0) = \left(\frac{8\sin^4\theta_W - 4\sin^2\theta_W + 1}{\cos^4\theta_W}\right) \frac{g^4 \cos^2(\beta - \alpha)\kappa^3}{192\pi\sqrt{s}\left[(s - m_Z^2)^2 + \Gamma_Z^2 m_Z^2\right]}, \tag{27}$$

where κ is the center-of-mass momentum of one of the final state Higgs bosons. For $\sigma(e^+e^- \to A^0 H^0)$, replace $\cos(\beta - \alpha)$ with $\sin(\beta - \alpha)$ in eq. (27). The detection of this process might be possible[50] if $m_{A^0} + m_{h^0}$ (or $m_{A^0} + m_{H^0}$) is not too large compared to the machine energy. For instance, if the $ZA^0 H^0$ coupling saturates the strength allowed by eq. (19), then the cross section for $e^+e^- \to Z^* \to A^0 + H^0$ can be as large as one-tenth of a unit of R when not suppressed by phase space.

Finally, let us consider charged Higgs production at an e^+e^- supercollider which proceeds via virtual γ and Z exchange. For $\sqrt{s} > m_Z$, the cross-section formula is a little different than the one given previously. The asymptotic result for $\sigma(e^+e^- \to H^+ H^-)$ in units of R (for $s \gg m_Z^2, 4m_{H^\pm}^2$) is given by:

$$\frac{1 + 4\sin^4\theta_W}{8\sin^4 2\theta_W} \sim 0.308. \tag{28}$$

The observability of the charged Higgs bosons in this production mode has been studied by Komamiya in ref. 51. He finds that a charged Higgs boson, whose decays are dominated by fermion pairs (or Wh) and with mass less than $0.4\sqrt{s}$, will be detectable at an e^+e^- collider with an integrated luminosity of 10^3 inverse units of R.

6. HIGGS BOSONS IN SUPERSYMMETRIC MODELS

We have seen above that it is relatively straightforward to go beyond the minimal one-doublet Higgs sector, even within the context of the Standard Model. However, there are a number of strong theoretical reasons to suppose that the Standard Model itself is, in fact, merely part of a larger structure. Indeed, it is the deep-rooted problems associated with the Higgs sector which suggest that we must ultimately look beyond the Standard Model. These are the problems of fine-tuning, naturalness, and hierarchy, referred to in section 1. Here, let us simply recall that, in the Standard Model, a calculation of the first order correction to the Higgs boson mass squared yields a quadratically divergent expression arising from Standard Model particle loop graphs. This implies that it is not "natural" to have a Higgs boson that is relatively light unless this divergence can be controlled by the structure of the theory. The Standard Model provides no mechanism for this. Many different ways of regulating the divergence have been proposed, including supersymmetry, technicolor, and composite Higgs models. In this section, I shall focus on supersymmetry. In a supersymmetric theory the quadratic divergence is naturally cancelled by related loop graphs involving the supersymmetric partners of the Standard Model particles which appear in the divergent loops. The result is that the tree level mass squared of the Higgs boson receives corrections that are limited by the extent of supersymmetry breaking. In order that the naturalness and hierarchy problems be resolved, it is necessary that the scale of supersymmetry breaking not exceed $\mathcal{O}(1 \text{ TeV})$. Supersymmetric theories are especially interesting in that, to date, they provide the only theoretical framework in which the problems of naturalness and hierarchy are resolved while retaining the Higgs bosons as truly elementary spin-0 particles.

In the minimal supersymmetric extension of the Standard Model,[53] one simply associates a supersymmetric partner to all Standard Model particles. In addition, one must enlarge the Higgs sector to contain two Higgs doublets: a $Y = -1$ Higgs doublet (H_1) which couples to down-type quarks and charged leptons, and a $Y = +1$ Higgs doublet (H_2), which couples to up-type quarks and neutrinos. In terms of the Higgs fields ϕ_1 and ϕ_2 introduced earlier,

$$
\begin{aligned}
H_1 &= \begin{pmatrix} H_1^1 \\ H_1^2 \end{pmatrix} = \begin{pmatrix} \phi_1^{0\,*} \\ -\phi_1^- \end{pmatrix} \\[2mm]
H_2 &= \begin{pmatrix} H_2^1 \\ H_2^2 \end{pmatrix} = \begin{pmatrix} \phi_2^+ \\ \phi_2^0 \end{pmatrix}
\end{aligned}
\tag{29}
$$

The requirement of two Higgs doublets with the specific Higgs-fermion coupling pattern mentioned above is dictated by the underlying supersymmetry of the model. (In addition, this structure is required in order to avoid gauge anomalies generated by the fermionic supersymmetric partners of the Higgs bosons.) Furthermore, supersymmetry imposes strong constraints on the form for the Higgs potential. Even allowing for the most general soft-supersymmetry breaking in the model, the dimension-four terms of the Higgs potential must respect the supersymmetry. These requirements impose relations among the λ_i of eq. (6). The resulting Higgs potential in the minimal supersymmetric extension of the Standard Model is:[54,52,42]

$$V = \left(m_1^2 + |\mu|^2\right) H_1^{i*} H_1^i + \left(m_2^2 + |\mu|^2\right) H_2^{i*} H_2^i - m_{12}^2 \left(\epsilon_{ij} H_1^i H_2^j + \text{h.c.}\right)$$
$$+ \tfrac{1}{8}\left(g^2 + g'^2\right)\left[H_1^{i*}H_1^i - H_2^{j*}H_2^j\right]^2 + \tfrac{1}{2}g^2|H_1^{i*}H_2^i|^2, \tag{30}$$

where μ is a supersymmetric Higgs mass parameter and m_1^2, m_2^2, m_{12}^2 are soft supersymmetry breaking masses. It is convenient to re-express the doublet fields in terms of the physical Higgs boson degrees of freedom and the Goldstone boson fields. The relations are:

$$H_2^1 = H^+ \cos\beta + G^+ \sin\beta$$
$$H_1^2 = H^- \sin\beta - G^- \cos\beta$$
$$H_1^1 = v_1 + \frac{1}{\sqrt{2}}(H^0\cos\alpha - h^0\sin\alpha + iA^0\sin\beta - iG^0\cos\beta) \tag{31}$$
$$H_2^2 = v_2 + \frac{1}{\sqrt{2}}(H^0\sin\alpha + h^0\cos\alpha + iA^0\cos\beta + iG^0\sin\beta),$$

where α is the mixing angle that arises in the process of diagonalizing the 2×2 neutral scalar Higgs mass matrix [see eqs. (13) and (14)], and $\tan\beta \equiv v_2/v_1$. It is possible to choose a phase convention in which v_1 and v_2 are real and positive. This implies that $0 \leq \beta \leq \pi/2$. In addition, due to the constraints imposed in the supersymmetric model, we find that $-\pi/2 \leq \alpha \leq 0$. The supersymmetric constraints also imply that two parameters are sufficient to fix the properties of the Higgs sector. Here I shall adopt $\tan\beta$ and $m_{H^\pm}$ as my two independent parameters. Once these two quantities are specified all the other (tree-level) Higgs masses can be computed according to

$$m_{A^0}^2 = m_{H^\pm}^2 - m_W^2$$

$$m_{H^0,h^0}^2 = \tfrac{1}{2}\left[m_{A^0}^2 + m_Z^2 \pm \sqrt{(m_{A^0}^2 + m_Z^2)^2 - 4m_Z^2 m_{A^0}^2 \cos^2 2\beta}\,\right], \tag{32}$$

while the mixing angle α can be computed using[55]

$$\cos 2\alpha = -\cos 2\beta \left(\frac{m_{A^0}^2 - m_Z^2}{m_{H^0}^2 - m_{h^0}^2}\right) \tag{33}$$

$$\sin 2\alpha = -\sin 2\beta \left(\frac{m_{H^0}^2 + m_{h^0}^2}{m_{H^0}^2 - m_{h^0}^2}\right). \tag{34}$$

There are several particularly crucial predictions following from eq. (32). We find: $m_{H^\pm} \geq m_W$, $m_{H^0} \geq m_Z$, $m_{A^0} \geq m_{h^0}$, and

$$m_{h^0} \leq m_Z |\cos 2\beta| \leq m_Z. \tag{35}$$

The spectrum of Higgs boson masses in the minimal supersymmetric model is illustrated in fig. 2. From this graph, it can be seen that in the limit where $m_{A^0} \to \infty$ (at

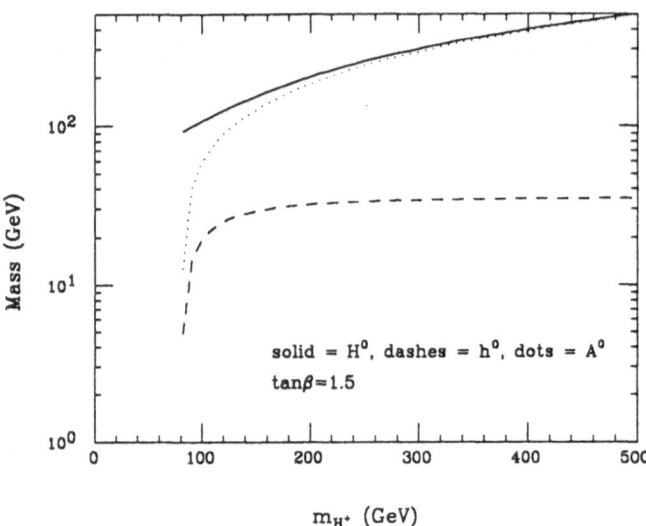

Figure 2. The masses of H^0, h^0 and A^0 are plotted as functions of $m_{H\pm}$ at fixed $\tan \beta = 1.5$ for the minimal supersymmetric model.

fixed $\tan \beta$), A^0, H^0 and $H^\pm$ decouple from the theory and we are left with a Higgs sector (consisting of a single physical CP-even scalar, h^0) which is identical to the Higgs boson of the minimal Standard Model. Moreover, in this limit, the interactions of h^0 with the Standard Model gauge bosons and fermions are equivalent to those of the minimal Higgs boson of the (nonsupersymmetric) Standard Model.

It is important to emphasize that the Higgs boson masses above are *tree-level* masses, which will be somewhat modified when radiative corrections are taken into account. In general, the relative size of the corrections will be small, except in the case of a very light scalar Higgs boson (h^0). For example, if $v_1 = v_2$ (*i.e.* $\tan \beta = 1$), one finds that $m_{h^0} = 0$ at tree-level, with radiative corrections of order 10 GeV. It is also of interest to investigate how stable the mass relations of eq. (32) are under one-loop renormalization. A numerical study was carried out in ref. 56, and more recently an analytical evaluation of the sum rule corrections has been obtained in ref. 57. A priori, one might expect that corrections to Higgs boson masses might be large when the superpartner masses become large; in this limit, the natural limitations coming from the supersymmetric structure of the theory become far removed from the tree-level Higgs boson mass scale. However, both calculations find very small corrections to the mass relations, once the standard infinities found in renormalization are absorbed into two physical parameters (*e.g.* the values of $\tan \beta$ and $m_{H\pm}$).

Let us now turn to the couplings of the various Higgs bosons of the minimal supersymmetric model. A complete list of all Higgs boson couplings in the minimal supersymmetric model is given in ref. 42. (See also Appendix A of ref. 1.) Here I shall discuss only couplings to Standard Model particles, in particular to quarks, W's and Z's. It is these couplings which are crucial in determining the production of the Higgs bosons, and in the absence of a light supersymmetric particles would also completely determine their decays.

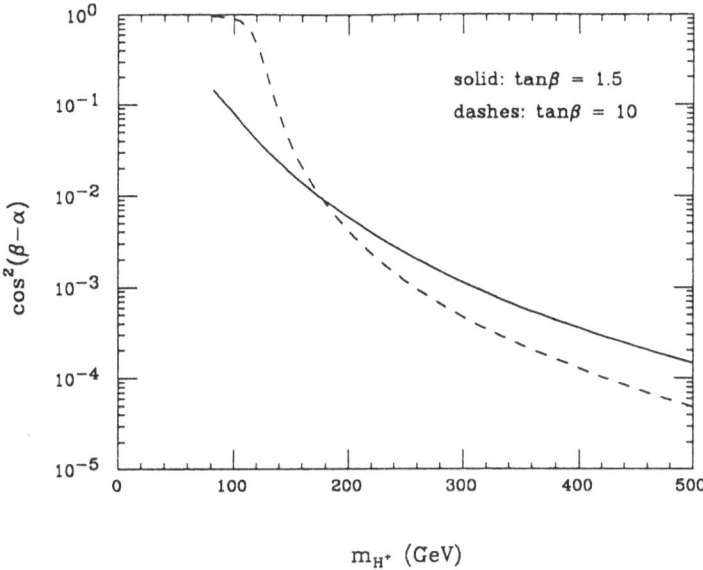

Figure 3. $\cos^2(\beta - \alpha)$ as a function of the charged Higgs mass, m_{H^+}, at $\tan\beta = 1.5$ and 10. Note that $m_{H^+} \geq m_W$ and $\cos^2(\beta - \alpha) = \cos^2 2\beta$ at $m_{H^+} = m_W$.

We have already seen that the A^0 has no VV couplings, while those of h^0 and H^0 are related by eq. (18) to the $\phi^0 VV$ coupling. In the supersymmetric model these important couplings can be computed in terms of two Higgs masses (and m_Z). We find[55]

$$\cos^2(\beta - \alpha) = \frac{m_{h^0}^2(m_Z^2 - m_{h^0}^2)}{(m_{H^0}^2 - m_{h^0}^2)(m_{H^0}^2 + m_{h^0}^2 - m_Z^2)}. \tag{36}$$

A close examination of this formula shows a dramatic suppression over a very large region of parameter space. It is easily verified that the maximum possible value for $\cos^2(\beta - \alpha)$ at fixed $\tan\beta$ (which occurs in the limit $m_{H^+} \rightarrow m_W$) is $\cos^2 2\beta$. Note that the lighter the value of m_{h^0} [eq. (35)], the smaller the maximum possible value of $\cos^2(\beta - \alpha)$. Furthermore, as m_{H^+} increases, so does m_{H^0}, and $\cos^2(\beta - \alpha)$ decreases as $1/m_{H^0}^4$. This behavior is illustrated in fig. 3 for the cases of $\tan\beta = 1.5$ and $\tan\beta = 10$. For example, when $\tan\beta = 1.5$, $\cos^2(\beta - \alpha)$ is $\lesssim 0.15$ at $m_{H^+} = m_W$, and is $\lesssim 0.01$ by the time $m_{H^0} > 2m_Z$. Thus, according to eq. (18), one should generally expect the coupling of W^+W^- and ZZ to the heavier Higgs scalar (H^0) to be greatly suppressed. Using eq. (17), this also implies that the coupling of W^+W^- and ZZ to the lighter Higgs scalar (h^0) should be roughly equal in strength to the corresponding couplings of the minimal Higgs boson.

The couplings of the various Higgs bosons to quarks were given in eqs. (22) and (23) as a function of α and β. A survey of the parameter space reveals that when $\tan\beta > 1$ the H^0 and h^0 couplings to $t\bar{t}$ are somewhat suppressed relative to the Standard Model ϕ^0, while the $b\bar{b}$ couplings are somewhat enhanced, and vice versa for $\tan\beta < 1$. In general, the suppression or enhancement is relatively mild. For instance, in the case of H^0, once $m_{H^0} \gtrsim 2m_Z$ the couplings agree with the corresponding Standard Model values to within (roughly) a factor of $\cot\beta$ or $\tan\beta$, for up and down quarks respectively, while in the case of the h^0 the couplings are very close to Standard Model values. See ref. 55 for detailed graphs.

131

With the above comments in mind, let us consider the new features of Higgs boson phenomenology in the context of the minimal supersymmetric model. Because there are fewer free parameters, the phenomenology is more constrained. We summarize some of the salient points below:

1. Perhaps the most dramatic prediction is that $m_h \leq m_Z |\cos 2\beta| \leq m_Z$. This implies that it is very likely that the h will be light enough to be detected at SLC, LEP or LEP-II in $Z \rightarrow Z^* h$ or $e^+ e^- \rightarrow Z^* \rightarrow Zh$. One must check, of course, that the $Z^* Zh$ coupling is not particularly suppressed from its Standard Model value. However, as discussed above,

$$g_{ZZh} = g_{ZZ\phi^0} \sin(\beta - \alpha) \qquad (37)$$

and $\sin(\beta - \alpha)$ is near 1 over a very large region of supersymmetric parameter space (see fig. 3). Thus, minimal supersymmetry strongly suggests the existence of a light Higgs boson with properties similar to that of the Standard Model minimal Higgs boson, which would be detected (if it exists) by $e^+ e^-$ colliders in the 1990s.

2. The converse of the above result is that the other physical Higgs particles H, A and $H^\pm$ will be very difficult to detect. In the same limit where the h properties approach those of the Standard Model Higgs minimal boson, the other physical Higgs particles tend to decouple. The pseudoscalar A can also be light (although there is no particular favored value for its mass). However, there is no tree-level coupling of A to vector bosons, which makes it impossible to observe using methods which depend on the $W^+ W^-$ and ZZ couplings to the Higgs boson. As mentioned in the previous section, the two most promising methods for A production are through toponium decay or $Z \rightarrow Ah^0$. The former requires a copious source of toponium. The latter is highly suppressed in the minimal supersymmetric model due to a suppression of the ZAh^0 coupling by the infamous $\cos(\beta - \alpha)$ factor, and a very limited region of parameter space over which the decay can occur. The charged Higgs boson $H^\pm$ is predicted to be heavier than the W. We have already seen that $H^+ H^-$ pair production is straightforward to detect at $e^+ e^-$ colliders, as long as $m_{H^+} \lesssim 0.4\sqrt{s}$. Thus, detection of H^+ will require the services of an $e^+ e^-$ supercollider. Detection of charged Higgs bosons at LHC or SSC is notoriously difficult as previously discussed.

3. As emphasized above, the coupling of the heavier scalar Higgs, H^0, to vector boson pairs is suppressed. Explicitly,

$$g_{W^+ W^- H} = g_{W^+ W^- \phi^0} \cos(\beta - \alpha) \qquad (38)$$

which, according to fig. 3, is very small over a large range of parameter space. In particular, for $m_H > 2m_W$, the $W^+ W^- H$ coupling is reduced by a factor of (at least) 10. This rules out all known techniques for Higgs detection at LHC or SSC, all of which depend on the decay of the Higgs boson to vector boson pairs.

The upshot of the above observations is twofold. In supersymmetric models, it should be rather straightforward to discover the lightest Higgs scalar of the model. However, it will be far more difficult to prove the existence of an extended Higgs sector, which is a necessary consequence of low energy supersymmetry. It could be argued that the above observations were very specific to the assumption of a minimal supersymmetric structure beyond the Standard Model. There has been some investigation in the literature of Higgs boson phenomenology in more complicated supersymmetric models.[58,59] Results in these models tend to confirm the general observations described above. Some interesting counterexamples are known to exist (*e.g.*, models with a charged Higgs boson which is lighter than the W); however these models tend to predict a rich structure of new physics at rather "low" energies (around 100 GeV) which should be rather easy to expose at colliders in the near future.[59] Thus, the verification of the supersymmetric scenario will require direct evidence of the supersymmetric particles. The decays of Higgs bosons into supersymmetric final states may also make up a major percentage of the Higgs branching ratio,[60] in which case, Higgs boson phenomenology would be absorbed into the general experimental exploration of the supersymmetric spectrum.

7. HOW MANY LIGHT HIGGS BOSON STATES ARE EXPECTED?

Most of the attention so far has been given to models with minimal Higgs structure. The minimal extension of the Standard Model to two Higgs doublets was examined in sections 4 and 5. In section 6, I focused almost entirely on the minimal supersymmetric extension of the Standard Model. I argued that the resulting Higgs phenomenology is also a general feature of many non-minimal supersymmetric models (although some interesting differences have been explored in the literature). Since there is no experimental information to guide us at present, the choice of minimal Higgs structures has usually been assumed to be a matter of convenience for the theorist. In this section, I will argue that minimality is more than a simplification—it is a feature to be expected in multi-Higgs models with a high energy scale. This section is based on recent work done in collaboration with Yosef Nir.[61]

In order to explicitly illustrate the mechanism we have in mind, let us return to the minimal supersymmetric model of section 6. As previously noted, the Higgs sector of this model is fixed once two parameters are specified. I shall choose these parameters to be $\tan \beta$ and the soft-supersymmetry-breaking mass m_{12} which couples H_1 and H_2 in the Higgs potential [eq. (30)]. It is easy to show that:

$$m_A^2 = m_{12}^2 (\tan \beta + \cot \beta).$$ (39)

Using this result, the other Higgs masses and mixing angle can be obtained using eqs. (32)–(34). If one now takes $m_{12} \gg m_W$, then it follows that: (i) $m_A \simeq m_{H^\pm} \simeq m_{H^0} \gg m_W, m_{h^0}$; (ii) $\cos(\beta - \alpha) \to 0$ [see eq. (36)]; and (iii) the coupling of h^0 to vector bosons and fermions approach the values of the Standard (nonsupersymmetric) Model with minimal Higgs structure. That is, A^0, $H^\pm$, and H^0 decouple, and the resulting Higgs sector consists of a single Higgs scalar, h^0, whose properties are precisely those of the minimal Higgs Model.* This tendency toward decoupling actually sets in for rather modest values of $m_{12} > m_W$. This explains why $\cos(\beta - \alpha)$ is small and

---

* It is interesting to note that one remnant of the underlying supersymmetry apparently remains—namely, $m_h \leq m_Z$. However, one must remember that this is only a tree-level relation, which presumably suffers very large corrections if m_{12} is too large.

the h^0 properties are nearly those of the minimal Higgs boson of the Standard Model over a very large region of supersymmetric parameter space, as noted in the previous section.

The above result is indicative of a more general result. The key ingredient in the above analysis was the existence of a new energy scale, m_{12}, which in the above example corresponds to the scale of supersymmetry breaking. In the limit of large m_{12}, all but one of the Higgs masses were driven to the large energy scale of the model, leaving only the minimal Higgs boson in the low energy effective theory at the electroweak breaking scale. Thus, it is of interest to consider an arbitrary multi-Higgs doublet model, assuming the existence of a new high energy scale (denoted by Λ) which characterizes the scale at which some unspecified new physics enters. We further assume that $\Lambda \gg v$, where $v \simeq 246$ GeV is the Higgs vacuum expectation value in the minimal model (corresponding to the scale of electroweak symmetry- breaking). Then, we can describe the effects of the new physics on the Higgs sector in the following way: all dimensionful parameters in the Higgs potential are of $\mathcal{O}(\Lambda)$, as long as the electroweak symmetry is maintained to low enough energies (of order v). That is, in order to permit electroweak breaking to occur at the low scale (v), there will be some combination of dimensionful parameters (which *a priori* are of order Λ) which must be of order v. To arrange this requires a fine-tuning of parameters. As a crucial aspect of our philosophy, we demand that only the minimal number of fine-tunings be performed in order that $v \ll \Lambda$.

This property is general enough to encompass a wide variety of theories beyond the Standard Model. In particular, we have in mind all those models where the $SU(2) \times U(1)$ symmetry is embedded in a larger symmetry group. The breaking of this larger symmetry occurs when a scalar $SU(2) \times U(1)$-singlet N assumes a vacuum expectation value: $\langle N \rangle \sim \Lambda$. Scalar non-singlets ϕ have couplings to N of the general form:

$$\eta N N \phi^\dagger \phi, \qquad (40)$$

where η is a quartic coupling constant. When N assumes a vacuum expectation value, this leads to a mass term for ϕ:

$$m^2 \phi^\dagger \phi = \eta \langle N \rangle^2 \phi^\dagger \phi. \qquad (41)$$

Thus, as long as all dimensionless scalar couplings are of $\mathcal{O}(1)$, we can conclude that $m^2 = \mathcal{O}(\Lambda^2)$ as indicated above.

Having stated our initial assumptions and philosophies, we now consider an arbitrary multi-Higgs-doublet model. It is convenient to parameterize the scale of new physics by introducing an $SU(2) \times U(1)$ singlet scalar field as indicated above. We perform the minimal number of fine tunings required to have electroweak symmetry breaking at a scale $v \ll \langle N \rangle$, but otherwise allow for the most general Higgs couplings.[†] When we diagonalize the Higgs mass matrices to obtain the physical states, the following results emerge:

[†] We do not impose artificial discrete symmetries to reduce the number of parameters. In this regard, a discrete symmetry is viewed as an additional fine-tuning of certain parameters of the model to zero, in violation of our stated dictum.

1. The low energy scalar sector contains precisely one light neutral Higgs boson, h^0, whose couplings differ from those of the minimal Higgs model by terms of order v^2/Λ^2.

2. We can allow all Higgs fields of the model to couple arbitrarily to the quarks and leptons, in violation of the conditions of the Glashow-Weinberg theorem for flavor conserving couplings (see section 3). Nevertheless, the h^0 couplings are approximately flavor conserving, with violations of $\mathcal{O}(v^2/\Lambda^2)$.

3. If we allow CP-violation to enter the model in an arbitrary fashion, then the only remnant of CP-violation (to leading order) in the low energy theory (at the electroweak scale) occurs in the form of a Kobayashi-Maskawa (KM) matrix in the charged current, with the usual CP-violating phase. The CP-violating h^0 interactions are suppressed by terms of $\mathcal{O}(v^2/\Lambda^2)$.

4. Unitarity constraints on h^0 interactions are the same as those of the minimal Higgs model, up to $\mathcal{O}(v^2/\Lambda^2)$ corrections. Furthermore, the coupling of the heavy Higgs sector (*i.e.*, all other Higgs scalars, apart from h^0) to W^+W^- and ZZ are likewise suppressed by $\mathcal{O}(v^2/\Lambda^2)$ terms.[30)]

5. The Linde-Weinberg bound for h^0 differs from the lower Higgs mass bound of the minimal Higgs model by terms of order v^2/Λ^2.

The last point is of some theoretical interest, so I shall briefly elaborate on its derivation in the two-Higgs-doublet model. Linde-Weinberg bound is derived by requiring that the $SU(2) \times U(1)$ breaking minimum is a global minimum of the one-loop corrected effective potential. In the minimal Higgs model, the result of this requirement is that $m_\phi \geq m_{LW}$ with

$$
\begin{aligned}
m_{LW}^2 &= \frac{1}{16\pi^2 v^2} \operatorname{Str} M_i^4 \\
&\simeq (7 \text{ GeV})^2 \left(1 - 1.09 \frac{m_t^4}{m_W^4} \right),
\end{aligned}
\tag{42}
$$

where we use the notation

$$
\operatorname{Str}\{\cdots\} \equiv \sum_i C_i (2J_i + 1)(-1)^{2J_i}\{\cdots\},
\tag{43}
$$

and the sum is taken over all particles of spin J_i and mass M_i, and C_i counts electric charge and color degrees of freedom. In multi-Higgs models, the value of m_{LW} is modified and no longer applies to a particular Higgs mass. In the two-doublet model,

$$
m_{H^0}^2 \cos^2(\beta - \alpha) + m_{h^0}^2 \sin^2(\beta - \alpha) \geq m_{LW}^2.
\tag{44}
$$

To compute m_{LW}, we must examine how particle masses depend on v. It is convenient to work in a basis in which only one scalar field possesses a nonzero vacuum expectation value, v. We may derive an analytic formula for m_{LW} if all particle masses are of the form[62)]

$$M_i^2 = \mu_i^2 + \lambda_i v^2 \,. \tag{45}$$

In accord with our philosophy above, the parameters μ_i are assumed to be of $\mathcal{O}(\Lambda)$. Then, we find[1,61)]

$$m_{LW}^2 = \mathrm{Str}\ \frac{\lambda_i^2 v^2}{16\pi^2} \left\{ 1 - \frac{2\mu_i^2}{\lambda_i v^2} \left[1 - \frac{\mu_i^2}{\lambda_i v^2} \log\left(\frac{\mu_i^2 + \lambda_i v^2}{\mu_i^2} \right) \right] \right\} \,. \tag{46}$$

It is instructive to examine this formula assuming that the particles i can be divided into two classes j and k such that $\mu_j = 0$ and $\mu_k \gg v$. Then, we find

$$m_{LW}^2 = \frac{1}{16\pi^2 v^2}\, \mathrm{Str}\, M_j^4 + \frac{v^4}{24\pi^2}\, \mathrm{Str}\, \frac{\lambda_k^3}{\mu_k^2} \,. \tag{47}$$

This is the desired result. The first term on the right hand side corresponds to the result of the minimal model [eq. (42)], and the second term is suppressed by a factor of order v^2/Λ^2.

We can extend many of the above considerations to supersymmetric models. As a simple example, consider the supersymmetric extension of the Standard Model, with *two pairs* of $Y = \pm 1$ Higgs doublet superfields and an $SU(2) \times U(1)$ singlet scalar superfield, $\hat{N}$. As above, we introduce a new high energy scale Λ such that $\langle N \rangle \sim \Lambda$. It is convenient to arrange the model such that a nonzero vacuum expectation value for N does not break supersymmetry. Supersymmetry is then broken at some scale Λ_{SUSY} when soft-supersymmetry breaking terms are added. Among such terms are squared-mass terms for the Higgs doublets and trilinear "A-terms" which mix the doublet and singlet scalars.[53)] Once again, we perform the minimal amount of fine-tuning of parameters to ensure the desired hierarchy of scales[*]

$$v \lesssim \Lambda_{SUSY} \ll \Lambda \,. \tag{48}$$

When the physical Higgs states and their interactions are extracted, one finds that at the low energy scale ($\ll \Lambda$), the scalar sector contains $H^\pm$, H^0, h^0, and A^0. This is precisely the Higgs content of the minimal supersymmetric extension of the Standard Model, with the expected couplings as described in section 6! All other Higgs bosons have masses of order the high energy scale, Λ. Specifically, we find:

$$\begin{aligned} \dot{m}_{h^0} &\sim \mathcal{O}(v) \,, \\ m_{H^0} \simeq m_{H^\pm} &\simeq m_{A^0} \sim \mathcal{O}(\sqrt{A\Lambda}) \,. \end{aligned} \tag{49}$$

---

[*] Although fine-tuning at tree-level is still required, supersymmetric models possess the theoretically pleasing feature that such fine-tuning is stable under radiative corrections.

If we wish to have $\Lambda_{SUSY} \sim v$, then the A-parameter must be chosen to be of order v^2/Λ. Note that if we break supersymmetry at the high scale ($\Lambda_{SUSY} \sim \Lambda$), then all Higgs bosons, except for h^0 move up to the high scale, in agreement with the analysis presented at the beginning of this section.

In summary, the main conclusion of this section is the following. In multi-Higgs models with a high energy scale, if the only low energy symmetry is the electroweak symmetry of the Standard Model, then there is only one light scalar. Its properties are identical to those of the minimal Higgs boson of the Standard Model, up to small corrections of order v^2/Λ^2. That is, effects of the additional scalar particles are suppressed by inverse powers of the high energy scale. They may induce flavor changing neutral currents which become significant if the scale is a few TeV or lower. They may provide us with several new sources of CP violation. The only source for CP violation which does not depend in magnitude on the high energy scale is the usual KM phase in the quark mixing matrix. Additional scalars contribute to the Linde-Weinberg bound. The modification of the bound can be significant only if the scale of new physics is a few TeV or less.

The existence of additional light scalars is associated with additional symmetries at low energies. The best known example is that of supersymmetry. If the only low energy symmetry is the supersymmetric $SU(2) \times U(1)$ symmetry, then there are two light neutral CP-even scalars, one light neutral CP-odd scalar and one light pair of charged scalars. The properties of these scalars are identical to those of the minimal supersymmetric extension of the Standard Model, up to small corrections of order v^2/Λ^2. If the scale of supersymmetry breaking is higher than that of electroweak breaking, then there will be one light scalar at the electroweak breaking scale, with properties identical to those of the minimal Higgs boson of the Standard Model, as before. The other above-mentioned light scalars have masses at the supersymmetry breaking scale.

Is it possible to have models which possess a large energy scale, Λ, and at the same time contain additional light scalar states beyond the possible minimal Higgs sectors described above? The answer is yes, but in all cases, the model is accompanied by additional symmetries at the low energy ($\ll \Lambda$) scale. Examples include additional low energy gauge theories and new spontaneously broken global symmetries. (See ref. 61 for further details.)

Most experimental Higgs searches are based on either the Standard Model with one physical Higgs boson or the minimal supersymmetric extension of the Standard Model. Based on the considerations above, one can conclude that these two models are not just toy models where calculations are simple and free of ambiguities; they indeed represent the low energy limit of a much larger class of models. The search for non-minimal Higgs bosons presents a challenge for the next generation of experiments. New difficulties are expected beyond those encountered in the search for the Standard Model Higgs boson. The discovery of non-minimal Higgs bosons will pose interesting constraints on the existence of low energy supersymmetry, as well as providing important insight into the origin of electroweak symmetry breaking. Moreover, if experiments find a light scalar sector which is richer or different from either of the minimal models mentioned above, it will be an important clue to the existence of new physics and additional symmetries at low energies.

Acknowledgements

I gratefully acknowledge my fellow Higgs Hunters: Sally Dawson, Jack Gunion, and Gordon Kane. Much of the discussion above (excluding section 7) is a consequence of our collaborative efforts over the past few years. In addition, I would like to thank Yosef Nir for his many insights and contributions to the material reported in section 7. Finally, I greatly appreciate the hospitality of CCSEM and applaud Ahmed Ali for an excellent workshop and an enjoyable ten days in Erice. This work was supported in part by the U.S. Department of Energy.

REFERENCES

1. For a comprehensive review of the physics of Higgs bosons and a guide to the relevant literature, see J.F. Gunion, H.E. Haber, G.L. Kane and S. Dawson, *The Higgs Hunter's Guide* (Addison-Wesley Publishing Company, Redwood City, CA, 1989).

2. G. 't Hooft, in *Recent Developments in Gauge Theories,* Proceedings of the NATO Advanced Summer Institute, Cargese 1979, edited by G. 't Hooft *et al.* (Plenum, New York, 1980) p. 135; L. Susskind, *Phys. Rep.* **104** (1984) 181.

3. E. Witten, *Nucl. Phys.* **B188** (1981) 513; S. Dimopoulos and H. Georgi, *Nucl. Phys.* **B193** (1981) 150; N. Sakai, *Z. Phys.* **C11** (1981) 153.

4. E. Farhi and L. Susskind, *Phys. Rep.* **74** (1981) 277; R.K. Kaul, *Rev. Mod. Phys.* **55** (1983) 449.

5. D.B. Kaplan, H. Georgi and S. Dimopoulos, *Phys. Lett.* **136B** (1984) 183; D.B. Kaplan and H. Georgi, *Phys. Lett.* **136B** (1984) 187, *Phys. Lett.* **143B** (1984) 152, and *Phys. Lett.* **145B** (1984) 216; M.J. Dugan, H. Georgi and D.B. Kaplan, *Nucl. Phys.* **B253** (1985) 299; R.S. Chivukula, and H. Georgi, *Phys. Lett.* **188B** (1987) 99; *Phys. Rev.* **D36** (1987) 2102.

6. D.A. Dicus and V.S. Mathur, *Phys. Rev.* **D7** (1973) 3111; B.W. Lee, C. Quigg and G.B. Thacker, *Phys. Rev. Lett.* **38** (1977) 883; *Phys. Rev.* **D16** (1977) 1519.

7. A.D. Linde, *JETP Lett.* **23** (1976) 64; *Phys. Lett.* **62B** (1976) 435; S. Weinberg, *Phys. Rev. Lett.* **36** (1976) 294.

8. S. Dawson, J.F. Gunion, and H.E. Haber, UCD-89-12 and SCIPP-89/22 (1989).

9. H. Baer *et al.*, in *Physics at LEP, Vol. 1,* edited by J. Ellis and R. Peccei, CERN 86-02 (1986) p.297.

10. G. Barbiellini *et al.*, in *Physics at LEP, Vol. 2,* edited by J. Ellis and R. Peccei, CERN 86-02 (1986) p. 1.

11. W. Buchmuller *et al.*, in *Physics at LEP, Vol. 1,* edited by J. Ellis and R. Peccei, CERN 86-02 (1986) p.203.

12. F. Abe *et al.*, preprint UPR-0172E (1989), to be published in *Physical Review Letters.*

13. P. Burchat, D.L. Burke and A. Petersen, *Phys. Rev.* **D38** (1988) 2735.

14. J.F. Gunion, H.E. Haber, F.E. Paige, Wu-Ki Tung, and S.S.D. Willenbrock, *Nucl. Phys.* **B294** (1987) 621.

15. R. Cahn, M. Chanowitz, M. Gilchriese, M. Golden, J. Gunion, M. Herrero, I. Hinchliffe, F. Paige, and E. Wang, *Proceedings of the 1987 Berkeley Workshop on "Experiments, Detectors and Experimental Areas for the Supercollider"*, edited by R. Donaldson and M. Gilchriese (World Scientific, Singapore, 1988), p. 20.

16. D.M. Atwood, J.E. Brau, J.F. Gunion, G.L. Kane, R. Madaras, D.H. Miller, L.E. Price, and A.L. Spadafora, "Intermediate Mass Higgs Boson(s)", *Proceedings of the 1987 Berkeley Workshop on "Experiments, Detectors and Experimental Areas for the Supercollider"*, edited by R. Donaldson and M. Gilchriese (World Scientific, Singapore, 1988), p. 728.

17. F. Paige, "ELMUD: An Electron Muon Detector for Higgs Physics at the SSC", *Proceedings of the 1987 Madison Workshop on "From Colliders to Supercolliders"*, edited by V. Barger and F. Halzen (World Scientific, Singapore, 1987), p. 103.

18. R. Kleiss and W.J. Stirling, *Phys. Lett.* **200B** (1988) 193.

19. J.F. Gunion, *"Probing Higgs Bosons/Electroweak Symmetry Breaking in Purely Leptonic Channels at Hadron Colliders"*, in these Proceedings.

20. R.N. Cahn and M.S. Chanowitz, *Phys. Rev. Lett.* **56** (1986) 1327.

21. J.F. Gunion, G. Kane, C.P. Yuan, H. Sadrozinski, A. Seiden and A.J. Weinstein, *Phys. Rev.* **D40** (1989) 2223.

22. M.S. Chanowitz and M.K. Gaillard, *Nucl. Phys.* **B261** (1985) 379.

23. M.S. Chanowitz, *Ann. Rev. Nucl. Part. Sci.* **38** (1988) 323.

24. U. Amaldi, A. Bohm, L.S. Durkin, P. Langacker, A. K. Mann, W.J. Marciano, *Phys. Rev.* **D36** (1987) 1385; G. Costa, J. Ellis, G.L. Fogli, D.V. Nanopoulos and F. Zwirner, *Nucl. Phys.* **B297** (1988) 244.

25. B.W. Lee, *Proceedings of the XVI International Conference on High Energy Physics*, Batavia, IL (1972), ed. J.D. Jackson, A. Roberts and R. Donaldson, Vol. 4, p. 249; D.A. Ross and M. Veltman, *Nucl. Phys.* **B95** (1975) 135.

26. H.-S. Tsao, in *Proceedings 1980 Guangzhou Conference on Theoretical Particle Physics*, ed. H. Ning and T. Hung-yuan (Science Press, Beijing, 1980) p. 1240.

27. H. Georgi and M. Machacek, *Nucl. Phys.* **B262** (1985) 463; M.S. Chanowitz and M. Golden, *Phys. Lett.* **165B** (1985) 105.

28. S. Glashow and S. Weinberg, *Phys. Rev.* **D15**, (1977) 1958.

29. H.E. Haber, G.L. Kane, and T. Sterling, *Nucl. Phys.* **B161** (1979) 493.

30. P. Langacker and H.A. Weldon, *Phys. Rev. Lett.* **52** (1984) 1377; H.A. Weldon, *Phys. Rev.* **D30** (1984) 1547; *Phys. Lett.* **146B** (1984) 59.

31. J.F. Gunion, H.E. Haber, and J. Wudka, work in progress.

32. N.G. Deshpande and E. Ma, *Phys. Rev.* **D18** (1978) 2574.

33. H. Georgi, *Hadronic Journal* **1** (1978) 155.

34. J.F. Donoghue and L.-F. Li, *Phys. Rev.* **D19** (1979) 945.

35. L.F. Abbott, P. Sikivie and M.B. Wise, *Phys. Rev.* **D21** (1980) 1393.

36. B. McWilliams and L.-F. Li, *Nucl. Phys.* **B179** (1981) 62.

37. J.A. Grifols and A. Mendez, *Phys. Rev.* **D22** (1980) 1725.

38. A.A. Iogansen, N.G. Ural'tsev, and V.A. Khoze, *Sov. J. Nucl. Phys.* **36** (1983) 717.

39. G. Keller and D. Wyler, *Nucl. Phys.* **B274** (1986) 410.

40. J.F. Gunion, G.L. Kane, and J. Wudka, *Nucl. Phys.* **B299** (1988) 231.

41. M. Capdequi Peyranere, P. Irulegui Gomes, and H.E. Haber, SCIPP preprint (1989).

42. J.F. Gunion and H.E. Haber, *Nucl. Phys.* **B272** (1986) 1.

43. H. Huffel and G. Pocsik, *Z. Phys.* **C8** (1981) 13; G. Pocsik and G. Zsigmond, *Z. Phys.* **C10** (1981) 367; *Phys. Lett.* **112B** (1982) 157; G. Zsigmond, *Acta Phys. Hung.* **56** (1984) 73.

44. L.J. Hall and M.B. Wise, *Nucl. Phys.* **B187** (1981) 397.

45. R.M. Barnett, H.E. Haber and D.E. Soper, *Nucl. Phys.* **B306** (1988) 697.

46. J.F. Gunion, H.E. Haber, S. Komamiya, H. Yamamoto, and A. Barbaro-Galtieri, "Probing the Non-Minimal Higgs Sector at the SSC", *Proceedings of the 1987 Berkeley Workshop on "Experiments, Detectors and Experimental Areas for the Supercollider"*, edited by R. Donaldson and M. Gilchriese (World Scientific, Singapore, 1988), p. 110.

47. See, for example, H.-J. Behrends *et al.*, (CELLO Collaboration), *Phys. Lett.* **193B** (1987) 376; W. Braunschweig *et al.*, (TASSO Collaboration), submitted paper to *XXIV International Conference on High Energy Physics*, Munich, West Germany, August 1988.

48. A. Chen *et al.* (CLEO Collaboration), *Phys. Lett.* **122B** (1983) 317.

49. J.F. Gunion *et al.*, *Phys. Rev.* **D38** (1988) 3444.

50. J. Alexander, D.L. Burke, C. Jung, S. Komamiya, and P. Burchat, in *Proceedings of the Summer Study on High Energy Physics in the 1990s,* June 27–July 15, 1988, Snowmass, CO, edited by S. Jensen (World Scientific, Singapore, 1989) p. 135.

51. S. Komamiya, *Phys. Rev.* **D38** (1988) 2158.

52. K. Inoue, A. Komatsu and S. Takeshita, *Prog. Theor. Phys.* **67** (1982) 927; [E: **70** (1983) 330]; **71** (1984) 413.

53. P. Fayet and S. Ferrara, *Phys. Rep.* **32** (1977) 249; H.P. Nilles, *Phys. Rep.* **110** (1984) 1; H.E. Haber and G.L. Kane, *Phys. Rep.* **117** (1985) 75.

54. R.A. Flores, and M. Sher, *Ann. Phys. (NY)* **148** (1983) 95.

55. J.F. Gunion and H.E. Haber, *Nucl. Phys.* **B278** (1986) 449.

56. S.P. Li and M. Sher, *Phys. Lett.* **140B** (1984) 339.

57. J.F. Gunion and A. Turski, *Phys. Rev.* **D40** (1989) 2325 and 2333.

58. H.E. Haber and M. Sher, *Phys. Rev.* **D35** (1987) 2206; J. Ellis, D.V. Nanopoulos, S.T. Petcov, and F. Zwirner, *Nucl. Phys.* **B283** (1987) 93; M. Drees, *Phys. Rev.* **D35** (1987) 2910; *Int. J. Mod. Phys.* **A4** (1989) 3635; J.F. Gunion, H.E. Haber, and L. Roszkowski, *Phys. Lett.* **189B** (1987) 409; *Phys. Rev.* **D38** (1988) 105; H. Baer, D. Dicus, M. Drees, X. Tata, *Phys. Rev.* **D36** (1987) 1363.

59. J. Ellis, J.F. Gunion, H.E. Haber, L. Roszkowski, and F. Zwirner, *Phys. Rev.* **D39** (1989) 844.

60. J.F. Gunion and H.E. Haber, *Nucl. Phys.* **B307** (1988) 405.

61. H.E. Haber and Y. Nir, SLAC-PUB-5089 and SCIPP-89/37 (1989).

62. E. Franco and A. Morelli, *Nuovo Cim.* **96A** (1986) 257.

AXIONS AND DILATONS: THE SEARCH FOR

VERY LIGHT SCALAR PARTICLES

W. Buchmüller

Institut für Theoretische Physik, Universität Hannover

Deutsches Elektronen-Synchrotron DESY, Hamburg

ABSTRACT

Axions, dilatons and other very light scalar particles are possible low energy manifestations of new interactions beyond the Standard Model. We discuss the connection between axion (a) and dilaton (σ) in supersymmetric theories and derive a prediction for their mass ratio, $m_a/m_\sigma \approx 10^{-4} - 10^{-5}$. We also review three recent proposals to search for light scalar particles in Bragg scattering, in a laser experiment and via the Mößbauer effect.

1. Light scalar particles and large mass scales

The success of the Standard Model as theory of strong and electroweak interactions [1] has reached an almost worrisome extent. Unless new, unexpected phenomena are found at the current generation of accelerators the conclusion seems unavoidable that the mass scale f of new interactions beyond the Standard Model is much larger than the Fermi scale of weak interactions, i.e., $f \gg G_F^{-1/2} \sim 300$ GeV.

How can such a new mass scale manifest itself at currently accessible energies ? One possibility are new "superweak" interactions which are suppressed by inverse powers of f. Their form is restricted by requiring invariance under the Standard Model gauge group. To order 1/f there is only one interaction, which yields Majorana mass terms for neutrinos:

$$L_{eff}^{(5)} = \frac{1}{f} (l_L \varphi)^T C (l_L \varphi) = \frac{v^2}{f} \nu_L^T C \nu_L + \dots , \qquad 1 = \binom{\nu}{e}_- ; \qquad (1)$$

here v = 174 GeV is the vacuum expectation value of the Higgs doublet φ. To order $1/f^2$ there are many interaction terms, in particular four-fermion operators with various flavour and Lorentz structures,

$$L_{eff}^{(6)} = \frac{1}{f^2} C_{ijkl}^{AB} \cdot \bar{\psi}_i \Gamma^A \psi_j \bar{\psi}_k \Gamma^B \psi_l , \qquad (2)$$

which give rise to deviations from weak interactions, flavour changing neutral currents, proton decay etc. The weakest lower bound on f stems from quark-lepton universality of the charged current weak interactions, which yields f > 5 TeV [2]. Other processes, such as flavour changing neutral currents or CP violating operators yield much larger bounds on f which almost reach the Planck mass $m_{PL} \sim 10^{18}$ GeV if baryon number violating interactions are allowed.

Another possibility is the existence of pseudo-Goldstone bosons, i.e., new, very light scalar particles whose mass is of order 1/f, e.g.,

$$m \sim \alpha_{S.M.} \frac{\Lambda_{S.M.}^2}{f} < \begin{cases} 100 \text{ MeV}, \Lambda_{S.M.} \sim G_f^{-1/2} \sim 300 \text{ GeV} \\ 1 \text{ keV}, \Lambda_{S.M.} \sim \Lambda_{QCD} \sim 300 \text{ MeV}, \end{cases} \qquad (3)$$

for f > 10 TeV.

Such light scalar particles have to exist if the mass scale f is associated with the spontaneous breaking of some global symmetry of the more fundamental theory which contains the Standard Model as its low energy limit. As in the case of the pion, the only known pseudo-Goldstone boson, these (pseudo)scalar particles are expected to have a 2-photon coupling, which may allow their experimental detection.

Many candidates [3] for light pseudo-Goldstone bosons have been advocated by theorists. From the point of view of the Standard Model, the axion [4] has the best motivation, since it is an unavoidable consequence of the Peccei-Quinn mechanism which has survived as the only viable solution of the strong CP-problem. Also interesting is the dilaton, a Brans-Dicke type scalar [5], which arises in theories with spontaneously

144

broken scale invariance [6] and which has recently been discussed in connection with the cosmological constant problem [7-13]. As we will see in the following section axion and dilaton occur together in supersymmetric theories.

2. Dilatons and axions: symmetries, couplings and masses

One of the most remarkable features of the Standard Model is that all particle masses are generated through spontaneous symmetry breaking. As a consequence the lagrangian

$$L = L_{gauge} + L_{scalar} + L_{fermion} \ , \tag{4a}$$

where

$$L_{gauge} = -\frac{1}{4} G^A_{\mu\nu} G^{A\mu\nu} - \frac{1}{4} W^I_{\mu\nu} W^{I\mu\nu} - \frac{1}{4} B_{\mu\nu} B^{\mu\nu}, \tag{4b}$$

$$L_{scalar} = -(D_\mu \varphi)^\dagger (D^\mu \varphi) - V_o(\varphi^\dagger \varphi) \ , \tag{4c}$$

$$V_o = \frac{\lambda}{2} (\varphi^\dagger \varphi + \frac{\mu^2}{2\lambda})^2 \ , \tag{4d}$$

contains only one parameter of dimension mass, μ, which sets the Fermi scale of weak interactions through the vacuum expectation value of the Higgs doublet φ. D_μ, $G^A_{\mu\nu}$, $W^I_{\mu\nu}$ and $B_{\mu\nu}$ are the gauge covariant derivative and the SU(3), SU(2) and U(1) field strengths respectively.

What is the origin of μ ? If the Fermi scale itself is dynamically generated through the vacuum expectation value f of some field Ψ, the Standard Model is part of a more fundamental theory whose effective lagrangian at distances larger than 1/f is obtained from eq. (4) by substituting

$$\mu \to \mu \ e^{\sigma/f} = g \ \Psi \ , \qquad \text{with}$$

$$\Psi = f \ e^{\sigma/f}, \ g = \frac{\mu}{f} \ . \tag{5}$$

Here the dilaton field σ corresponds to the fluctuations of Ψ around the vacuum expectation value f.

The new effective action is well known to be invariant under dilatations [14]:

$$\delta\sigma = f + x^\mu \partial_\mu \sigma, \qquad \delta\varphi = (1 + x^\mu \partial_\mu) \, \varphi, \ldots \qquad (6)$$

This symmetry of the action, $\delta\Gamma = 0$, leads to a conserved dilatation current:

$$\partial^\mu D_\mu = 0 \; . \qquad (7)$$

Associated with this invariance under scale and conformal transformations is the existence of a flat direction of the scalar potential which, as we shall see, is lifted by quantum corrections which also generate a small mass for the dilaton. The kinetic term for the dilaton can be choosen as

$$e^{2\sigma/f} \, (\partial_\mu \sigma)^2 \quad \text{or} \quad (\partial_\mu \sigma)^2 \; , \qquad (8)$$

depending on whether scale invariance is treated as fundamental symmetry of the theory or just as accidental symmetry of a certain part of the lagrangian.

It is interesting that in this latter case, if gravity is encluded by adding the Einstein-Hilbert action, one obtains precisely the Brans-Dicke theory of gravity with the ordinary Standard Model as matter sector [12]. In order to see this let us use the field variables σ and $\phi = e^{-\sigma/f}\varphi$ in terms of which the lagrangian (4) reads:

$$L_{\text{S.M.}} (\hat{g}_{\mu\nu}, \phi, \ldots) = - \sqrt{\hat{g}} \left[\hat{g}^{\mu\lambda} \hat{g}^{\nu\tau} \frac{1}{4} G^A_{\mu\nu} G^A_{\lambda\tau} + \ldots \right.$$

$$\left. + \hat{g}^{\mu\nu} (D_\mu \phi)^\dagger (D_\nu \phi) + \frac{\lambda}{2} (\phi^\dagger \phi + \frac{\mu}{2\lambda})^2 \right] , \qquad (9)$$

where

$$\hat{g}_{\mu\nu} = \eta_{\mu\nu} \, e^{2\sigma/f}, \qquad \hat{g}^{\mu\nu} \hat{g}_{\nu\lambda} = \delta^\mu_\lambda \; . \qquad (10)$$

This is the Standard Model lagrangian in a conformally flat background metric, where the conformal factor is given by the dilaton field. Let us now add the gravitational field in the standard manner, i.e.,

$$L = -\sqrt{g} \left(\frac{\kappa}{2} R + \frac{1}{2} g^{\mu\nu} \partial_\mu \sigma \partial_\nu \sigma \right) + L_{\text{S.M.}} (g_{\mu\nu} e^{2\sigma/f}, \phi, \ldots), \qquad (11)$$

$$\kappa = 1/8\pi G_N \, ,$$

and perform a Weyl transformation

$$\bar{g}_{\mu\nu} = g_{\mu\nu}\, e^{2\sigma/f} \;, \tag{12}$$

$$R = e^{2\sigma/f}\left[\,\bar{R} - 6\,\frac{1}{\sqrt{\bar{g}}}\,\partial_\mu\!\left(\bar{g}^{\mu\nu}\sqrt{\bar{g}}\,\partial_\nu\frac{\sigma}{f}\right) + 6\,\bar{g}^{\,\mu\nu}\partial_\mu\frac{\sigma}{f}\,\partial_\nu\frac{\sigma}{f}\,\right]\;. \tag{13}$$

We then obtain

$$L = -\sqrt{\bar{g}}\,\Big\{\,\chi\,\bar{R}\;+\;\frac{\omega}{\chi}\,\bar{g}^{\mu\nu}\partial_\mu\chi\,\partial_\nu\chi$$

$$-\,L_{\text{S.M.}}(\bar{g}_{\mu\nu},\phi,\dots)\Big\}\;, \tag{14}$$

$$\chi = \frac{\kappa}{2}\,e^{-\sigma/f}\;,\quad \omega = \frac{f^2-6\kappa}{4\kappa}\;,$$

which is precisely the Brans-Dicke lagrangian. We note that the strong experimental bounds on the parameter ω only apply if the dilaton mass vanishes, which is usually assumed, but which is not the case for the theory defined by the lagrangian (9), where in general quantum corrections lead to a nonvanishing mass.

Let us briefly digress and discuss the possible implication of the dilaton field for the cosmological constant problem. The field equations of the Brans-Dicke theory (14) read:

$$R = -\frac{1}{\kappa}\,(g^{\mu\nu}\partial_\mu\sigma\,\partial_\nu\sigma) + T^\mu{}_\mu\;, \tag{15a}$$

$$f\,\sigma = T^\mu{}_\mu\;, \tag{15b}$$

and imply that the vacuum state, which is characterized by constant fields $\sigma = \sigma_0 = $ const. and $\varphi = \varphi_0 = $ const., has vanishing curvature, i.e., $R_0 = 0$ [12]. Unfortunately, this cannot be regarded as solution of the cosmological constant problem, since the existence of the constant solution of eq. (15b) requires fine-tuning of a constant in the scalar potential. Non-linearly realized scale invariance is certainly not sufficient [9,11,12] to obtain flat space-time as uniquely determined ground state, but the presence of the dilaton field does have some intriguing aspects which may turn out to be essential for understanding the observed flatness of space-time. We note that the Brans-Dicke theory emerges from one particular definition of "dilatons" in curved space-time. Other options are discussed in refs. [9] and [12].

Classical scale invariance is not preserved in the full quantum theory where one has:

$$\delta\Gamma = M \frac{\partial}{\partial M} \Gamma = \int d^4x \, A_D + O(\hbar^2) \ , \tag{16a}$$

$$\partial^\mu D_\mu = A_D \ . \tag{16b}$$

Here Γ and M are effective action and renormalization mass, and A_D is the well known conformal anomaly which is given by the various operators appearing in the lagrangian multiplied by the appropriate β-functions [15]:

$$A_D = \frac{\beta(g_s)}{2g} \, G^A_{\mu\nu} G^{A\mu\nu} + \quad \cdots \quad - \frac{\beta(\lambda)}{\lambda} \frac{\lambda}{2} (\varphi^\dagger \varphi)^2 \ . \tag{17}$$

The presence of the dilaton, the Goldstone boson of spontaneously broken scale invariance, allows to restore scale invariance by adding a Wess-Zumino term [16], satisfying

$$\delta\Gamma = \delta\Gamma_{WZ} + O(\hbar^2) \ . \tag{18}$$

In the case of dilatons one has

$$\Gamma_{WZ} = \int d^4x \, A_D \frac{\sigma}{f} \ . \tag{19}$$

If the theory is renormalized such that the conformal anomaly (16) is kept, the dilaton aquires a mass which is determined by the Wess-Zumino term (19) [8]:

$$m_\sigma = - \frac{4}{f^2} < A_D >_o$$

$$= \frac{1}{8\pi^2 f^2} \, \text{Str } M^4 = \frac{1}{8\pi^2 f^2} \, (\sum_{bosons} m_b^4 - \sum_{fermions} m_f^4) \ . \tag{20}$$

The ground state of the theory is now uniquely determined.

The coupling of the dilaton to the conformal anomaly is reminiscent of the coupling of the axion to the chiral anomaly of a Peccei-Quinn current. A possible connection between "dilaton" and "axion" in superstring theories has been particularly emphasized by Nilles [17]. In supersymmetric theories the analogue of conformal invariance is superconformal invariance, and in the nonlinearly realized version of this symmetry [18] dilaton- and axion-like degrees of freedom occur together.

In these supersymmetric theories dilaton and axion form together a complex scalar field,

$$\chi = \sigma + ia \quad , \tag{21}$$

which is part of a chiral superfield Σ:

$$\Sigma(x,\Theta,\bar{\Theta}) = \chi(y) + \sqrt{2}\Theta\psi(y) + \Theta\Theta F(y) \quad , \qquad y^{\mu} = x^{\mu} + i\Theta\sigma^{\mu}\bar{\Theta} \quad . \tag{22}$$

Instead of the scalar potential with dilaton one now has the superpotential

$$W = e^{3\Sigma/f} \left[g_U QH_1 U_c + g_D QH_2 D_c + g_E LH_2 E_c \right.$$

$$\left. + \lambda_1 S (H_1 H_2 + \mu_1^2) + \frac{1}{2} \mu_2 S^2 + \frac{1}{3} \lambda_2 S^3 \right] \quad , \tag{23}$$

where Q, U_c, D_c, L and E_c are the familiar quark and lepton superfields, and H_1, H_2 and S are Higgs superfields (cf. ref. [19]). The corresponding action is invariant under superconformal transformations [20] which include dilatations and chiral R-transformations which play the role of the Peccei-Quinn symmetry. The kinetic term may be chosen as

$$\frac{1}{64} D^2\bar{D}^2 \, \bar{\Sigma} \, \Sigma + c.c. \tag{24a}$$

$$\text{or} \quad \frac{f^2}{64} D^2\bar{D}^2 \, \exp\left(\frac{\Sigma + \bar{\Sigma}}{f} \right) + c.c. \tag{24b}$$

The second option is easily identified as the superconformally invariant lagrangian first obtained by Kobayashi and Uematsu who studied nonlinear realizations of the superconformal group [18]:

$$\frac{f^2}{64} D^2\bar{D}^2 \, \exp\left(\frac{\Sigma + \bar{\Sigma}}{f} \right) + c.c.$$

$$= - e^{2\sigma/f} \left(\frac{1}{2} (\partial_\mu \sigma)^2 + \frac{1}{2} (\partial_\mu a)^2 - \frac{1}{2} (\bar{F} - \frac{1}{2f} \overline{\psi\psi})(F - \frac{1}{2f} \psi\psi) \right.$$

$$\left. + \frac{i}{2f} \psi\sigma^\mu \partial_\mu \bar{\psi} + \frac{1}{2f} \psi\sigma^\mu\bar{\psi}\partial_\mu a \right) \quad . \tag{25}$$

Quantum corrections do not respect this classical invariance. The variation of the effective action with respect to R-transformations yields the superconformal anomaly [20,21]:

$$\delta^R \Gamma = \int d^4x \, A + O(\hbar^2) \; , \tag{26a}$$

$$A = \partial^\mu J_\mu = \frac{1}{2} (\bar{D}^2 \bar{S} - D^2 S) \; . \tag{26b}$$

Here $J_\mu = (R_\mu, Q_{\mu\alpha}, \bar{Q}_{\mu\dot\alpha}, \Theta_{\mu\nu})$ is the supercurrent which contains R-current, supersymmetry currents and energy-momentum tensor, and the chiral superfield S is again given by the various operators appearing in the lagrangian multiplied by the appropriate β-functions:

$$S \sim \begin{cases} \dfrac{\beta(\lambda)}{\lambda} \, \bar{D}^2 \bar{\Phi} \Phi \; , & \text{chiral models} \\[2ex] \dfrac{\beta(g)}{g} \, tr\left[\lambda^\alpha \lambda_\alpha\right] \; , & \text{gauge theories} \quad , \\[2ex] \lambda_\alpha = -\dfrac{1}{4} D_\alpha \bar{D}^2 e^{-V} D_\alpha e^V \qquad . \end{cases} \tag{27}$$

In the supersymmetric case, instead of eq.(16b) the following trace identities hold:

$$\bar{D}^\alpha J_{\alpha\dot\alpha} = D_\alpha S \; , \qquad J_{\alpha\dot\alpha} = \frac{1}{2} \sigma^\mu_{\alpha\dot\alpha} J_\mu \; , \tag{28}$$

which include eq. (26b). The supersymmetric extension of the Wess-Zumino term (19) reads ($\delta^R \Sigma = i\frac{2}{3}f$):

$$\delta^R \Gamma = \delta^R \Gamma_{WZ} + O(\hbar^2) \; ,$$

$$\Gamma_{WZ} = \frac{3}{4f} \int d^4x \left[\bar{D}^2(\bar{S} \, \bar{\Sigma}) + D^2(S \, \Sigma) \right] \quad . \tag{29}$$

Γ_{WZ} contains couplings of dilaton and axion to the conformal anomaly and the chiral anomaly of the R-current, respectively, and also interaction terms of the "dilatino" ψ.

What is left of this connection between axion and dilaton after supersymmetry breaking ? This question can be answered in a model independent way by means of "secret supersymmetry", i.e., nonlinear realizations of local supersymmetry which were systematically studied by

Samuel and Wess [22]. The starting point is the goldstino λ_α, the Goldstone fermion of spontaneously broken supersymmetry, which transforms as

$$\delta_\xi \lambda_\alpha = \kappa \xi_\alpha - \frac{2i}{\kappa} \lambda \sigma^\mu \bar{\xi} \partial_\mu \lambda_\alpha \quad . \tag{30}$$

Here δ_ξ and κ denote supersymmetry transformation and supersymmetry breaking scale, respectively. From λ_α one then obtains a superfield Λ_α which allows to construct supersymmetric lagrangians for fields which have no superpartners. Physically, they correspond to low energy effective lagrangians which are valid at distances larger than $1/\kappa$. In the "unitary gauge" the goldstino is "eaten" and one obtains a massive gravitino. In this way one can construct a supersymmetric Standard Model without scalar quarks and leptons, gauginos and higgsinos. The only remnant of supersymmetry at low energies is an enlarged Higgs sector with complex scalars H_1, H_2 and S and some relations among coupling constants which are determined by the superpotential of the original lagrangian.

In a theory with nonlinearly realized superconformal invariance one finds, that the low energy effective lagrangian contains dilaton and axion together. Their couplings to quarks and leptons follow from eq. (23):

$$L_P = - \frac{3\sigma}{f} (m_{q_i} \bar{q}_i q_i + m_{l_i} \bar{l}_i l_i)$$

$$- \frac{3a}{f} (m_{q_i} \bar{q}_i i \gamma_5 q_i + m_{l_i} \bar{l}_i i \gamma_5 l_i) + O\left(\frac{1}{f^2}\right) \quad . \tag{31}$$

The Wess-Zumino term yields (cf. eqs. (17), (29)):

$$L_{WZ} = \frac{1}{f} (\sigma A_D + a A_R) \quad , \tag{32a}$$

$$A_D = \partial^\mu D_\mu \sim \frac{\beta(g_s)}{2g_s} G^A_{\mu\nu} G^{A\mu\nu} + \ldots \quad , \tag{32b}$$

$$A_R = \partial^\mu R_\mu \sim \frac{\beta(g_s)}{2g_s} G^A_{\mu\nu} \tilde{G}^{A\mu\nu} + \ldots \quad . \tag{32c}$$

In eq. (29), which leads to (32a), we have assumed that the superfield Σ couples to the complete superconformal anomaly S (cf. (28)). This is plausible but not necessary. The breaking of superconformal invariance through the anomaly would also be realized if Σ would couple to different contributions to S, which are indicated in (27), with different strength.

Given the axion and dilaton couplings (32a) it is straightforward to compute axion and dilaton masses (cf. [23], [12]). We find:

axion:
$$m_a^2 = \frac{16}{9} N_g^2 m_\pi^2 \left(\frac{f\pi}{f} \right)^2 \frac{m_u m_d}{(m_u + m_d)^2} \quad , \qquad (33a)$$

$$m_a = 3 \text{ keV} \left(\frac{10 \text{ TeV}}{f} \right) , \quad N_g = 3 \quad ; \qquad (33b)$$

dilaton:
$$m_\sigma^2 = \frac{1}{8\pi^2 f^2} (6 m_W^4 + 3 m_Z^4 + \sum_i m_{H_i}^4 - 12 m_t^4) , \qquad (34a)$$

$$m_\sigma \approx 250 \text{ MeV} \left(\frac{10 \text{ TeV}}{f} \right) \quad . \qquad (34b)$$

The numerical factor in (33a) is due to the R-charges of quarks; in (34b) we have assumed that the contributions from Higgs particles and t-quark to the dilaton mass approximately cancel. Furthermore we have assumed that the strengths of QCD and electroweak contributions to the superconformal anomaly are determined by the particle content of the Standard Model. Eqs. (33) and (34) are based on the one-loop contributions to the anomaly. To this order the different currents which are used in connection with the superconformal anomaly [21] all give the same result. From eqs. (33) and (34) we conclude:

$$\frac{m_a}{m_\sigma} \approx 10^{-4} - 10^{-5} \quad . \qquad (35)$$

This result is a generic feature of theories in which superconformal invariance is broken by the superconformal anomaly and does not depend on the value of the symmetry breaking scale f.

How large is f ? On phenomenological grounds, as discussed in section 1, we expect f > 10 TeV. A simple, theoretically motivated guess is that f has the same order of magnitude as the supersymmetry breaking scale κ which is a function of the gravitino mass $m_{3/2}$ and the Planck mass $m_{PL} \sim 10^{18}$ GeV, i.e., $f \sim \kappa = (m_{3/2} m_{PL})^{1/2}$ [22]. If $m_{3/2}$ should be related to the Fermi scale, e.g., $m_{3/2} \sim G_F^{-1/2} \sim 300$ GeV one would have $f \sim \kappa \sim 10^{10}$ GeV and, correspondingly, $m_\sigma \sim 200$ eV and $m_a \sim 10^{-3}$ eV. This is the range of masses and couplings discussed for "invisible" axions [3].

3. The search for very light scalar particles

In this section we will discuss three recent suggestions how to produce and detect very light scalar particles in laboratory experiments. The first two, Bragg scattering and a laser experiment, are based on the two-photon coupling of light scalars which, like the two-gluon coupling, is generated by the superconformal anomaly (cf. eqs. (29), (32a)). The third experiment, which makes use of the Mößbauer effect, is based on the coupling of light scalars to nucleons.

(1) Bragg scattering [24]

In order to be specific we will restrict ourselves to the pseudoscalar axion in the following. The result which we will finally obtain holds also for the scalar dilaton. The effective lagrangian for axions and photons reads

$$L = -\frac{1}{4} F_{\mu\nu}F^{\mu\nu} + \frac{1}{2} (\partial_\mu a)^2 - \frac{m_a^2}{2} a^2 - \frac{1}{4M} F_{\mu\nu}\tilde{F}^{\mu\nu} a \quad , \tag{36}$$

and the corresponding field equations are

$$(\Box + m_a^2)a = -\frac{1}{M} \vec{E}\cdot\vec{B} \quad , \tag{37a}$$

$$\vec{\nabla}\times\vec{B} - \frac{\partial}{\partial t}\vec{E} = -\frac{1}{M} (\vec{B} \frac{\partial}{\partial t} a - \vec{E}\times\vec{\nabla}a) \quad , \tag{37b}$$

$$\vec{\nabla}\cdot\vec{E} = \frac{1}{M} \vec{B}\cdot\vec{\nabla}a \quad . \tag{37c}$$

They show that in an external electromagnetic field electromagnetic waves generate axion waves and vice versa. This "Primakoff-effect" (cf. Fig. 1) is the basis of Sikivie's methods [25] for the detection of axions in strong external magnetic fields.

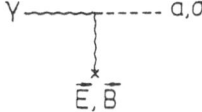

Fig. 1. Primakoff process: photon – axion (dilaton) conversion
in an external electromagnetic field

In crystals X-rays penetrate strong electric fields. For instance, the strength of the electric field at a distance d = 1 Å away from a nucleous with charge Z = 10 is

$$|\vec{E}| = \frac{Ze}{4\pi d^2} \approx 10^6 \text{ eV}^2 \approx 10^4 \text{ Tesla} \quad . \tag{38}$$

This suggests that Bragg scattering may be an efficient way to produce axions. An experiment for production and detection of axions is shown in Fig. 2: After the first Bragg scattering the reflected beam contains photons and axions; only the axions penetrate the absorber and produce in a second Bragg reflection, again via the Primakoff effect, photons which are detected.

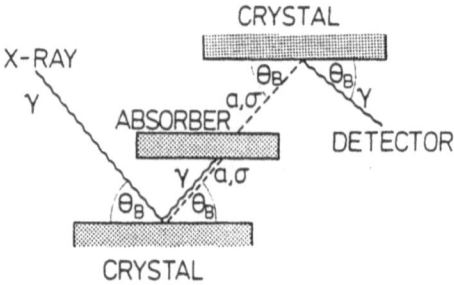

Fig. 2. Experimental setup to search for light scalar particles in Bragg scattering (from ref. [24])

In order to calculate the intensity of the final outgoing electromagnetic wave one first has to calculate the scattering amplitudes for photon-axion and axion-photon conversion. From the field equations (37) one finds that incoming plane waves

$$\vec{B}(t,\vec{x}) = \vec{B}^{(o)} e^{i(\omega t - \vec{k}\vec{x})} \quad , \tag{39a}$$

$$\dot{a}(t,\vec{x}) \equiv \frac{\partial}{\partial t} a(t,\vec{x}) = \dot{a}^{(o)} e^{i(\omega t - \vec{k}\vec{x})} \quad , \qquad \dot{a}^{(o)} = i\omega \, a^{(o)} \quad , \tag{39b}$$

generate the outgoing spherical waves

$$\dot{a}(t,\vec{x}) = \frac{F_a(2\Theta)}{4\pi M} \; \vec{e}_r \cdot \vec{B}^{(o)} \; \frac{1}{r} \; e^{i(\omega t - kr)} \quad , \tag{40a}$$

$$\vec{B}(t,\vec{x}) = - \frac{F_a(2\Theta)}{4\pi M} \; \vec{e}_r \times (\vec{e}_r \times \overset{\wedge}{\vec{k}}) \dot{a}^{(o)} \frac{1}{r} e^{i(\omega t - kr)} \quad , \tag{40b}$$

where

$$F_a(2\Theta) = k^2 \int d^3x \; \phi(\vec{x}) \; e^{i\vec{q}\vec{x}} \quad , \qquad \vec{E} = -\vec{\nabla}\phi \quad ,$$

$$\vec{q} = \vec{k}' - \vec{k}, \quad \vec{k}' = k\vec{e}_r, \quad \vec{e}_r = \frac{\vec{x}}{r} \; , \quad 2\Theta = \sphericalangle (\vec{k},\vec{k}') . \tag{40c}$$

Here we have neglected the axion mass for simplicity. Comparison with eq. (37a) shows that the formfactor $F_a(2\Theta)$ corresponds to an average electric field seen by the X-rays,

$$\bar{E}(k,2\Theta) = \frac{1}{kd^3} F_a(2\Theta) \quad , \tag{41}$$

which, as we will see, determines the probability for photon-axion conversion in a crystal.

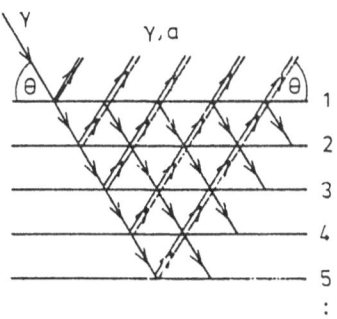

Fig.3.Coherent production of axions in a crystal

Given the scattering amplitudes for photon-axion transitions and elastic photon (Thomson) scattering by a single atom one can calculate intensity and opening angle of the outgoing beam in a Bragg reflection on a crystal by means of Darwins dynamical theory [26]. Multiple scattering inside the crystal (cf. Fig. 3) leads to transmitted and reflected photon and axion plane waves, which at the n-th layer of atoms satisfy the linear system of equations (cf. Fig. 4):

$$A_{n+1} = (1+i\rho_o)A_n + i\rho \; e^{2in\phi} B_{n+1} + i\xi \; e^{2in\phi}D_{n+1} \quad , \tag{42a}$$

155

$$B_n = i\rho e^{-2in\phi} A_n + (1+i\rho_o)B_{n+1} + i\xi\, e^{-2in\phi}\, C_n \ , \tag{42b}$$

$$C_{n+1} = i\xi\, e^{2in\phi}\, B_{n+1} + C_n \ , \tag{42c}$$

$$D_n = i\xi\, e^{-2in\phi}\, A_n + D_{n+1}, \tag{42d}$$

where

$$\phi = k\, d\, \sin\Theta \ , \tag{43}$$

$$\rho = \rho(2\Theta) = \frac{\alpha F_\gamma(2\Theta) N_s \lambda}{m\,\sin\Theta} \ , \qquad \rho_o = \rho F_\gamma(0)/\, F_\gamma(2\Theta) \ , \tag{44}$$

$$\xi = \xi(2\Theta) = \frac{F_a(2\Theta) N_s \lambda}{4\pi M \sin\Theta}\ \sin 2\Theta \ . \tag{45}$$

Here α, m, λ, F_γ, N_s and d are fine structure constant, electron mass, photon wave length, atomic structure factor (cf. [26]), density of scattering centers per unit area and lattice spacing, respectively. For $\phi = \pi$ the Bragg condition is fullfilled and one has constructive interference. In the case $\xi=0$ eqs. (42) reduce to a system of equations familiar from Thomson scattering [26].

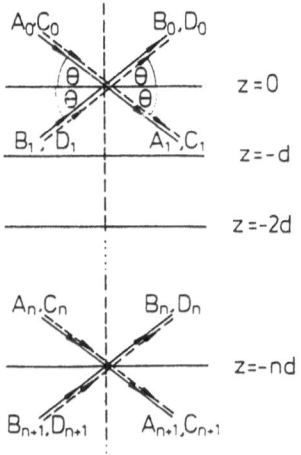

Fig. 4. Beams of photons and scalar particles in the crystal: A(B) = transmitted (reflected) photon beam, C(D) = transmitted (reflected) axion (dilaton) beam (from ref. [24])

Since $\xi \ll \rho$ eqs. (42) can be solved as for Thomson scattering and obtains for the axion-photon and photon-axion transition propabilities up to terms of order $\rho\xi$, ξ^2:

$$P(a \rightarrow \gamma) = \left| \frac{D_o}{A_o} \right|^2 = \left| \frac{B_o}{C_o} \right|^2$$

$$= P(\gamma \rightarrow a) = \left(\frac{\xi}{\rho} \right)^2 = \left(\frac{\bar{E}}{2} \frac{l}{M} \sin 2\Theta \right)^2 , \qquad (46)$$

where

$$l = \frac{md}{\alpha N_s \lambda F_\gamma (2\Theta)} \qquad (47)$$

is the penetration depth of the X-ray into the crystal. Eq. (46) is very similar to the transition probability for photon-axion conversion in an external magnetic field.

In a realistic experiment [24, 27], which makes use of the very intense source of X-rays with a brightness of $\Phi \sim 10^{18}$/sec(0.1% BW) soon available at the European Synchrotron Radiation Facility the following lower bound on the mass scale M can be achieved:

$$M > 1 \cdot 10^3 \text{ GeV} \left(\frac{\bar{E}}{1 \text{keV}^2} \frac{l}{1 \mu m} \frac{\sin 2\Theta}{0.34} \right) \left(\frac{\Phi}{10^{18}/\text{s} \ 0.1\% \text{BW}} \frac{T}{100 d} \frac{10}{N^{\text{obs}}} \right)^{1/4} . \qquad (48)$$

Here we have normalized the running time to T = 100 days and the number of observed photons to $N^{\text{obs}} = 10$. This lower bound will be slightly decreased if finite detection efficiency and temperature effects, i.e. the Debye-Waller factor, are taken into account. In the case of nonvanishing axion mass the axions are emitted under an angle $\bar{\Theta}_B < \Theta_B$. In principle the experiment is sensitive to masses $m_a < \omega \sin\Theta_B$, where ω is the photon energy.

The lower bound on M is linear in the penetration depth l which is only 1 μm for a typical Bragg scattering process. In the case of Laue scattering (cf. Fig. 5) a much larger penetration depth, up to 1 cm for 100 keV photons and scattering angle $\Theta \sim 1^\circ$, can be achieved [27]. If eq. (46) for the transition probability $P(a \rightarrow \gamma) = P(\gamma \rightarrow a)$ also holds for Laue scattering, which remains to be seen, the lower bound on M given in (48) could be improved by three orders of magnitude up to $\sim 10^6$ GeV.

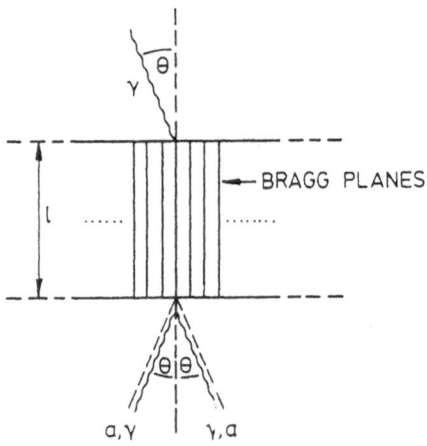

Fig.5. Axion production in Laue scattering

(2) Laser experiment [28]

The experiment described in the last subsection is based on the idea to discover axions by "shining light through walls", which was first put forward in connection with the laser experiment shown in Fig. 6 [28]. Here the photon-axion transitions take again place via the Primakoff effect, now with an external magnetic field. The transition probability in the homogeneous magnetic Field B of a dipole magnet with length L can be computed from eqs. (37). For zero mass axions one finds [25, 28]:

$$P(a \to \gamma) = P(\gamma \to a) = \left(\frac{BL}{2M}\right)^2 , \qquad (49)$$

which is the analog of the transition probability (46) in an external electric field.

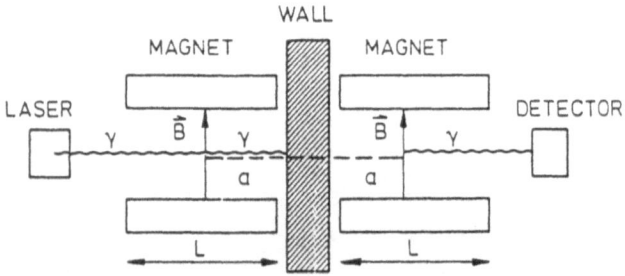

Fig.6. "Shining light through walls": photon-axion and axion-photon conversion in an external magnetic field (cf. [28])

In a realistic experiment a rather large lower bound on the mass scale M can be achieved (cf. [28]):

$$M > 4 \cdot 10^8 \text{ GeV} \left(\frac{B}{10\text{TeV}} \frac{L}{10\text{m}} \right) \left(\frac{\Phi}{2.5 \cdot 10^{21} \text{s}^{-1}} \frac{T}{100\text{d}} \frac{10}{N^{\text{obs}}} \right)^{1/4}. \qquad (50)$$

In comparison with Bragg scattering one looses 3 orders of magnitude in the field strength but one gains 7 orders in the penetration length L. Furthermore the photon flux is larger: $\Phi = 2.5 \cdot 10^{21} \text{s}^{-1}$ corresponds to a 1 kW laser with frequency $\omega = 2.5$ eV. However, due to this small energy the laser experiment is sensitive only to axions with masses below ~ 1 eV, which is 4 - 5 orders of magnitude below the sensitivity of X-ray experiments.

An alternative laser experiment, where the change in polarization due to the photon-axion coupling is measured, has been suggested by Maiani, Petronzio and Zavattini [29] and is currently carried out at Brookhaven. First results yield the lower bound $M > 10^5$ GeV for $m_a < 10^{-3}$ eV [30].

(3) Mößbauer effect [31]

Another variant of the "shining light through walls"-idea is the recent suggestion by de Rújula and Zioutas to search for axions by means of the Mößbauer effect [31]. The experimental set-up is sketched in Fig. 7: An intense Mößbauer source, which emits M1 photons in the process $N^* \to N + \gamma$, is used as source for axions through the transition $N^* \to N + a$; the source is shielded; only the axions penetrate the wall and resonantly excite the nuclei of the absorber, i.e., $N + a \to N^*$; the deexcitation of these nuclei by emission of photons or internal conversion is the axion signal.

A realistic experiment[1] can reach the following upper bound on the axion-nucleon coupling [29]:

---

[1] The required specifications are [31]: 200 mCi of ^{119m}Sn (produced from 0.1 gr of ^{118}Sn exposed for 200 days to a flux of $3 \cdot 10^{14}$ neutrons/cm^{2}s) as source; 10 layers of 100 mg/cm^2 ^{119}Sn as resonant absorber; the frequency of the γ-signal is $\omega = 24$ keV.

$$\frac{\alpha_a}{\alpha_{em}} < 5 \cdot 10^{-8} \quad , \quad \alpha_a = \frac{g^2_{aNN}}{4\pi} \quad . \tag{51}$$

With $g_{aNN} \sim \frac{m_N}{f} \sim 1/f \, [\text{GeV}]$ [32] the corresponding lower bound on the mass scale f is

$$f > 10^4 \text{ GeV} \quad . \tag{52}$$

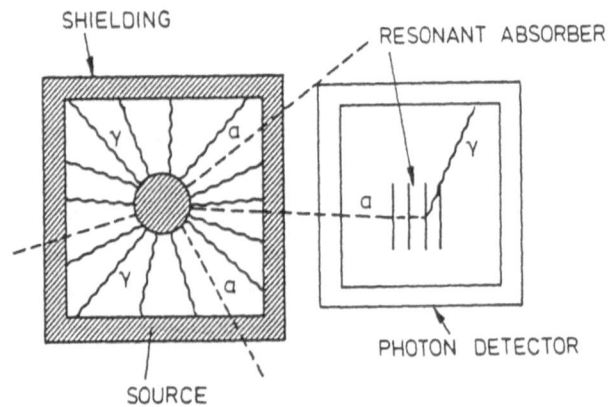

Fig. 7. Sketch of experimental setup to produce and detect axions via the Mössbauer effect

Comparing this bound on f with the bounds (48) and (50) on M, one has to take into account that in models where the two-photon coupling is generated by the anomaly, the connection between f and M is given by $f \sim 10^{-2}$ M. Hence the bound on f from the Mößbauer effect appears more stringent than the bound achievable in Bragg scattering, less stringent than the bound obtainable with the laser experiment and possibly comparable to Laue scattering.

The lower bounds on the axion - two photon coupling, which can be obtained in the three experiments discussed in this section, have to be compared to the astrophysical bounds [33]. Fig. 7, which is partly taken from ref. [28], shows that the range accessible by the simplest version of the laser experiment is already excluded by bounds derived from the allowed energy loss of the sun. The advantage of the X-ray and Mößbauer experiments is that the sensitivity with respect to the axion mass extends to about 10 keV, i.e., about five orders of magnitude beyond the sensitivity of the laser experiment. However, if bounds from lifetimes of

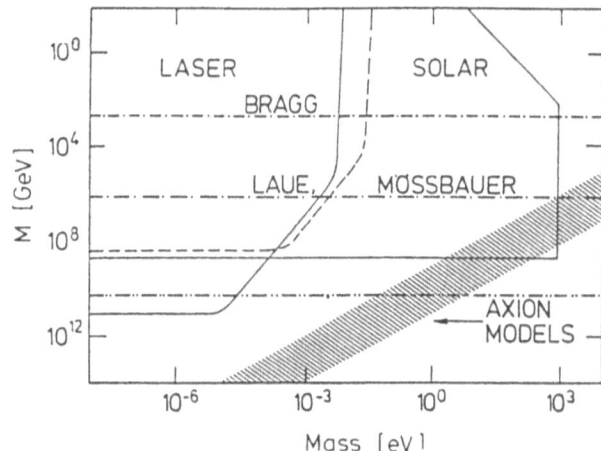

Fig. 8. Comparison of the sensitivity of the three laboratory experiments with astrophysical bounds. The solar and laser bounds are taken from ref. [28]. The dashed double-dotted line is obtained from helium burning red giants [34].

helium burning red giants [34] are included, no "window" in the $M - m_a$ plane remains for the laboratory experiments, which is not already excluded by astrophysical considerations.

4. Summary

New interactions beyond the Standard Model, which are characterized by a mass scale f much larger than the Fermi scale of weak interactions, can manifest themselves at low energies through pseudo-Goldstone bosons, i.e., light, weakly interacting (pseudo)scalar particles.

Symmetry arguments based on the Standard Model single out two spin-0 particles, dilaton and axion, which are related to spontaneously broken symmetries – scale invariance and a chiral, Peccei-Quinn type symmetry. In supersymmetric extensions of the Standard Model, dilaton and axion belong to the same supermultiplet. If their masses are generated by the superconformal anomaly, their mass ratio is $m_a/m_\sigma \approx 10^{-4} - 10^{-5}$, independent of the mass scale f.

Several novel experiments have been suggested which can significantly extend the bounds on masses and couplings of light scalars obtained from current laboratory experiments. So far, however, none of these new experiments can improve the present astrophysical bounds. We

161

leave it to the reader to decide whether such experiments may nevertheless be sufficiently interesting to be carried out.

The proposed experiments to search for axions and dilatons all require production $\underline{\text{and}}$ detection of the scalar particles, i.e., the rate of observed events is proportional to $(1/f)^4$. The sensitivity on the mass scale f might be improved if some new indirect signature for the production of light scalars could be found, since the corresponding signal would then only be proportional to $(1/f)^2$.

On the theoretical side it appears most important to obtain a prediction of, or restrictions on the mass scale f. This may be possible in theories where f and the Fermi scale $G_F^{-1/2}$ are related, as it is the case in some supergravity models.

Acknowledgements: It is a pleasure to thank N. Dragon and F. Hoogeveen for collaboration on the topics discussed in this lecture, and G. Raffelt and C. Wetterich for helpful discussions. I also thank A. Ali and the staff at Erice for organizing a pleasant and stimulating workshop.

References

[1]   P. Langacker, in Proc. of the XXIV Int. Conf. on High Energy Physics, eds. R. Kotthaus and J.H. Kühn (Munich, 1988) p. 190

[2]   W. Buchmüller and D. Wyler, Nucl. Phys. B268 (1986) 621

[3]   For a review, see J.E. Kim, Phys. Rep. 150 (1987) 1

[4]   R.D. Peccei and H. Quinn, Phys. Rev. Lett. 38 (1977) 1440;
S. Weinberg, Phys. Rev. Lett. 40 (1978) 223;
F. Wilczek, Phys. Rev. Lett. 40 (1978) 279

[5]   P. Jordan, Z. Phys. 157 (1959) 112;
C. Brans and R.H. Dicke, Phys. Rev. 124 (1961) 925

[6]   G. Mack, Nucl. Phys. B5 (1968) 499;
P.G.O. Freund and Y. Nambu, Phys. Rev. 174 (1968) 1741

[7]   R.D. Peccei, J. Solà and C. Wetterich, Phys. Lett. B 195 (1987) 183

[8]   W. Buchmüller and N. Dragon, Phys. Lett. B 195 (1987) 417

[9]   C. Wetterich, Nucl. Phys. B302 (1988) 668

[10]   S. Weinberg, Rev. Mod. Phys. 61 (1989) 1

[11]   G.D. Coughlan, I. Kani, G.G. Ross and G. Segrè, Nucl. Phys. B316 (1989) 469

[12]   W. Buchmüller and N. Dragon, Nucl. Phys. B321 (1989) 207

[13]   E. T. Tomboulis, preprint UCLA/88/TEP/42 (1988)

[14]   For a review, see S. Coleman, Aspects of Symmetry
         (Cambridge University Press, 1985) p. 67

[15]   S. L. Adler, J. C. Collins and A. Duncan, Phys. Rev. D15 (1977)
         1712;
         J. C. Collins, A. Duncan and S. D. Joglekar, Phys. Rev. D16
         (1977) 438

[16]   J. Wess and B. Zumino, Phys. Lett. B37 (1971) 95

[17]   H. P. Nilles, in Proc. of the 9th Johns Hopkins Workshop,
         Florence, 1985;
         J. P. Derendinger, L. E. Ibañez and H. P. Nilles, Nucl. Phys. B267
         (1986) 365

[18]   K. Kobayashi and T. Uematsu, Nucl. Phys. B263 (1986) 309

[19]   H. P. Nilles, Phys. Rep. 110 (1984) 1

[20]   For a review, see O. Piguet and K. Sibold, Renormalized
         Supersymmetry, Birkhäuser Inc. (Boston 1986)

[21]   O. Piguet and K. Sibold, Int. J. Mod. Phys. A (1986) 913;
         M. T. Grisaru and P. C. West, Nucl. Phys. B254 (1985) 249;
         M. A. Shifman and A. I. Vainshtein, Nucl. Phys. B277 (1986) 456

[22]   S. Samuel and J. Wess, Nucl. Phys. B233 (1984) 488

[23]   W. A. Bardeen, R. D. Peccei and T. Yanagida, Nucl. Phys. B279
         (1987) 401

[24]   W. Buchmüller and F. Hoogeveen, Hannover preprint ITP-UH 9/89
         (1989)

[25]   P. Sikivie, Phys. Rev. Lett. 51 (1983) 1415; 52 (1984) 695 (E);
         Phys. Rev. D32 (1985) 1988

[26]   See, for instance, B. E. Warren, X-Ray Diffraction, Addison-
         Wesley Comp. (Reading, 1969)

[27]   G. Materlik, private communication

[28]   K. van Bibber et al., Phys. Rev. Lett. 59 (1987) 759

[29]   L. Maiani, R. Petronzio and G. Zavattini, Phys. Lett. B 175
         (1986) 359

[30]   A. C. Melissinos, private communication

[31]   A. de Rujula and K. Zioutas, Phys. Lett. B217 (1989) 354

[32]   T. W. Donnelly et al., Phys. Rev. D18 (1987) 1607

[33]   M. Yoshimura, in Proc. of the XXIII Int. Conf. on High Energy
         Physics, Berkeley, ed. S. C. Loken (1986) p. 189;
         G. G. Raffelt, in Proc. of the XIV Int. Conf. on High Energy
         Physics, Munich, eds. R. Kotthaus and J. H. Kühn (1988) p. 1519

[34]   G. G. Raffelt, private communication; G. G. Raffelt and
         D. S. P. Dearborn, Phys. Rev. D36 (1987) 2211

HIGGS PARTICLES AND DARK MATTER SEARCHES

Graciela Gelmini*

International School for Advanced Studies
Trieste, Italy

INTRODUCTION

Over the past few years, a consensus has developed in the astrophysics community that the mass in luminous objects (stars, dust and gas,
the traditional object of study of astronomy and astrophysics) can account
for only a minor part of the mass in galaxies, small groups, clusters and
superclusters of galaxies, and even at scales smaller than a galaxy, in
globular clusters of stars. Most of the matter of the universe, up to 90%
of the total, is revealed only through its gravitational effect. This is
the "dark matter" (DM) [1].

One of the outstanding questions in cosmology and astrophysics is the
nature of the "dark matter" (DM), the most abundant form of matter in the
universe. An important recent development in this respect is the attempt
to detect DM particles from the halo of our galaxy in laboratory experiments. Only some of the particle candidates can be so tested, in particular some of the WIMP's, the candidates heavier than protons and with interactions of weak strength. DM searches are already going on and new better detections techniques are being developed.

I will concentrate on the possible connection between DM detection
and Higgs bosons, in particular "the" Higgs. By this I mean the neutral
Higgs particle expected in the standard model, or in extensions of the
standard model, which high energy experiments attempt to detect now. In
some of the current DM particle models the relation is given by the Higgs
particles being the mediators of the interaction of the DM candidates with
normal matter. Thus, the existence of a light Higgs boson would mean that
the DM searches could be successful while negative results in searches for
the Higgs field yield bounds on DM models.

Other ways of incorporating scalars in DM models have been proposed,
but these scalars are not Higgs bosons (i.e. they are not the scalars necessary for the Higgs mechanism) and, thus, they are not included here.
For example, a scalar boson has been recently proposed as a DM candidate
[2]. It is the lightest boson of a scalar sector decoupled from all the

---

* Also ICTP, Trieste, Italy. On leave from Dip. di Fisica, Uni-
versità di Roma II, "Tor Vergata", Rome Italy and associated to
INFN, Italy. From November 1989, Phys. Dept.,Univ. of California at Los Angeles.

fermions, in which there is a exactly conserved global symmetry. The mentioned model [2] is supersymmetric, but the same idea could be equally applied to any model, included the standard model, by just assuming arbitrarily the existence of this ad-hoc scalar sector. Thus, scalars could constitute the DM.

Among the DM models recently studied there are two in which a relatively light Higgs plays an essential role: a heavy neutrino proposed as a candidate for the "cosmion" (the DM particle that would constitute the halo of our galaxy and solve the solar neutrino problem) and the lightest neutralino as a DM candidate. I will present the bounds on these models obtained from direct and indirect searches of DM particles in the galactic halo, after a brief introduction to the evidences for DM and the DM searches underway.

THE DARK MATTER

The first evidence for DM was found in galaxy clusters: the velocities of galaxies in a cluster implied that the mass of the cluster exceeded the mass of the stars in the galaxies that made the cluster. This discrepancy between the mass in clusters found through the virial theorem and the luminous mass has been confirmed. Also most of the mass of a spiral galaxy is in a spheroidal dark "halo", that extends as far as stellar components can be found to trace its existence [3]. In fact, the orbital velocity as function of the distance R from the center (what is called the "rotation curve") is constant beyond the distance where the light of the disk dies exponentially out, while if the mass of the galaxy would end where the light ends, the velocity should decrease as $R^{-1/2}$. A constant velocity means that the mass grows linearly with R. The ratio of this virial mass to light at R=20 Kpc (1pc is approximately 3 light yrs) is 10^3 times larger than the ratio of dynamically detected mass to light in the disk of the galaxy. Observations of hot X-ray emitting gas in elliptical galaxies indicate the dominant presence of DM also there. The DM in the halos of all galaxies amounts to a fraction Ω=0.05 to 0.1 of the critical density, while at the largest scales observed the DM amounts to Ω=0.2-0.3 . The critical density is now ρ_c=10.5 h^2 KeV cm^{-3} (h is the Hubble constant in units of 100 Km sec^{-1}Mpc^{-1}; it is a number between 0.4 and 1). It is possible that there is not just one DM problem but many, i. e. that the DM at different scales is different.

Our own spiral galaxy, the Milky Way, has a dark halo. Through the simultaneous analysis of a large amount of data of our galaxy, astrophysicists have determined the local ("local" meaning "in the vicinity of the solar system") density and characteristic velocity of our galactic halo : 0.2-0.4 GeV cm^{-3} [4] and 200 - 400 Km sec^{-1} (and the escape velocity from the galaxy is roughly twice as much)[4,5]. We are immersed in the DM of our galaxy. If the DM consists of DM particles of mass m, the flux on earth is large $\simeq 10^7$ cm^{-2} sec^{-1} (m/ 1 GeV)$^{-1}$.

The nature of the DM is a hotly debated topic among astrophysicists. Some suggest that it may consist of very low mass stars (the brown dwarfs) or remnants of dead stars, or primordial black holes. Others suggest more exotic matter, such as new neutral massive particles. There is no compelling argument one way or the other, and none of the possible candidates have been found.

There are two main arguments in favor of non baryonic matter. One is that primordial nucleosynthesis suggests that baryonic matter (that formed by protons and neutrons) accounts for only 10% of the critical density, while the observed amount of DM in the universe is larger in some cases. Some people support the theoretical prejudice that the universe has precisely the critical density, because any other value at present (Ω close to one, as it is observed) would require a severe fine tuning (Ω only infinitesimally different from one) at some point in the past. In both cases, the remaining mass should be in non-baryonic form. The other main argument in favor of non-baryonic matter is that it can better help explain the formation of galaxies and the large scale structure of the universe (superclusters of galaxies with the form of filaments and walls and large voids in between). With respect to galaxy formation, any DM candidate belongs to one of three classes: hot DM, which is relativistic at the time galaxies should start forming (when the temperature of the universe is of order 1 KeV); cold DM, which is non relativistic at that moment; warm DM, which is becoming non relativistic then [6].

Assuming that a non baryonic component of the DM exists, we may ask ourselves if elementary particle physicists have preferred particle candidates for the DM and if the DM particles in the halo of our galaxy could be detected in the laboratory. This question has been extensively discussed in the last years [7]. There are two main types of DM candidates that particle physicists may offer:

- particles that appear in popular models proposed to solve elementary particles problems completely independent of the DM, that are found to be good DM candidates for some phenomenologically allowed range of values of the parameters of the model. Among these are neutrinos, the axion, the lightest supersymmetric particle (if it is stable).

- candidates provided ad-hoc to have particular desirable properties, such as candidates for the "cosmions" [8-12], the type of DM that may solve the solar neutrino problem [13]. As we will see below cosmions must have mass and cross section in the sun in narrow ranges: 5-10 GeV and 10^{-36} cm^2.

These candidates must have the density required to account for all, or part, of the DM. The relic abundance of particles that were in equilibrium in the early universe can be computed reliably [14]. It depends on the type of interactions of these particles, on their mass and on their asymmetry in number (i.e. the difference in number between particles and antiparticles) in the early universe. Interactions of weak strength, such as those of standard neutrinos or of supersymmetric particles, yield just two ranges of masses for particles with a relic density Ω from 0.10 to 1: a few times 10 eV, thus hot DM, or between a few to some 10 GeV (or even a few TeV, as for some supersymmetric particles, for example), thus cold DM . Because particle physics is exploring the scale of weak interactions these ranges of masses are of particular interest. Particles in the heavy range have a special name: WIMPs, Weakly Interacting Massive Particles.

For WIMP's with no cosmic asymmetry, the relic density Ωh^2 turn out to be

$$\Omega h^2 \simeq \frac{1 \cdot 10^{-37} \text{cm}^2}{<\sigma v>}, \tag{1}$$

where $<\sigma v>$ is the thermal average of the annihilation cross section by the velocity of WIMPs. Using order of magnitude and dimensional arguments on the annihilation cross section, it is easy to obtain the dependence of the density on the mass, and the range of masses for which WIMP's could have the abundance adequate to be DM candidates. For nonrelativistic particles lighter than the Z and W weak gauge bosons, the annihilation cross section is roughly $\sigma \simeq G^2_{\text{Fermi}} m^2 N$ (with a number N of annihilation channels), what is $\sigma \simeq 0.4 N 10^{-37} (m / 1 \text{GeV})^2$ cm^2. Thus,

$$\Omega h^2 \simeq \frac{O(1 - 10)}{(m/1\text{GeV})^2}, \tag{2}$$

the density decreases with increasing mass and weakly interacting WIMP's must have masses in the GeV range to account for the DM. For WIMPs much heavier than the weak gauge bosons, or with electromagnetic cross sections, $\sigma \simeq \alpha^2 N / m^2 \simeq 0.2 N 10^{-37}$ cm^2 (m / 1 TeV)$^{-2}$. Thus

$$\Omega h^2 \simeq O(1 - 10)(m/1\text{TeV})^2, \qquad\qquad (3)$$

the density increases with increasing mass and DM candidates with this interactions must have masses of order TeV.

DARK MATTER SEARCHES

Some of the WIMP's could be detected either in direct or indirect searches if they are in the halo of our galaxy.

Direct searches [15] look for the energy deposited by a halo DM particle in a collision with a nucleus within a detector. This energy can be seen through ionization in a Ge or Si crystal or through the increase in temperature in a bolometric detector, or by still other techniques. Only the first has already given results [16,17]. There are two experimental groups, devoted originally to the search of double beta decay, which are carrying out the search through ionization. These ionization detectors count the number of electrons which jump into the conduction band, i.e., the number of "shallow" electron-hole pairs as a result of energy deposited (which is of the order of 10 KeV) by the halo DM in a nucleus of the Ge or Si crystal . The predicted rate R_P depends on the local density in number ρ/m of the dark matter particles of mass m, its scattering cross-section σ on Ge or Si nuclei and the distribution of their velocity in the halo v, $R_P \sim (\rho\sigma \text{ v}/ \text{ m})$. It is necessary to integrate over velocities to obtain an upper bound on $(\rho\sigma)$ by requiring $R_P < R_O$, the observed rate. To fix ideas the σ of standard heavy neutrinos was used in Ref. 74. By fixing σ an area in a $(\rho, \text{ m })$ space is excluded. By fixing $\rho = \rho_{halo}$ an area in a $(\sigma, \text{ m})$ space is excluded.

This type of research was initiated by the PNL/USC group with Ge [16]. A Spanish group from Zaragoza has recently started a similar experiment in Spain and joined the collaboration. The second group, which has now the best data available, is that of the LBL/UCSB collaboration [17]. Besides continuing the search with Ge, they have recently initiated a search with Si. This technique, using the ionization power of the energy deposited on a nucleus, will soon reach a saturation. To improve bounds on DM new techniques that are under development will have to be used.

Indirect searches look for the products of the annihilation of the DM candidates, either in the halo of our galaxy [18] or within a body such as the sun or the earth [19], which may capture the particles from the halo. The products of DM particles in the halo may be found in cosmic rays, as an excess of positrons or antiprotons or photons. The only annihilation products that may escape from the sun or the earth are neutrinos with large energies. At energies larger than a few GeV, the background of neutrinos produced in the atmosphere by cosmic rays dies out [20]. Thus there are already interesting bounds on DM candidates from the non observation of an excess of neutrinos coming from the sun and from the earth in the large proton decay experiments.

If the mean free path of the halo WIMPs is smaller that the radius of the earth or sun most WIMPs will be trapped, because, in most cases, the energy lost by a WIMP in one collision is enough to bound it to the sun or earth. The WIMPs captured by the earth or the sun remain trapped provided they are heavy enough not to evaporate. The critical evaporation mass is $m_{ev} \simeq 12$ GeV for the earth and $m_{ev} \simeq 4$ GeV for the sun [21]. The number of trapped WIMPs with $m \lesssim m_{ev}$ decreases exponentially with decreasing WIMP mass, and no appreciable neutrino signal comes from them.

The age of the earth or the sun is long enough to achieve equilibrium between capture and annihilation of WIMPs. Thus, the total annihilation rate is given by half the capture rate. The neutrino production rate is proportional to the branching fraction of the annihilation into neutrinos by the scattering cross section. Some WIMPs annihilate into a pair of neutrinos. Some times this direct production is suppressed, for Majorana

fermion (non relativistic) WIMPs , for example. Then there are intermediate steps in the production of neutrinos. Fermions produced in the annihilations of WIMPs fragment and decay. The energy distributions of neutrino yield per fermion in the rest frame of the decaying fermion in the
sun and earth have been studied by Ritz and Seckel [22] with Montecarlo
simulations. They have to be boosted to the frame of the sun or earth. As
discussed in [22], the only fermions relevant in these computations are τ,
c and b (and t for heavy enough WIMPs), since solar and terrestrial media
stop lighter fermions before they decay and thus they do not contribute to
the high energy neutrinos. The signals in the underground "proton decay"
experiments of the high energy neutrinos resulting from annihilations
are either contained (or vertex contained) neutrino event or through going muons, produced by interactions of the incoming neutrinos in the rock
surrounding the experiment. Muons from the sun are observed at night, to
avoid the background of down going atmospheric muons (this is a really
"indirect" search, it consists in trying to see the sun underground at
night!)

For a WIMP to be observed in either direct or indirect searches its
non relativistic elastic scattering cross section with nuclei has to be
large enough. The difference in scattering cross-sections between WIMPs
with spin dependent and those with spin independent interactions is very
relevant for direct and indirect detection. For example, the rate of collisions with heavy nuclei is expected to be (10^2 to 10^4)/(kg day) for Dirac
neutrinos and sneutrinos (which have weak spin independent interactions),
while it is expected to be (10^{-1} to 10)/(kg day) for pure photinos (that
have only spin dependent interactions). This difference implies that
there are good bounds on the first ones and no bound on the second ones
from, both, direct and indirect searches. The difference is given by the
nuclear coherence factor in the scattering cross section of WIMPs on heavy
nuclei.

Given that the typical velocities of DM particles in the halo are v $\simeq$
10^{-3} c, a halo WIMP transfers at most an energy of order 100 KeV in a collision with a nucleus. The inverse of the momentum transferred to nuclei is
larger than (or at most of the order of) their radius, i. e. the interaction of WIMPs with nuclei is coherent (the small loss of coherence present
in some cases is incorporated through the nuclear form factor). The non-
relativistic cross-section for a projectile of velocity v mass m against a
target of mass M at rest is of the form

$$\sigma = \frac{m^2 M^2}{\pi (m + M)^2} |A|^2, \qquad (4)$$

where A is a reduced amplitude which depends on the dynamics of the collision.

For coherent interactions A scales with the atomic number of the target or some number in relation to it, like the number of neutrons for weak
interactions. For example, the factor $|A|^2$ for heavy standard Dirac neu-
trinos is proportional to $T_3{}^2[(1 - 4\sin^2 \theta_W)Z - N]^2$ where T_3 is the third
component of the weak isospin of the neutrino, Z and N are the number of
protons and neutrons in the target. Since $\sin^2 \theta_W = 0.23$ the contribution
of protons is negligible. This is a large factor of order $10^3 - 10^4$.

For spin-dependent interactions A depends on the total spin of the
target $|A|^2 \sim \lambda^2 J(J + 1)$, where J is the spin of the nucleus, and λ is a
nucleus dependent factor; $\lambda(J + 1)$ is a number between 0.1 and 1 for most
nuclei with nonzero spin. Thus the spin dependent cross-section is never
much larger for scattering on a heavy nucleus than for the scattering on a
proton.

The spin dependence (or independence) of the interactions of WIMPs
depend on their couplings. For example, for fermionic WIMPs, the parti-

cle exchanged that mediates the interaction with quarks, may be a vector boson or a scalar boson. In the first case the couplings may contain γ_μ or $\gamma_\mu\gamma_5$, that, in the non relativistic limit, go respectively to 1 and $\vec{s}$ in a scattering (and to $\vec{s}$ and 1, respectively, in annihilations). In the second case the coupling can be proportional to 1 or to γ_5, that go respectively to 1 and $\vec{s}\cdot\vec{v}$ in the non relativistic limit in a scattering (to $\vec{s}\cdot\vec{v}$ and 1 respectively in annihilations). Here $\vec{s}$ is the spin and $\vec{v}$ the velocity of the WIMP. A standard Dirac neutrino has both γ_μ and $\gamma_\mu\gamma_5$ in the coupling through Z boson exchange and the spin independent part of the coupling dominates in the scattering. A Majorana standard neutrino has only a $\gamma_\mu\gamma_5$ coupling. Thus it has only a spin dependent coupling in a scattering.

A HEAVY STANDARD NEUTRINO AS THE COSMION

Faulkner, Gilliland, Press and Spergel [13] have shown that the solar neutrino problem and the dark matter in the halo of our galaxy can be simultaneously explained assuming the existence of a particular kind of WIMP, the cosmion, with mass and cross section in a narrow range. Cosmions should constitute the dark halo of our galaxy, they would be captured from the halo of our galaxy by the sun, and once trapped inside would lower the temperature in the neutrino producing core, by improving the heat conduction, thus reducing the rate of production of ^{8}Boron solar neutrinos by the desired factor of 1/3. To obtain this result cosmions should be present in the sun with a significant concentration, more than 10^{-12} per nucleus. As a surprizing cosmic coincidence, that allows the whole picture to work, this is just the amount of cosmions that would be accumulated in the sun if all capturable cosmions are effectively captured and remain in the sun. Thus cosmions have to be efficiently captured, they have to be stable in the lifetime of the sun and their annihilation rate in the sun must be very small.

In order to have both an efficient capture of cosmions by the sun and an efficient conduction of heat by cosmions in the sun, the effective scattering cross section of cosmions per nucleus in the sun should be about $\sigma_c = 4\cdot10^{-36}$ cm^2. This is the critical value of the cross section for which the mean free path in the sun is approximately the solar radius. Also the cosmion mass should be in a narrow range, between 4 and 10 GeV. If lighter, cosmions evaporate from the sun and if heavier, they do not move far enough from the center of the sun to produce the necessary heat transport away from the core.

There are mainly three ways to suppress cosmion annihilations in the sun: i) the non relativistic annihilation cross section is suppressed (this is the only alternative to have a Majorana fermion as a cosmion candidate); ii) the trapping rate of cosmions and anticosmions in the sun are different, yielding an asymmetry in their numbers only in the sun; iii) a cosmic asymmetry is assumed, i. e., the number of cosmions and anticosmions in the universe are different. In the last two cases, the annihilation rate in the sun is determined by the minority component. A very remarkable property of these bounds is that they can be made compatible with a cosmion relic abundance suitable to account for the dark matter.

Krauss et al. [13] showed that none of the "conventional" dark matter candidates have the expected properties, one of the reasons being that the standard scattering weak cross section is some orders of magnitude smaller than required. New ad hoc cosmion candidates have been proposed (in chronological order): a spin 1/2 neutral particle interacting with nucleons via the exchange of a colour triplet scalar of mass around 100 GeV [8]; the magnino [9], a neutral fermion which interacts mainly through an anomalous magnetic moment; the $EXon$ [10], a fourth generation neutrino interacting through a very light (0.5-1 GeV) standard Higgs boson; a spin 1/2 neutral particle with interactions mediated by an extra Z' neutral gauge boson, arising from an E_6 symmetry breaking into the standard model and an extra $U(1)$ factor [11]. A suitable combination of the fermionic supersymmetric partners of the neutral bosons in the supersymmetric standard model (i.e. a special case of the lightest supersymmetric particle), interacting through a light (1-2 GeV) Higgs boson [12].

We want to concentrate here on the $EXon$. A light Higgs scalar is essential to obtain the required large scattering cross section in the sun. The exchange of a scalar boson give rise to a spin independent scattering cross section, as we mentioned above. Thus there is a large coherence factor in the cross section. In fact, the non relativistic $EXon$ scattering cross section from a nucleus N_i is mainly due to the Higgs boson exchange [23]:

$$\sigma_i = \frac{8G_F^2}{(27)^2 \pi} \frac{n_H^2}{m_h^4} \frac{(M_i m)^4}{(M_i + m)^2},$$ (5)

m, m_h and M_i are the cosmion, Higgs boson and the nucleus N_i masses respectively and n_H is the number of heavy quarks that couple to the Higgs boson, corresponding in this model to c, b, t and the two new quarks of the fourth generation, thus $n_H = 5$. The coupling of a Higgs field to a nucleon is dominated by the coupling to gluons through a loop of heavy quarks [23].

The effective scattering cross section in the sun is obtained as a weighted sum $\sigma_{eff} = \sum x_i \sigma_i$ with the solar number abundances x_i, tabulated in ref. [24]. σ_{eff} is dominated by scattering off of Helium, and in order to get $\sigma_{eff} \simeq \sigma_c$ the Higgs boson mass should be in the range between approximately 0.5 GeV (for m=4 GeV) and 1 GeV (for m=10 GeV).

The $EXons$ are Dirac neutrinos with a cosmic asymmetry in number that suppresses their annihilation in the sun. Enough particles of the minoritary component remain, to obtain bounds from the non observation of the neutrinos resulting in the annihilation of $EXons$ and anti-$EXons$ in the sun. Bounds on the cosmic asymmetry of $EXons$ are obtained from their indirect detection in underground proton decay detectors and from cosmological requirements. These constraints result in severe bounds for the parameter space (cosmion asymmetry and mass).

Griest and Seckel [25] have considered in detail the effect of a cosmic asymmetry for massive "standard" neutrinos (with scattering due mainly to Z^0 exchange). They have analysed the relic abundances of heavy neutrinos, their capture by the sun and the possibility of observing their annihilation products. The main difference in the $EXon$ model is the small value assumed for the Higgs boson mass, which yields larger scattering cross sections. Hence, the rate of capture by the sun of cosmions from the halo is larger than in [25], and the neutrino annihilation signal at the earth increases proportionally [26].

The relic densities Ω_xs of $EXons$ and $\Omega_{\bar{x}}$ of anti-$EXons$ depend on the cosmic asymmetry and can be determined by following their evolution in the early stages of the universe.

The relevant bounds are presented in fig. 1, taken from [26] Cosmological bounds (continuous lines) and constraints from underground detection (dashed lines) as a function of the cosmion asymmetry α and the cosmion mass are shown there. The continuous lines show the values of $\alpha = (n_x - n_{\bar{x}})/2s$, the asymmetry scaled to the entropy density s (n is the number density) that yield $\Omega h^2 \equiv (\Omega_x + \Omega_{\bar{x}})h^2 = 1$, 0.25 and 0.1, for all the mass range of interest. As it is expected, the relic density of cosmions increases with the asymmetry, for a fixed mass. For the curves shown, it holds approximately that $\Omega h^2 \propto n_x m \propto \alpha m$ (because $n_x >> n_{\bar{x}}$) for any cosmion mass above a threshold, below which the chosen Ωh^2 value is too small to be obtained, even for $\alpha = 0$ (recall that in the usual case of a standard heavy Dirac neutrino with $\alpha = 0$, for masses of some GeV's Ωh^2 increases with decreasing mass, and $\alpha \neq 0$ can only make Ωh^2 larger).

The observationally allowed values of Ωh^2 are mainly restricted by the determinations of the age of the universe. For a fixed h a universe

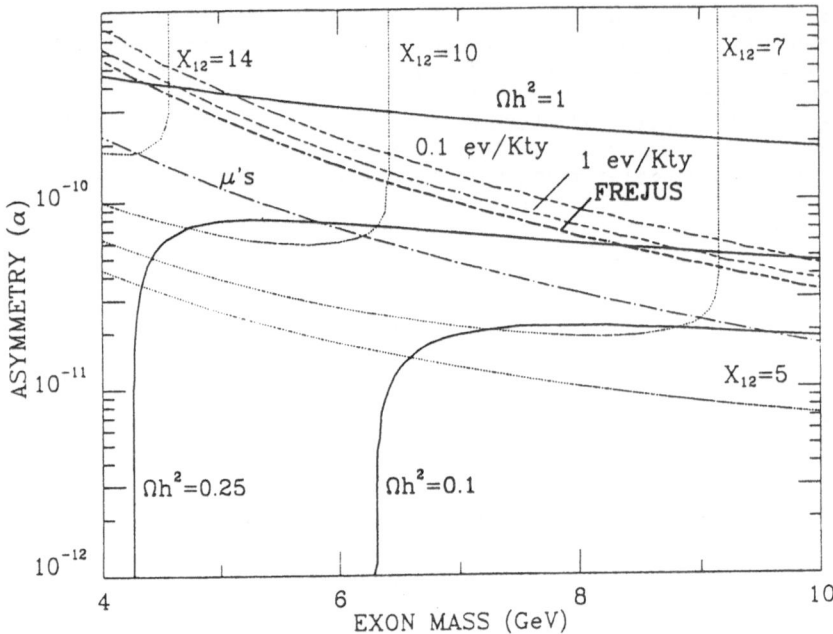

FIGURE 1 Cosmological bounds (continuous lines) and constraints from underground detection (dashed lines) on the $EXon$ model as a function of the cosmion asymmetry α and the cosmion mass [26]. Only the wedge between them is allowed (but largely excluded by direc searches see the text).

with larger total energy density Ω is younger. For a fixed Ω or a fixed Ωh^2, a universe with larger h is younger. Independent measures of the age of the universe τ suggest that 10 Gyr$\leq \tau \leq$20 Gyr [27]. For a matter dominated universe, the lifetime of the universe $\tau \geq 10$ Gyr imposes α to lay below the isoabundance curve with $\Omega h^2 \equiv (\Omega_x + \Omega_{\bar{x}})h^2 = 1$. The region above $\Omega h^2 = 0.25$ is, however, only possible for a rather young universe, $\tau \leq 15$ Gyr allowing for Ω larger than one or $\tau \leq 13$ Gyr if $\Omega = 1$.

The number of cosmions x (N_x) and the number of anticosmions $\bar{x}$ ($N_{\bar{x}}$) trapped in the sun, depend on both, the capture of x and $\bar{x}$ from the halo and their annihilation in the sun.

To solve the solar neutrino problem, a particular value of N_x is needed for each specific value of σ_{eff} and of the cosmion mass. In general cosmion concentrations larger than 10^{-12} (around 10^{-11}) per nucleus are needed [13]. It is shown in figure 1 that these concentrations are easily achievable in the cosmologically allowed region of the α-m space (below $\Omega h^2 = 1$ or $\Omega h^2 = 0.25$, as explained above): the dotted lines are the values of α and m yielding concentrations (in units of 10^{-12}) $X_{12} \equiv N/N_{\odot} \times 10^{12} =$ 5, 7, 10 and 14.

Cosmion-anticosmion pairs annihilate into pairs of light neutrinos of type k with a branching ratio $f_k \simeq .07$, giving rise to a flux of monoenergetic neutrinos (or antineutrinos) at the earth surface.

The resulting rate of contained electron and muon neutrino events is very large (several thousands of events per Kton year) with respect to the experimental bounds (few events per Kton year), for most of the range of interest. However, for large asymmetries (10^{-10}-10^{-9}) the relic anticosmion abundance starts decreasing exponentially with α [25], reducing the signal well below the experimental bound. The Frejus 90% C. L. upper limit of 3.3 event/Kt y for contained neutrino events (for $\nu_e + \bar{\nu}_e + \nu_\mu + \bar{\nu}_\mu$ from the direction of the sun, with $E_\nu > 2.3$ GeV) [28]), exclude the asymmetries below the bold dashed line in fig. 1. Also shown are the curves of 1 and 0.1 in the same units. At these neutrino energies only muons produced at less than a few meters outside the detector can reach it so that their signal yields weaker bounds than those from contained events. For instance, the 90 % C.L. IMB bound of 8.4×10^{-14} muons cm^{-2} s^{-1} [28] on the excess flux of muons with energy $E_\mu \geq 2$ GeV should be improved by about one order of magnitude to yield comparable bounds on the cosmion parameters as those from contained events. The region excluded by the IMB bound on up-going μ's is the one below the dot-dashed line in fig. 1.

These bounds were obtained [26] using the most unfavorable halo parameters (those corresponding to the lowest expected neutrino rate): $\rho_h \simeq 0.2$ GeV/cm^3 and $\bar{v} \simeq 400$ Km/s.

As a summary, indirect DM searches rule out the EXon parameter space for a universe of age $\tau \geq 15$ Gyr or strongly constrain it for a younger universe, $\tau \geq 10$ Gyr. The upper-limit on the cosmion asymmetry (the $\Omega h^2 = 0.25$ or 1 curve respectively) is related to the age of the universe because the cosmion relic density increases with the asymmetry and a too large relic density would imply a too young universe. On the contrary, to obtain a neutrino signal from the sun at underground detectors below the experimental upper-bound requires an extremely low anticosmion relic density, achievable only with large cosmion asymmetries (above the FREJUS curve of fig. 1). The wedge between these two bounds is the only allowed region left by indirect DM searches.

The allowed area, completely or almost, disappears when also the direct DM searches are taken into account. The best bounds on cosmions with coherent interactions with nuclei, as is the case of $EXons$, are the preliminary ones obtained by the LBL/UCSB collaboration with a Silicon spectrometer [29]. They reject the mass region near 10 GeV, where the allowed wedge in fig. 1 is larger. Actually, they only allow cosmions with masses near the lower limit of the allowed range, i.e. near 4 GeV.

THE LIGHTEST NEUTRALINO

The lightest neutralino is one of the most likely candidates for the lightest supersymmetric particle (LSP) and, as such, a good dark matter candidate.

A general neutralino, not a pure photino or higgsino, but an admixture of a gaugino and higgsino may have coherent interactions with matter [30,31] The dominant coherent interactions are mediated by the lightest neutral Higgs field [31] (the scalar called H$_2$ below). The enhancement of the scattering cross section improves the capture of these particles by the sun or the earth and helps increase their direct interaction with the nuclei in detectors. Thus, DM searches yield interesting bounds on (and could detect soon) general neutralinos [30,31,32,33] but not on pure photinos or higgsinos. These have to wait for the new detection techniques under development.

I will consider here the lightest neutralino in the MSSM, the Minimal Supersymmetric Standard Model. This is the simplest phenomenologically acceptable supersymmetric version of the Standard Model of elementary particles [34]. In this model there is one new particle, the supersymmetric partner, for each particle of the Standard Model, with the additional difference that there are two Higgs doublets and their supersymmetric partners, instead of just one.

The model has a N=1 supergravity spontaneously broken at a large energy scale in a hidden sector (a sector that communicates with the observable fields only through gravitational interactions). At lower energy scales the remaining theory has a global supersymmetry broken by soft terms. With some simplifying assumptions (among these the existence of a grand unification that equates the masses of the three gauginos at a large scale [34]) the theory at this stage depends on just five new unknown parameters (besides those already present in the Standard Model such as the Yukawa couplings, the standard gauge couplings and the vacuum expectation values of the Higgs fields). Sometimes, a particular assumption on the hidden sector further reduces the new parameters to just four. This theory, valid at a still large energy scale, has to be rescaled down to low energies according to the renormalization group equations. In so doing, the electroweak spontaneus breaking is achieved when, through radiative corrections, one of the masses of the scalar Higgs fields becomes negative. This indicates the appearance of a new minimum of the scalar potential. The two minimization conditions on the scalar potential, that insure the correct breaking of the Standard Model, further reduce the number of independent new parameters to three, that here will be chosen to be the parameters usually called M_2, μ and m_{H_2} presented below.

In the MSSM there are four neutralinos (the Majorana fermions that are supersymmetric partners of the neutral bosons present in the model). They are the $\tilde{B}$, the $\tilde{W}_3$, the $\tilde{H}_1^0$ and the $\tilde{H}_2^0$, the partners of the neutral U(1) and SU(2) gauge bosons and of the neutral components of the two Higgs doublets. Their four by four mass matrix depends on the parameters M_2, μ and v_2/v_1, which are respectively the $\tilde{W}_3$ mass parameter (from the supersymmetry breaking terms), a mixing mass parameter between the two Higgs superfields in the superpotential, and the ratio of the two vacuum expectation values $\langle H_1^0 \rangle = v_1$ and $\langle H_2^0 \rangle = v_2$. In the base $(\tilde{B}, \tilde{W}_3, \tilde{H}_1^0, \tilde{H}_2^0)$, their four by four mass matrix is

$$
\begin{bmatrix}
M_1 & 0 & -M_Z c_\beta s_W & M_Z s_\beta s_W \\
0 & M_2 & M_Z c_\beta c_W & -M_Z s_\beta c_W \\
-M_Z c_\beta s_W & M_Z c_\beta c_W & 0 & -\mu \\
M_Z s_\beta s_W & -M_Z s_\beta c_W & -\mu & 0
\end{bmatrix},
\tag{6}
$$

with the abbreviations c_β, s_β, c_W and s_W for $\cos\beta$, $\sin\beta$, $\cos\theta_W$ and $\sin\theta_W$ respectively. Through the usual assumption (mentioned above) of equating the masses of all the gauginos at some grand unification scale, the $\tilde{B}$ mass parameter M_1, results $M_1 = \frac{5}{3} tg^2 \theta_W M_2$.

The lightest neutralino, here called χ, is in general a combination of different current eigenstates (photino, zino and higgsinos):

$$
\chi = \gamma_1 \tilde{\gamma} + \gamma_2 \tilde{Z} + \gamma_3 \tilde{H}_1^0 + \gamma_4 \tilde{H}_2^0,
\tag{7}
$$

where the γ_i depend only on M_2, μ and v_2/v_1. When $M_2 \to 0$ the χ is the photino $\tilde{\gamma} = \cos\theta_W \tilde{B} + \sin\theta_W \tilde{W}_3$ with mass proportional to M_2. When $\mu \to 0$ the χ is the higgsino $\tilde{h} = \sin\beta \tilde{H}_1^0 + \cos\beta \tilde{H}_2^0$, with mass proportional to μ. Here $\sin^2\theta_W = 0.23$ and $\tan\beta \equiv v_2/v_1$. A general χ as in eq. (1) is usually heavier than the pure $\tilde{\gamma}$ or $\tilde{h}$. Disregarding any physical bounds, the mass of the lightest neutralino can be anything between zero and a few TeV. The upper bound is a matter of consistency with the motivation to introduce supersymmetry in the Standard Model as a solution to the hierarchy problem; there is no reason to consider supersymmetry if the mass of all supersymmetric particles is much larger than the weak scale.

In what follows I will present the results of references 31 and 33. There all parameters of the lagrangian were taken real, neglecting possible small CP violation effects. Without loss of generality M_2 was chosen non negative. This can be done due to a symmetry under a simultaneous change of the sign of μ and M_2 [35] of all the equations used in both papers.

A correct electroweak breaking requires $v_2/v_1 \geq 1$ [35]. The values that v_2/v_1 may have, depend on the top quark mass through the renormalization group equations that determine the electroweak spontaneous symmetry breaking in the MSSM [36]. Only for a light top, $m_t \lesssim 50$ GeV (no longer

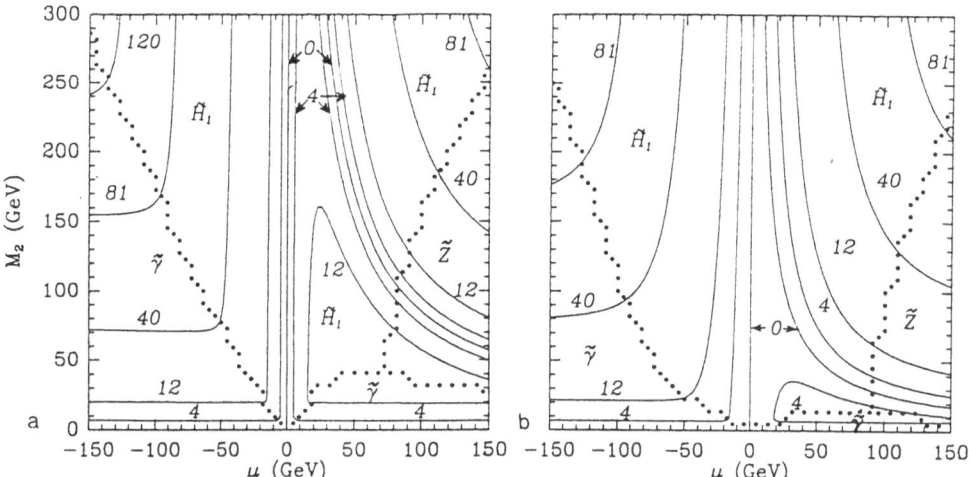

FIGURE 2 Contour maps of the mass of the lightest neutralino χ in GeV and its dominant component (in regions separated by dotted lines) in the $\mu - M_2$. for v_2/v_1 equal to 2 (fig. a, left) and 8 (fig. b right) [33]. For explanations see the text.

acceptable phenomenologically), the solution $v_2/v_1 \simeq 1$ is favoured; for larger top masses, larger values of v_2/v_1 are permitted (see [36]) and for $m_t \simeq 120$ GeV a value as large as 20 may be allowed. Υ-decay experiments rule out large values of v_2/v_1 for $m_{H_2} \lesssim m_\Upsilon \simeq 9.4$ GeV, but if $m_{H_2} \gtrsim m_\Upsilon$ even the most stringent B-decays experiments still allow for $v_2/v_1 \lesssim 20$ [37].

Let us consider the region -150 GeV $< \mu < 150$ GeV, $0 < M_2 < 300$ GeV and $1 \lesssim v_2/v_1 \lesssim 20$. Contour maps of m_χ and the dominant neutralino component of the χ (in regions separated by dotted lines) in the $\mu - M_2$ plane are shown in figs. 2.a and b, for $v_2/v_1 = 2$ and $v_2/v_1 = 8$ respectively [33].

The same three parameters determine the masses of the two charginos, the Dirac fermions mass eigenstates that are linear combinations of the $\tilde{W}^-$, the $\tilde{W}^+$, the $\tilde{H}_1^-$ and the $\tilde{H}_2^+$. These are the superpartners of the $W^\pm$ gauge bosons and of the charged components of the two Higgs fields.

The mass matrix of the charginos in the base $(\tilde{W}^+, \tilde{H}_2^+, \tilde{W}^-, \tilde{H}_1^-)$ is given by

$$\begin{bmatrix} 0 & 0 & M_2 & M_W\sqrt{2}c_\beta \\ 0 & 0 & M_W\sqrt{2}s_\beta & \mu \\ M_2 & M_W\sqrt{2}s_\beta & 0 & 0 \\ M_W\sqrt{2}c_\beta & \mu & 0 & 0 \end{bmatrix}. \tag{8}$$

Since the χ is the LSP, the region of parameter space where the lightest chargino is lighter than the χ is excluded. This region is anyhow within that excluded by the experimental lower bound on the chargino mass $m_{\chi^+} > 25.5$ GeV at the 95% C.L. from TRISTAN [38], shown in figs. 4 and 7.

In the MSSM there are three physical neutral Higgs particles, two scalars (usually called H_1 and H_2) and one pseudoscalar (H_3). Their masses depend only on two parameters, that we choose to be the mass of the lightest one, m_{H_2}, and v_2/v_1. They are given by $m_{H_a}^2 = r_a m_Z^2$, where $r_2 \leq c^2$, $c \equiv \cos 2\beta$, $r_1 = c^2(1 - r_2)/(c^2 - r_2)$, $r_3 = r_2(1 - r_2)/(c^2 - r_2)$. These equations imply $m_{H_2} \leq m_{H_3} \leq m_{H_1}$, with the lightest neutral scalar, H_2, lighter (and the heaviest, H_1, heavier) than the Z. As previously mentioned, m_{H_2} can be chosen to be an independent parameter together with M_2, μ and v_2/v_1. For values of v_2/v_1 near one, the mass m_{H_3} is much larger than m_{H_2}, thus H_3 could be ignored, but for large values of v_2/v_1 both masses become equal (see fig. 4). H_3 does not contribute significantly to the scattering of non relativistic neutralinos with matter because its coupling is proportional to the velocities, but it is important for the annihilation of neutralinos. The annihilation of non relativistic χ's through H_3 is velocity independent and, thus, is important if H_3 is light. Also the annihilation into a pair $H_2 H_3$ (contrary to that into $H_2 H_2$ and $H_3 H_3$ pairs) is not velocity suppressed and, for the ranges of χ masses considered here, is the dominant channel when it is kinematically allowed. Using the relation $m_{H_1}^2 = m_Z^2 + m_{H_3}^2 - m_{H_2}^2$, derived from eq. (2), it is evident that when m_{H_3} becomes equal to m_{H_2} then $m_{H_1} \simeq m_Z$. Thus, H_3 can be ignored in direct DM searches, but it has to be taken into account in indirect searches.

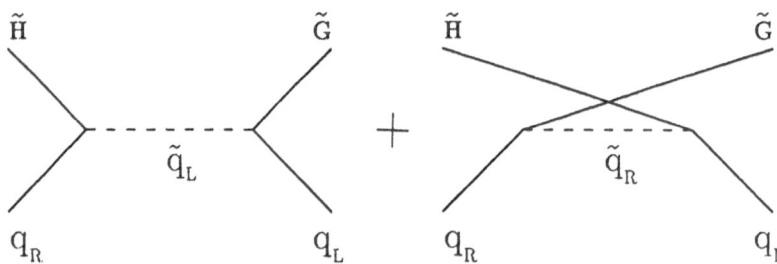

FIGURE 3 Feynman diagrams for the spin-independent part of the elastic scattering of a neutralino with a quark, through squark exchange

The lightest neutralino (when it is not a pure photino or pure higgsino) may have coherent interactions with nuclei through the exchange of scalar quarks [30] and Higgs bosons [31]. The coherence enhancement in the scattering cross section may be of several orders of magnitude for heavy nuclei. This enhancement improves the chances of detecting neutralinos that may be present in the dark halo of our galaxy, in both direct and indirect searches.

The scattering of a neutralino with a quark may proceed through a squark formed in the s-channel. Only if there is a chirality flip for the quark, the Fierz transformed interaction contains a scalar term. This term originates the same type of coherent interaction with a nucleus as that generated by the exchange of a scalar boson (the same we mentioned for the $EXon$ above). If there are no mass mixing terms between right and left squarks, at the tree level the only way of changing the chirality of the quark when a squark is exchanged in the s channel, is by picking a higgsino component in one of the vertices and a gaugino (photino or zedino) in the other, as shown in fig 3 (remember that a higgsino couples a right-quark to a left-squark and a left-quark to a right-squark while a gaugino changes a right-quark into a right- squark and a left quark into a left-squark). Thus, this interaction is zero if the interacting neutralino is a pure higgsino or a pure gaugino.

The two contributions shown in fig 3 have opposite sign because the right squarks carry (by convention) the quantum numbers of the anti right-quark. If the right and left squark are degenerate both amplitudes cancel

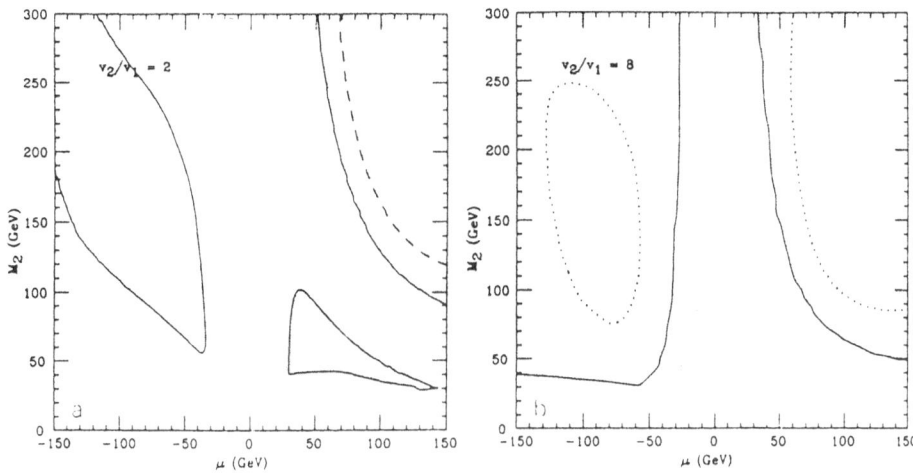

FIGURE 4 Direct dark matter searches using a ^{76}Ge-spectrometer: excluded regions in the μ-M_2 plane for $m_{H_2} = 10$ GeV (solid line), 20 GeV (dashed line), 30 GeV (dotted line) and $v_2/v_1 = 2$ in 4.a (left) and 8 in 4.b (right) [31].

exactly when the gaugino is a photino. Thus, for degenerate squarks there is a coherent squark mediated interaction only between the zedino and the higgsino components of the interacting neutralino.

In more general models there can be, usually small, mixing mass terms between right and left squarks. By inserting this mixing mass terms in the squark propagator, the chirality of the interacting quark can be flipped even when two gauginos or two higgsinos are attached to the vertices of the Feynman diagram. Thus in these models coherent squarks mediated interactions are possible for a pure photino, or zedino or higgsino.

Squarks are usually too heavy to yield detectable scattering cross sections of neutralinos. The exchange of the lightest scalar Higgs field might be more important. This Higgs field only couples a zedino component to a higgsino component of the interacting neutralino. Again this coupling is not present for a pure higgsino or a pure gaugino.

In the following the squark mediated scattering interactions are neglected because squarks are heavier than the Z for the relevant parameter region. Bounds from CDF [39] say that $m_{\tilde{q}} \gtrsim 74$ GeV. This limit depends on the gluino mass, and becomes stronger for smaller M_2. These bounds are, however, based on the same assumptions as the previous bounds from colliders due to the UA1 collaboration: the LSP is taken to be a massless pure photino (which in the MSSM amounts to take $M_2 = 0$) and all the sfermion masses are assumed equal. This last assumption is valid only for $\cos 2\beta \simeq 0$ ($v_2/v_1 \simeq 1$) and for small values of M_2, because the mass splitting among different squarks consist of terms of order $M_2^2/m_{\tilde{q}}$ and $\cos 2\beta\ m_Z^2/m_{\tilde{q}}$ [35]. For the general values of M_2 and β considered here, the above mentioned assumptions do not hold. Anyhow, for $M_2 \gtrsim 30$ GeV and given the upper bound from TRISTAN of 28 GeV on the selectron mass [38], the squarks (when left and right squark mixings are small) become much heavier than the Z so that their contribution to neutralino scattering is negligible. The bounds from DM searches come out in that region. The only relevance of sfermion

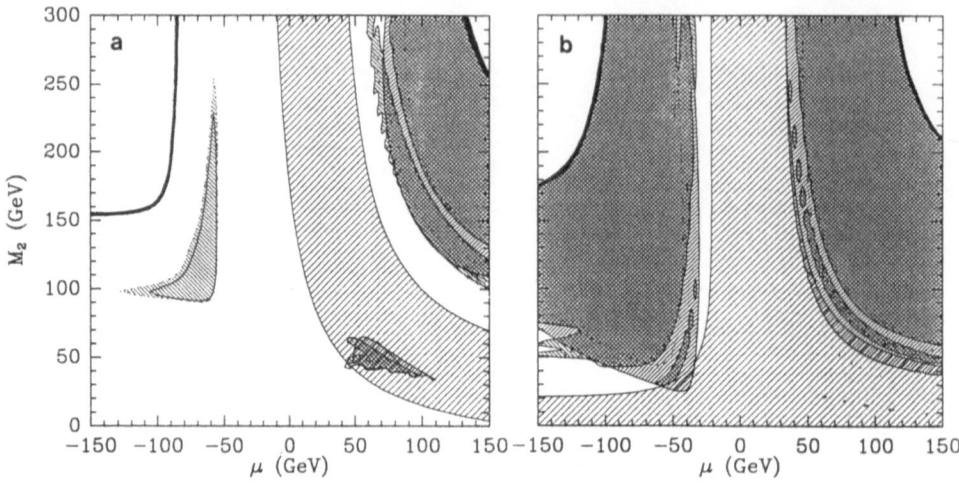

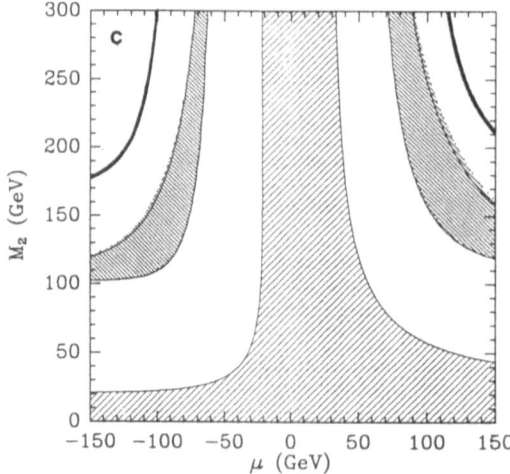

FIGURE 5 (.a left, .b right and .c center) Bounds from indirect dark matter searches, under the assumption that only neutralinos constitute the
halo of our galaxy. The hatched central region is excluded by accelerator
bounds on chargino masses [33]. For explanations see the text

interactions is in the computation of the relic density of χ's that are almost pure photinos.

Assuming that our galactic halo consists of the lightest neutralino,
bounds were obtained [31], from direct searches of dark matter with a Ge
spectrometer [16], on the parameters of the MSSM. These bounds are shown
in fig.4.a and b for for values of $v_2/v_1 = 2, 8$ and m_{H_2}=10 GeV respectively.
The excluded regions reflect the dependence of the Higgs mediated interaction with a Ge nucleus on the neutralino mass and the zino-higgsino mixing.

In ref. 33, the expected event rate of neutrinos from the sun and the
earth (neutrinos and antineutrinos of the electron and the muon type with

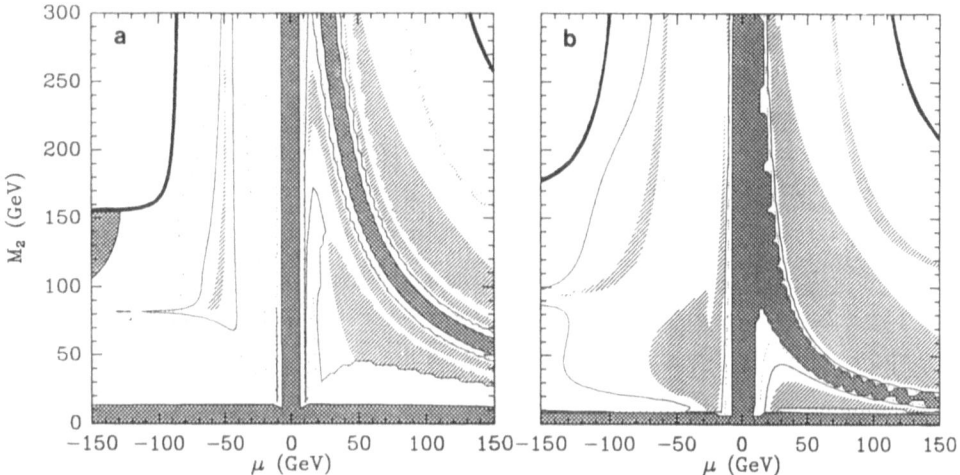

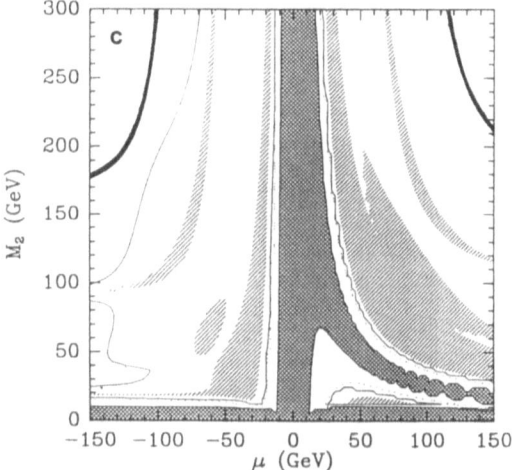

FIGURE 6 (.a left, .b right and .c center) Relic density of the lightest neutralino. Areas of $m_\chi > m_W$ avoided (thick full lines). The darker regions are excluded because $\Omega_\chi h^2 \geq 1$. The levels $\Omega_\chi h^2 = 0.05$ (dotted lines), 0.01 (full lines) and the regions of $\Omega_\chi h^2 < 0.001$ (hatched areas) are shown [33].

energy $E_\nu \geq 2$ GeV), was compared with the corresponding upper limits at the 90% C.L. of 4.1 events kton^{-1} yr^{-1} (for the sun) and 6.4 events kton^{-1} yr^{-1} (for the earth) resulting from charged current contained and vertex contained events at Fréjus [40]. Also the limits on the flux of up-going muons with energy larger than 2 GeV coming from the sun, $8.4 \, 10^{-14} \, \mu$ cm^{-2} s^{-1}, and the conservative limit of $2.65 \, 10^{-13} \, \mu$ cm^{-2} s^{-1} from the earth, both obtained by IMB [41] (at the 90% C.L.) were applied.

To minimize the expected signals and obtain conservative upper bounds on them, the values of the local halo density and velocity most unfavorable for the capture rate, $\rho_h = 0.20$ GeV/cm^3 and $v = 400$ Km/s, were chosen in 33 (to compare with fig. 4 notice that the values 0.40 GeV/cm^3 and 300 Km/s were used in 31) Bounds were obtained under two assumptions: either that the halo of our galaxy consists of neutralinos (fig. 5), or that

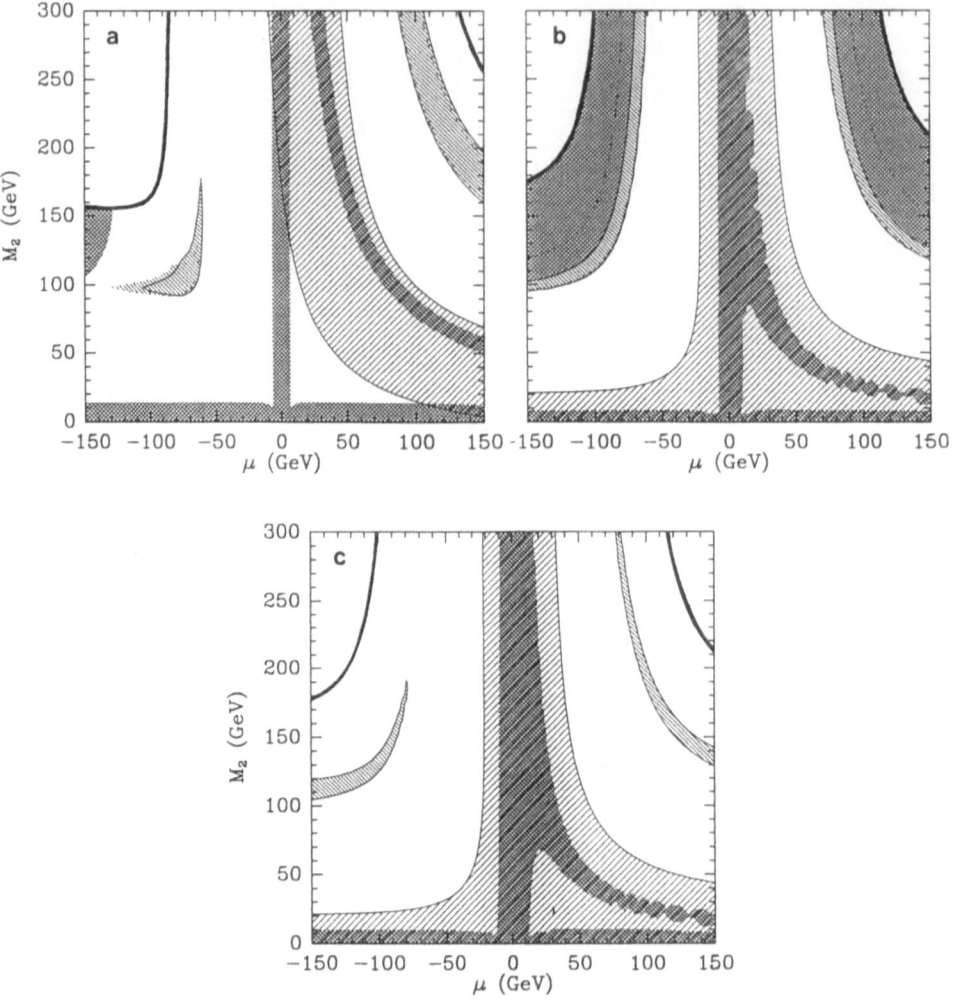

FIGURE 7 As in figs. 5 but relaxing the assumption that the neutralinos constitute the halo by themselves: if $\Omega_\chi h^2 < 0.05$ it is assumed that neutralinos contribute a fraction $\Omega_\chi h^2/0.05$ of the local halo density [33].

for abundances smaller than some value (0.05 of the critical density), the neutralinos only constitute a fraction of the halo, proportional to their relic cosmic abundance (fig. 7). The results are given in the figures 5 to 7 in a $\mu - M_2$ plane for three representative choices of v_2/v_1 and m_{H_2}: 2 and 10 GeV in the figures a, 8 and 10 GeV in the figures b, 8 and 30 GeV in the figures c. In these figures χs heavier than the $W^\pm$ (the regions at the top-left and top-right corners, indicated by thick full lines), are not considered.

The areas of the $\mu - M_2$ plane excluded by the lower bound on the chargino mass and by the limits of Fréjus and IMB, under the assumption that the neutralino constitutes the halo of our galaxy, are presented in figs. 5. The large central up-right to down-left hatched area is excluded by the TRISTAN bound on the chargino mass. The bounds from the sun are indicated with up-right to down-left hatched areas (with a larger density of lines than for the chargino mass limit). Those from the earth exclude the areas hatched with up-left to down-right lines. In both cases re-

gions excluded by contained neutrino events are enclosed by a full line
and those excluded by through-going muons are enclosed by a dotted line.
For $v_2/v_1 = 2$ (or 8) the maximum value of m_{H_2} is 56 GeV (or 91 GeV), but the
bounds from indirect searches disappear for $m_{H_2} > 20$ GeV (or > 40 GeV,
because the couplings increase with v_2/v_1). No bounds from Fréjus and IMB
remain for $m_{H_2} > 60$ GeV even for the largest coherent coupling, i. e. for
$v_2/v_1 = 20$.

The assumption that the neutralinos constitute the halo of our galaxy
is not tenable if their cosmological abundance Ω_χ is not large enough to
account for all the dark matter in the halos of galaxies, $\Omega_g \simeq 0.05$. What
results from the computations of the relic abundance is the combination
$\Omega_\chi h^2$ where h. In figs. 6(a-c) we show $\Omega_\chi h^2$ for the same values of v_2/v_1
and m_{H_2} as in figs. 5(a-c).

Only in the regions of $0.01 \leq \Omega_\chi h^2 \leq 1$ neutralinos do not overclose
the universe and may constitute by themselves the halos of galaxies, i.e.
the relic density may be $\Omega_\chi \geq 0.05$ for an allowed value of h. Notice in
figs. 6.b and c how small these areas are for large v_2/v_1.

Figures 7 show the results if in the regions of the parameter space
where $\Omega_\chi h^2 \lesssim 0.05$ (see figs. 6) it is assumed that the χ's constitute only
a fraction $\Omega_\chi h^2/0.05$ of the halo, namely that their local density is only
that fraction of ρ_{halo}. This assumption is reasonable if halos consist of
cold dark matter with cosmological density $\Omega_g h^2 = 0.05$ and with a spatial
distribution similar to that of the χ's. The areas excluded by the lower
bound on the chargino mass and those excluded by the limits of Fréjus and
IMB are shown with the same conventions as in figs. 5. These last bounds
disappear for $m_{H_2} > 15$ GeV for $v_2/v_1 = 2$ and > 30 GeV for $v_2/v_1 = 8$.
In figs. 7.a and 7.c only the bounds from the earth survive, because of
the kinematical enhancement of the scattering cross section in the earth
around $m_\chi = m_{Fe} = 56$ GeV. In fig. 5.b the couplings are larger and the
Higgs field is lighter. Also the areas excluded because $\Omega_\chi h^2 > 1$ are shown
as in figs. 7.

As a summary, figures 5 to 7 show the relation between the values of
the Higgs masses and the possibility of having a detectable neutralino
present in the halo of our galaxy. Only in the regions between the bor-
ders of the dark areas and the dotted lines in figs. 6 the halo may consist of neutralinos. As can be seen in figures 6 just the requirement of an
appropriate relic abundance for the neutralino excludes large areas (the
dark areas) of the parameter space. In figs. 5 and 7 the parameter ranges
excluded by the TRISTAN bound on the chargino mass are shown. The areas
excluded by indirect searches in figs. 5, where we assume $\rho_\chi = \rho_h$, correspond therefore to regions of the parameter space in which the lightest
neutralino cannot be alone the dark matter in the halo of our galaxy, but
the points excluded by the IMB and Fréjus data in these figures might still
correspond to a viable model.

In figs. 7 the bounds are more ambitious but less certain. The areas
excluded in these figures, under the assumptions specified above, correspond to excluded parameter ranges of the MSSM: a point excluded corresponds to a model which predicts the existence of a lightest neutralino
with such a mass, relic abundance and interactions that should have been
seen if it existed.

Only near the boundaries of the regions excluded by indirect searches
could we expect that halo neutralinos might soon be discovered. The larger
the lightest Higgs mass and the smaller the ratio v_2/v_1, the more dificult
it is to detect neutralinos. For Higgs masses above 30 GeV, only neutralinos with masses similar to that of a Iron nucleus could be seen (through
neutrinos from the earth).

We have not yet considered the case of a lightest neutralino heavier
than the weak gauge bosons. Disregarding experimental bounds, the upper
bound on the lightest supersymmetric particle is given by the solution

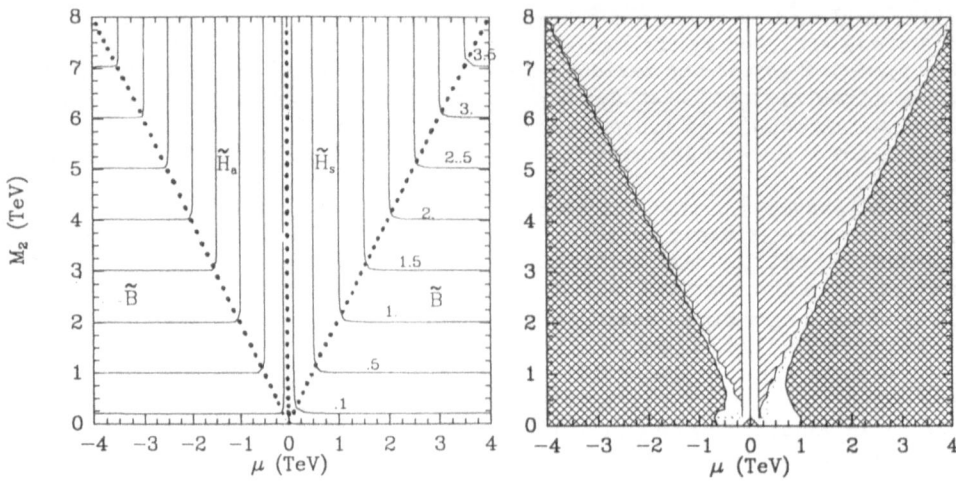

FIGURE 8 a. (left) Contour map of the mass of the lightest neutralino in TeV and its dominant components (in regions separated by dotted lines). b. (right) Relic density of the same neutralinos, with the same conventions used in the fig. 6.

of the gauge hierarchy problem. A mass larger than a few TeV, let us say 3 TeV, would invalidate supersymmetry as the solution to this important problem. From the matrix in eq. 6, it is easy to see that for most values of the parameter space the χ is a pure bino or a pure higgsino (see fig. 8.a), and thus, do not have coherent interactions with nuclei. Moreover this heavy neutralinos are in most of the parameter space either excluded because too abundant (the dark area in fig 8.b) or too poorly abundant to constitute the dark matter [42]. Still regions of the parameter space are left where a lightest neutralino with a mass of some 100 GeV could be the dark matter in our galactic halo. The questions of how and if they could be detected are still open.

REFERENCES

[1] For a review see Dark Matter in the Universe, eds. J.Knapp and J. Kormendy (reidel,1986), Proceeding of the International Union Symposium N° 117, Princeton June 24-28, (1985).

[2] K. Griest and M. Sher, CfPA-TH-89-007 (or WM-89-107) preprint, July (1989)

[3] S. M. Faber and J. S. Gallagher, Ann. Rev. Astron. and Astrophys.,17 135 (1979); A. Bosma, Astron.J. 220, 1825 (1981); V. C. Rubin, W. K. Ford and N. Thonnard Ap. J. 238,471 (1980); T. S. Van Albada and R. Sancisi, Phil. Trans. Royal Soc. Lond. A320, 447 (1986).

[4] J.P.Ostriker and J.A.R. Caldwell, Ap. J.251 (1981) 61 and ''Kinematics, Dynamics and the Structure of the Milky Way'' (W. L. H. Shuter ed., Reidel Dordrecht, (1983) pp. 249-257; J. N. Bahcall, M. Schmidt and R.M. Soneira, Ap. J.265 (1983) 760; R.A.Flores, CERN preprint TH 4736 (1987).

[5] D.N. Spergel and D.O. Richstone, Proc. of the Moriond Astrophysics Meeting on Dark Matter, March 1988 and references therein.

[6] J Bond and A. Szalay, Ap.J.274, 443 (1983); J. R. Primack and G. R. Blumenthal, Moriond conference: Formation and Evolution of Galaxies and the Large Structures in the Universe, Reidel-Dordrecht, 1983.

[7] Some reviews are P. F. Smith, Proc. of the ''2nd ESO/CERN Symposium on Cosmology, Astronomy and Fundamental Physics", Garching (1986); B. Sadoulet, Proc. of the 13th Texas Symposium on Relativistic Astrophysics, Chicago (1986), LBL preprint 23468 (1987); B Sadoulet, lectures at the ''School of Astroparticle Physics, Dark Matter in the Universe", Erice, Italy, 4-14 May 1988; G Gelmini, lectures at the same last mentioned school; Proceedings of the Particle Astrophysics Workshop- Forefront Experimental Issues-, LBL, Berkeley, 8-10 December 1988; J. Primack, D. Spergel and B. Sadoulet, Ann. Rev. Nucl. Sci. 38 (1988) 751

[8] G.B. Gelmini, L.J. Hall and M.J. Lin, Nucl. Phys. B281 (1987) 726;

[9] S. Raby and G.B. West, Nucl. Phys. B292 (1987) 793; Phys. Lett. B194 (1987) 557;

[10] S. Raby and G.B. West, Phys. Lett. B202 (1988) 47;

[11] G.G. Ross and G. Segré, Phys. Lett. B197 (1987) 45;

[12] G.F. Giudice and E. Roulet, FERMILAB preprint PUB-88/129-T, sept. 1988, to appear in Phys. Lett. B;

[13] D.N. Spergel and W.H. Press, Ap. J. 294 (1985) 663; W.H. Press and D.N. Spergel, Ap. J. 296 (1985) 679; J. Faulkner and R.L. Gilliland, Ap. J. 299 (1985) 994; R.L. Gilliland, J. Faulkner, W.H. Press and D.N. Spergel, Ap. J. 306 (1986) 703; L.M. Krauss, K. Freese, D.N. Spergel and W.H. Press, Ap. J. 299 (1985) 1001.

[14] For reviews see G. Steigman, Ann. Rev. Nucl. Part.Sci. 29,313 (1979); S. Wolfram, Phys. Lett. 82B, 65 (1979); A. D. Dolgov and Ya. B. Zel'dovich, Rev. of Mod. Phys. 53, 1 (1981).

[15] M.W. Goodman, and E. Witten, Phys. Rev. D 31, 3059 (1985); I.Wasserman, Phys. Rev. D, 33, 2071 (1986); 2295 (1984); A.K. Drukier, K. Freese, and D.N. Spergel, Phys. Rev. D, 33, 3495 (1986).

[16] F.T. Avignone, S. Ahlen, R. Brodzinski, S. Dimopoulos, A. Drukier, G.Gelmini, B. Lynn, H. Miley, J. Reeves, D. Spergel and G. Starkman, Proceedings of the ''7th. International Conference", Vanderbilt University, 15-17 May (1986); G.B. Gelmini, '' Proceedings of the Theoretical Workshop on Cosmology and Particle Physics", 28 July - 15, August 1986, Lawrence Berkeley Laboratory, Berkeley, California; S. Ahlen, F. Avignone, R. Brodzinski, A. Drukier, G. Gelmini and D. Spergel, Phys. Lett. 195B, 603 (1987).

[17] D. O. Caldwell, R. M. Eisberg, D. M. Grumm, M. S. Witherell, B. Sadoulet, F.S. Goulding and A. R. Smith, Phys. Rev. Lett. 61 (1988) 510.

[18] J.Silk and M. Srednicki, Phys. Rev. Lett.53 (1984) 624; J. Hagelin, G.L. Kane, Nucl. Phys.. B263 (1986) 399; etc.

[19] J. Silk, K.A. Olive and M. Srednicki, Phys. Rev. Lett.55 (1985) 257; K. Freese, Phys. Lett.167B (1986) 295; L. Krauss, M. Srednicki and F. Wilczek, Phys. Rev.D33 (1986) 2079; T.K. Gaisser, G. Steigman and S. Tilav, Phys. Rev.D 34 (1986) 2206; J. Hagelin, K.W. Ng and K. Olive, Phys. Lett.180B (1986) 375; M. Srednicki, K.A. Olive and J. Silk, Nucl. Phys.. B279 (1987) 804; K.W. Ng, K. Olive and M. Srednicki, Phys. Lett.B188 (1987) 138; J. Ellis, R.A. Flores and S. Ritz, Phys. Lett.198B (1987) 393; K. Olive and M. Srednicki, Phys. Lett. B (1988) 553; G.F. Giudice and E. Roulet, Nucl. Phys.. BB316 (1989) 429; K. Griest and D. Seckel, Nucl. Phys. B283 (1987) 681. E: B296 (1988) 1034; A. Gould, Ap. J. 321 (1987) 560 and 571, and Ap. J. 328 (1988) 919; S. Ritz and D. Seckel, Nucl. Phys. B304 (1988) 877.

[20] T.K. Gaisser and T. Stanev, Phys. Rev. D30 (1984) 985; T.K. Gaisser and T. Stanev, Phys. Rev. D31 (1985) 2770; T.K. Gaisser, T. Stanev, S.A. Bludman and H. Lee, Phys. Rev. Lett. 51 (1983) 223.

[21] A. Gould, Ap. J. 321 (1987) 560 and 571.

[22] S. Ritz and D. Seckel, Nucl. Phys. B304 (1988) 877.

[23] M.A. Vainshtein and V.I. Zakharov, Phys. Lett. 78B (1978) 443.

[24] J.E. Ross and L.H. Aller, Sci. 191 (1976) 1223.

183

[25] K. Griest and D. Seckel, Nucl. Phys. B283 (1987) 681. E: B296 (1988) 1034.
[26] G. Gelmini and E. Roulet, Nucl. Phys B325 (1989), 733.
[27] I. Iben and A. Renzini, Phys. Rep. 105C (1984) 329; W.A. Fowler, Q. J. R. Astr. Soc. 28 (1987) 87; F. Caputo, V. Castellani and M.L. Quarta, Astron. Astrophys. 138 (1984) 457; for a review see The Early Universe, E. Kolb and M. Turner, book in preparation.
[28] Frejus Collab., B. Kuznik contribution at the Moriond Astrophysics Meeting on Dark Matter, March 1988; and private communication; L. Moscoso contribution at the International Workshop on Neutrino Telescopes, Venice, nov. 1988; IMB Collab., R. Svoboda contribution at the Moriond Astrophysics Meeting on Dark Matter, March 1988.
[29] D. Caldwell, talk at the "Worshop on the Physics of th Gran Sasso", L'Aquila, september 1989
[30] K. Griest, Phys. Rev. D38 (1988) 2357.
[31] R. Barbieri, M. Frigeni and G. F. Giudice, Nucl. Phys.B313 (1989) 725.
[32] G.F. Giudice and E. Roulet, Nucl. Phys. B316 (1989) 429.
[33] ''Bounds on the Minimal Supersymmetric Standard Model from Neutralino Annihilations in the Sun and the Earth",G. Gelmini, P. Gondolo and E. Roulet, SISSA 88 EP89 preprint, 1989.
[34] H. E. Haber and G. L. Kane, Phys. Rep.117 (1985) 75; J. F. Gunion and Haber, Nucl. Phys. B272 (1986) 1; R. Barbieri, Pisa preprint IFUP-TH 33/87 (1987), La Rivista del Nuovo Cimento, vol. 11, Ner. 4 (1988) and references therein.
[35] L.E. Ibáñez and C. López, Nucl. Phys. B233 (1984) 511; L.E. Ibáñez, C. López and C. Muñoz, Nucl. Phys. B256 (1985) 218.
[36] G.F. Giudice and G. Ridolfi, Z. PHYS. C41 (1988) 447.
[37] S. Pokorsky, Rutherford Appleton Labs preprint RAL--88--069.
[38] TOPAZ collaboration, I. Adachi et al., Phys. Lett. B218 (1989) 105.
[39] CDF collaboration, F. Abe et al., Phys. Rev. Lett. 62 (1989) 1825.
[40] Fréjus Collaboration, Topical Seminar on Astrophysics and Particle Physics, San Mininato, Italy, 1989, contribution presented by H. J. Daum
[41] IMB Collaboration: J. Lo Secco et al. Phys. Lett. B188 (1987) 388; R. Svoboda et al., Ap. J. 315, (1987) 420.
[42] G. Gelmini and P. Gondolo, in preparation.

HADRONIC AND MUONIC BRANCHING RATIOS FOR DECAYS OF LIGHT HIGGS-BOSONS

R. S. Willey

Department of Physics and Astronomy
University of Pittsburgh
Pittsburgh, PA 15260

INTRODUCTION

A Higgs boson of mass less than m_B can be produced with large probability in the decays of a $B-$meson[1]. The Higgs boson would be detected by observing its decay products. For $2m_\mu < m_H < 2m_\tau$ (3.6 GeV.) the cleanest signal may be the $H \rightarrow \mu\bar\mu$ mode. But then one has to know that the BR for $H \rightarrow \mu\bar\mu$ is not too small. At the most naive quark level, for $m_H < 2m_\tau$, the H will decay primarily into $s\bar s$ and $\mu\bar\mu$, with a ratio of about three to one (the color factor for $q\bar q$) or somewhat more, depending on what one takes for m_s. It was pointed out[2] that heavy quarks could also contribute substantially through a virtual heavy quark loop giving $H \rightarrow gg$ with the gluons then materializing as light hadrons - in particular, as $\pi\pi$. This observation was eventually refined into a low energy theorem[3] (LET) for the $H \rightarrow \pi\pi$ amplitude. The crucial feature of this result is that the amplitude contains a term proportional to m_H^2 as well as one proportional to m_π^2. But even if one extrapolates this result well beyond its range of validity, it still leads to a BR for $H \rightarrow \mu\bar\mu$ which ranges from 40% at $m_H = 0.3$ GeV. to 7% at $m_H = 2$ GeV. It was then observed by Raby and West [4] that strong final state interactions could further enhance the rate for $H \rightarrow \pi\pi$, and hence decrease the BR for $H \rightarrow \mu\bar\mu$. They made an estimate and claimed very large enhancements for $H \rightarrow \pi\bar\pi$. We find that their estimate is a substantial overestimate for several reasons, the most important of which is that they treated the $f_o(975)$ (formerly, and in this paper, S*) resonance as an elastic $\pi\pi$ resonance. But the S* is strongly coupled to the $K\bar K$ channel and cannot be reasonably treated in a single channel framework, even below the $K\bar K$ threshold (which is the range treated by Raby and West). We[5] have carried out a coupled channel analysis which incorporates the constraints of unitarity on the S* resonance parametrization both above and below the $K\bar K$ threshold. We also find a large enhancement of the rate for $H \rightarrow \pi\pi$ very close to the $K\bar K$ threshold, but not as large as the estimate of ref. 4; and, more importantly, there is no large enhancement for m_H below 0.95 GeV.

Higgs Particle(s)
Edited by A. Ali
Plenum Press, New York, 1990

CALCULATION

The amplitude for the decay $H \to \pi\pi$ defines a pion scalar form factor[5]

$$M(H \to \pi\pi) = < \pi\pi_{out} \mid \mathcal{L}'_{eff}(0) \mid 0 > \equiv -(G\sqrt{2})^{1/2} F_\pi(q^2). \qquad (2.1)$$

$$\mathcal{L}'_{eff} = \frac{\partial \mathcal{L}_{eff}}{\partial H}$$

where H is the Higgs field. A similar definition holds for $F_K(q^2)$ with $\pi\pi$ replaced by $K\bar{K}$.

The coupled channel unitarity equations for F_π, F_K are

$$Im \; F_i = F_j^* \rho_j t_{ji} \qquad (2.2)$$

where

$$\rho_i = \frac{k_i}{W}, \; k_i = (\frac{W^2}{4} - m_i^2)^{1/2}, \; W^2 = s = q^2 (= m_H^2) \qquad (2.3)$$

and below the inelastic threshold

$$t_{11} = t = \frac{1}{\rho} e^{i\delta} sin \; \delta \qquad s < 4m_\pi^2 \qquad (2.4)$$

Neglecting the left-hand cut singularities of the t_{ij}, the coupled unitarity equations are satisfied by the construction

$$F_i(s) = C_j(s) t_{ji}(W)/W \qquad (2.5)$$

The $C_j(s)$ are real polynomials to be determined by the LET, and the $t_{ji}(W)$ are determined by a fit to the low energy $\pi\pi, K\bar{K}$ scattering and reaction data.

The inverse matrix t^{-1} satisfies a simple matrix unitarity equation

$$Im t^{-1} = \rho. \qquad (2.6)$$

A simple parametrization of t^{-1}, which satisfies (2.6), is

$$t^{-1} = \begin{pmatrix} \frac{M_1^2 - s - 2ik_1\Gamma_1}{2W\Gamma_1} & \frac{\lambda}{W} \\ \frac{\lambda}{W} & \frac{M_2^2 - s - 2ik_2\Gamma_2}{2W\Gamma_2} \end{pmatrix} \qquad (2.7)$$

We use (all in GeV units)

$$m_\pi = 0.138, \quad m_K = 0.496$$

$$M_1 = 0.87, \quad \Gamma_1 = 0.7, \quad M_2 = 0.92, \quad \Gamma_2 = 1.0, \quad \lambda = 0.1 \qquad (2.8)$$

Note that the output parameters ('observed resonances') can be substantially different from the input m_i, Γ_i - because of the coupling ($\lambda \neq 0$) and the requirement to invert (2.7). In fact, these parameters and (2.7) provide a good representation of the known elastic $\pi\pi$ and inelastic $\pi\pi \to K\bar{K}$ data (see ref. 5 for details). We also emphasize that the five parameters of (2.8) are overdetermined by the quantitative and qualitative features of that data. We list seven such features: the scattering length, the energy at which δ rises through $\pi/2$, the slope there, the position and slope of the rapid rise of δ near the $K\bar{K}$ threshold, and the rapid rise and then turnover of the inelasticity.

The low energy theorem[3,6] which we use to determine the $C_j(s)$ is

$$\frac{1}{(G\sqrt{2})^{1/2}} < M_a M_b | \mathcal{L}'_{eff}(0)|0> = \delta_{a,b}(\frac{11}{9}m_a^2 + \frac{2}{9}q^2 + \cdots) \qquad (2.9)$$

With the C_j and t_{ji} determined, the form factors F_π, F_K are computed from (2.5). Then the decay rates for $H \to \mu\bar{\mu}, \pi\pi, K\bar{K}$ are

$$\Gamma(H \to \mu\bar{\mu}) = \frac{G\sqrt{2}}{8\pi}m_H m_\mu^2 \left(1 - \frac{4m_\mu^2}{m_H^2}\right)^{3/2} \qquad (2.10)$$

$$\Gamma(H \to \mu\bar{\mu}) = \frac{G\sqrt{2}}{32\pi}\frac{|F_{1,2}|^2}{m_H}(1 - \frac{4m_{1,2}^2}{m_H^2})^{1/2}. \qquad (s = m_H^2) \qquad (2.11)$$

The ratio of hadronic to muonic decay rates is, below the $K\bar{K}$ threshold

$$f = f_\pi = \frac{\Gamma(H \to \pi\pi)}{\Gamma(H \to \mu\bar{\mu}} \qquad (2.12a)$$

and above the $K\bar{K}$ threshold

$$f = \frac{\Gamma(H \to \pi\pi) + \Gamma(H \to K\bar{K})}{\Gamma(H \to \mu\bar{\mu}} \qquad (2.12b)$$

The BR for $H \to \mu\bar{\mu}$ is $1/(f+1)$.

RESULTS AND DISCUSSION

In figure 1, we give the values of $f\pi, f_K$ as functions of the Higgs mass ($W = m_H$). Note the dip in f_π just below the S^* resonance, which is the result of destructive interference between the two terms in (2.5). Note also that f_π never exceeds 4 until m_H is within 50 MeV of the S^* resonance and $K\bar{K}$ threshold i.e. the BR for $H \to \mu\bar{\mu}$ is never less the 20% for $m_H < 950$ MeV.

We have already remarked that Raby and West[4] have overestimated the enhancement of $H \to \pi\pi$ by the S^* by treating it as an elastic $\pi\pi$ resonance, ignoring the constraint of coupled channel unitarity. In addition, even in the context of a single channel resonance parametrization, we have found two other errors in the treatment of RW, each of which also goes in the direction of overestimating the (dominant) S^* enhancement. RW use a resonance parametrization of a single-channel Omnés-Muskeleshvili representation (RW (64))

$$F(s) = P(s)\Omega(s).$$

(In the single-channel approximation this is entirely equivalent to the single-channel version of (2.5)). For $P(s)$ they take the LET result (2.9) (RW (95)). However this does not satisfy the LET because $\Omega'(0) \neq 1$. When the contribution of $\Omega'(0)$ is taken out to satisfy the LET, the value of F at the S^* is decreased by almost $1/2$. (And F is squared in the rate). A second error in RW is that for the denominators in their resonance forms they use $M^2 - s - ik\Gamma$ (RW (67), (96)). The conventional form is $M^2 - s - 2ik\Gamma \approx 2M(M - W - i\Gamma/2)$. (See for example the Particle Data Group review). Thus when they take a (total) S^* width of 33 GeV from the PDG and use it in their formulas, they are affectively using a width of 16.5 GeV. Thus

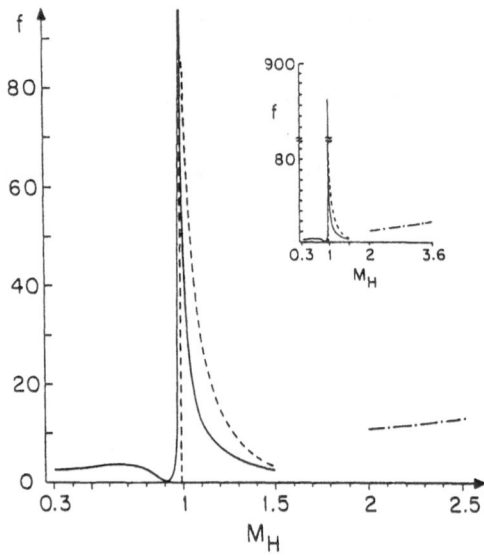

Figure 1. The ratios f_π (solid line), f_k (dashed line), and f_{QCD} (dash-dot line).

they overestimate the peak height by another factor of two in amplitude and four in rate. (This also explains why they obtain $\Gamma_\epsilon = 1.3$ GeV. from fitting to elastic $\pi\pi$ phase shift, while we obtain 0.63 GeV., fitting to the _same_ phase shift, with the conventional resonance form, or its equivalent $\Gamma = 2/(d\delta/dW)$). Compounding all of these factors leads to a large overestimate of the S^* contribution.

Somewhere above the $K\bar{K}$ threshold, additional channels ($\eta\eta, 4\pi, \rho\rho, ..$) become important and our two-channel formalism is no longer adequate. At still higher energies, where many channels are open, for the inclusive hadronic ratio $f_{had} = \Gamma(H \rightarrow hadrons)/\Gamma(H \rightarrow \mu\bar{\mu})$, one can use the quark-gluon QCD description[7].

$$\Gamma_{\mu\bar{\mu}} : \Gamma_{s\bar{s}} : \Gamma_{gg} \approx m_\mu^2 : 3m_s^2 : (\frac{\alpha_s}{\pi})^2(\frac{N_h^2}{9})m_H^2 \qquad (3.1)$$

For m_H in the range 2 to 3.6 GeV. we take $\alpha_s/\pi \approx 0.1$ and $N_h = 3$ (or 2). Then roughly

$$f_{had} \approx \frac{\Gamma_{ss} + \Gamma_{gg}}{\Gamma_{\mu\bar{\mu}}} \approx 7 + m_H^2 \qquad (3.2)$$

This is also plotted in figure 1.

In the range above 1.2 or 1.3 GeV where our two-channel description breaks down. but below some several GeV where the inclusive QCD description becomes accurate; it is very difficult to make any detailed calculation. But we remark that the large enhancement associated with the S^* is a very special case. The S^* is a very narrow resonance, and, more importantly, it is located very close to a threshold to which it is strongly coupled. This special set of circumstances is not repeated at higher masses. S-wave resonances at higher energy are broad, and there are no sharply defined thresholds. The most reasonable guess is just to extrapolate f_{QCD} (eq. 3.2) back to smaller q^2 with some smooth broad bumps ($f_0(1400), f_0(1590), ?$) superposed on it. A conservative estimate is that $B_{\mu\mu}$ is not less than 1% for m_H in the range from 1.1 to 3.6 GeV.

We briefly consider the application of these results to the experimental searches for the H in B−decays. In the standard model with a single physical H and just three generations of fermions, the BR for the decay of a b−meson into H plus anything is[1]

$$\frac{B(B - HX)}{B(B - \ell\nu X)} = \frac{27\sqrt{2}}{64\pi^2} G_F m_b^2 \frac{|V_{tb}V_{ts}^*|^2}{|V_{cb}|^2} (\frac{m_t}{m_b})^4 \frac{(1 - \frac{m_H^2}{m_b^2})^2}{f(m_c^2/m_b^2)} \qquad (3.3)$$

In order to evade the Linde-Weinberg bound[8] and have an H with mass less than m_B, in the minimal standard model, it is required that the top mass be ≥ 80 GeV. Then, with the $b \to c\ell\nu$ phase space factor $f(m_c^2/m_b^2) = 0.5$ and $B(B \to \ell\nu X) = 0.12$, eq. (3.3) gives the theoretical value

$$B(B \to HX) \geq 0.26(1 - \frac{m_H^2}{m_b^2})^2. \qquad (m_t \geq 80GeV,\ m_H < m_B) \qquad (3.4)$$

There are a series of experiments[9] which give

$$B(B - \mu\bar{\mu}X)_{exp} \leq 0.008 \qquad (for\ m_H > 0.3GeV) \qquad (3.5)$$

In the range $0.3 \leq m_H \leq 0.95$ our calculation gives $B(H \to \mu\bar{\mu}) > 0.20$, hence

$$B(B \to HX) \cdot B(H \to \mu\bar{\mu} > 0.05 \qquad (0.3 \leq m_H \leq 0.95) \qquad (3.6)$$

which exceeds (3.5) by a factor of six. In the range 0.95 to 1.1 Gev. the strong S^* enhancement of the $\pi\pi$ and $K\bar{K}$ decay modes of the H depresses the BR for $H \to \mu\bar{\mu}$ so much that no conclusion can be drawn from this mode. For $m_H > 1.1$ GeV., up to 3.6 GeV., the recent CLEO experiment[10] provides stronger bounds. In Fig. 2 we have reproduced the CLEO limits for $B(B \to \mu\bar{\mu}X)$ (fig. 6 of ref. 14) and superimposed the theoretical lower limit following from (3.4) (with $m_b = 4.9$ GeV.) and our conservative estimate $B(H \to \mu\bar{\mu}) > 0.01$ in this range.

$$B(B \to HX) \cdot B(H \to \mu\bar{\mu}) > 26 \times 10^{-4} \Big(1 - \frac{m_H^2}{m_{b^2}}\Big)^2 \qquad (1.1 \leq m_H \leq 3.6) \qquad (3.7)$$

we see that the theoretical lower limit substantially exceeds the experimental upper limit everywhere in this range, except just at the position of the J/ψ resonance, where the experiment observes $B \to \psi X$ followed by $\psi \to \mu\bar{\mu}$ at the level 8×10^{-4}.

Acknowledgement. This is a report of work done with Tran Truong. A detailed account is published in Phys. Rev. D40, 3635 (1989).

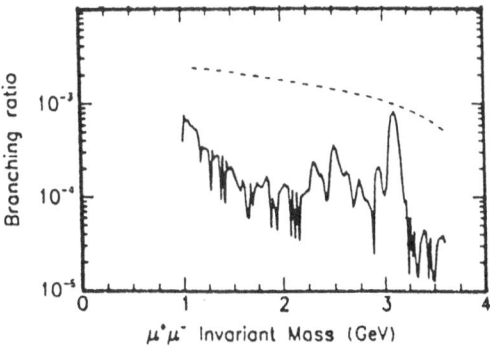

Figure 2. The CLEO[10] upper limit on $B(B \to \mu\bar{\mu}X)$ (solid line) and the theoretical lower limit on $B(B \to HX) \cdot B(H \to \mu\bar{\mu})$ under the conditions stated in the text (dashed line).

REFERENCES

[1] R. S. Willey and H. L. Yu, Phys. Rev. $D26$, 3086 (1982).

[2] F. Wilczek, Phys. Rev. Lett. 39, 1304 (1977); M. A. Shifman, A. I. Vainshtein, and V.I. Zakharov, Phys. Lett. $78B$, 443 (1978).

[3] M. Voloshin and V. Zakharov, Phys. Rev. Lett. 45, 688 (1980); V. A. Novikov and M. A. Shifman, Z. Phys. $C8$, 43 (1981); M. B. Voloshin, Sov. J. Nucl. Phys. 44, 478 (1986).

[4] S. Raby and G. B. West, Phys. Rev. $D38$, 3488 (1988).

[5] Tran N. Truong and R. S. Willey, Phys. Rev. D, to be published.

[6] Alternative derivations of the low energy theorem have recently been given. R. S. Chivukula, A. Cohen, H. Georgi, and A. V. Manohar, Phys. Lett. B (to be published); L. S. Brown, Phys. Rev. $D39$, 3085 (1989). S. Dawson and H. E. Haber, report SCIPP-89/14.

[7] B. Grinstein, L. Hall, and L. Randall, Phys. Lett. $211B$, 363 (1988).

[8] A. D. Linde, JETP Lett. 23, 64 (1976). S. Weinberg, Phys. Rev. Lett. 36, 294 (1976).

[9] A CLEO experiment, Phys. Rev. Lett. 53, 1309 (1984) gives the combined $BR < 0.008$ for $m_H > 0.5$ GeV. The lack of sensitivity to $m_H < 0.5$ GeV is because of cuts on the muon momentum. Two other experiments, done at high energy in the continuum, looking directly for $b \to \mu\bar{\mu}X$, do not have this problem. However, their efficiencies were evaluated assuming three body spectrum for each μ rather than the sequence $b \to Hs$ followed by $H \to \mu\bar{\mu}$. JADE, Phys. Lett. $B132$, 241 (1983); MARKJ, Phys. Lett. 50, 799 (1983).

[10] CLEO collaboration, Phys. Rev. D40, 712 (1989).

RULING OUT THE LIGHT HIGGS BOSON BY KAON DECAY*

Hoi-Lai Yu

Institute of Physics, Academia Sinica
Nanking, Taipei, Taiwan

ABSTRACT

We re-examine the theoretical estimates of the decay $K \to \pi H$ and the experimental constraints on the existence of a light Higgs boson from this process. We find that: (i) pole diagrams generated from the Higgs-gluon coupling via a loop of heavy quarks do contribute to $K \to \pi H$, (ii) there is an additional contribution to the $K \to \pi H$ amplitude coming from the effective KHW and πHW couplings, (iii) even if B , the unknown parameter in the chiral-Lagrangian description of $K \to \pi H$ transitions, is nonzero and even if the real part of the $K \to \pi H$ amplitude is canceled accidentally, the imaginary contribution alone suffices to rule out a Higgs boson lighter than $2m_{\pi}$, and (iv) whether Higgs bosons in the mass range $2m_{\pi} < m_H < 350\ MeV$ are excluded by the imaginary part of the $K \to \pi H$ amplitude depends on the branching ratio of $H \to \mu^+\mu^-$ and the top-quark mass. Decay modes $K_L \to \pi^+\pi^- H$ and $K^+ \to l^+\nu H$ are briefly discussed.

1. Introduction

Higgs boson searching is one of the main topic of this workshop. Unfortunately, the problem is obscured by the absence of any theoretical constraint on the number or masses of the Higgs bosons within the framework of the Standard Model (SM). For the minimal model, there is a vacuum stability bound[1], $m_H > 7$ GeV, but this argument fails when there exists heavy fermions, or when there are more than one doublet of Higgs bosons. Recently, the ARGUS and CLEO Collaborations[2] have reported a large $B_d^0 - \bar{B}_d^0$ mixing. This, when combining with the null signal of the top quark search by the UA1 Collaboration[3], indicates that the top quark is likely to be heavy. Thus, it is

---

*Talk based on works with H.Y. Cheng.

important to consider possible experimental limits on the existence of the light Higgs bosons.

Within the minimal model, it has been pointed out earlier[4] that the branching ratio for $B \to HX$ could be as large as several percents for a light Higgs boson (*i.e.* $m_H \lesssim 4.5 \ GeV$) and a heavy top quark. This provides a good laboratory for the light Higgs boson searching. However, results from various experimental graups[5] seem to rule out the existence of a light Higgs boson within the mass renge of $0.3 \ GeV \lesssim m_H \lesssim 5 \ GeV$. But as pointed out by the authors of refs.[6] and [7] that even though the branching ratio for $B \to HX$ is fairly large, the theoretical uncertainties[14] in the $H \to \mu^+ \mu^-$ branching ratio may ruin any definite conclusions.

In this talk, we update arguments[6-13] of using various K decays to exclude light Higgs bosons with mass $m_H \lesssim 360 \ MeV$. In an earlier paper, Vainshtein et al.[8] have pointed out that $m_H \lesssim 350 \ MeV$ is ruled out by the $K^\pm \to \pi^\pm H$ mode. However, their calculation assumed a momentum-independent $K \pi$ transition amplitude which is inconsistent with the underlying chiral symmetry. Later, Willey and one of us (H. L. Y.)[9] presented a quark model calculation which includes a one loop $s \to d + H$ transition, but neglected the contributions from the nonspectator diagram Fig.1e. Pham and Sutherland[10] advocated that both the nonspectator contributions and $\Delta I = \frac{1}{2}$ enhancement are important and may partially cancel the spectator term, therefore raise doubt in the conclusion of Willey and Yu. Willey[11], in a following paper, has made a detailed re-analysis of this problem by using Fermi statistics and Bethe-Salpeter equation techniques to demonstrate that the spectator and nonspectator contributions are in fact constructive and hence rule out a Higgs boson of a mass between $50 \ MeV \lesssim m_H \lesssim 211 \ MeV$. On the other hand, Chivukula and Manohar[13] (CM) used chiral perturbative theory and vacuum insertion to obtain an expression which indicates a destructive interference between the spectator and nonspectator diagrams and which depends on an unknown parameter B in the effective chiral Lagrangian. Despite the above complications, CM still managed to conclude that $m_H < 360 \ MeV$ from the $K_L \to \pi^0 e^+ e^-$ and $K_L \to \pi^0 \mu^+ \mu^-$ decay modes. On taking a rather conservative attitude, Raby, West and Hoffman[7] argued that because B is unknown, many of the claims in the literature excluding Higgs bosons from kaon decays are not valid.

In light of the above-mentioned confusing status we shall present in this talk an updating version on the calculation of the $K \to \pi H$ decay amplitude[15]. We claim that: (1) pole diagrams which include the Higgs interaction with gluons via a triangular loop of heavy quarks do contribute to $K \to \pi H$, (2) there is an additional contribution to the $K \to \pi H$ amplitude coming from the effective KHW and πHW couplings, and (3) the imaginary contributions alone to $K \to \pi H$ suffice to rule out a Higgs boson lighter than 270 MeV. Finally, to complete our discussion we also calculate the branching ratio of $K_L \to \pi^+ \pi^- H$ and $K^+ \to e^+ \nu H$, though the present experimental limits on the corresponding $K_L \to \pi^+ \pi^- e^+ e^-$ and $K^+ \to e^+ \nu e^+ e^-$ decay modes cannot give any conclusion on the Higgs boson masses.

2. Theoretical Evaluation of $K \to \pi H$ Amplitudes

We shall re-examine in this section the theoretical estimate of the $K \to \pi H$ rate. To begin with we write down the relevant effective Lagrangian for $K \to \pi H$ [13]

$$\mathcal{L} = i \sum_i \bar{q}_i \gamma_\mu \partial^\mu q - (1 + \frac{H}{v}) \sum_i m_i \bar{q}_i q_i + (1 + \frac{H}{v})^{-2} \mathcal{L}^{\Delta S = 1} + \mathcal{L}_{1\text{-loop}} - \frac{n_h \alpha_s}{12\pi} H G^a_{\mu\nu} G^{\mu\nu}_a \tag{2.1}$$

where we have included effective Higgs-gluon interactions via a heavy-quark triangle diagram for later purposes, n_h is the number of heavy quarks, $v = 1/(\sqrt{2}G_F)^{1/2} = 246 \ GeV$, $\mathcal{L}^{\Delta S = 1}$ is the effective $\Delta S = 1$ weak Lagrangian

$$\mathcal{L}^{\Delta S = 1} = -\sqrt{2} G_F V_{us} V^*_{ud} \sum_{i=1}^{6} c_i(\mu) \mathcal{O}_i(\mu) \tag{2.2}$$

where V_{ij} are the Kobayashi-Maskawa mixing matrix elements, $\mathcal{O}_i$ are four-quark operators (we follow the notation of Eq.(3.9) of ref.[16]), and $c_i(\mu)$ are Wilson coefficient functions in which perturbative QCD corrections from M_W down to the renomalization scale, typically chosen at 1 GeV, are taken into account. The characteristic values of Wilson coefficients are [17]

$$c_1 = -2.11, \ c_2 = 0.12, \ c_3 = 0.09, c_4 = 0.45, \ c_5 = -0.025, \ c_6 = -0.003 \tag{2.3}$$

The term $\mathcal{L}_{1\text{-loop}}$ in Eq.(2.1) is the flavour-changing $\Delta S = 1$ two-quark interaction induced at the one loop level

$$\mathcal{L}_{1\text{-loop}} = \frac{3\alpha}{32\pi \sin^2 \theta_W} (\sum_i V_{is} V^*_{id} \frac{m_i^2}{M_W^2}) \frac{H}{v} [m_s \bar{d}(1+\gamma_5)s + m_d \bar{d}(1-\gamma_5)s] + \text{h.c.} \tag{2.4}$$

which was first obtained by Willey and Yu [9].

193

We first consider the pole diagrams Figs.1a and 1b. At the quark level, pole diagrams correspond to Higgs emission from the bound-state quarks of initial and final mesons. It has been argued in the literature[10] that diagrams 1a and 1b compensate as the Higgs coupling is proportional to the meson mass squared, i.e. $g_{H\pi\pi}/g_{HKK} = m_\pi^2/m_K^2$. However, as we shall see shortly, the Higgs-meson coupling does not vanish even in the chiral limit, and hence the pole contributions are not necessarily zero. To see this, let us consider the matrix element

$$\langle \pi^+\pi^- |H\rangle = \frac{1}{v}\langle \pi^+\pi^-| \sum_i m_i \bar{q}_i q_i |0\rangle \tag{2.5}$$

It is well known that heavy quarks contribute indirectly to the Higgs-pion coupling by virture of the triangle diagram[17] [i.e. the last term in Eq.(2.1)]

$$\langle \pi^+\pi^- |H\rangle = \frac{1}{v}\langle \pi^+\pi^-| \sum_l m_l \bar{q}_l q_l - \frac{n_h \alpha_s}{12\pi} G_{\mu\nu}^a G_a^{\mu\nu} |0\rangle \tag{2.6}$$

where the subscript l denotes light quarks. The momentum dependence of the matrix element $\langle \pi^+(p_1)\pi^-(p_2)|H\rangle$ is easily seen from the approach of chiral Lagrangian. In the chiral limit the general chiral representation for the HGG interaction given in Eq.(2.1) reads[6]

$$\frac{f_\pi^2}{16}\frac{H}{v}(a\mathrm{Tr}\partial_\mu U \partial^\mu U^\dagger + b\mathrm{Tr}U\Box U^\dagger) + \mathrm{h.c.} \tag{2.7}$$

This chiral Lagrangian will contribute not only to the Higgs-meson coupling but also to the effective $H\phi W$ vertex responsible for the nonspectator diagrams Figs.1c and 1d, as we shall see later. It is straightforward to check that[18]

$$\langle \pi^+(p_1)\pi^-(p_2)|H\rangle = \frac{(a-b)}{v}p_1 \cdot p_2 \tag{2.8}$$

The unknown coefficient $(a-b)$ in Eq.(2.8) is intimately related to the trace of energy-momentum tensor

$$\theta^\mu{}_\mu = \sum_l m_l \bar{q}_l q_l - \frac{(33-2n_l)\alpha_s}{24\pi}G_{\mu\nu}^a G_a^{\mu\nu} \tag{2.9}$$

where n_l is the number of light quarks, and use of heavy-quark operator-product

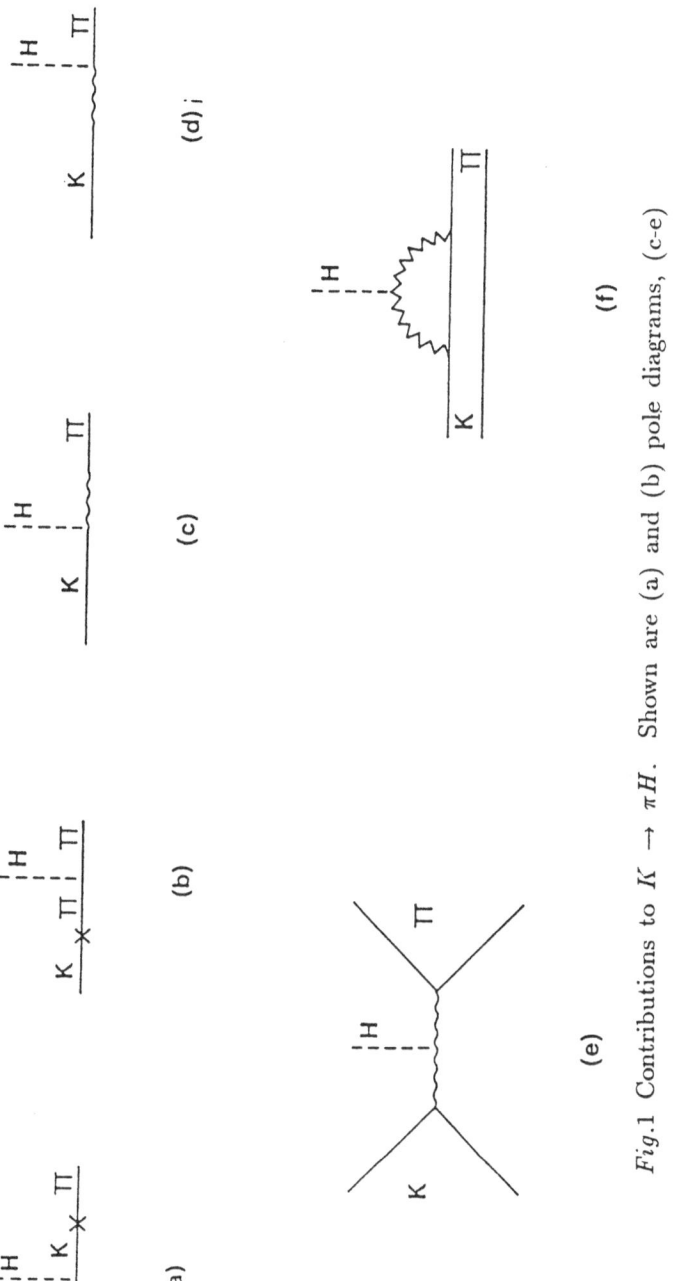

Fig.1 Contributions to $K \to \pi H$. Shown are (a) and (b) pole diagrams, (c-e) nonspectator diagrams, and (f) the one-loop spectator diagram.

195

expansion[18] has been made. Therefore,

$$\langle \pi^+\pi^-|H\rangle = \frac{1}{v}\frac{2n_h}{33-2n_l}\langle\pi^+\pi^-|\theta^\mu{}_\mu\,|0\rangle + \frac{1}{v}(1 - \frac{2n_h}{33-2n_l})\langle\pi^+\pi^-|\sum m_i\bar{q}_i q_i|\,|0\rangle \tag{2.10}$$

The lowest order chiral Lagrangian implies

$$\langle\pi^+(p_1)\pi^-(p_2)|\,\theta^\mu{}_\mu\,|0\rangle = 2p_1\cdot p_2 + 4m_\pi^2 \tag{2.11}$$

Hence, $(a - b)$ is fixed to be

$$a - b = \frac{4n_h}{33 - 2n_l} \tag{2.12}$$

and

$$\langle\pi^+(p_1)\pi^-(p_2)|H\rangle = \frac{1}{v}(\frac{4n_h}{33-2n_l}p_1\cdot p_2 + \frac{6n_h}{33-2n_l}m_\pi^2 + m_\pi^2) \tag{2.13}$$

for small $(p_1 + p_2)^2$. Likewise,

$$\langle K^+(p_1)K^-(p_2)|H\rangle = \frac{1}{v}(\frac{4n_h}{33-2n_l}p_1\cdot p_2 + \frac{6n_h}{33-2n_l}m_K^2 + m_K^2) \tag{2.14}$$

Raby and West[20] have emphasized recently that the $H \to \pi\pi$ decay is further enhanced by final-state interactions via a possible resonance in the $\pi\pi$–scattering amplitude. However, this is not rel4evant for the pole contributions discussed here.

From Eqs.(2.13) and (2.14) it is easily seen that the pole diagrams with momentum-dependent Higgs-pion couplings (i.e. the $p_1\cdot p_2$ terms) do not cancel. We find the pole amplitude to be

$$A(K^+ \to \pi^+H)_{\text{pole}} = \frac{1}{v}(\frac{4n_h}{33-2n_l})p_K\cdot p_\pi$$
$$\cdot\,\frac{\langle\pi^+(p_K)|\mathcal{L}^{\Delta S=1}|K^+(p_K)\rangle - \langle\pi^+(p_\pi)|\mathcal{L}^{\Delta S=1}|K^+(p_\pi)\rangle}{m_K^2 - m_\pi^2} \tag{2.15}$$

The $K - \pi$ transition can be evaluated in several different methods (for a detailed discussion, see Secs. 6.1 and 7.3 of ref.[16]). Here we focus on the chiral-Lagrangian approach. The lowest-order chiral representation for $\mathcal{L}^{\Delta S=1}$ has the form (using the notation of chapter 7 of ref.[16])

$$\mathcal{L}^{\Delta S=1}_{\text{chiral}} = g_8\text{Tr}(\lambda_6\partial_\mu U\partial^\mu U^\dagger) + g_{27}^{(1/2)}\Theta^{(27,1/2)} + g_{27}^{(3/2)}\Theta^{(27,3/2)} \tag{2.16}$$

where $U = \exp(2i\phi/f_\pi), \phi = (1/\sqrt{2})\phi^a\lambda^a$, $\text{Tr}(\lambda^a\lambda^b) = 2\delta^{ab}$. Total derivative terms like $Tr\partial^\mu(\lambda_6 U\partial_\mu U\dagger)$ are prohibited because the $\Delta S = 1$ weak Hamiltonian at the quark level respects an additional discrete CPS symmetry[21], which is the product of ordinary CP with a switching symmetry S, which switches the s and d quarks[22]. The coupling constants are determined from the experimental $K \to \pi\pi$ rates to be[16]

$$|g_8 + g_{27}^{(1/2)}| = 0.26 \times 10^{-8} m_K^2, \quad |g_{27}^{(3/2)}| = 0.86 \times 10^{-10} m_K^2 \qquad (2.17)$$

but their signs undertermined. To fix the sign we see that a direct application of factorization yields (cf. Eq.(6.40) of ref.[16])

$$\langle \pi^+(p_\pi)|\mathcal{L}^{\Delta S=1}|K^+(p_K)\rangle = \frac{\sqrt{2}}{6} G_F V_{us} V_{ud}^* \left[c_1 - 2c_2 - 2c_3 - 2c_4 \right.$$
$$\left. + \frac{32}{3}(c_5 + \frac{16}{3}c_6)\frac{\sigma^2}{\Lambda_\chi^2} \right] f_\pi^2 (p_K \cdot p_\pi) \qquad (2.18)$$

where $\sigma = m_\pi^2/(m_u + m_d) = m_{K^+}^2/(m_u + m_s)$ characterizes the spontaneous breaking of chiral symmetry, and $\Lambda_\chi \sim 1$ GeV[23] sets the scale of higher order chiral terms. With the Wilson coefficients Eq.(2.3) it turns out that the sign of $\langle \pi^+|\mathcal{L}^{\Delta S=1}|K^+|$ and hence g_8 is fixed to be negative. It follows from Eq.(2.16) that

$$\langle \pi^+(p_\pi)|\mathcal{L}_{chiral}^{\Delta S=1}|K^+(p_K)\rangle = \frac{4}{f_\pi^2}(g_8 + g_{27}^{(1/2)} + g_{27}^{(3/2)})p_K \cdot p_\pi \qquad (2.19)$$

The final result for the pole contribution has the form

$$A(K^+ \to \pi^+ H)_{pole} = \frac{1}{v}(\frac{8n_h}{33 - 2n_l}) \times \frac{g_8}{f_\pi^2}(m_K^2 + m_\pi^2 - m_H^2) \qquad (2.20)$$

We turn next to the nonspectator diagram Fig.1e, in which the spectator u quark in K^+ participates directly in the Higgs production. The amplitude of Fig.1e governed by the interaction $-2(H/v)\mathcal{L}^{\Delta S=1}$ reads

It is worth stressing here that in the chiral Lagrangian approach the flavor-changing $\Delta S = 1$ Higgs interactions are given by[13]

$$A(K^+ \to \pi^+ H)_e = \frac{2}{v}\langle \pi^+(p_\pi)H|H\mathcal{L}^{\Delta S=1}|K^+(p_K)\rangle \qquad (2.21)$$

Because of the Higgs-gluon interaction arising from the heavy-quark triangle diagram, there are two additional contributions to the $K \to \pi H$ decay, namely Figs.1c and 1d. At the quark level, the effective $H\phi W$ vertex is depicted in Fig.2. The $H\phi W$ coupling can be obtained by coupling the chiral Lagrangian for HGG interactions [Eq.(2.7)] to external gauge fields A_μ^L and A_μ^R

$$\partial_\mu U \to D_\mu U = \partial_\mu U + A_\mu^L U - U A_\mu^R \qquad (2.22)$$

In the present case the external fields are identified with the left-handed $W_\mu^\pm$ boson fields

$$A_\mu^L = -i\frac{g}{\sqrt{2}}W_\mu\Gamma, \quad A_\mu^R = 0 \qquad (2.23)$$

where Γ is identical to the KM matrix element V_{ij} for the pseudoscalar meson constructed from $\bar{q}_i q_j$ and vanishes otherwise. For example, $\Gamma_{ij} = 0$ except for $\Gamma_{12} = V_{ud}$ for $H\pi^+ W$ coupling. It is easily seen from Eqs.(2.7), (2.22) and (2.23) that

$$\mathcal{L}_{H\phi W} = -\left(\frac{4n_h}{33-2n_l}\right)\frac{gf_*}{2\sqrt{2}}\frac{H}{v}W^\mu \mathrm{Tr}(\Gamma\partial_\mu\phi) \qquad (2.24)$$

Recalling that

$$\mathcal{L}_{\phi W} = -\frac{g}{2\sqrt{2}}f_* W^\mu \mathrm{Tr}(\Gamma\partial_\mu\phi) \qquad (2.25)$$

we find[24]

$$\begin{aligned} A(K^+ \to \pi^+ H)_c &= A(K^+ \to \pi^+ H)_d \\ &= -\frac{1}{v}\left(\frac{4n_h}{33-2n_l}\right)\langle\pi^+ H|H\mathcal{L}^{\Delta S=1}|K^+\rangle \end{aligned} \qquad (2.26)$$

Summing over the nonspectator amplitude of Figs.1c, 1d and 1e, we obtain

$$A(K^+ \to \pi^+ H)_{\mathrm{NS}} = \frac{2}{v}\left(2 - \frac{4n_h}{33-2n_l}\right)\langle\pi^+ H|H\mathcal{L}^{\Delta S=1}|K^+\rangle \qquad (2.27)$$

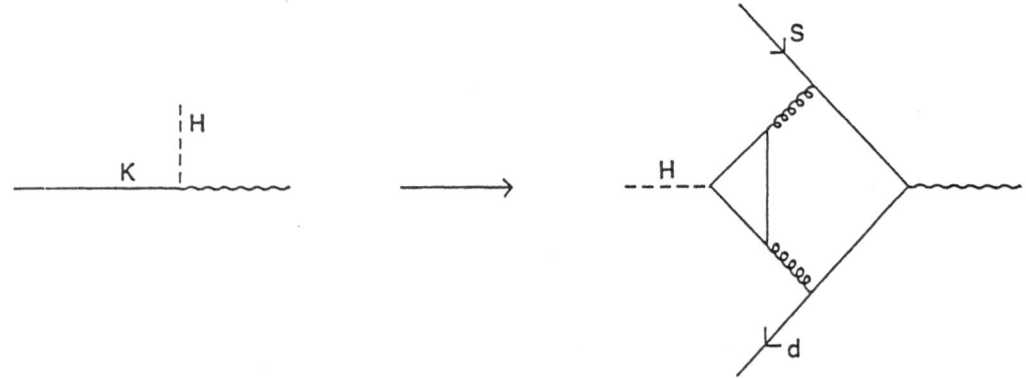

*Fig.*2 The effective HKW vertex at the quark level.

To evaluate the $K \to \pi H$ matrix elements we note that in the chiral Lagrangian approach the flavor-changing $\Delta S = 1$ interactions involving Higgs are given by [13,26]

$$(1 - 2\frac{H}{v})\mathcal{L}^{\Delta S=1} = g_8[\text{Tr}(\lambda_6 \partial_\mu U \partial^\mu U^\dagger) - B\text{Tr}\partial^\mu(\lambda_6 U \partial_\mu U^\dagger)](1 - 2\frac{H}{v})$$
$$+ C(1 - 2\frac{H}{v})\text{Tr}[\lambda_6 M U^\dagger(1 + \frac{H}{v})] + \text{h.c.} \tag{2.28}$$

where we have neglected the 27-plet contributions (recall that $g_{27}^{(1/2)} \approx g_{27}^{(3/2)}/5$). As noticed in ref.[25], CPS symmetry eliminates the B term as before, but it does not remove away the $C\text{Tr}(\lambda_6 M U^\dagger)H$ contribution. The chiral rotation which diagonalizes the quark mass only deletes the Higgs-independent piece of the C term. To eliminate the remaining C term via a Higgs-dependent chiral rotation will reintroduce the B term [26]. This means that the $K - \pi H$ transition receives an additional contribution depending on the unknown parameter B. We find the nonspectator amplitude to be

$$A(K^+ \to \pi^+ H)_{NS} = \frac{4}{v}(2 - \frac{4n_h}{33 - 2n_l})\frac{g_8}{f_\pi^2}[m_K^2 - m_\pi^2 - m_H^2 + \frac{1}{2}B(m_K^2 - m_\pi^2)] \tag{2.29}$$

We now turn to the spectator diagram Fig.1f in which $s \to dH$ occurs at the level. The hadronic matrix element of the $\bar{s}d$ density is given by [28]

$$\langle \pi^+(p_\pi)| \bar{s}d |K^+(p_K)\rangle = \sigma \tag{2.30}$$

with σ being defined in Eq.(2.18). It follows from Eq.(2.4) that

$$A(K^+ \to \pi^+ H)_{\text{1-loop}} = \sum_i^{u,c,t} \frac{m_K^2}{v} \frac{3\alpha}{32\pi \sin^2 \theta_W} V_{id} V_{is}^* \frac{m_i^2}{M_W^2} \equiv \sum_i^{u,c,t} h_i \tag{2.31}$$

Note that the sign of h_i relative to $A(K^+ \to \pi^+ H)_{\text{NS}}$ is unambiguously fixed and is independent of the quark phase convention. For example, h_c is of opposite sign to the nonspectator amplitude due to the negativity of the coupling g_8, in agreement with ref.[13]. To evaluate the t quark contribution we recast the KM matrix element $V_{td}V_{ts}^*$ in terms of the Wolfenstein parametrization [29]

$$V_{td}V_{ts}^* = -\lambda^5 A^2(1 - \rho - i\eta) \tag{2.32}$$

where $\lambda = |V_{us}| = 0.22$, A is close to unity and η measures CP violation. The spectrum of leptons in semileptonic B decay implies $(\rho^2 + \eta^2)^{1/2} \leq 0.9$. The recent ARGUS and CLEO observations of $B_d^0 - \bar{B}_d^0$ mixing [2] strongly suggest a negative ρ and a heavy top quark, $m_t \geq 60 \; GeV$. The parameter η can be determined from the recent NA31 measurement[30] of ϵ'/ϵ. In the standard KM model ϵ'/ϵ has the expression[31]

$$\frac{\epsilon'}{\epsilon} = 3.3(\frac{300\ MeV}{m_s})^2 B_K'\ \text{Im}(V_{td}V_{ts}^*) \tag{2.33}$$

where

$$B_K' \equiv \frac{\langle \pi^+\pi^-|\mathcal{O}_5|K^0\rangle}{\langle \pi^+\pi^-|\mathcal{O}_5|K^0\rangle^{(1/N_c)}} = \frac{\langle \pi^+\pi^-|\mathcal{O}_5|K^0\rangle}{0.055\ GeV^3} \tag{2.34}$$

measures the deviation of the $K - \pi\pi$ penguin matrix element from the $1/N_c$ calculation (for a detailed discussion, see ref.[31]), analogous to the parameter B_K defined in $K^0 - \bar{K}^0$ mixing. Using $m_s = 150\ MeV$, $B_K' = 1$ and the NA31 result[30] $\epsilon'/\epsilon = (3.3 \pm 1.1) \times 10^{-3}$, we find

$$\eta = 0.57 \pm 0.19 \tag{2.35}$$

Now we have

$$h_t = 4.3 \times 10^{-10} \text{GeV}(\frac{m_t}{M_W})^2(1 - \rho - i\eta) \tag{2.36}$$

Because $\rho < 0$ inferred from the $B_d^0 - \bar{B}_d^0$ mixing data, a conservative lower bound for the real part of h_t can be set by putting $\rho = 0$. Since $h_c = 0.73 \times 10^{-10}\ GeV$ for $m_c = 1.5\ GeV$, it is evident that the dominant contribution to the spectator amplitude arises from the top quark. As we shall see in the next section, even the imaginary part of h_t alone suffices to rule out light Higgs bosons within certain mass ranges.

Summing over the contributions from Figs.1a-1f, we find the amplitude for $K \rightarrow \pi H$ to be (for three generations $i.e.$ $n_l = n_h = 3$)[32]

$$A(K^+ \rightarrow \pi^+ H) = [-1.5 \times 10^{-10}(1 - \frac{2}{9})(1 + \frac{m_\pi^2 - m_H^2}{m_K^2})$$
$$+ 0.73 \times 10^{-10} + h_t + B(0.39 \times 10^{-10})]GeV \tag{2.37}$$
$$A(K_L \rightarrow \pi^0 H) = \text{Re}A(K^+ \rightarrow \pi^+ H)$$

From Eq.(2.37) it is obvious that the $K \rightarrow \pi H$ amplitude is at least of order $10^{-10}\ GeV$ and is dominated by the t quark contribution. The branching ratios are then given by

$$Br(K^+ \rightarrow \pi^+ H) = 7.57 \times 10^{-6} (\frac{2p_H}{m_K}) |\frac{A}{10^{-10} GeV}|^2,$$

$$Br(K_L \rightarrow \pi^0 H) = 3.15 \times 10^{-5} (\frac{2p_H}{m_K}) |\frac{A}{10^{-10} GeV}|^2,$$

(2.38)

where the p_H is the momentum of the Higgs boson.

3. Limits from $K \rightarrow \pi H$ decays

To set a limit on a light Higgs boson from existing experimental data, it is important to take the experimental situation into consideration, for instance, the experimental decay vertex requirement. For this purpose, we follow the analysis of Raby, West and Hoffman (RWH)[7] to summarize the available data for $K \rightarrow \pi H$

1) ref.[33], $Br(K^+ \rightarrow \pi^+ H) Br(H \rightarrow e^+ e^-) < 2.7 \times 10^{-7}$ for $100~MeV < m_H < 2m_\mu$,

2) ref.[34], $Br(K^+ \rightarrow \pi^+ H) Br(H \rightarrow e^+ e^-) < 3.5 \times 10^{-7}$ for $140~MeV < m_H < 2m_\mu$,

3) ref.[35], $Br(K^+ \rightarrow \pi^+ H) Br(H \rightarrow \mu^+ \mu^-) < 2.1 \times 10^{-7}$ for $2m_\mu < m_H < 360~MeV$,

4) ref.[36], $Br(K^+ \rightarrow \pi^+ H) < 1.5 \times 10^{-6}$ for $0 < m_H < 80~MeV$,

5) ref.[37], $Br(K^+ \rightarrow \pi^+ X) < 1.4 \times 10^{-6}$ for $5~MeV < m_H < 100~MeV$,

6) ref.[38], $Br(K_L \rightarrow \pi^0 H) Br(H \rightarrow e^+ e^-) < 2.3 \times 10^{-6}$ for $80~MeV < m_H < 2m_\mu$,

7) ref.[39], $Br(K_L \rightarrow \pi^0 H) Br(H \rightarrow e^+ e^-) < 2.0 \times 10^{-7}$ for $10~MeV < m_H < 2m_\mu$,

8) ref.[38], $Br(K_L \rightarrow \pi^0 H) Br(H \rightarrow \mu^+ \mu^-) < 1.2 \times 10^{-6}$ for $2m_\mu < m_H < 360~MeV$

(3.1)

When $m_H < 2m_\mu$, $H \rightarrow e^+ e^-$ is the dominant decay mode, while for $2m_\mu < m_H < 2m_\pi$, the Higgs boson will predominately decay into $\mu^+ \mu^-$. It was argued recently by RWH[7] that because of the undetermined B parameter in the chiral Lagrangian(2.28), no unambiguous and definite limits on the existence of light Higgs bosons can be drawn from kaon decays. However, we have shown in Sec.II that the imaginary contribution to the $K^+ \rightarrow \pi^+ H$ amplitude coming from the intermediate top quark in the $s \rightarrow dH$ loop is large and suffices to rule out a Higgs boson lighter than $2m_\pi$ as long as $m_t > 45 GeV$. Hence, Even if B is non-zero and even if the real part is canceled accidently, the imaginary contribution alone is enough to exclude Higgs bosons with $m_H < 2m_\pi$.

For $2m_\pi < m_H < 360 GeV$, the branching ratio of $H \rightarrow \mu^+ \mu^-$ is suppressed due to the existence of the deacy mode $H \rightarrow \pi\pi$. An estimate of $H \rightarrow \pi\pi$ rates by Eq.(2.13) leads to a branching ratio of 40% for $H \rightarrow \mu_+ \mu^-$ at $m_H = 300 MeV$. Raby and West[20] claimed a large enhancement of $H \rightarrow \pi\pi$ by final-state interactions and they concluded that $Br(H \rightarrow \mu^+\mu^-) \approx 1/24$ at the same Higgs mass. A recent reanalysis of this issue by Truong and Willey[38], found, however, no large enhancement of $H \rightarrow \pi\pi$ for $m_H < 950 MeV$; they estimated $Br(H \rightarrow \mu^+\mu^-)$ to be of 30% at $m_H = 300 MeV$. As a result, whether Higgs bosons in this mass range can be excluded by the imaginary contribution to the $K_+ \rightarrow \pi^+ H$ amplitude depends strongly on the branching ratio of $H \rightarrow \mu^+\mu^-$ and te top-quark mass. For $(H \rightarrow \mu^+\mu^-) \approx 0.3$, we find that light Higgs bosons with $2m_\pi < m_H < 350 MeV$ do not exist if $m_t > 65 GeV$. (Recall that $m_t \geq 80 GeV$ is required to evade the Linde-Weinberg constraint[1].) But if $Br(H \rightarrow \mu^+\mu^-) \approx 1/24$, the top quark must be heavier than 105 GeV in order to implement the job.

4. $K_L \rightarrow \pi^+\pi^- H$ and $K^+ \rightarrow l^+\nu H$ decays

Processes, e.g. $K \rightarrow \pi\pi H$ and $K \rightarrow l\nu H$ can in principle be used to constraint the existence of a light Higgs boson. Here we confine our attention to the decay modes, $K_L \rightarrow \pi^+\pi^- H$ and $K^+ \rightarrow l^+\nu H$ because of the availability of the experimental measurements of $K_L \rightarrow \pi^+\pi^- e^+ e^-$ and $K^+ \rightarrow e^+\nu e^+ e^-$.

From Eq.(2.1) it is clear that the decay $K_L \rightarrow \pi\pi H$ is prohibited in the limit of CP symmetry. Since the imaginary and real parts of the KM matrix element $V_{td}V_{ts}^*$ are comparable, the dominant contribution to the CP-violating $K_L \rightarrow \pi\pi H$ obviously arises from the t-quark loop diagram. From the chiral representation of the quark density[41]

$$\bar{q}_{Rj} q_{Li} = -\frac{1}{4} f_\pi^2 \sigma U_{ij} \qquad (4.1)$$

it follows[42]

$$\langle \pi^+\pi^- | \bar{s}\gamma_5 d | K^0 \rangle = -\langle \pi^+\pi^- | \bar{d}\gamma_5 s | \bar{K}^0 \rangle = -i\frac{2}{3}\frac{\sigma}{f_\pi} \qquad (4.2)$$

Consequently,

$$A(K_L \rightarrow \pi^+\pi^- H) \cong i\frac{\sqrt{2}\alpha}{16\pi \sin^2\theta_w}(\frac{m_K^2 - m_\pi^2}{f_\pi v})(\frac{m_t}{M_w})^2 \text{Im}(V_{td}V_{ts}^*)$$
$$\approx i(1.64 \pm 0.55) \times 10^{-9} \frac{(m_t}{M_w}^2) \qquad (4.3)$$

where use of Eq.(2.33) has been made. After integrating over the phase space we find

202

$$Br(K_L \rightarrow \pi^+\pi^- H) \lesssim (1.7 \pm 1.2) \times 10^{-6} (\frac{m_t}{m_W})^2 \tag{4.4}$$

for $m_H \neq 0$. Consequently, the current experimental upper bound[43] $Br(K_L \rightarrow \pi + \pi - e + e-) < 2.5 \times 10^{-6}$ is not strong enough to provide evidence against the existence of light Higgs bosons unless $m_t \gtrsim 100 \ GeV$.

We have also calculated the branching ratio for the process $K^+ \rightarrow e^+\nu H$. Since the pole contribution vanishes in the limit $m_e \rightarrow 0$ due to vector current conservation, Higgs production comes from the emission from the virtual W–boson and from the K–W vertex. We find[43]

$$Br(K^+ \rightarrow e^+\nu H) = \frac{\sqrt{2}G_F m_K^4}{96\pi^2 m_\mu^2 (1 - \frac{m_\mu^2}{m_K^2})^2} Br(K^+ \rightarrow \mu^+\nu)_{\text{expt}} f(x)$$
$$= 6.4 \times 10^{-8} f(x) \tag{4.5}$$

where

$$f(x) = [(1 - 8x + x^2)(1 - x^2) - 12x^2 \ln x](1 - \frac{2n_h}{33 - 2n_l})^2 \tag{4.6}$$

with $x = m_H^2/m_K^2$. The available experimental measurement[44] $Br(K^+ \rightarrow e^+\nu e^+e^-) = (2.1^{+2.1}_{-1.1}) \times 10^{-7}$ indicates that nothing can be learned from this decay mode. Similarly, for $\pi^+ \rightarrow e^+\nu H$ we obtain

$$Br(\pi^+ \rightarrow e^+\nu H) = 3.3 \times 10^{-9} f(y) \tag{4.7}$$

with $y = m_H^2/m_\pi^2$. It seems to us that the branching ratio obtained in ref.[7] is too large by a factor of two. Recently, the SINDRUM Collaboration[45] has obtained upper limits on the branching ratio $Br(\pi^+ \rightarrow e^+\nu H)$ ranging from 10^{-9} to 10^{-11} depending on the mass and the lifetime of the Higgs boson. A mass range $10 \ MeV < m_H < 100 \ MeV$ is clearly ruled out[45].

5. Conclusions

We have re-explored the Higgs boson production in the process $K \rightarrow \pi H$. Effects such as heavy quark contributions to the pole and nonspectator diagrams via the triangle diagram with external gluons and CP violating effects on the $K \rightarrow \pi H$ decay mode, which were not considered by previous calculations, are elaborated on.

We find the $K \rightarrow \pi H$ amplitude to be dominated by the t-quark contribution in the $s \rightarrow dH$ loop. Unfortunately, there is an undetermined parameter B in the chiral-Lagrangian description of the $K \rightarrow \pi H$ transition. Nevertheless, the imaginary contribution to the $K^+ \to \pi^+ H$ amplitude coming from the imaginary part of the Kobayashi-Maskawa mixing angles in $s \to dH$ is large. We conclude that even if B is nonzero and even if the real part of the $K^+ \to \pi^+ H$ amplitude is conceled accidently, the imaginary contribution alone with $m_t \geq 45\ GeV$ will suffice to rule out a Higgs boson lighter than $2m_\pi$. Whether the Higgs bosons with $2m_\pi < m_H < 350\ MeV$ can be excluded by the CP-violating contribution depends strongly on the branching ratio of $H \to \mu^+\mu^-$. For $Br(H \to \mu^+\mu^-) \approx 30\%$, a top quark heavier than 65 GeV will rule out a light Higgs boson in the above-mentioned mass range. But if $Br(H \to \mu^+\mu^-) \approx 1/24$, the top quark should not be lighter than 105 GeV in order to put constraints on the light Higgs boson in the mass range $2m_\pi < m_H < 350\ MeV$.

ACKNOWLEDGEMENT

This work was supported in part by the National Science Council of the Republic of China. We wish to thank R. S. Chivukula, Y. C. Lin and R. S. Willey for helpful discussions.

REFERENCES

1. A. D. Linde, *JETP Lett.* **23** (1976) 73; S. Weinberg, *Phys. Rev. Lett.* **36** (1976) 294.

2. ARGUS Collaboration, H. Albrecht *et al.*, *Phys. Lett.* **B192** (1987) 245;CLEO Collaboration, M. Artuso *et al.* CLNS-89/889(1988).

3. UA1 Collaboration, C. Albajar *et al.*, *Phys. Lett.* **B186** (1986) 247.

4. R.S. Willey and H.L. Yü, *Phys. Rev.* **D26** (1982) 3086.

5. CLEO Collaboration, P. Avery *et al.*, *Phys. Lett.* **B183** (1987) 429; Mark J Collaboration, B. Adeva *et al.*, *Phys. Rev. Lett.* **50** (1983) 799;TASSO Collaboration, M. Althoff *et al.*, *Z. Phys.* **C22** (1984) 219;JADE Collaboration, W. Bartel *et al.*, *Phys. Lett.* **B132** (1983) 241.

6. B. Grinstein, L. Hall and L. Randall, *Phys. Lett.* **B211** (1988) 363.

7. S. Raby, G.B. West and C.M. Hoffman, *Phys. Rev.* **D39** (1989) 828.

8. A.I. Vainshtein, V.I. Zakharov and M.A. Shifman, *Sov. Phys. Usp.* **23** (1980) 429.

9. R.S. Willey and H.L. Yu, *Phys. Rev.* **D26** (1982) 3287.

10. T.N. Pham and D.G. Sutherland, *Phys. Lett.* **B151** (1985) 444.

11. R.S. Willey, *Phys. Lett.* **B173** (1986) 480.

12. R. Ruskov, *Phys. Lett.* **B187** (1987) 165.

13. R.S. Chivukula and A.V. Manohar, *Phys. Lett.* **B207** (1988) 86;, **B217** (1989) (E)568.

14. R.S. Willey had presented to this workshop, a coupled channel analysis on the final state enhencement of $H \to \pi\pi$ and found no large final state enhencement.

15. See also R.S. Willey, *Phys. Rev.* **D39** (1989) 2784 for a recent updating limit on the existence of a light Higgs boson implied by rare K decays.

16. H.Y. Cheng, *J. Math. Phys.* **A4** (1989) 495.

17. F.J. Gilman and M.B. Wise, *Phys. Rev.* **D20** (1979) 2392;*ibid.* **D27**(1983)1128.

18. M.A. Shifman, A.I. Vainshtein and V.I. zakharov, *Phys. Lett.* **78B** (1978) 443.

19. In the literature Eq.(2.8) is often written as $\frac{1}{2}(a - b)(p_1 + p_2)^2$, which is valid only in both *chiral* and *soft − pion* limits. In the framework of current algrbra, Eq.(2.8) is derived by applying the soft-pion theorem and the commutator relation $[\theta^\mu{}_\mu, Q_5^\pm] = 0$ valid in the chiral limit.

20. S. Raby and G.B. West, *Phys. Rev.* **D38** (1988) 3488.

21. C. Bernard, T. Draper, A. soni, H. D. Politzer and M. Wise, *Phys. Rev.* **D32** (1985) 2343.

22. Even without CPS symmetry, total derivative chiral terms do not contribute to momentum-conserving processes involving only pseudoscalar mesons.

23. A. Manohar and H. Georgi, *Nucl. Phys.* **B234** (1984) 189.

24. To obtain Eq.(2.28) we note that the amplitude of Fig.1c or 1d has the form

$$-\frac{1}{v}(\frac{4n_h}{33 - 2n_l})\langle\pi^+ H|\mathcal{L}_{tree}|K^+\rangle$$

with

$$\mathcal{L}_{tree} = (G_F/\sqrt{2})V_{us}V_{ud}^* H \bar{s}\,\gamma^\mu(1 - \gamma_5)\mu\bar{\mu}\gamma_\mu(1 - \gamma_5)d + \text{h.c.}$$

being the effective $\Delta S = 1$ lagrangian at the electroweak scale. The evolution of the $\Delta S = 1$ weak Lagrangian from M_W down to the renormalization done in ref.[24] is actually our pole contribution.

25. R.S. Chivukula and A.V. Manohar, *Phys. Lett.* **B217** (1989) (E)568.

26. H. Haber had also demonstrated this point by an explict rotating in this workshop. See also SCIPP-19/14 preprint.

27. We adopt the convention of ref.[13] putting a minus sign in front of the parameter B.

28. Contributions of higer order chiral Lagrangians are of order $(p_K - p_\pi)^2/\Lambda_\chi^2 = m_H^2/\Lambda_\chi^2$.

29. L. Wolfenstein, *Phys. Rev. Lett.* **51** (1984) 1945.

30. H. Burkhardt *et al.*, *Phys. Lett.* **B206** (1988) 169.

31. H.Y. Cheng, IP-ASTP-10-88.

32. The first term in Eq.(2.37) is in agreement with ref.[25] obtained in a different method.

33. R.J. Cence *et al.*, *Phys. Rev.* **D10** (1974) 776.

34. P. Bloch *et al.*, *Phys. Lett.* **B56** (1975) 201.

35. BNL-787 Experiment, talk presented by D. Marlow at the I2 International Workshop on Weak Interactions and Neutrinos, Ginosar, Isreal, April 1989.

36. T. Yamazaki *et. al.*, *Phys. Rev. Lett.* **52** (1984) 1089.

37. Y. Asano *et. al.*, *Phys. Lett.* **B107** (1981) 159; **B113** (1982) 195.

38. A. S. Carroll *et. al.*, *Phys. Rev. Lett.* **44** (1980) 525.

39. NA31 Collaboration, talk presented by H.G. Sander at the XII International Workshop on Weak Interactions and Neutrinos, Ginosar, Israel, April 1989.

40. T.N. Truong and R.S. willey, PITT-89-05.

41. See *e.g.* H. Y. Cheng, *Phys. Rev.* **D36** (1987) 2056.

42. However, it should be stressed that $\langle \pi^+\pi^- | \bar{s}\gamma_5 d | K^0 \rangle = 0$ for the physical $K \to \pi^+\pi^-$ amplitude (see Sec.7.1 of ref.[15]).

43. Our result for $Br(K^+ \to e^+\nu H)$ is numerically different from the branching ratio obtained in ref.[13] by a factor of 1.5.

44. A. M. Diamant-Berger *et. al., Phys. Lett.* **B62** (1970) 485.

45. SINDRUM Collaboration, S. Egli *et. al., Phys. Lett.* **B222** (1989) 533.

DISPOSING OF THE LIGHT HIGGS BOSON:

THEORETICAL ISSUES IN $K \to \pi H$

Howard E. Haber

Santa Cruz Institute for Particle Physics
University of California, Santa Cruz, CA 95064

ABSTRACT

The absence of K decay into a pion and a scalar particle would rule out the existence of a Standard Model Higgs boson with mass less than about 350 MeV. I review aspects of the theoretical calculation of $K \to \pi H$ and briefly compare the theoretical expectation with the present experimental observations.

1. INTRODUCTION

The mass of the Higgs boson is not predicted in the Standard Model.[1] It is generally accepted that the Standard Model Higgs boson must have mass less than of order 1 TeV.[2] In addition, a lower limit for the Higgs mass of 7 GeV was obtained by Linde and Weinberg.[3] The latter limit was derived under the assumption that all fermion masses are much lighter than m_W. It is now known that the t-quark mass violates this assumption; the most recent measurement at the Tevatron by the CDF Collaboration finds that $m_t > 77$ GeV at 95% confidence level.[4] Allowing for a heavy t-quark with mass of order m_W, the modified Linde-Weinberg bound reads:

$$m_H^2 \gtrsim (7 \text{ GeV})^2 \left(1 - 1.09 \frac{m_t^4}{m_W^4} \right) . \tag{1}$$

For example, if $m_t \simeq 78$ GeV, the lower bound vanishes, and the Standard Model can support an arbitrarily light Higgs boson.* Furthermore, in models with more than one Higgs doublet, the Linde-Weinberg bound applies only to one diagonal element of the scalar Higgs boson squared-mass matrix. In such models, arbitrarily light Higgs bosons cannot be theoretically excluded.

⋆ For $m_t > 78$ GeV, the Linde-Weinberg bound does not apply. Other arguments can be invoked, based on renormalization group scaling and stability of the scalar potential, which imply once again the existence of a Higgs mass lower bound.[5]

As a result of the above considerations, experimentalists in search of the Higgs boson must conduct their searches over *all* mass ranges below 1 TeV. In this paper, I shall consider whether it is possible to rule out very light Higgs bosons with mass below 350 MeV. It is remarkable that as recently as last year (1988), the only reliable Higgs mass limits were $m_H < 11.5$ MeV based on nuclear physics experiments! [6,7] During the past two years, both experimentalists and theorists have been actively engaged in improving this limit. Searches for Higgs bosons in K-decay are sensitive to Higgs masses up to 350 MeV, and searches in B-decay and Υ-decay could extend the mass limit beyond 5 GeV. All Higgs searches rely on theoretical calculations in order to set mass limits if no evidence for Higgs bosons is seen. It is here where controversy has arisen, since in each case the theoretical computations have not been totally clean. Some theoretical assumptions must be made in order to come to a final conclusion.

In this paper, I shall focus on the theoretical calculation of $K \to \pi H$. I will emphasize the theoretical uncertainties involved in the calculation and problems which may arise in the comparison with experiment. Noting that the experimental searches for $K \to \pi H$ have (to date) all been negative, I will finally discuss to what extent these data can be used to rule out the existence of light Higgs boson with $m_H \lesssim 350$ MeV.

2. HIGGS BOSON LOW-ENERGY THEOREMS

In the decay $K \to \pi H$, the momenta of the final state particles are small compared to the scale of strong interactions and chiral symmetry breaking. Thus, low-energy theorems can be used to obtain the leading term in the $K \to \pi H$ amplitude. In this section, we review the low-energy theorems for Higgs boson interactions which have been studied in great detail by Vainshtein, Voloshin, Zakharov and Shifman [8,9,10]. These theorems relate the amplitudes of two processes which differ by the insertion of a zero momentum Higgs boson. In the next section, I will employ the method of chiral Lagrangians which automatically reproduces soft-pion and soft-kaon theorems. Combining the two approaches allows one to compute rigorously the leading low-momentum behavior of the $K \to \pi H$ amplitude.

The Higgs boson low-energy theorems can be derived by observing that the Higgs interactions in the standard electroweak theory is of the following form

$$\mathcal{L}_{\text{int}} = -\left(1 + \frac{H}{v}\right) \sum_f m_f \bar{f} f - \left(1 + \frac{H}{v}\right)^2 (m_W^2 W^{\mu+} W_\mu^- + \tfrac{1}{2} m_Z^2 Z_\mu Z^\mu), \quad (2)$$

where $v \equiv (\sqrt{2} G_F)^{-1/2} = 2m_W/g \simeq 246$ GeV, and the sum runs over all fermions f in the theory. Consider a Higgs field with zero four-momentum: $[P_\mu, H] = i\partial_\mu H = 0$. This implies that H is a constant field. From eq. (2), it follows that the effect of a constant field H is equivalent to redefining all mass parameters of the theory

$$m_i \to m_i \left(1 + \frac{H}{v}\right). \quad (3)$$

This immediately implies the following low-energy theorem [11,9]

$$\lim_{p_H \to 0} \mathcal{M}(A \to B + H) = \frac{1}{v} \left(\sum_f m_f \frac{\partial}{\partial m_f} + \sum_V m_V \frac{\partial}{\partial m_V} \right) \mathcal{M}(A \to B), \quad (4)$$

where the sum over V includes both the W and Z bosons. This theorem is rather trivial when applied to the elementary particles of the model. But its range of applicability is much wider. As a demonstration, let us derive a low-energy theorem for the Hgg interaction (g = gluon). At one-loop, the transition amplitude $\mathcal{M}(g \to g)$, which is just the gluon two-point function, does have m_f dependence, due to an intermediate quark loop. One can show that the effect of heavy fermion loops is to add the following piece to the effective QCD Lagrangian

$$\delta\mathcal{L} = \frac{-\alpha_s}{24\pi} \, G^a_{\mu\nu}G^{\mu\nu a} \sum_f \log\left(\frac{\Lambda^2_{UV}}{m^2_f}\right) , \tag{5}$$

where Λ_{UV} is the ultraviolet cutoff. Using eq. (4), the following effective Lagrangian governing the Higgs–gluon interaction is obtained

$$\mathcal{L}_{Hgg} = \frac{N_H \alpha_s}{12\pi v} \, HG^a_{\mu\nu}G^{\mu\nu a} , \tag{6}$$

where N_H is the number of heavy quark flavors. Here, "heavy" means that N_H is the number of quarks heavier than H and the scale of QCD (denoted by Λ_{QCD}).[*] As expected, eq. (6) gives precisely the same answer as the one obtains by directly computing the triangle diagram in a theory of quarks and gluons; namely, the Hgg matrix element is constant in the limit of $m_q \to \infty$.

Consider now the application of the low-energy theorems to the study of Higgs interactions with mesons at low energy. The mesons are complicated bound state systems made up of light quarks and gluons. The Higgs bosons can interact with these systems in three distinct ways: (i) interaction with the gluons via eq. (6); (ii) direct interactions with the light constituent quarks; and (iii) via a weak interaction process, where the quarks exchange a W or Z boson, and the Higgs interacts with the exchanged vector boson. It is convenient to develop low-energy theorems which separate out the interactions via gluons from the direct interactions with fermions and vector bosons. First, I divide up the quarks into "light" ($m_q < \Lambda_{QCD}$) and "heavy" ($m_q > \Lambda_{QCD}, m_H$). The heavy quarks are important in that they are responsible for the Hgg interaction; hence, we remove the heavy quarks from eq. (4). Then, I can derive a new low-energy theorem by observing that eq. (6) can be combined with the gluon kinetic energy term

$$\mathcal{L} = \frac{-1}{4g^2_s}(\partial_\mu A^a_\nu - \partial_\nu A^a_\mu - f_{abc}A^b_\mu A^c_\nu)^2 \left(1 - \frac{N_H \alpha_s}{3\pi v}H\right) , \tag{7}$$

where $\alpha_s \equiv g^2_s/(4\pi)$. Note that the gluon field above has been rescaled, $A^a_\mu \to g^{-1}_s A^a_\mu$. In the zero momentum limit where H is a constant field, it is apparent that the H

---

[*] Λ_{QCD} is expected to be of the order of a few hundred MeV. This means that the strange quark is neither heavy nor light. For simplicity, I shall henceforth regard the strange quark as being light.

interactions can be reproduced simply by rescaling α_s. Thus, if the corresponding change is denoted by $\alpha_s \to \alpha_s + \delta\alpha_s$, then to first order,

$$\delta\alpha_s = \frac{N_H \alpha_s^2}{3\pi v} H .$$

(8)

The following low-energy theorem is thereby obtained[10)]

$$\lim_{p_H \to 0} \mathcal{M}(A \to B + H)|_{\text{gluons}} = \frac{N_H \alpha_s^2}{3\pi v} \frac{\partial}{\partial\alpha_s} \mathcal{M}(A \to B) ,$$

(9)

where the subscript "gluons" indicates that I have exhibited the partial contribution to $\mathcal{M}(A \to B + H)$ due to the Hgg interactions induced by the heavy quark loops. By dimensional analysis, it is often possible to deduce the dependence of $\mathcal{M}(A \to B)$ on the intrinsic scale of QCD, Λ. In defininig Λ, I shall use the following normalization in the definition of the QCD β-function

$$\mu \frac{\partial\alpha_s}{\partial\mu} = \alpha_s \beta(\alpha_s) ,$$

(10)

with

$$\beta(\alpha_s) = \frac{-b\alpha_s}{2\pi} + \mathcal{O}(\alpha_s^2) ,$$

(11)

where $b \equiv 11 - \frac{2}{3} N_f$, and N_f is the number of quark flavors. Then, Λ is defined as[12)]

$$\Lambda = \mu \exp\left\{ -\int \frac{d\alpha_s}{\alpha_s \beta(\alpha_s)} \right\} .$$

(12)

Note that by using eq. (11), $d\Lambda/d\mu = 0$, which implies that Λ is a physical parameter of the theory. It then follows that

$$\frac{\partial\Lambda}{\partial\alpha_s} = \frac{-\Lambda}{\alpha_s \beta(\alpha_s)} .$$

(13)

Thus, in the one-loop approximation, I can replace eq. (9) with:

$$\lim_{p_H \to 0} \mathcal{M}(A \to B + H)|_{\text{gluons}} = \frac{2N_H}{3bv} \Lambda \frac{\partial}{\partial\Lambda} \mathcal{M}(A \to B) ,$$

(14)

where b should be computed in a theory where the heavy flavors are decoupled, namely $b = 11 - \frac{2}{3} n_L$, with n_L being the number of light flavors. Since I have taken the charmed quark to be the lightest of the "heavy" quarks, the error made in the analysis by using the one-loop approximation for the β-function is the neglect of terms of order $\alpha_s(m_c)$, i.e. α_s evaluated at the last heavy quark which has been decoupled.[13)]

The remaining contributions due to the interactions with the light constituent quarks and weak vector bosons are obtained by deleting the heavy quarks from eq. (4). Thus, the final low-energy theorem is:

$$\lim_{p_H \to 0} \mathcal{M}(A \to B + H)|_{q,W,Z} = \frac{1}{v}\left(\sum_q m_q \frac{\partial}{\partial m_q} + \sum_V m_V \frac{\partial}{\partial m_V}\right)\mathcal{M}(A \to B), \quad (15)$$

where the sum runs over the *light* quarks and $V = W$ and Z.

3. HIGGS BOSON INTERACTIONS AND THE CHIRAL LAGRANGIAN

I shall now apply the Higgs low-energy theorems to the study of low energy interactions of a light Higgs boson and the pseudoscalar mesons. The presentation here follows work done in collaboration with Sally Dawson.[14] I will employ the chiral Lagrangian approach; my notation is that of ref. 15. To briefly review the approach, consider the chiral Lagrangian which describes $\Delta S = 0$ and $\Delta S = 1$ processes involving the pseudoscalar mesons. Keeping only the terms with the fewest number of derivatives,

$$
\begin{aligned}
\mathcal{L} = {} & \tfrac{1}{4}f^2 \operatorname{Tr} \partial^\mu \Sigma \partial_\mu \Sigma^\dagger + \tfrac{1}{2}f^2 \left[\operatorname{Tr} \mu M \Sigma^\dagger + \text{h.c.}\right] \\
& + \tfrac{1}{4}f^2 \left[\lambda \operatorname{Tr} h \partial^\mu \Sigma \partial_\mu \Sigma^\dagger + \text{h.c.}\right] \\
& + \tfrac{1}{2}f^2 \left[A \operatorname{Tr} h\mu(M\Sigma^\dagger + \Sigma M) + \text{h.c.}\right] \\
& + \tfrac{1}{4}f^2 \left[a T_{k\ell}^{ij}(\Sigma \partial^\mu \Sigma^\dagger)_i^k (\Sigma \partial_\mu \Sigma^\dagger)_j^\ell + \text{h.c.}\right],
\end{aligned}
\quad (16)
$$

where $f \equiv f_\pi = 93$ MeV, $\Sigma \equiv \exp(2i\Pi^a T^a/f)$, with the SU(3) generators normalized to $\operatorname{Tr} T^a T^b = \tfrac{1}{2}\delta_{ab}$ and $\Pi \equiv \Pi^a T^a$ denoting the pseudoscalar octet

$$\Pi = \frac{1}{\sqrt{2}}\begin{pmatrix} \pi^0/\sqrt{2} + \eta/\sqrt{6} & \pi^+ & K^+ \\ \pi^- & -\pi^0/\sqrt{2} + \eta/\sqrt{6} & K^0 \\ K^- & \overline{K}^0 & -2\eta/\sqrt{6} \end{pmatrix}, \quad (17)$$

and M is the quark mass matrix. The matrix h appears in a $\Delta S = 1$ term which transforms as $(8,1)$ under $SU(3)_L \times SU(3)_R$; it is a pure $\Delta I = 1/2$ operator. The tensor T appears in an operator which transforms as $(27,1)$; it contains both $\Delta I = 1/2$ and $\Delta I = 3/2$ pieces. Specifically,

$$M = \begin{pmatrix} m_u & 0 & 0 \\ 0 & m_d & 0 \\ 0 & 0 & m_s \end{pmatrix}, \qquad h = \begin{pmatrix} 0 & 0 & 0 \\ 0 & 0 & 1 \\ 0 & 0 & 0 \end{pmatrix}, \quad (18)$$

and T is a traceless tensor, symmetric in its upper and lower indices. For example, the nonzero components of T corresponding to $\Delta S = 1$ and $\Delta I = 3/2$ are

$$T_{13}^{12} = T_{13}^{21} = T_{31}^{12} = T_{31}^{21} = -T_{23}^{22} = -T_{32}^{22} = \tfrac{1}{2}. \quad (19)$$

If CP-violating effects are neglected, then the coefficients λ, A and a are real.

The parameter μ is adjusted in order that the meson masses come out correctly. For example, if we focus on the charged meson sector, then the chiral Lagrangian takes the form

$$
\begin{aligned}
\mathcal{L} = {} & \partial^\mu \pi^+ \partial_\mu \pi^- + \partial^\mu K^+ \partial_\mu K^- + \tfrac{1}{2}\lambda(\partial^\mu \pi^+ \partial_\mu K^- + \partial^\mu \pi^- \partial_\mu K^+) \\
& - m_\pi^2 \pi^+ \pi^- - m_K^2 K^+ K^- - m_{K\pi}^2(\pi^+ K^- + \pi^- K^+) + \text{interactions} ,
\end{aligned}
\tag{20}
$$

where

$$
\begin{aligned}
m_\pi^2 &= \mu(m_u + m_d) \\
m_K^2 &= \mu(m_u + m_s) \\
m_{K\pi}^2 &= \tfrac{1}{2}A\mu(m_d + m_s) .
\end{aligned}
\tag{21}
$$

At this point, one can remove λ and $m_{K\pi}^2$ from the quadratic part of the Lagrangian [eq. (20)] by performing wave function renormalization and diagonalizing the pseudoscalar mass matrix, respectively. Wave function renormalization in this context means diagonalization of the pseudoscalar kinetic energy, followed by a rescaling of the fields to obtain canonically normalized terms. To diagonalize the pseudoscalar mass matrix, I perform an $\mathrm{SU}(3)_L \times \mathrm{SU}(3)_R$ transformation, $\Sigma \rightarrow L\Sigma R^\dagger$, and find

$$
\mathrm{Tr}[1 + Ah + A^* h^\dagger]\mu M \Sigma^\dagger \rightarrow \mathrm{Tr}\, \mu M_D \Sigma^\dagger ,
\tag{22}
$$

where

$$
M_D = L^\dagger[1 + Ah + A^* h^\dagger]MR
\tag{23}
$$

is the diagonalized mass matrix. Since I shall always work to first order in λ and A, this rotation has no affect on the other (interaction) terms of the Lagrangian. Thus, I can simply set $m_{K\pi} = 0$. On the other hand, when wave function renormalization is performed (as described above), interaction terms are generated which depend on λ. Thus, λ is a physical parameter, which can be measured in the $\Delta S = 1$, $\Delta I = 1/2$ weak interactions of pseudoscalar mesons. Fitting to data, Cohen and Manohar[16] find* $|\lambda| = 3.2 \times 10^{-7}$. They also have fit the parameter a to the measured data on $\Delta I = 3/2$ transitions, and find $|a| = 1.0 \times 10^{-8}$. Thus, for most purposes, the $\Delta I = 3/2$ piece can be neglected.

Let us now consider coupling the Higgs boson into this system. By the rules of the chiral Lagrangian technique, the most general coupling (with the fewest derivatives) is[17]

$$
\begin{aligned}
\Delta\mathcal{L} = {} & \tfrac{1}{4}f^2 c_1 \frac{H}{v} \mathrm{Tr}\, \partial^\mu \Sigma \partial_\mu \Sigma^\dagger + \tfrac{1}{2}f^2 c_2 \frac{H}{v} \left[\mathrm{Tr}\, \mu M \Sigma^\dagger + \text{h.c.} \right] \\
& + \tfrac{1}{4}f^2 c_3 \frac{H}{v} \left[\lambda \, \mathrm{Tr}\, h \partial^\mu \Sigma \partial_\mu \Sigma^\dagger + \text{h.c.} \right] \\
& + \tfrac{1}{2}f^2 c_4 \frac{H}{v} \left[A \, \mathrm{Tr}\, h\mu(M\Sigma^\dagger + \Sigma M) + \text{h.c.} \right] \\
& + \tfrac{1}{4}f^2 c_5 \frac{H}{v} \left[a T^{ij}_{k\ell}(\Sigma \partial^\mu \Sigma^\dagger)^k_i (\Sigma \partial_\mu \Sigma^\dagger)^\ell_j + \text{h.c.} \right] ,
\end{aligned}
\tag{24}
$$

---

* The sign of λ is not directly determined by current experiment. However, since λ is implicitly proportional to $V^*_{ud}V_{us} \equiv \cos\theta_c \sin\theta_c$, the sign of λ is well-defined once the usual convention is adopted where $\sin\theta_c \approx 0.22$ is positive. Then, if one matches the chiral Lagrangian to the $\Delta S = 1$ weak Lagrangian of the full theory at 1 GeV, one finds that λ is *negative* in the vacuum insertion approximation.

where the c_i are strong interaction real parameters which cannot be fixed on the basis of chiral symmetry alone. In fact, these parameters can be determined by applying the Higgs low-energy theorems, as I will show later in this section. When studying the full Lagrangian $\mathcal{L} + \Delta\mathcal{L}$ [eqs. (16) and (24)], one must again perform wave function renormalization and diagonalize the pseudoscalar mass matrix, as described earlier. Working to first order in the small parameters μ/v, λ, A and a, I find

$$
\begin{aligned}
\mathcal{L} = {} & \tfrac{1}{4}f^2\left(1 + c_1\frac{H}{v}\right)\mathrm{Tr}\,\partial^\mu\Sigma\partial_\mu\Sigma^\dagger + \tfrac{1}{2}f^2\left(1 + c_2\frac{H}{v}\right)\left[\mathrm{Tr}\,\mu M\Sigma^\dagger + \mathrm{h.c.}\right] \\
& + \tfrac{1}{4}f^2(c_3 - c_1)\frac{H}{v}\left[\lambda\,\mathrm{Tr}\,h\partial^\mu\Sigma\partial_\mu\Sigma^\dagger + \mathrm{h.c.}\right] \\
& + \tfrac{1}{2}f^2(c_4 - c_2)\frac{H}{v}\left[A\,\mathrm{Tr}\,h\mu(M\Sigma^\dagger + \Sigma M) + \mathrm{h.c.}\right] \\
& + \tfrac{1}{4}f^2 c_5\frac{H}{v}\left[aT^{ij}_{k\ell}(\Sigma\partial^\mu\Sigma^\dagger)^k_i(\Sigma\partial_\mu\Sigma^\dagger)^\ell_j + \mathrm{h.c.}\right] ,
\end{aligned}
\tag{25}
$$

where I have written M for the diagonal mass matrix M_D [see eqs. (22) and (23)]. We see explicitly that the term proportional to A is not completely removed by an $\mathrm{SU}(3)_L \times \mathrm{SU}(3)_R$ transformation, unlike in the case in which no Higgs field is present. That is, A is a *new* parameter which arises when Higgs boson interactions are incorporated, and thus it cannot be determined from experiment at present.

Before proceeding to apply eq. (25) to obtain the amplitude for $K \to \pi H$, a comment should be made on the generality of the chiral Lagrangian being used here. I have restricted the above considerations to terms with at most two derivatives; nevertheless, some terms have been apparently left out of eq. (24)

$$
\begin{aligned}
&(i)\ \frac{\partial_\mu H}{v}\,\mathrm{Tr}\,\Sigma\partial^\mu\Sigma^\dagger , \\
&(ii)\ \tfrac{1}{2}f^2\frac{H}{v}\left[A'\,\mathrm{Tr}\,h\mu(M\Sigma^\dagger - \Sigma M) + \mathrm{h.c.}\right] \\
&(iii)\ \tfrac{1}{4}f^2\frac{\partial_\mu H}{v}\left[B\lambda\,\mathrm{Tr}\,h\Sigma\partial^\mu\Sigma^\dagger + \mathrm{h.c.}\right] .
\end{aligned}
\tag{26}
$$

Term (i) above is actually identically equal to zero, since $\Pi^a\,\mathrm{Tr}\,T^a = 0$. Terms (ii) and (iii) violate CPS symmetry,[18] which is a discrete symmetry that combines CP and the interchange of s and d quarks. The CPS symmetry is respected (in the chiral symmetry limit) by all quark operators of the full effective $\Delta S = 1$ electroweak Lagrangian (including gluonic corrections), even when CP-violation is taken into account. [19] Since chiral symmetry breaking in the fundamental theory is due to $\sum_q m_q q\bar{q}$, I can still make use of the CPS symmetry to restrict possible terms in the chiral Lagrangian if m_s and m_d are interchanged when performing the S symmetry operation. For example, an explicit computation of the quadratic terms contained in term (ii) above results in a CPS-even expression multiplied by m_s–m_d. Thus, assuming that A' and B are mass independent constants, I can thus omit terms (ii) and (iii) above from the chiral Lagrangian.

In the discussion above, I noted that if no Higgs field is present, then the term proportional to A can be removed by an $\mathrm{SU}(3)_L \times \mathrm{SU}(3)_R$ rotation. This is no longer the case when $\Delta\mathcal{L}$ [eq. (24)] is included. However, one could remove the A term completely by performing an appropriate *H-dependent* $SU(3)_L\times SU(3)_R$ transformation, $\Sigma \to L\Sigma R^\dagger$, in eq. (25). Clearly, L and R will depend on the coordinate (through $H(x)$); hence the kinetic energy term in eq. (25) will not be invariant under this transformation. (Other terms with derivatives are also not invariant, but the extra terms generated are second order in the small parameters and thus negligible.) It is easy to compute L and R; to first order I find

$$
\begin{aligned}
L &\simeq I + \left(1 + \frac{(c_4 - c_2)H}{v}\right)\frac{m_s^2 + m_d^2}{m_s^2 - m_d^2}\left(Ah - A^*h^\dagger\right), \\
R &\simeq I + \left(1 + \frac{(c_4 - c_2)H}{v}\right)\frac{2m_s m_d}{m_s^2 - m_d^2}\left(Ah - A^*h^\dagger\right),
\end{aligned}
\tag{27}
$$

where I is the 3×3 identity matrix and I have assumed that $(m_s + m_d)|A| \ll |m_s - m_d|$. The extra terms in $\mathcal{L}$ which are generated are

$$
\begin{aligned}
\mathcal{L}_{\text{extra}} = \frac{1}{4}\frac{\partial_\mu H}{v} \\
\times \left[B_1\lambda\,\text{Tr}\,h(\Sigma\partial^\mu\Sigma^\dagger + \Sigma^\dagger\partial^\mu\Sigma) + B_2\lambda\,\text{Tr}\,h(\Sigma\partial^\mu\Sigma^\dagger - \Sigma^\dagger\partial^\mu\Sigma) + \text{h.c.}\right]
\end{aligned}
\tag{28}
$$

with[*]

$$
\begin{aligned}
\lambda B_1 &= A(c_4 - c_2)\left(\frac{m_s + m_d}{m_s - m_d}\right), \\
\lambda B_2 &= A(c_4 - c_2)\left(\frac{m_s - m_d}{m_s + m_d}\right).
\end{aligned}
\tag{29}
$$

The ratio of quark masses can be rewritten in terms of meson masses by using eq. (21), with $m_u \approx m_d$. Using the arguments analogous to those below eq. (26), it is clear that $\mathcal{L}_{\text{extra}}$ is CPS-even since the coefficients B_1 and B_2 are odd under interchange of m_d and m_s. This is no surprise since these terms have been generated from the CPS-even "A term". If I now expand out the two terms proportional to B_1 and B_2, I find that the former contains only even powers of the pseudoscalar field Π, while the latter contains only odd powers of Π. Thus, if one wishes to parameterize the amplitude for $K \to \pi H$, it is a matter of convenience whether one works with an A-term or a B-term; between them there is only one independent parameter.[†] Furthermore, these parameters are nonzero since $c_2 \neq c_4$ (as I will now demonstrate) and hence *cannot be rotated away*.

Finally, I turn to the parameters $c_1, \ldots, c_5$. These parameters cannot be determined by chiral symmetry. However, they can be determined by applying the Higgs low-energy theorems [in particular eqs. (9) and (15)] discussed above [20-22]. To apply

---

[*] Here, I have used $\Sigma^\dagger\partial^\mu\Sigma = -(\partial^\mu\Sigma^\dagger)\Sigma$, which follows from taking the divergence of the equation, $\Sigma^\dagger\Sigma = 1$.

[†] It is important to appreciate that B_2 is predicted once B_1 is known [see eq. (29)] which is a consequence of the CPS symmetry of the quark operators of the underlying theory.

these equations, one must investigate the m_q, m_W and α_s dependence of the terms appearing in eq. (16). It is easy to argue that

$$f, \mu \sim \Lambda; \qquad M \sim m_q; \qquad A, \lambda, a \sim \frac{\Lambda^2}{m_W^2}, \tag{30}$$

where the α_s dependence of Λ has been given in eq. (12). In the case of the dimensionless parameters (A, λ, a), the m_W^{-2} behavior is clear, since these terms arise from the charged-current weak interactions. By dimensional analysis, the Λ^2 behavior of A, λ and a can be ascertained. The Higgs boson low-energy theorems of section 2 can now be applied. From eq. (14), it follows that a term in eq. (16) which behaves like Λ^p will be multiplied by

$$\left(1 + \frac{2pN_H}{3b} \frac{H}{v}\right). \tag{31}$$

From eq. (15), it follows that all terms in eq. (16) which are proportional to the quark mass matrix will be multiplied by $(1 + H/v)$, and those terms proportional to A, λ or a will be multiplied by $(1 - 2H/v)$. (It is sufficient to work to leading order in H/v.) The latter factor is easily understood as being the result of summing a geometric series of multiple Higgs emission from the exchanged W. This leads to a factor of $(1 + H/v)^{-2}$, which is the expected result. Combining these results, I find

$$\begin{aligned}
c_1 &= \frac{4N_H}{3b}, \\
c_2 &= 1 + \frac{2N_H}{b}, \\
c_3 &= c_5 = -2 + \frac{8N_H}{3b}, \\
c_4 &= -1 + \frac{10N_H}{3b}.
\end{aligned} \tag{32}$$

The values of $c_1, \ldots, c_5$ have also been obtained in refs. 17 and 22 (I have corrected their value of c_4). The above results also disagree slightly with those of ref. 21.

In the above derivation, there is one sleight of hand which needs to be justified. Namely, there is a factor of f hidden in the definition of $\Sigma = \exp(2i\Pi^a T^a / f)$. Nevertheless, even though I stated that $f \sim \Lambda$, this dependence was neglected in Σ when we applying the low-energy theorem [eq. (14)]. In fact, one can show that this is the correct procedure [23]. The low-energy theorems of section 2 can be recast into a theorem involving the divergence of the scale current [11,24,22]

$$\theta_\mu^\mu = \partial_\mu s^\mu = -\left(\Lambda \frac{\partial}{\partial \Lambda} + \sum_{f=u,d,s} m_f \frac{\partial}{\partial m_f} + \sum_V m_V \frac{\partial}{\partial m_V}\right) \mathcal{L}, \tag{33}$$

where I have used the same technique as discussed below eq. (4) to replace the effects of the heavy quarks by $\Lambda \, \partial / \partial \Lambda$. However, as shown in ref. 13, the field Σ (and likewise the Π field) do not scale under a dilatation transformation; i.e., they have scale dimensions equal to zero. It follows that for the leading term of the chiral Lagrangian, $\mathcal{L} = \frac{1}{4}f^2 \operatorname{Tr} \partial^\mu \Sigma \partial_\mu \Sigma^+$, one can derive $\partial_\mu s^\mu = -2\mathcal{L}$. As a result, for consistency, we must use $\partial\Sigma/\partial\Lambda = 0$ in the low-energy theorem, which is the justification for the procedure described above.

4. CALCULATION OF THE AMPLITUDE FOR $K \to \pi H$

Before using the formalism of section 3, it is useful to consider the various mechanisms for $K \to \pi H$ on the more fundamental quark level. There are two classes of diagrams to consider. First, the direct quark decay $s \to dH$ generates a Higgs coupling to an effective two-quark operator. Second, there are contributions in which the Higgs boson couples to effective four-quark operators. An example of such a process is $u\bar{s} \to u\bar{d}H$, where H is emitted from the exchanged W. The process $s \to dH$ has been computed by several groups,[25-28,24,17] and can be summarized by the effective Lagrangian

$$\mathcal{L}_{sdH} = \frac{H}{2v} \left[\zeta m_s \bar{d}(1 + \gamma_5)s + \zeta^* m_d \bar{s}(1 + \gamma_5)d + \text{h.c.} \right] , \tag{34}$$

where

$$\zeta = \frac{3\alpha}{16\pi m_W^2 \sin^2 \theta_W} \sum_i m_i^2 V_{id}^* V_{is} , \tag{35}$$

and the sum is taken over up-type quarks (u,c,t). The m_i and V_{ij} are the corresponding quark masses and Kobayashi-Maskawa mixing angles. When one evaluates $\langle \pi | \mathcal{L}_{sdH} | K \rangle$, the γ_5 pieces give no contribution. Thus, following ref. 17, one may simply make the standard chiral Lagrangian replacement

$$\frac{1}{2} \sum_{ij} [N_{ij} \bar{\psi}_i \psi_j + \text{h.c.}] \to -\frac{1}{2}f^2 \operatorname{Tr} \mu N \Sigma^\dagger + \text{h.c.}, \tag{36}$$

to represent the two-quark operator contribution to $K \to \pi H$. Thus, in the chiral Lagrangian, the effect of the two-quark operator is[*]

$$\mathcal{L}_{2q} = -\frac{1}{2}f^2 \frac{H}{v} \left[\zeta \operatorname{Tr} h\mu(M\Sigma^\dagger + \Sigma M) + \text{h.c.} \right] , \tag{37}$$

where the matrices M and h are defined in eq. (18). Note that this term is precisely of the same form as the "A-term" of eq. (24).

Next, let us consider the effects of the four-quark operators. Here, I will simply invoke the full chiral Lagrangian given in eq. (25). For convenience, I will keep separate the contributions of the two-quark and four-quark operators. That is, I add eqs. (25) and (37), so that A (as well as λ and a) is by definition a consequence of the four-quark

---

[*] QCD corrections can be incorporated by using running masses defined at the chiral Lagrangian scale[17].

operators. It is then straightforward to extract the full amplitude for $K \to \pi H$. The result is

$$\mathcal{M}(K^- \to \pi^- H) = \mathcal{M}^*(K^+ \to \pi^+ H)$$

$$= \frac{-\lambda(m_K^2 + m_\pi^2 - m_H^2)}{2v}\left(1 - \frac{2N_H}{3b}\right)$$

$$+ \frac{Am_K^2}{v}\left(1 - \frac{2N_H}{3b}\right) + \frac{m_K^2\zeta}{2v}, \qquad (38)$$

$$\mathcal{M}(K_L^0 \to \pi^0 H) = -\operatorname{Re}\mathcal{M}(K^- \to \pi^- H),$$

$$\mathcal{M}(K_S^0 \to \pi^0 H) = \frac{im_K^2}{2v}\operatorname{Im}\left[2A\left(1 - \frac{2N_H}{3b}\right) + \zeta\right].$$

In deriving these results, I have omitted the effect of the $\Delta I = 3/2$ contribution [the term proportional to a in eq. (25)]. I have also neglected the effects of ϵ (which measures the CP-violation in the physical states of the K^0–$\overline{K^0}$ system) and Im λ (which is related to ϵ'/ϵ), as both these effects are numerically small. It is interesting to note that if A is of order λ, then all the terms in eq. (38) are of order 10^{-10}, and I would then expect $BR(K \to \pi H) \sim 10^{-5}$. However, a precise prediction cannot be made without precise knowledge of the value of the parameter A (with some additional uncertainty due to presently unknown t-quark mass and mixing angles).

Even in the absence of knowledge of the value of A, there is some theoretical information which can be extracted. Consider the CP-violating effects in eq. (38). I argued above that it was numerically safe to neglect ϵ and Im λ. Nevertheless the CP-violating effects can be significant because Im ζ can be phenomenologically important, as first suggested in ref. 29. The key point to observe is the presence of m_t^2 in the definition of ζ in eq. (35). In particular, the top-quark contribution is greatly enhanced in the sum, and this allows Im ζ/Re ζ to be of $\mathcal{O}(1)$. The m_t^2 amplitude growth is a feature which is unique to the two-quark operator. Simple dimensional reasoning suggests that the contributions to the amplitude from the four-quark operators can at best approach a constant in the large m_t limit. It follows that the ratio of the imaginary to the real part of the four-quark amplitudes can be no larger than (roughly) Im $V_{td}^* V_{ts}/(V_{ud}V_{us})$ which is of order 10^{-3}. Hence, Im $A \sim \mathcal{O}(10^{-3})$ Re A. That is, the dominant contribution of the CP-violating part of the amplitude for $K^+ \to \pi^+ H$ (and for $K_S^0 \to \pi^0 H$) is proportional to Im ζ.

5. IMPLICATIONS FOR RARE K-DECAY EXPERIMENTS

Finally, I shall briefly extract numbers from the above predictions and indicate the implications of current rare K-decay experiments. First, consider charged kaon decay. For $N_H = 3$ heavy quarks and $b = 11 - 2n_L = 9$ (for $n_L = 3$ light quarks), the results of the previous section yield:

$$\mathcal{M}(K^- \to \pi^- H) = G_F^{1/2} 2^{1/4}\left[\frac{7Am_K^2}{9} - \frac{7\lambda(m_K^2 + m_\pi^2 - m_H^2)}{18} + \tfrac{1}{2}m_K^2\zeta\right], \qquad (39)$$

where ζ is defined in eq. (35). In the absence of information about A, one can obtain a lower limit by setting the real part of the total amplitude to zero (which would happen for some choice of the parameter A). Since the real parts of A, λ and ζ should be of the same order, the arguments presented at the end of the last section imply that the term proportional to Im ζ is the dominant part of Im $\mathcal{M}$, which therefore cannot be accidentally small due to a cancellation among terms. It is convenient to use the Wolfenstein parametrization[30] of the Kobayashi-Maskawa matrix, where

$$V_{ud}^* V_{us} \simeq -V_{cd}^* V_{cs} \simeq \sin\theta_c \cos\theta_c$$
$$V_{td}^* V_{ts} \simeq -A_w^2 \sin^5\theta_c(1 - \rho + i\eta). \tag{40}$$

Experimentally, $\sin\theta_c \simeq 0.22$, the parameter A_w is close to unity ($A_w = 1.05 \pm 0.17$ according to ref. 31), and $\rho \lesssim 0$ (based on the observed B-$\overline{B}$ mixing). In addition, values in the range $0.1 \lesssim \eta \lesssim 0.6$ have been obtained in the literature [29,32]. In the analysis below, I shall take a somewhat conservative value, $\eta = 0.2$. Thus, keeping only the term proportional to Im ζ in the total amplitude,

$$|\text{Im}\,\mathcal{M}(K^\pm \rightarrow \pi^\pm H)| \gtrsim \frac{G_F^{3/2}}{2^{1/4}} \frac{3 m_K^2 m_t^2 \eta A_w^2 \sin^5\theta_c}{16\pi^2} \gtrsim 7.9 \times 10^{-11} \text{ GeV}, \tag{41}$$

where I have taken $m_t \approx 80$ GeV, as required in order to have a very light Standard Model Higgs boson [see eq. (1)]. Using $\Gamma(K \rightarrow \pi H) = B_H |\mathcal{M}|^2/(16\pi m_K)$, where $B_H = 2 p_H/m_K$, and normalizing to the total decay rate, $\Gamma(K^\pm) = 5.32 \times 10^{-17}$ GeV, I find

$$BR(K^\pm \rightarrow \pi^\pm H) \gtrsim 4.3 \times 10^{-6} B_H. \tag{42}$$

Let us next turn to the decay $K_L^0 \rightarrow \pi^0 H$. The one important change in the analysis results from the fact that the matrix element for this process is real [see eq. (38)]. Thus, one cannot obtain such a definitive theoretical bound for $K_L^0 \rightarrow \pi^0 H$ without the explicit knowledge of A, since (unlike in the case above) there exists a particular value of A for which this amplitude vanishes. Nevertheless, it is useful to examine the branching ratio as a function of A (which is presumably of order λ) and compute the decay width as a function of $m_t^2 |\text{Re}\, V_{td}^* V_{ts}|$. The resulting width is normalized with respect to the total width $\Gamma(K_L^0) = 1.27 \times 10^{-17}$ GeV. According to the discussion above, I expect $m_t^2 |\text{Re}\, V_{td}^* V_{ts}| \gtrsim 2.5$ GeV^2 (for $m_t \approx 80$ GeV), with large uncertainty. In fig. 1, $BR(K_L^0 \rightarrow \pi^0 H)$ vs. $m_t^2 |\text{Re}\, V_{td}^* V_{ts}|$ is plotted for three different values of A. We see that only for a very narrow range of $m_t^2 |\text{Re}\, V_{td}^* V_{ts}|$, and only for positive values of A does the branching ratio for $K_L^0 \rightarrow \pi^0 H$ ever fall below 10^{-5}. Therefore, barring a very unlikely conspiracy of parameters, the predicted branching ratio satisfies

$$BR(K_L^0 \rightarrow \pi^0 H) \gtrsim 10^{-5} B_H. \tag{43}$$

Much of the above uncertainty can be eliminated if a reliable calculation of A becomes available. In this regard, I would like to cite a recent computation by Leutwyler and Shifman[33] in which the parameter A is estimated using the vacuum saturation approximation in the evaluation of certain hadronic matrix elements. They find a rather small value for A, roughly $A \sim -0.1\lambda$ (with about a factor of two

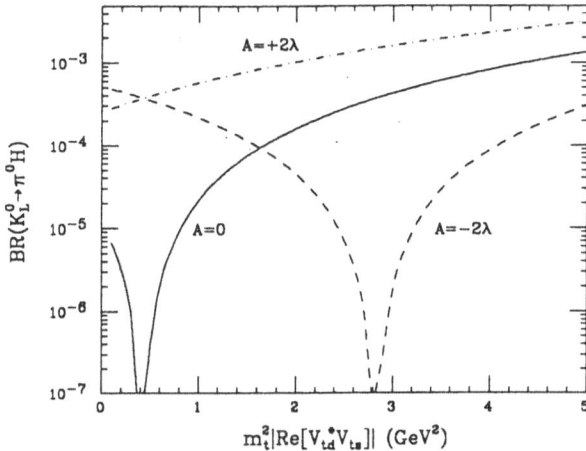

Fig. 1. The $K_L^0 \rightarrow \pi^0 H$ branching ratio as a function of $m_t^2 |\text{Re}\, V_{td}^* V_{ts}|$, assuming $m_H \ll m_K$. The three curves shown are: $A = 0$ (solid), $A = 2\lambda$ (dot-dashed), and $A = -2\lambda$ (dashes), where $\lambda = -3.2 \times 10^{-7}$.

uncertainty). Thus, numerically, the kaon decay amplitudes are well approximated by simply neglecting A! This leads to predictions for $BR(K \rightarrow \pi H)$ somewhat larger than the estimates quoted above. It should also be noted that as the limits on the t-quark mass become larger, the relative importance of ζ increases, in which case the uncertainty in A (and λ) becomes moot.

A report on Higgs searches in kaon decays has been presented at this meeting by M.E. Zeller. Sally Dawson, Jack Gunion and I have also summarized the recent searches for the Higgs boson in rare K-decays in ref. 34. Using the theoretical results quoted above, we concluded that a Standard Model Higgs boson with $m_H < 320$ MeV can presently be excluded. Although some theoretical uncertainty remains, we are convinced that the existence of a very light Standard Model Higgs boson is unlikely. Certainly, the improved limits from rare K-decay experiments now in progress (and additional information from the search for Higgs bosons in B-decay) should further reduce the impact of the remaining theoretical uncertainties.

ACKNOWLEDGMENTS

I gratefully acknowledge Sally Dawson and Jack Gunion for a very enjoyable and fruitful collaboration as we tried to dispose of the light Higgs boson. In addition, conversations with C. Bernard, A. Cohen, D. Kaplan, A. Manohar, A. Nelson, L. Randall, S.D. Rindani, M. Shifman and M. Voloshin were vital to my understanding of some of the theoretical issues described here. Finally, I greatly appreciate the hospitality of CCSEM and applaud A. Ali for an excellent workshop and an enjoyable ten days in Erice. This work was supported in part by the Department of Energy.

REFERENCES

1. For a comprehensive review of Higgs boson physics, see J.F. Gunion, H.E. Haber, G.L. Kane and S. Dawson, *The Higgs Hunter's Guide*, (Addison-Wesley Publishing Company, Redwood City, CA, 1989).

2. D.A. Dicus and V.S. Mathur, *Phys. Rev.* **D7** (1973) 3111; B.W. Lee, C. Quigg and G.B. Thacker, *Phys. Rev. Lett.* **38** (1977) 883; *Phys. Rev.* **D16** (1977) 1519.

3. A.D. Linde, *JETP Lett.* **23** (1976) 64; *Phys. Lett.* **62B** (1976) 435; S. Weinberg, *Phys. Rev. Lett.* **36** (1976) 294.

4. F. Abe *et al.*, preprint UPR-0172E (1989), to be published in *Physical Review Letters*.

5. N. Cabibbo, L. Maiani, G. Parisi, R. Petronzio, *Nucl. Phys.* **B158** (1979) 295; E. Gross and E. Duchovni, *Phys. Rev.* **D38** (1988) 2308; M.J. Duncan, R. Philippe, and M. Sher, *Phys. Lett.* **153B** (1985) 165; M. Lindner, M. Sher and H.W. Zaglauer, *Phys. Lett.* **228B** (1989) 139.

6. S.J. Freedman, J. Napolitano, J. Camp, and M. Kroupa, *Phys. Rev. Lett.* **52** (1984) 240.

7. S. Raby, G. West and C.M. Hoffman, *Phys. Rev.* **D39** (1989) 828.

8. A.I. Vainshtein, M.B. Voloshin, V.I. Zakharov, and M.A. Shifman, *Sov. J. Nucl. Phys.*, **30** (1979).

9. A.I. Vainshtein, V.I. Zakharov, and M.A. Shifman, *Sov. Phys. Usp.* **23** (1980) 429.

10. M. B. Voloshin, *Sov. J. Nucl. Phys.* **44** (1986) 478.

11. J. Ellis, M.K. Gaillard, and D.V. Nanopoulos, *Nucl. Phys.* **B106** (1976) 292.

12. E. de Rafael, in *Quantum Chromodynamics,* Lecture Notes in Physics No. 118, ed. by J.L. Alonso and R. Tarrach (Springer-Verlag, Berlin, 1980), p. 1.

13. R.S. Chivukula, A. Cohen, H. Georgi, B. Grinstein, and A.V. Manohar, *Ann. Phys.* **192** (1989) 93.

14. S. Dawson and H.E. Haber, SCIPP-89/23 (1989).

15. H. Georgi, *Weak Interactions and Modern Particle Theory* (Benjamin-Cummings, New York, 1984).

16. A. Cohen and A.V. Manohar, *Phys. Lett.* **143B** (1984) 481.

17. R.S. Chivukula and A. V. Manohar, *Phys. Lett.* **207B** (1988) 86, [erratum: **217B** (1989) 568].

18. C. Bernard, T. Draper, A. Soni, H.D. Politzer and M.B. Wise, *Phys. Rev.* **D32** (1985) 2343.

19. J. Flynn and L. Randall, UCB-PTH-88-29 (1988).

20. M.B. Voloshin, *Sov. J. Nucl. Phys.* **45** (1987) 122.

21. R. Ruskov, *Phys. Lett.* **187B** (1987) 165.

22. R. Chivukula, A. Cohen, H. Georgi and A.V. Manohar, *Phys. Lett.* **222B** (1989) 258.

23. A.V. Manohar, private communication.

24. B. Grinstein, L. Hall, and L. Randall, *Phys. Lett.* **211B** (1988) 363.

25. R. Willey and H. Yu, *Phys. Rev.* **D26** (1982) 3287.

26. R. Willey, *Phys. Lett.* **173B** (1986) 480.

27. B. Grzadkowski and P. Krawczyk, *Z. Phys.* **C18** (1984) 43.

28. F. Botella and C. Lim, *Phys. Rev. Lett.* **56** (1986) 1651.

29. H.-Y. Cheng and H.-L. Yu, preprint IP-ASTP-02-89 (1989).

30. L. Wolfenstein, *Phys. Rev. Lett.* **51** (1984) 1945.

31. G. Altarelli and P. Franzini, *Z. Phys.* **C37** (1988) 271.

32. J.R. Cudell, F. Halzen and S. Pakvasa, Wisconsin preprint MAD/PH/491 (1989).

33. M.A. Shifman, private communication.

34. S. Dawson, J.F. Gunion, and H.E. Haber, UCD-89-12 and SCIPP-89/22 (1989).

SEARCHES FOR LIGHT SCALAR PARTICLES IN K MESON DECAYS

Michael E. Zeller

Department of Physics
Yale University
New Haven, CT 06511

1. Introduction

One of the most perplexing yet fundamental problems in particle physics is the origin of mass. The very successful Standard Model of QCD and electroweak interactions begins to address this problem by incorporating the Higgs mechanism of spontaneous symmetry breaking to give masses to the vector bosons. In its simplest formulation, the minimal model, there is a single doublet of scalar Higgs fields which, by its Yukawa coupling to fermion fields, can also generate masses for the fermions.

The Higgs mechanism, however, is not understood at a fundamental level. While the Higgs fields appear in the Lagrangian of the minimal model in a way that would suggest that Higgs particles might exist, their masses are not uniquely specified. It is also not known if the minimal model is the correct one, placing uncertainty on the number of such particles if they were physically realized. There is theoretical prejudice that if Higgs particles do exist they are quite massive, but this too has no fundamental underpinning. Thus, examination of processes wherein low mass Higgs bosons might be created is of value, if only to rule out the possibility of the existence of such particles.

Since they decay only weakly and since they can be produced in abundance in favorable laboratory environments, K mesons provide an excellent vehicle for sensitive searches for low mass Higgs or other low mass scalar particles. Also because of recent interest in very rare decay modes of K's, experiments have been and are being performed which have significantly more sensitivity than was heretofore achieved. It is therefore an opportune time to review the status of scalar particles resulting from K decays, and to reexamine the limits of Higgs particle production via this mechanism.

In this article we will refer to the general class of low mass particles with the symbol ϕ^0, and when referring specifically to Higgs particles we will use H^0.

Higgs Particle(s)
Edited by A. Ali
Plenum Press, New York, 1990

2. Rate estimates

In the minimal model with decays of low mass Higgs boson, $H^0 \rightarrow x\bar{x}$, where x is a photon, electron, muon, or pion, decay widths of the H^0 as a function of the invariant mass of the $x\bar{x}$ pair, M_{xx}, are predicted fairly unambiguously. (The uncertainty arises in the pion decay mode due to strong interaction effects, and in the photon mode due to the fact that the photons do not couple directly to the Higgs.) For decays to fundamental fermions (f) or photons, predictions are shown in Fig. 1 in the invariant mass range of interest.[1] We have noted the kinematic limit of Higgs mass, M_H, for $K \rightarrow \bar{f}f$ decays on the figure. The widths for decays to fermions originate from the formula:

$$\Gamma(H \rightarrow \bar{f}f) = \frac{G_F M_H m_f^2}{4\pi\sqrt{2}} C_f (1 - \frac{4m_f^2}{M_H^2})^{3/2},$$

where m_f is the mass of the fundamental fermion, and C_f is 3 for quarks and 1 for leptons.

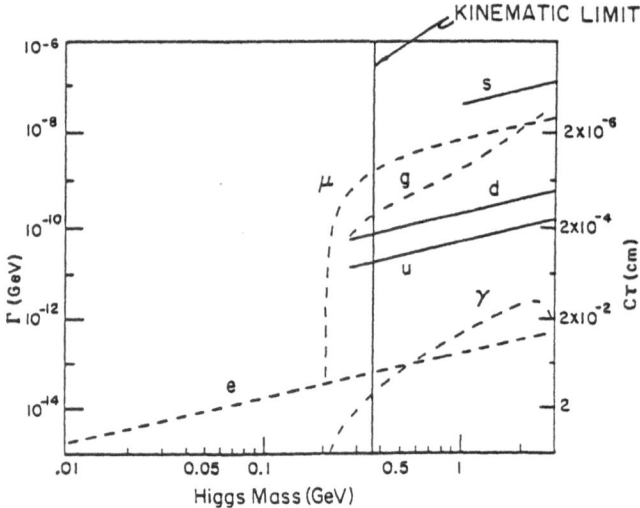

Fig. 1.   Decay width *vs.* Higgs mass for $H^0 \rightarrow \bar{f}f$, where f is a fundamental fermion or photon as noted on the figure. Approximate values of $c\tau$ are given on the left ordinate.

Below the mass of two muons, 211 MeV/c^2, the dominant decay mode is to an electron positron pair, and the lifetime ranges from 3.8×10^{-9} s at M_H=1.8 MeV/c^2 to 1.8×10^{-11} s at 211 Mev/c^2. Above this Higgs mass the lifetime drops by four orders of magnitude, and thus becomes inconsequential when considering measurable effects.

For leptonic decays the above formula gives accurate predictions of widths, while for pion decays of low mass Higgs particles corrections must be made which involve considerable theoretical uncertainties. There seems to be agreement, however, that there is an enhancement of the decay to two pions due to final-state interactions,[2] and that the ratio of pion to muon decays is between 5 and 25 in the mass range under consideration, $280 < M_H < 354$ Mev/c^2.[3,4] Predicting the properties of Higgs particle decays is only one component of describing $K \to \pi H^0; H^0 \to \bar{x}x$. The first piece, $K \to \pi H^0$, is a subject of a great deal of theoretical controversy. This has been thoroughly examined in "The Higgs Hunter's Guide".[5] To estimate the rates we will simply use a constant amplitude, $\mid A(K \to \pi H^0) \mid= 10^{-10}$ GeV, for an upper limit, and place an order of magnitude uncertainty on the lower bound of the rate.[6] It should be noted, however, that in the case of K_L^0 decays, as opposed to $K^\pm$'s, possible cancellation between different components of the matrix elements could lead to even lower rates.

Figure 2 shows our expected branching ratios for K decays to Higgs particles which subsequently decay according to the minimal model. The shaded region displays the uncertainty mentioned above. With this figure we now have an expectation with which to compare the results of various experiments.

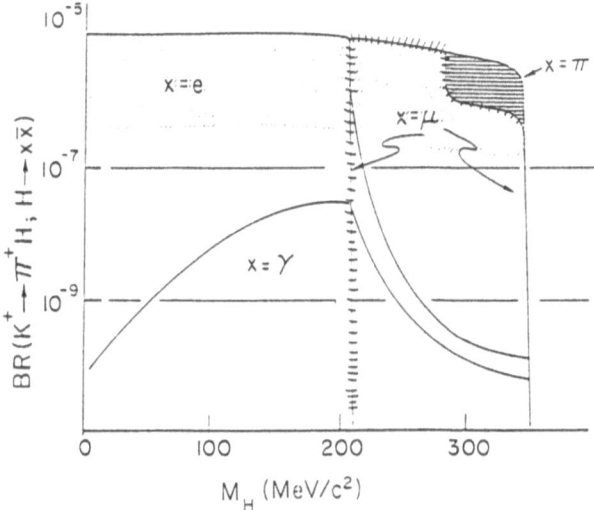

Fig. 2.   Expected branching ratios for $K \to \pi H^0$; $H^0 \to \bar{x}x$ with x being γ, e, μ, or π as noted. The cross-hatched line represents $x = \mu$.

3. Experimental Review

We now will examine several experiments which have searched for light scalar particles in K decays. This is not an exhaustive review of all such experiments, but a presentation of those with the greatest sensitivity of or those which evoke special interest.

3.a. An unbiased search for $K^+ \to \pi^+ \phi^0$

While this is not the most sensitive search, it has special significance because it does not place any assumptions on the properties of the particle recoiling against the π^+.

In the experiment performed at KEK with the apparatus shown in Fig. 3, a K^+ beam was stopped in an active target and allowed to decay.[7] Resulting charged decay products emitted into the acceptance of the spectrometer were tracked and momentum analyzed by proportional chambers MWPC1-4 in conjunction with field of the "C" magnet. In order to identify pions, the time of flight (TOF) of these particles from the T0 scintillation counter to the TOF counters was measured. The range necessary to stop the particles was measured by the range counters, and the pulses in these counters were examined to determine that a $\pi \rightarrow \mu \rightarrow e$ decay chain occurred.

Given the acceptance of the apparatus and the number of stopped kaons, the authors show the branching ratio limits at the 90% confidence level (90% CL) for $K^+ \rightarrow \pi^+ \phi^0$, see Fig. 4.

One sees for masses less than 120 MeV/c^2 the limits are below or near those expected in our naive calculation. Above this mass, due to the increased number of events from 2π and 3π decays of the K^+, the limits are not as good. As noted, however, these limits are unbiased by assumptions about the lifetime, decay modes, or interaction strength of the hypothetical missing particle. They are the best measurements of their kind, and thus provide unique information about this possible decay mode.

3.b. "Search for a Rare Decay Mode $K^+ \rightarrow \pi^+ \nu\bar{\nu}$ and Axion"

This experiment, also preformed at KEK, was designed to look for signs of weakly interacting particles, possibly axions, accompanying a π^+ in K^+ decays.[9] As opposed to the experiment of ref. 7, a premium was placed on good photon and charged particle rejection capability surrounding the pion detector. This was necessary because without such rejection rather common final states such as $\mu^+\nu\gamma$, $\pi^+\pi^0$, and $\pi^0\mu^+\nu$ can simulate the desired signal. Whereas experiments looking for π^+ plus 'anything' seek a signal in a momentum spectrum containing many events, experiments searching for π^+ plus 'nothing' attempt to observe signal in an otherwise event free environment.

A plan view of the detector is shown in Fig. 5. The K^+ beam was stopped in target counters T1-T7 which were surrounded by veto counters A1-A6. The π^+ exited into the detector region, was tracked with proportional chambers PC1-PC4, and was brought to rest in a range stack of scintillation counters M1-M7. There was no spectrometer field, with pions being identified by their range and the required $\pi \rightarrow \mu \rightarrow e$ decay chain. On the side opposing the pion detector was a set of lead-glass Čerenkov counters to veto photons.

Because of the abundant number of $\pi^+\pi^0$ decays, with a π^+ range of $\sim$33 g/cm^2, the search was constrained to those missing masses for which the range of the π^+ was greater than 35 g/cm^2, corresponding to $0 < M_{\phi^0} < 106$ MeV/c^2. No events were observed, so a limit was placed on decays containing 'axions': BR($K^+ \rightarrow \pi^+\phi^0$) < 3.8×10^{-8}. This, however, assumes that the particle recoiling against the π^+ does not decay into ionizing particles before passing through the veto counters. Thus the sensitivity of the experiment is function of the lifetime and mass of the 'axion'. This dependence has been investigated and the limit for a 1.8 MeV/c^2 particle as a function of its lifetime is shown in Fig. 6, curve (a).[11]

By looking for photons in the lead-glass for events containing a π^+ with appropriate range, the experimenters could also place limits on the branching ratio times decay probability for $K^+ \rightarrow \pi^+\phi^0$; $\phi^0 \rightarrow \gamma\gamma$. With a similar analysis of the geometry

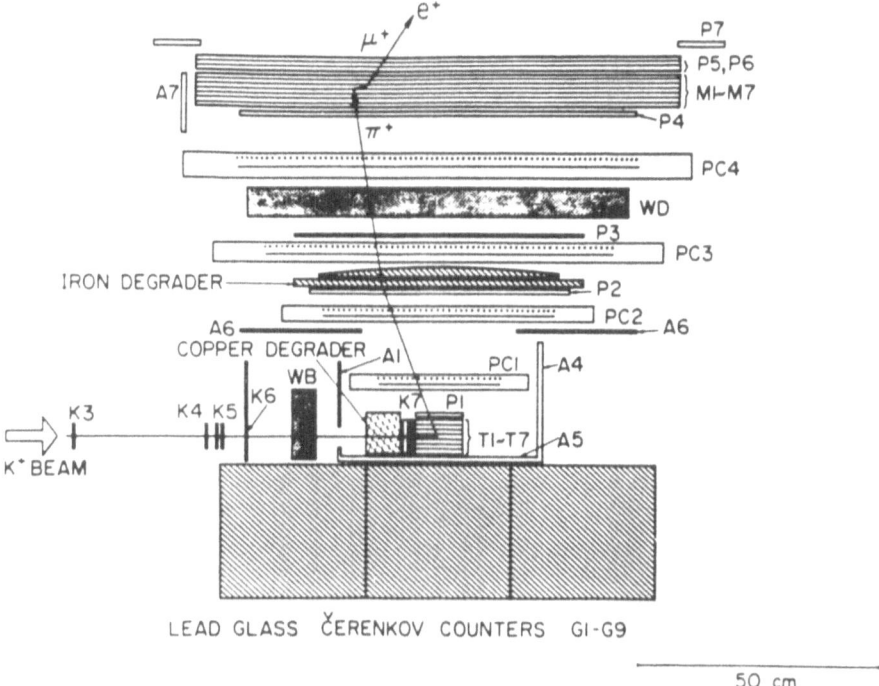

Fig. 3.   Experimental apparatus for a "Search for a Neutral Boson in a Two-Body Decay of $K^+ \to \pi^+ X^0$.[7]

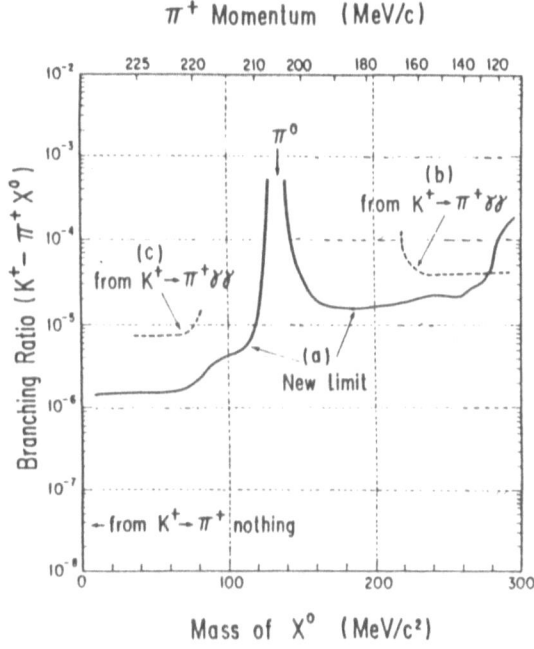

Fig. 4.   (a) Branching ratio limits (90%CL) for $K^+ \to \pi^+ \phi^0$ as a function of the mass of the supposed ϕ^0 from ref. 7. (b) and (c) are limits from refs. 8 and 10, while the zero mass limit is from ref. 9.

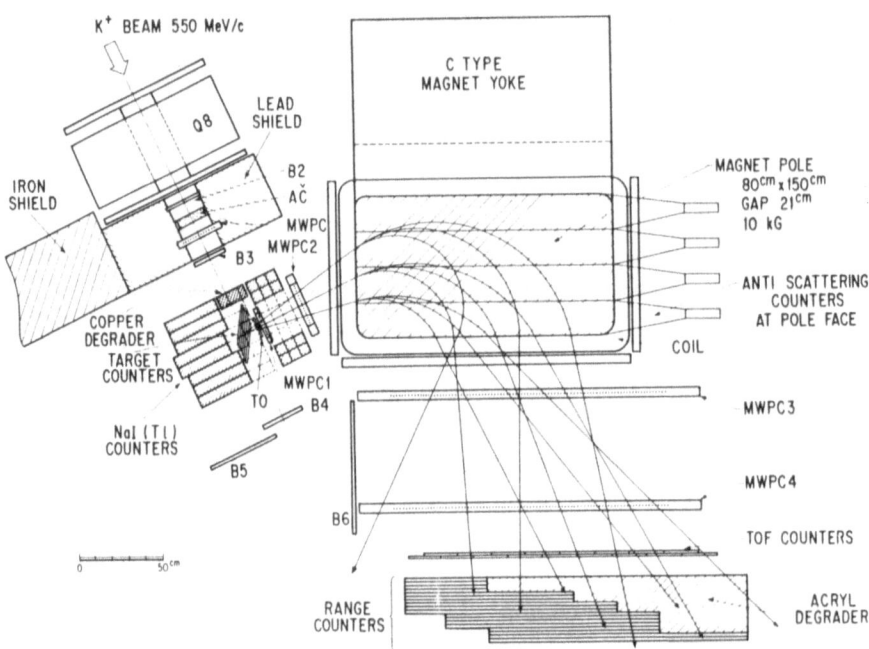

Fig. 5.   Experimental apparatus used in a search for $K^+ \to \pi^+ \bar{\nu}\nu$ and
Axions from ref. 9.

of the system, and requiring the ϕ^0 to decay before the last half of the lead-glass, the limits for a 1.8 MeV/c^2 axion are also shown in Fig 6, curve (b).

These limits can also be applied to Higgs decays since the neutral Higgs particle will not be detected by the veto counters. In fact, since the expected lifetime for a light (1.8 MeV/c^2) Higgs is rather long (3.8×10^{-9} s) this experiment provides the best limit for low mass Higgs production in K decays. And the limit is significantly below our expected lower bound of 10^{-6}.

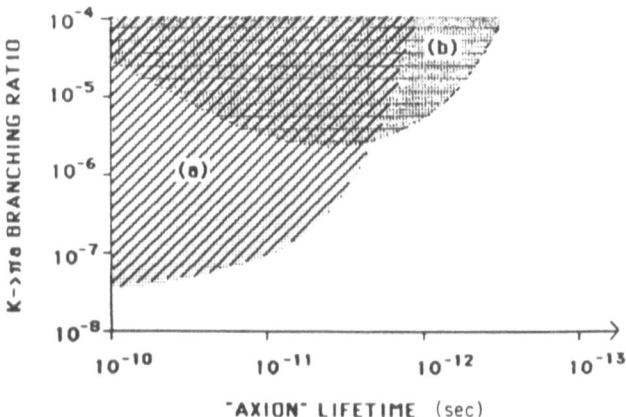

Fig. 6. Limits on the branching ratio for the decay $K^+ \rightarrow \pi^+\phi^0$ as a function of lifetime, interpreted from data of ref. 9, from ref. 11.

3.c. Search for low mass Scalars amid the decay $K^+ \rightarrow \pi^+e^+e^-$

Measurements of the branching ratio and Dalitz plot distributions for the decay $K^+ \rightarrow \pi^+e^+e^-$ ($K^+_{\pi ee}$) have been made[12] as a byproduct of a sensitive search for $K^+ \rightarrow \pi^+\mu^+e^-$ ($K^+_{\pi\mu e}$) which was recently completed at Brookhaven.[13] As opposed to previously discussed experiments, this one examines K^+ decays in flight.

The apparatus is shown in Fig. 7. A 6 GeV/c beam of K^+ mesons entered a 5 m long evacuated decay region where approximately 10% decayed. The decay products were deflected out of the beam region with magnet M1 and were momentum analyzed with proportional wire chambers P1-P4 surrounding the spectrometer magnet M2. Particle identification was accomplished with two atmospheric pressure Cerenkov counters on either side or the apparatus, the left counters filled with H$_2$ and the right with CO$_2$, followed by a lead-scintillator shower detector. The probability that a pion would be identified as an electron with this arrangement was about 10^{-5} on each side.

Events were selected based on the quality of the reconstructed tracks, the presence of an identified πee final state, the presence of an acceptable vertex within the decay volume for the three tracks, and a satisfactory extrapolation of the reconstructed initial particle trajectory back to the production target.

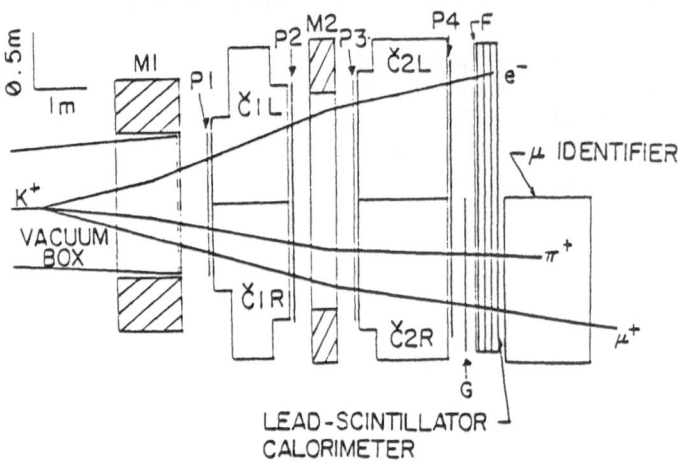

LEFT SIDE

RIGHT SIDE

Fig. 7. Experimental apparatus for the Brookhaven experiment studying $K^+ \rightarrow \pi^+ e^+ e^-$ decays.

Candidate events are displayed on a plot of the invariant mass of the electron positron pair *vs.* the invariant mass of the three final state particles, πee. In this display, signal will appear in the band with $M_{\pi ee} = 494 \pm 20$ MeV/c². Figure 8 shows such a plot, along with various projections, for a very early sample of data which was used to place an upper limit on low mass $e^+ e^-$ states.[14] In Fig. 8a the large number of events with $M_{e^+ e^-}$ less than 120 MeV/c² is due to the sequential process $K^+ \rightarrow \pi^+ \pi^0$; $\pi^0 \rightarrow \gamma e^+ e^-$ (Dalitz decays). Looking above $M_{ee} = 150$ MeV/c² one sees a small signal from the direct πee decay, Fig 8b. A projection of the $M_{\pi ee}$ spectrum for $M_{ee} < 15$ MeV/c², Fig. 8c, shows the Dalitz spectrum rapidly falling near the K^+ mass. Finally, Fig. 8d depicts the $M_{e^+ e^-}$ spectrum for events with $M_{\pi ee}$ between 492.5 and 520 MeV/c².

Figure 8d was then used to place the limits on low mass $e^+ e^-$ states, in particular the possible state suggested by the peaks in the electron positron sum energy spectrum observed in heavy ion collisions at GSI.[15] A Monte Carlo simulation of a signal with $M_{ee} = 1.8$ MeV/c² with branching ratio times decay probability of 10^{-6} is also shown in this figure, for a short lived state.

Because this is an in flight experiment, it is sensitive to only short lived particles. This comes about mainly because a vertex must be formed between the reconstructed particle trajectories for an event to be retained. The sensitivity thus becomes a function of the lifetime of the state in the laboratory rest frame; higher mass states will be accepted with shorter lifetimes. The 90% confidence limit for branching ratio times decay probability to a 1.8 MeV/c² state as a function of lifetime is shown in Fig. 9, along with the limit from ref. 9. Also shown in this figure is the result for a 100 MeV/c² particle. As can be seen, these results significantly limit the possibility of short lived particles coming from K^+ decays, but do not address the standard Higgs whose lifetime is $\sim 4 \times 10^{-9}$ s.

232

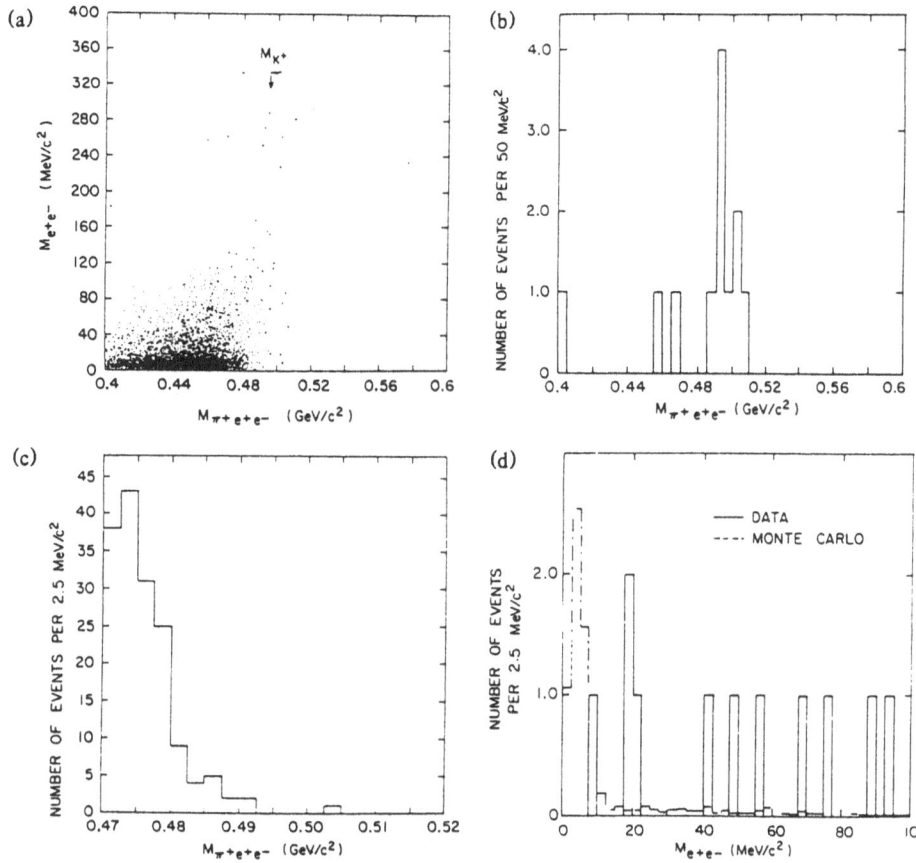

Fig. 8. (a) Scatter plot of $M_{e^+e^-}$ vs. $M_{\pi^+e^+e^-}$ for events where a $\pi^+e^+e^-$
final state was observed from a K^+ decay, the K^+ mass is de-
noted by the arrow. (b) Distribution of $M_{\pi^+e^+e^-}$ for $M_{e^+e^-} \geq$
150 MeV/c^2. (c) Distribution of $M_{\pi^+e^+e^-}$ for $M_{e^+e^-} \leq$ 15 MeV/c^2.
(d) Distribution of $M_{e^+e^-}$ for $(.493 \leq M_{\pi^+e^+e^-} \leq .52)$ GeV/c^2.
The dashed line indicates the signal expected from a 1.8 MeV/c^2
e^+e^- short lived state originating from a K^+ decay with a branching ratio times decay probability of 10^{-6}. From ref.[14]

This experiment has continued for two runs beyond that from which the above
data was accumulated. In the spring of 1988 approximately 800 $K^+_{\pi ee}$ events with
$M_{ee} \geq 120$ MeV/c^2 where acquired. The scatter plot of M_{ee} vs. $M_{\pi ee}$ from that run is
displayed in Fig. 10a; the lower M_{ee} bound of 100 MeV/c^2 is imposed because of the
multitude of Dalitz decay events at lower masses. One can see a clear signal at the K^+
mass of events with M_{ee} above 120 MeV/c^2. It should be noted that the diminution
of events between $\sim$ 130 and $\sim$ 180 MeV/c^2 is an artifact of the acceptance of the
apparatus imposed by the necessity for an efficient $K^+_{\pi\mu e}$ trigger. Figure 10b shows
the M_{ee} distribution of events with $(.44 \leq M_{\pi^+e^+e^-} \leq .52)$GeV/c^2. Also shown in this
figure is the spectrum for a vector matrix element, normalized to total events.

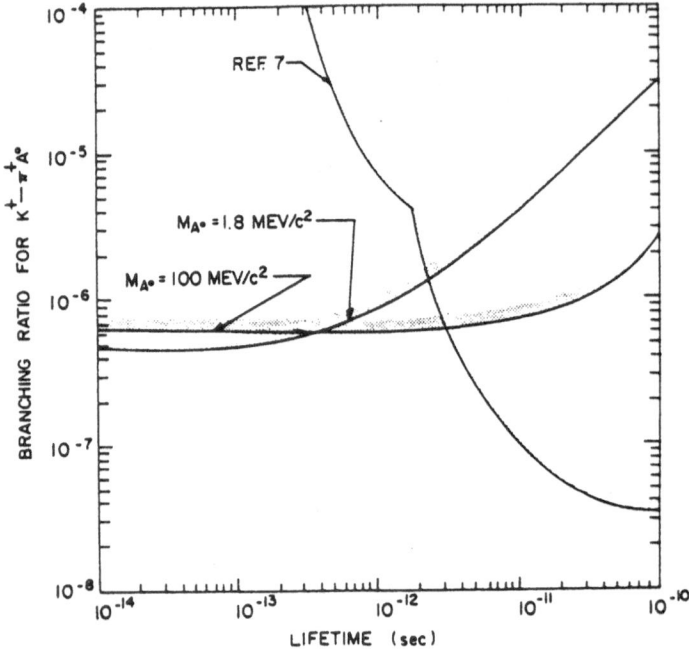

Fig. 9. Upper limits (90% C.L.) for branching ratio times decay probability as a function of lifetime for $K^+ \to \pi^+ \phi^0$; $\phi^0 \to e^+ e^-$ from E777 and ref. 9.

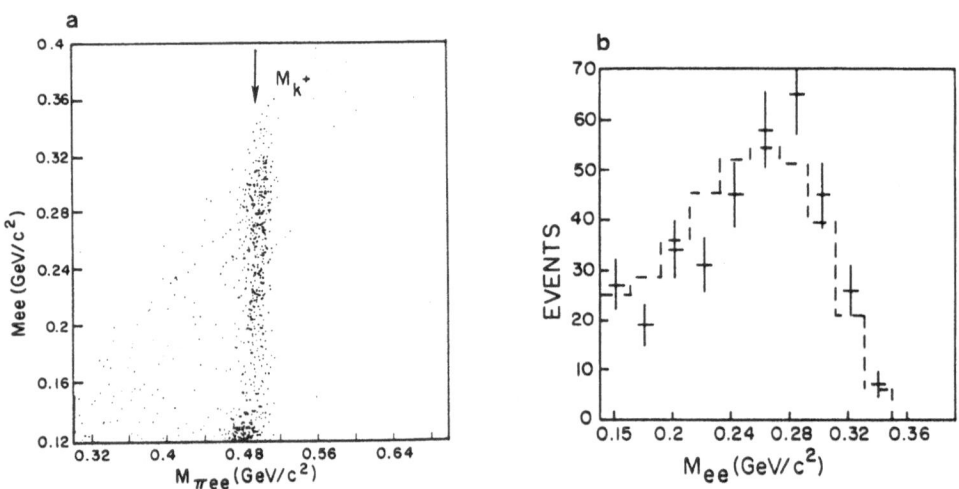

Fig. 10. (a) Scatter plot of $M_{e^+ e^-}$ vs. $M_{\pi^+ e^+ e^-}$ from the 1988 run of E777. (b) M_{ee} distribution of events with $(.44 \leq M_{\pi^+ e^+ e^-} \leq .52)$ GeV/c^2. The histogram is the expected distribution for a vector decay matrix element.

Figure 10b can be used to place upper limits on possible decays to $\pi^+ H^0$ with the H^0 decaying to $e^+ e^-$. Since there is a large 'background' under any potential signal, one must establish a criterion for there being no statistically significant number of signal events. By assuming the data of Fig. 10b contains no H^0 signal, and that such a signal would be seen as a Gaussian peak with a width which was the resolution of the apparatus (~ 8 MeV/c^2), an upper limit at the 90% confidence level for the branching ratio times decay probability as a function of the mass of the H^0 can be determined. This calculation takes into account the acceptance of the apparatus as a function of the $e^+ e^-$ invariant mass, and assumes the lifetime of the H^0 as a function of mass to be that described above. The resulting distribution is shown in Fig. 11. Recall that the abundance of Dalitz decays for ee invariant masses below M_{π^0} significantly increases the 'background' in this region, and thus decreases the sensitivity to a possible H^0. The group has not yet performed the analysis of this region.

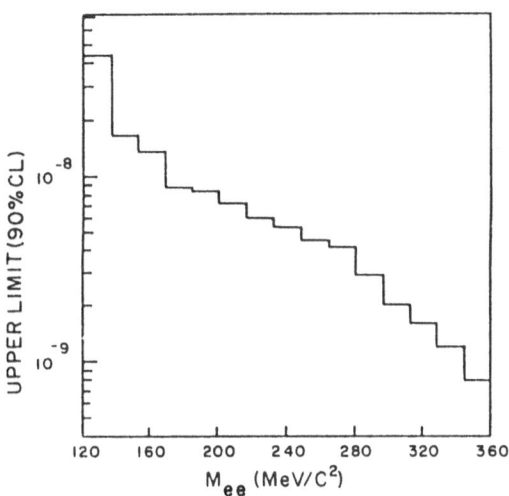

Fig. 11. Distribution of the upper limit (90% C.L.) of the branching ratio times decay probability for $K^+ \to \pi^+ H^0$; $H^0 \to e^+ e^-$ as a function of the mass of the H^0, from Brookhaven E777.

3.d. Search for low mass Scalars while searching for the decay $K_L^0 \to \pi^0 e^+ e^-$

As a result of experiments designed to measure $\epsilon\prime/\epsilon$ in the K^0 system, an improved search for the decay $K_L^0 \to \pi^0 e^+ e^-$ has been conducted.[17] In the Standard Model this mode violates CP when it proceeds through a single vector boson mediator (either a photon or a Z^0), and is expected to have a branching ratio less that 10^{-10} when two photons are involved. Of course one of the possible modes through which this decay might occur is $K_L^0 \to \pi^0 \phi^0$; $\phi^0 \to e^+ e^-$. Because of the suppressed nature of the former mode, the latter would be present in an environment of greatly reduced background compared to that of K^+ decays.

The CERN group has performed this search, has presented preliminary results in seminars and conferences, but as of the time this is being written it has not yet published its final results. Therefore, what follows is only the author's interpretation.

and with the knowledge of the energy of each photon as determined in the calorimeter, the longitudinal position of the decay point of the π^0 is found. Where that point intersects the line of flight of the K^0 is taken to be the decay point of the primary K^0. This also establishes a plane perpendicular to the K^0 flight path. Finally, the momentum vector of the candidate ϕ^0 is extrapolated back to the this plane. The distance in the plane between this trajectory and the decay point is then a measure of the quality of the fit to the $\pi^0\phi^0$ hypothesis.

With this construction, the distance traveled by the by the ϕ^0 from the K^0 decay point to its decay point can be determined. Since the momentum four-vectors of the electron and positron are measured, both the rest mass and the proper decay time of the ϕ^0 can then be calculated.

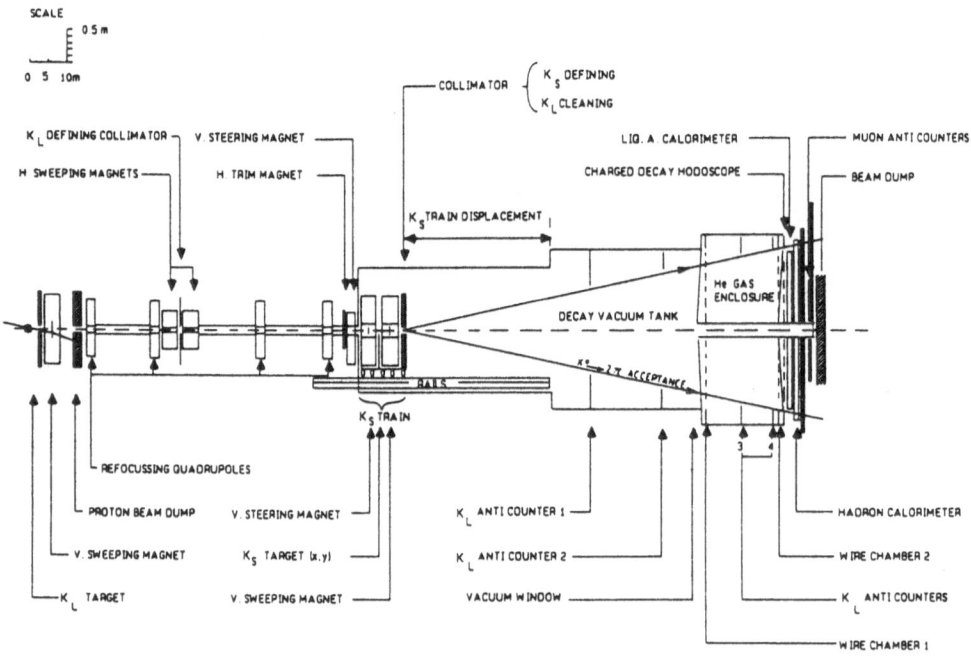

Fig. 12.   Elevation view of the NA31 experimental apparatus from ref. 18.

The experimental apparatus is shown in Fig. 12.[18] For $\pi^0\phi^0$ mode K_L^0's which decay between the collimator and anticounter 1 are expected to produce two photon showers from the π^0 decay, and two electron showers in the liquid argon calorimeter. The electrons will register in the charged decay hodoscope and the wire chambers, while the photons will not.

Reconstruction of events occurs as follows. The center of gravity of the energy deposited in the calorimeter, in conjunction with the K^0 production target, establishes the line of flight of the incident K^0. By means of the hit points of the electron and positron trajectories in wire chambers 1 and 2, and the energies deposited in the calorimeter, the vector momentum and decay point of the candidate ϕ^0 can be determined. Under the assumption that the two photons are emitted from a π^0 decay,

For events satisfying the desired quality criteria, the group displays the invariant mass of the e^+e^- pair as a function of the flight path of the possible ϕ^0, see Fig. 13a. Assuming that the three events in this plot are signal with zero lifetime, the 90% confidence level upper limit on the process is then shown in Fig. 13b. And since the acceptance of the apparatus is a function of the lifetime of the ϕ^0, a contour plot of upper limits for the lifetime $vs.$ e^+e^- invariant mass can be made, see Fig. 14.

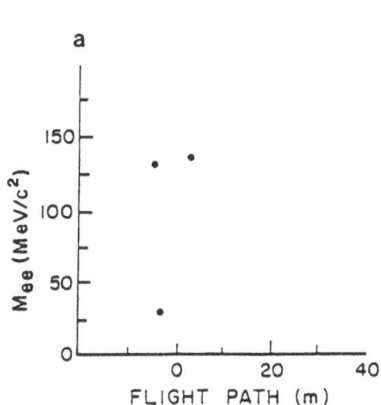 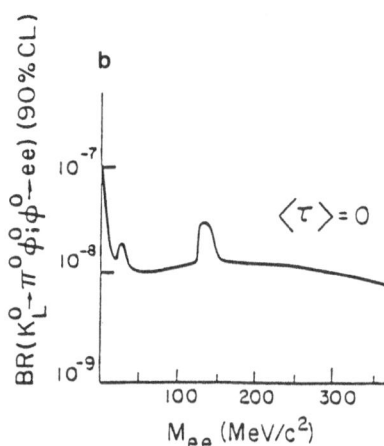

Fig. 13. (a.) Scatter plot of the e^+e^- invariant mass $vs.$ Flight path of candidate events from NA31. (b.) Upper limits (90% C.L.) for the decay $K_L^0 \to \pi^0 H^0$; $H^0 \to e^+e^-$ as a function of the H^0 mass, assuming zero lifetime.

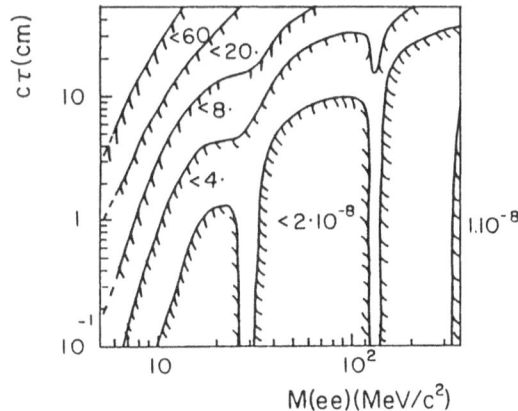

Fig. 14. Contour plot of upper limits for the process $K_L^0 \to \pi^0 H^0$; $H^0 \to e^+e^-$ for lifetime $vs.$ e^+e^- invariant mass.

3.e. Search for low mass Scalars in the decay $K^+ \to \pi^+\mu^+\mu^-$

A more recent experiment dedicated primarily to a search for the decay $K^+ \to \pi^+\nu\bar{\nu}$ has been under way at the Brookhaven AGS since 1988.[19] Because this experiment will have a greater sensitivity to the $\pi\nu\bar{\nu}$ mode than heretofore achieved, its results will eventually supercede those of the previous search for the $\pi^+\phi^0$ mode with low ϕ^0 masses. At the time of this review, analyses of possible new limits have not yet been made. From a short run in 1988, however, results from a search for $K^+ \to \pi^+\mu^+\mu^-$ have been obtained.[20] $K^+ \to \pi^+\mu^+\mu^-$ is analogous to the $\pi^+e^+e^-$ decay of the K^+ discussed above, but will be reduced in branching ratio by about a factor of 5 due to kinematic considerations.[21] Thus this mode should have a branching ratio of roughly 5×10^{-8} if no decays such as $K^+ \to \pi^+\phi^0$ are present. Recalling the fact that the coupling of Higgs particles to fermions is proportional to the fermion mass, we expect that if a π^+H^0 mode exists it will be dominated by $H^0 \to \mu^+\mu^-$ for H^0 masses above twice the mass of the muon, see Fig. 2.

Since this is a search for the same decay as that of ref. 9, the experimental problems to be solved are those described above in section 3.b.. The apparatus, shown in Fig. 15, is quite different, however, since the stopping target is surrounded by a solenoid magnetic field. This allows the momentum of the π^+ to be determined over a much larger solid angle. In addition to the larger acceptance of the apparatus, this technique, coupled with a range measurement and observation of the $\pi \to \mu \to e$ decay chain gives a cleaner determination that the charged particle emitted was indeed a pion.

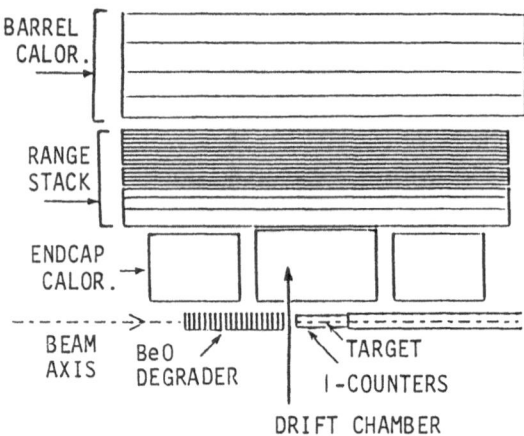

Fig. 15. Schematic drawing of the apparatus used in the $K^+ \to \pi^+\nu\bar{\nu}$ search. The system has axial symmetry about the beam line.

Many technological innovations have been incorporated in the apparatus to insure that any observed events originated from K^+ decays, and no particles other than those detected were emitted from the decay. The stopping target is composed of scintillating fibers which yield a pattern characteristic of a stopping kaon and the correct number of charged decay products; the central drift chamber tracks and measures the momentum of charged particles leaving the target; and the scintillation counter range stack measures the range, energy deposition, and decay chain using transient digitizers to record the time and pulse shape evolution of the process. Surrounding these components of the apparatus are lead-scintillator detectors to observe any photons emitted in the decay.

Measurement of the $\pi^+\mu^+\mu^-$ branching ratio is a natural byproduct of an experiment employing a detector such as this. The signature of such events is the observation of three charged particles - one of which is identified as a pion, the presence of no other particles, a good decay vertex, and an invariant mass of the observed decay products equal to that of the kaon assuming the other two particles are muons.

For approximately 10^{10} stopped kaons, and with a total average acceptance of about 0.34%, the experiment has observed three candidate events. The $\mu\mu$ invariant masses of these are 283.6, 255.4, and 256.0 MeV/c^2. Of these events the first two are very clean and convincing, while the third could be explained as a $K^+ \rightarrow \pi^+\pi^-e^+\nu$ decay. The group chooses to not call this a measurement of the $\pi\mu\mu$ branching ratio, but rather places an 90% C.L. upper limit of 2.1×10^{-7} on the process . They also place a 90% C.L. upper limit on the the branching ratio times decay probability for $K^+ \rightarrow \pi^+\phi^0$; $\phi^0 \rightarrow \mu^+\mu^-$ as a function of $\mu\mu$ invariant mass as shown in Fig. 16. For these ϕ^0 masses, if the ϕ^0 were a Standard Model Higgs particle its lifetime would be less than 10^{-12}s. Thus the limits shown in Fig. 16 apply to Higgs decays as well, without need of correction for decay pathlength.

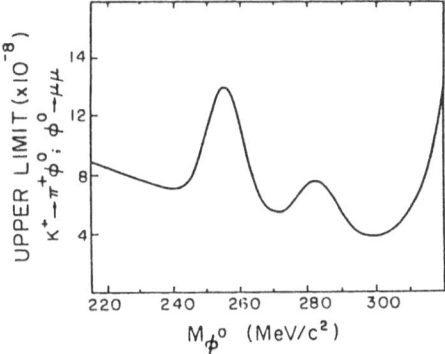

Fig. 16. 90% C.L. upper limit of the branching ratio times decay probability for $K^+ \rightarrow \pi^+\mu^+\mu^-$ as a function of $M_{\mu\mu}$.

4. Conclusion

In this review we have compiled upper limits for K decays into a pion and a neutral scalar particle. If one uses the Higgs lifetime as a function of Higgs mass specified by the formula in section 2, and the acceptance versus ϕ^0 lifetime relationships of the various experiments, a composite of all these results as applied to $K \rightarrow \pi H^0$ can be constructed. We display this in the plot shown in Fig. 17.

As one can see, the limits established by existing experiments are low enough to definitively rule out a minimal Standard Model Higgs particle in the mass range below 320 MeV/c². The only caveat to this is in the mass range covered by $K_L^0 \to \pi^0 H^0$, ~ 40 to ~ 120 MeV/c². As pointed out above and in ref. 5, the theoretical lower bound for K^0 decays is less certain than that for K^+ decays, and may involve cancellations that would make the branching ratio to $\pi^0 H^0$ significantly smaller.

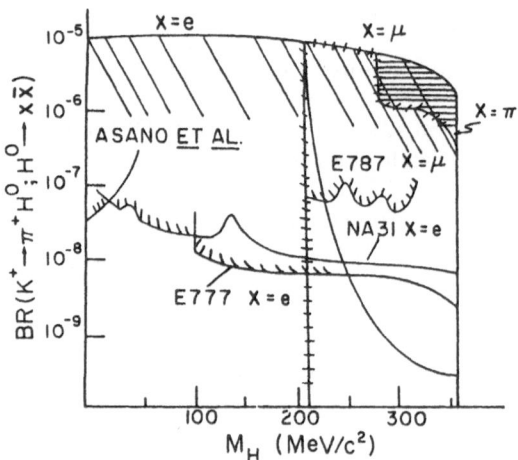

Fig. 17. Upper limits for the branching ratio times decay probability for $K \to \pi H^0$; $H^0 \to x\bar{x}$ vs. M_H^0, superimposed on the theoretical limits of Fig. 2.

5. Acknowledgements

I thank David Coward and Harry Nelson for sharing the NA31 results before publication, Stu Smith for providing information on E787, and Ahmed Ali and Antonio Zichichi for their warm hospitality.

This research was supported in part by the Department of Energy under contract No. DE-AC02-76ER03075.

References

1. R. N. Cahn, Rep. Prog. Phys. **52**, 389 (1989).
2. R. Chivukula, A. Cohen, H. Georgi, and A. Manohar, Phys. Lett. **222B**, 258 (1989).
3. S. Raby and G. West, Phys. Rev. **D38**, 3488 (1988).
4. R. S. Willey, This workshop (1989).
5. J. F. Gunion, H. E. Haber, G. L. Kane, and S. Dawson, UCD-89-4 (1989).
6. R. Chivukula and A. Manohar, Phys. Lett. **207B**, 86 (1988).
7. T. Yamazaki *et al.*, Phys. Rev. Lett. **52**, 1089 (1984).
8. R. Abrams *et al.*, Phys. Rev. **D15**, 22 (1977).

9. Y. Asano *et al.*, Phys. Lett. **107B**, 159 (1981).

10. Y. Asano *et al.*, Phys. Lett. **113B**, 195 (1982).

11. L. M. Krauss and M. E. Zeller, Phys. Rev. **D34**, 3385 (1986).

12. C. Alliegro *et al.*, Brookhaven Experiment 777/851 (1989).

13. C. Campagnari, *et al.*, Phys. Rev. Lett. **61**, 2062 (1988); A. Lee *et al.* "A New Search for the Decay $K^+ \rightarrow \pi^+ \mu^+ e^-$" (submitted to Phys. Rev. Lett.) (1989).

14. N. J. Baker *et al.*, Phys. Rev. Lett. **59**, 2832 (1987).

15. T. Cowan *et al.*, Phys. Rev. Lett. **56**, 444 (1986).

16. P. Bloch *et al.*, Phys Lett. **56B**, 201 (1975).

17. CERN experiment NA31. CERN, Edinburgh, Mainz, Orsay, Pisa, Siegen collaboration (1989), with thanks to D. Coward and H. Nelson for private communications.

18. H. Burkhardt *et al.*, Nucl. Inst. and Meth. **A268**, 116 (1988).

19. Brookhaven AGS experiment E787. BNL, LANL, Princeton, TRIUMF collaboration.

20. M.A.Selen, Princeton Ph.D. Thesis (1988)(unpublished).

21. D.S.Beder and G.V.Dass, Phys.Lett. **59B**,444(1975).

HIGGS BOSON(S) AND RARE B PROCESSES :

THEORETICAL EXPECTATIONS

Stefano Bertolini

DESY, Theory Division
Notkestrasse 85
D-2000 Hamburg 52

INTRODUCTION

Rare weak processes have always represented a source of sensitive and crucial tests for the Glashow-Weinberg-Salam model of electroweak interactions (hereafter referred to as the standard model, or SM), supplying us, at the same time, with strict guidelines for modelling viable extensions. After the successes linked to Kaon physics, of which the most recent is the first possible observation of "direct" CP violation in $K_{L,S} \to \pi\pi$ decays at CERN [1], the physics of $B-$mesons opens a window on the structure of the third generation quark sector (for a review on the present status of experimental results see ref. [2]). The discovery of an "anomalously" large $B_d - \bar{B}_d$ mixing, made two years ago by the ARGUS collaboration and subsequently confirmed by the CLEO collaboration, was in this respect quite remarkable. In fact, this has been the first substantial indication in favour of a "heavy" top quark, likely to have a mass above 100 GeV. An indication which is now corroborated by the comparison of the recent precise measurements of the Z^0 mass with neutral-current phenomenology data (for a "status of the art" review see ref. [3]).

In this talk we will elaborate upon the information that we can extract from our present knowledge of rare B processes with regard to the existence and detectability of the standard Higgs boson and/or other physical Higgs scalars, present in a number of extensions of the SM. The existence of a neutral scalar boson in the standard electroweak theory is a consequence of the spontaneous breaking of the $SU(2)_L \times U(1)_Y$ gauge symmetry, achieved through the Higgs mechanism [4]. Of the four original scalar degrees of freedom, assembled in a $SU(2)_L$ doublet, three are "eaten", in the symmetry breaking process, by the weak gauge bosons $W^{\pm}$ and Z^0. The latter acquire a longitudinal component and become massive, leaving as a result, together with a massless photon, a physical massive neutral scalar: *the* Higgs boson. The mass of such a scalar is a free parameter of the model. Upper bounds below the 1 TeV scale are usually called for on the grounds of unitarity violation and/or failing of perturbation theory in the Higgs sector (for a review see ref. [5]). These results follow from a self-consistency ansatz on the validity of the perturbative approach for the whole electroweak theory. Aside from our wishful thinking, there is no a priori argument for that to be the case. Nothing physically "wrong" is associated with a heavy Higgs boson, from the aforementioned point of view, other than the "unpleasantness" of having to cope with a non-perturbative problem. Here, the lattice regularization approach and the "triviality" of the ϕ^4 theory come crucially into play: lattice results show in fact that the minimal Higgs sector becomes non-interacting when the cut-off is removed (with correspondingly vanishing Higgs mass), thus suggesting that the SM is an effective theory valid up to a finite cut-off scale Λ. An upper bound on m_H of about 640 GeV, then follows for Λ larger than $2m_H$ (for a review see ref. [6]). This result therefore justifies the perturbative treatment of the Higgs sector, although it does not yet include the effects of large Yukawa couplings (heavy top).

Since we are here interested in the interplay between Higgs scalars and the B system, our attention will focus on the lower range of Higgs masses. In this respect, use of perturbation theory allows us to derive a lower bound on the standard Higgs boson mass, by requiring that the symmetry breaking vacuum is stable in the presence of higher order corrections. For a light top quark ($m_t \ll m_W$), the bound amounts to about 7 GeV, as first derived, independently, by Linde and S. Weinberg [7]. This value is however crucially related to the assumption of negligeable fermion masses. For the remainder of this talk it is important to recall that the Linde-Weinberg bound vanishes as the top quark mass approaches m_W, and that a second sharply rising lower bound arises when $m_t > m_W$, due to the requirement that a true minimum of the potential exists [8,9] (for an extension to non-minimal Higgs scenarios, see ref. [10]). In the limit of large cutoff, the bound is roughly given by [9]: $m_H > 1.85(m_t - 80)\ GeV$. As a consequence, whenever we will consider the possibility of Higgs production in B-meson decays, within the SM framework, we will assume for consistency that the top quark mass lies in the close neighbourood of 80 GeV.

As already mentioned, the aim of this talk is to summarize the status of theoretical work which has been done in the area of rare B processes, with special attention to the role plaied by the standard and/or exotic Higgs sector. For this purpose I have chosen a "preferred" set of processes and examined them in the context of the SM and two paradigmatic Higgs extensions: a two Higgs doublet model with natural suppression of flavour changing neutral currents (FCNC), and a minimal supersymmetric scenario, which originates from spontaneously broken N = 1 supergravity. This talk does not provide an exhaustive review on the topic, which would be rather extensive, but aims at singling out some general features which determine the sensitivity of these low energy processes to the structure of the Higgs sector. Particular attention is also given to the effects of QCD corrections, which, in spite of the "heaviness" of the B-mesons, play in some instances a relevant role, and to the top dependence. An effort will be made to reach a qualitative understanding of the various numerical results.

The remainder of this talk is organized in five sections. We will first examine the issue of whether $b-$related processes provide constraints on the existence of a light Higgs scalar. Then, we will review the status of the theoretical predictions for a class of rare decays and mixings in SM, with special attention to the role of strong interactions. These aspects are then reexamined in two popular extensions of the standard scenario, which involve enlarged Higgs sectors. In the last section, the main results are summarized.

THE STANDARD HIGGS BOSON

Within the standard context, the connection between B-meson physics and the minimal neutral Higgs boson relies on the possibility that the latter may be directly produced in decays involving b-quark states. In the following, we will consider two such processes: the radiative Υ decay, $\Upsilon \to H + \gamma$, and the effective FCNC transition $b \to s + H$. Absence of signals for these decays may raise the lower bound on the Higgs mass, from the few hundred MeV's of Kaon physics (see ref. [11]), up to about 7 GeV.

a) $\Upsilon \to H + \gamma$

It was pointed out in 1977 by Wilczek [12] that the radiative decay of "heavy" $q\bar{q}$ vector bound states, such as J/Ψ and Υ, provides a promising way of detecting a light Higgs scalar. The presence of a monochromatic photon allows for an inclusive tagging of the process, thus avoiding the theoretical uncertainties related to exclusive Higgs decay modes. The lowest order (tree level) prediction for the decay rate can be written as

$$R_0 \equiv \frac{\Gamma_0(q\bar{q} \to \gamma H)}{\Gamma_0(q\bar{q} \to \mu^+\mu^-)} = \frac{G_F m_q^2}{\sqrt{2}\pi\alpha} \left(1 - \frac{m_H^2}{M_{q\bar{q}}^2}\right) \tag{1}$$

By plainly applying the Wilczek formula to the data obtained by the CUSB collaboration [13], who searched for monochromatic photon signals at the $\Upsilon(3s)$, a lower bound on the Higgs mass of about 7 GeV follows (see fig. 1). However, $O(\alpha_s)$ gluon corrections have been shown [14] to suppress the tree level transition by more than a factor of two. When analogous QCD corrections are included in the denominator of eq. (1) [15] one finds

$$R^{(1)} = R_0 \frac{\left[1 - \frac{4}{3}\frac{\alpha_s}{\pi} f\left(\frac{m_H^2}{M_\Upsilon^2}\right)\right]}{\left[1 - \frac{16}{3}\frac{\alpha_s}{\pi}\right]} \simeq 0.23\, R_0 \tag{2}$$

or, by expanding at first order in the strong coupling,

$$R^{(2)} = R_0 \left[1 - \frac{4}{3}\frac{\alpha_s}{\pi} f\left(\frac{m_H^2}{M_\Upsilon^2}\right) + \frac{16}{3}\frac{\alpha_s}{\pi}\right] \simeq 0.49\, R_0 \tag{3}$$

In the previous equations, $f(0) \simeq 10$ and $\alpha_s = 0.2$ have been used to derive the numerical results. The factor two of discrepancy between eq. (2) and eq. (3), shows a problem in the perturbative calculation. due to the large size of the corrections to the Wilczek formula. An alternative evaluation based on the modified minimal subtraction scheme ($\overline{MS}$) has been presented by the authors of ref. [16]. It is argued that a value for the t'Hooft parameter $\mu \simeq 2m_b$ is preferred by the stability of the corrections, thus fixing the scale at which the strong coupling constant should be evaluated, and leading, as a net effect, to a reduction of their magnitude ($\alpha_{s\overline{MS}}(2m_b) \simeq 0.155$). The result of ref. [16] is reported in fig.1 (long-dashed-dotted line): although

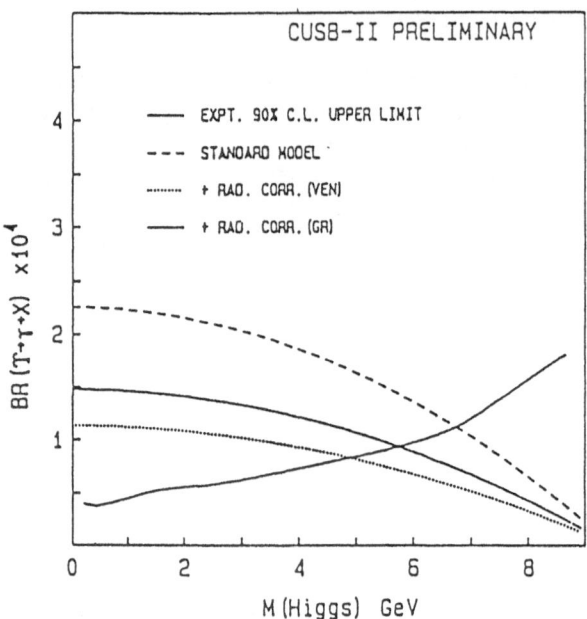

Figure 1. The 90% c.l. upper limit from CUSB on the inclusive $BR(\Upsilon \to \gamma + X)$ is shown (solid line) and compared with various theoretical expectations as explained in the text. The figure is taken from ref. [13].

still somewhat suppressed with respect to the uncorrected prediction, a bound on the Higgs mass of about 6 GeV would nevertheless follow.

In addition to the theoretical uncertaintes related to the evaluation of short distance QCD effects, bound state and relativistic corrections appear to be large and critically affect the magnitude of the process [17,18]. Bound state effects become important when $m_H \sim M_{q\bar{q}}$, since after emitting the soft photon, the propagating quark still feels the bounding potential. These corrections suppress further the rate for the process and spoil the possibility of achieving a sensitivity to the Higgs mass above 6-7 GeV in the present CUSB experiment. The relativistic corrections to the Υ wavefunction have a major impact on the above results. These corrections become important in the region of light Higgs masses and are due to soft gluon effects. They have been evaluated in ref. [18] and shown to lead for $m_H \ll M_\Upsilon$ to a further suppression of about a factor of two in the rate, on top of the hard gluon effects. When all these effects and the uncertainty related to the QCD corrections are combined, it is apparent from fig. 1 that no reliable bound on a light Higgs boson mass can be presently extracted from the data.

b) $b \to s + H$

Another possibility of detecting the presence of a Higgs boson with a mass up to about 4 GeV is offered by the electroweak induced transition $b \to s + H$. The strong sensitivity of the process to the top quark mass was pointed out originally by Willey and Yu [19] and by Grzadwkowsky and Krawczyk [20], who showed that the sum of the various loop contributions, with W and top exchange, reduce to a simple (m_t^2/m_W^2) dependence in the final amplitude (thus leading to an unscreened m_t^4 dependence of the rate). This result, perhaps due to its "accidental" simplicity, has since then gone under the scrutiny of a number of groups and its validity has been at times questioned and reaffirmed [21,22]. A relatively recent criticism was raised on the grounds of a low-energy theorem regarding the structure of the effective $qq'H$ vertex [24], making the original result once again controversial [25]. The issue has been finally settled by Grinstein, Hall and Randall [23], who proposed a calculation based on current algebra methods (the Higgs boson is viewed as the pseudo-Goldstone boson associated with the breaking of scale invariance in the electroweak lagrangian). This alternative and elegant derivation confirms the original result, providing also a deeper understanding for the cancellation of logarithmic terms in the amplitude. We refer the reader to ref. [23] for details.

The branching ratio for the inclusive $B \to X_s + H$ decay can be written as

$$BR(B \to HX_s) \simeq 0.32 \left(\frac{m_t}{m_W}\right)^4 \left(1 - \frac{m_H^2}{m_b^2}\right)^2 \qquad (4)$$

where use of $|V_{ts}| \simeq |V_{cb}|$ and of $BR(b \to ce\bar{\nu}) \simeq 0.11$ has been made. Since, due to the Linde-Weinberg bound on the minimal Higgs boson, we have to assume $m_t \simeq 80\ GeV$ for the process to be kinematically allowed, we obtain, neglecting phase space suppression, $BR(B \to X_s H) \simeq 30\%$!

After the discussion on the radiative Υ decay, a question naturally arises: do QCD corrections affect the large one-loop electroweak transition ? The answer is in this case negative. By working in the framework of an effective low-energy theory we may write the amplitude for the process $(B \to X_s H)$ as

$$C(\mu, m_t^2/m_W^2)h_b(\mu) \langle X_s| \bar{s}_L b_R |B \rangle (\mu) \qquad (5)$$

where $h_b = m_b/v$ is the bottom quark Yukawa coupling, and μ is the 't Hooft mass parameter. At the m_W scale, $C(m_W, m_t^2/m_W^2) = (3/64\sqrt{2}\pi^2)V_{ts}^* V_{tb} g^2 (m_t/m_W)^2$, thus fixing the boundary condition for the renormalization problem. When QCD is turned on one expects that no large logarithms appear in the hadronic matrix element in eq. (5) if evaluated at $\mu \simeq m_b$, since only momenta at that scale are involved. Thus, all the leading QCD effects are "transferred" in the renormalization of the coefficients C and h, which is controlled by perturbation theory through renormalization group (RG) methods. We therefore hope that by evaluating the amplitude of the process at a convenient scale, $\mu = m_b$. we may minimize the uncertainties related to our ignorance on the μ dependence of the hadronic matrix element (the full amplitude must be independent on μ).

The evolution of the coefficient $C(\mu)h(\mu)$ (the explicit separation in two parts is just matter of convenience) is determined by the anomalous dimension induced for the operator $O_H \equiv \bar{s}_L b_R H$ by gluon exchange. It is a simple exercise (especially if performed in the Landau gauge where the fermion wavefunction renormalization is absent) to show that O_H has the same anomalous dimension of a mass operator. This is, on the other hand, just an obvious consequence of the fact that gluons do not "feel" flavour. The effect of gluon renormalization consists therefore in replacing $m_b(\mu = m_W)$ by $m_b(\mu = m_b)$, the scale at which it is defined. The coefficient C is therefore, at the leading level, not affected. In addition, it is fairly simple to show that any non-multiplicative renormalization of O_H, v.i.z. any mixing with low-energy effective four-quark operators such as $O_1 \equiv (\bar{s}_{L\alpha}\gamma^\mu b_{L\alpha})(\bar{c}_{L\beta}\gamma_\mu c_{L\beta})$ or $O_2 \equiv (\bar{s}_{L\alpha}\gamma^\mu b_{L\beta})(\bar{c}_{L\beta}\gamma_\mu c_{L\alpha})$, is either absent, at one loop [22], or, at the two-loop level, always suppressed by powers of light quark masses (this conclusion can be simply drawn by inspection of the relevant two-loop diagrams, on dimensional grounds, by counting the needed helicity flips on the fermion lines).

At variance with $\Upsilon \rightarrow H + \gamma$, the transition $b \rightarrow s + H$ is free from problems related to strong corrections. On the other hand, in order to compare the theoretical prediction with experiment, one must here deal with the evaluation of the hadronic matrix elements for the relevant exclusive modes, as well as the calculation of the various Higgs branching fractions ($H \rightarrow \mu\mu$, $\pi\pi$, KK, etc.), which are a function of the Higgs mass. For a "heavy" enough Higgs, let us say $m_H > 2\,GeV$, one can thrust the use of perturbative QCD in estimating the hadronic decay modes, while non-relativistic quark models should give sensible estimates for the quark hadronization. At the other end, for $m_H < 1\,GeV$ the use of chiral Lagrangian and low-energy theorems may give a handle on the relevant Higgs branching fractions (for reviews and further references on these aspects see refs. [26], [27] and [28]). However, for such a light Higgs boson, the estimates of the hadronic matrix elements, based on extrapolation of non-relativistic quark models, become unreliable. As an example, the branching fraction $\Gamma(B \rightarrow KH)/\Gamma(B \rightarrow X_s H)$ is estimated to be about 0.07, for $m_H^2/m_b^2 \ll 1$, by Haber, Schmidt and Schnider [29], whereas it can be as low as 0.009 using the Grinstein, Isgur and Wise approach [30,23]. In the intermediate mass gap one can only afford "reasonable" extrapolations.

These uncertainties are minimized by convenient semiinclusive studies in the different mass regimes. Detailed experimental studies for $2m_\mu < m_H < 2m_\tau$ have been recently presented by the CLEO collaboration [31] (however the most stringent constraints on $B \rightarrow X_s \mu^+\mu^-$ in the $2m_\mu - 2m_\pi$ mass region, where $H \rightarrow 2\mu$ is dominant come from the data of the TASSO collaboration [32]). On the basis of these results, when $m_t \simeq 80\,GeV$ is assumed, a light Higgs in the above mass regime has to be considered excluded. This is summarized in fig. 2, where the various bounds on the *inclusive* $B \rightarrow X_s + H$, as derived in ref. [31], are compared with the theoretical prediction (solid line on top). It is clear that even allowing for large theoretical uncertainties in the exclusive to inclusive branching ratios used in [31], the occurrence of a Higgs boson with mass between $2m_\mu$ and $2m_\tau$ can be excluded with good confidence (for a detailed discussion we refer the reader to [28,33]).

For $2m_e < m_H < 2m_\mu$, where $H \rightarrow e^-e^-$ is the dominant channel open, no relevant bounds can presently be extracted from experimental data on B decays, contrary to the claim in ref. [23] (see the discussion of ref. [33]).

The above results hold in the standard electroweak scenario, where only one Higgs doublet is present. These conclusions are somewhat modified when the Higgs sector is extended and additional neutral and/or charged scalars appear in the physical spectrum. In particular, the presence of charged scalars can affect loop-induced processes, mediated in the SM by the charged gauge bosons. In fact, whereas at the tree level charged Higgs induced transitions are generally suppressed by small Yukawa couplings, at the loop level we can take advantage of the possibility of top exchange and avoid the strong suppression in the scalar couplings.

In the next sections we will consider the effects due to additional scalars on a class of rare B-processes and examine the information that we can extract on the structure of the Higgs sector. In order to do that, it is convenient, before leaving our standard and "preferred" scenario, to summarize the status of the theoretical predictions on the relevant processes.

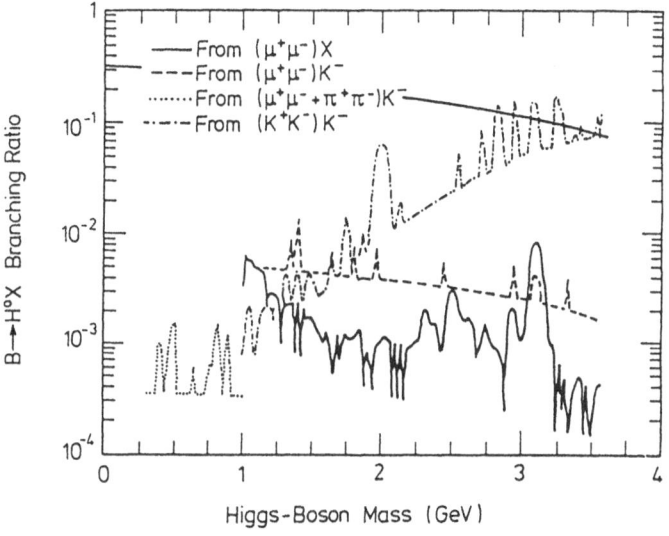

Figure 2. A summary of the experimental limits on $B \to X_s + H$, from the study of various semiinclusive and exclusive channels, is shown as reported in ref. [31], and compared to the SM prediction for $m_t = 80 \ GeV$ (top solid line).

RARE LOOP-INDUCED B-TRANSITIONS IN THE SM

Loop-induced B processes can open a window on mass scales which are not yet directly probed by existing machines. In particular, in the context of the SM, they may give us an indication of where the top mass lies. Theoretical calculation are generally expected to be cleaner than for the case of Kaon processes, since long-distance effects should not sizeably affect transitions which are dominated by heavy quark exchange. Non-spectator contributions are also suppressed by a small wavefunction factor ($f_B^2/m_b^2 \simeq 1/400$) and often disfavoured by mixing angles: as a consequence, they can be safely neglected relative to the spectator quark contributions. Because of asymptotic freedom, also the effects of QCD corrections can be expected to be smaller than in the Kaon system. As we will see, this is in fact true when we consider the *multiplicative* scaling of the effective operators responsible for the various transitions. However, important effects may arise due to operator mixing. In this respect, it is important to take into account the detailed mechanism of flavour- changing (FC) suppression, before drawing wrong conclusions. A typical example is provided by the $b \to s\gamma$ decay.

a) $b \to s\gamma$

The one-loop electroweak amplitude for the quark transition can be written as

$$A(b \to s\gamma) = A_0 \, F(x_t) \, O_\gamma \qquad (6)$$

where $A_0 \equiv \sqrt{2} G_F V_{ts}^* V_{tb}$, $O_\gamma = (e/16\pi^2) m_b \bar{s}_L \sigma^{\mu\nu} b_R F_{\mu\nu}$, and $x_t \equiv m_t^2/m_W^2$. The function $F(x_t)$ can be found for instance in ref. [34]. What is important for the present discussion is that, for $x \ll 1$, one finds $F(x) \sim O(x)$ ("hard" GIM [35] suppression). Already in 1982, this fact prompted Campbell and O' Donnel [36] to emphasize the relevance of this decay as a test for a heavy top quark. Amusingly enough, it turns out that it is just the accidental cancellation of large $\log(m_t^2/m_c^2)$ terms ("soft" GIM suppression) in the one-loop electroweak amplitude, that paves the way for sizeable QCD corrections at higher orders, when gluons are exchanged between internal and external fermion lines [37–40]. Following the notation of eq. (6), the QCD corrected amplitude can in fact be written as

$$A(b \to s\gamma) = \eta^{-16/23} A_0 \left\{ F(x_t) + X \left[\frac{3}{10} \left(\eta^{10/23} - 1 \right) + \frac{3}{28} \left(\eta^{28/23} - 1 \right) \right] \right\} O_\gamma \qquad (7)$$

where $\eta \equiv \alpha_s(m_b)/\alpha_s(m_W)$ and $X = (4/3)[3Q_u - (4/9)Q_d]$ [39], $Q_{u,d}$ being the electric charges of the up and down type quarks respectively. A few comments are in order. The mass of the top quark does not explicitely appear in the term in square brackets (mixing of O_γ with O_1 and O_2) since the top quark field has been integrated away at the same scale as the W gauge boson ($m_t = m_W$). The result thus obtained is expected to hold with an accuracy of 10% for $m_W/2 < m_t < 2m_W$, due to the slow running of $\alpha_s(\mu)$ in that range. The resummation of all leading $(\alpha_s \log)^n$ terms, which eq. (7) provides, is already well approximated by the $O(\alpha_s \log)$ result, as reported in refs. [37.38] (based on the results of an earlier analysis on the $s \to d\gamma$ transition [41]). At first order in α_s, the expression in square brackets in eq. (7) reduces indeed to $(\alpha_s/\pi)\log(m_t/m_b)$. This shows explicitly the replacement of the power flavour-changing suppression, present in the one-loop electroweak transition, with a softer logarithmic cancellation, when a gluon is exchanged. In refs. [37,38,41] only the term proportional to Q_u was kept in the mixing coefficient X, being alone responsible for about 90% of the final result. The naive treatment of γ_5 in obtaining the results of refs. [39,37,38,41], where use of dimensional regularisation is made, has been criticized by the authors of ref. [40], who, using dimensional reduction, did find a smaller effect. Recently, however, the issue has been cleared in favour of the original dimensional regularization results [42].

In terms of the decay rate, the addition of QCD corrections to $b \to s\gamma$ leads to a dramatic effect for $m_t < m_W$ (see fig. 3); however still for $m_t = 80 \; GeV$ almost an order of magnitude enhancement is present, and reduces to about a factor 3 for $m_t = 140 \; GeV$. Indeed, a "strong" effect ! As a counterpart for this experimentally welcomed enhancement, the original sensitivity to the top mass is now softened. The variation of the decay rate in terms of the top mass is in fact smaller than the present uncertainties in the evaluation of the branching fractions for the exclusive decay modes. The channel $K\gamma$ being forbidden by angular momentum conservation, the lightest vector resonance that can appear in the final state is $K^*(892)$ (the interested reader may find a discussion on the allowed two body decay modes in ref. [43], where all Kaon resonances lighter than $D^*(2010)$ are considered). Predictions for $R_{K^*} \equiv \Gamma(B \to K^*\gamma)/\Gamma(B \to X_s\gamma)$ vary, in the literature, between 4% and 40% [43,38,44,45]. The present available upper limits on $BR(B \to K^*\gamma)$ are 2.8×10^{-4} (CLEO [46]) and 4.2×10^{-4} (ARGUS [47]), which, for $R_{K^*} \simeq 10\%$, bound $BR(B \to X_s\gamma)$ to be less than a few parts in a thousand: about one order of magnitude above the QCD enhanced prediction. It is however likely that CLEO II, with the improved resolution in photon detection, may reach the sensitivity needed to test the SM prediction within the end of next year.

b) $b \to se^+e^-$

The amplitude for the transition $b \to se^+e^-$ can be written in terms of effective current-current interactions as

$$
\begin{aligned}
A(b \to se^+e^-) &= A_1(x_t, x_c, q^2)(\bar{s}_L\gamma^\mu b_L)(\bar{e}\gamma_\mu e) \\
&+ A_2(x_t)\frac{q_\nu}{q^2}m_b(\bar{s}_L i\sigma^{\mu\nu} b_R)(\bar{e}\gamma_\mu e) \\
&+ A_3(x_t)(\bar{s}_L\gamma^\mu b_L)(\bar{e}_L\gamma_\mu e_L)
\end{aligned} \tag{8}
$$

where $x_c = m_c^2/m_W^2$ and q is the momentum transferred. The amplitude A_1 receives contributions both from the electromagnetic and Z^0 penguin diagrams. Its dependence on x_c and q^2 is related to the presence of a soft (logarithmic) FC suppression in the electromagnetic penguin. This fact has noteworthy consequences: i) for $m_t < 100 \, GeV$ photon exchange is the dominant mechanism for the decay (however, it is the full A_1 that is gauge-invariant); ii) at variance with $b \to s\gamma$, the relevance of operator mixing, due to QCD effects, is sizeably softened. The second term in eq. (8) is induced by the same operator that is responsible for real photon emission and exhibits the features discussed in the previous subsection. The third term, A_3, is generated by the box-diagram with W exchange and by the $SU(2)_L$ component of the Z^0 penguin.

It is interesting to compare the pattern of QCD corrections with the case of the radiative decay. As a first remark, the components A_1 and A_3 do not receive any *multiplicative* renormalization. This is simply a consequence of the fact that the current $\bar{s}_L\gamma_\mu b_L$ is conserved, in the limit of massless quarks. For conserved currents, Ward identities imply that no divergent counterterm arises. Since the ultraviolet behaviour is independent of the quark masses (only gauge couplings are present), the assertion follows. Sizeable corrections may, nevertheless, come into play through operator mixing (additive effect), as in the case of $b \to s\gamma$ (resp. the amplitude A_2). The analysis of the QCD corrections to $b \to se^+e^-$ has been carried out in the leading-log approximation in refs. [48,49] (it is worth recalling that some of the results in [49] have to be modified, for the reasons mentioned in the previous subsection). At one-

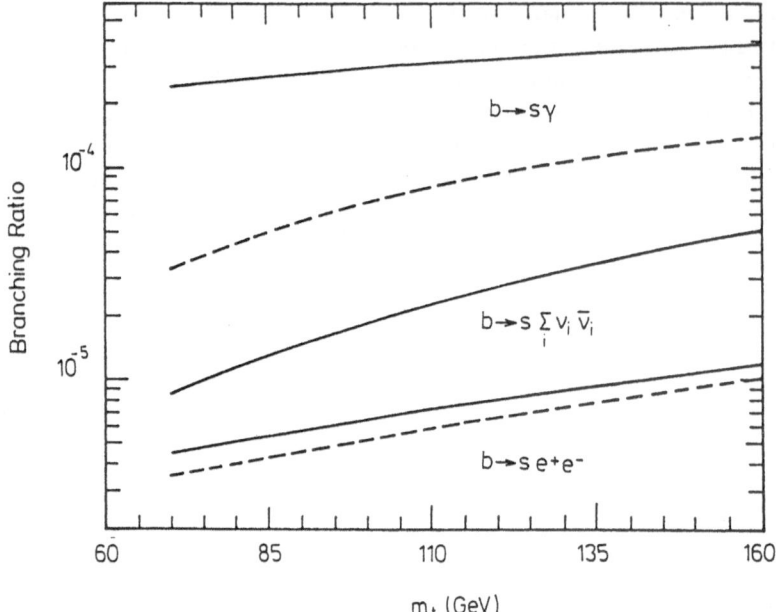

Figure 3. The SM predictions for three loop-induced $b \to s$ transitions, with (solid line) or without (dashed line) QCD corrections, are shown as a function of the top quark mass. The process $b \to s\bar{\nu}\nu$ is not affected by leading QCD effects.

loop, a mixing with the four quark operators O_1 and O_2 arises in the electromagnetic penguin. Its size is dominated by the *electroweak* $\log(m_b^2/m_W^2)$ mentioned above, which, in the effective Hamiltonian approach, appears when the amplitude A_1 is scaled down to the m_b scale. As a consequence, a consistent calculation of the QCD effects should also include the subleading two-loop mixings (see ref. [48] for a discussion on this aspect). For the present purposes it is enough to observe that, because of the large logarithmic term present in the electroweak amplitude, QCD corrections turn out to be relatively small. The largest renormalization effect appears, as expected, in A_2. On the other hand, A_2 is numerically suppressed with respect to A_1, and, once again, the net effect on the final decay rate is small (we recall that the component A_3 is not affected by QCD renormalization at the leading level). These considerations are summarized in fig. 3, which shows the expected branching ratio for the inclusive $B \to X_s e^+ e^-$ process as a function of the top mass, with and without inclusion of QCD corrections. The effect of QCD rescaling is not larger than a 40% increase of the decay rate.

At the exclusive level, modes like Ke^+e^- or $K^*e^+e^-$ are expected to contribute a $10\% - 20\%$ fraction of the inclusive rate respectively [50,51]. Present experimental bounds (see for instance ref. [46]) are still more than one order of magnitude away from testing the SM predictions.

c) $b \to s\bar{\nu}\nu$

A process closely related to the previous one is $b \to s\bar{\nu}\nu$. Although there are not experimental bounds available yet (vertex detectors are needed in order to screen

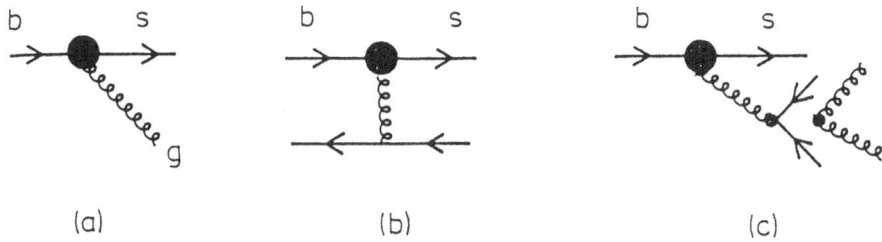

Figure 4. The different contributions to $b \to s$ "$gluc$", as described in the text.

the charmed induced bakground), it is worth considering it for the specific features which differentiate this decay from $b \to se^+e^-$. The amplitude in the SM is readily written as

$$A(b \to s\bar{\nu}\nu) = A_3'(x_t)(\bar{s}_L\gamma^\mu b_L)(\bar{\nu}_L\gamma_\mu\nu_L) \qquad (9)$$

where $A_3'(x_t)$ receives contributions from the W-mediated box-diagram and the Z^0 penguin. From what we have learned in the previous two subsections, two features can be immediately deduced from the form of eq. (9) : i) since photon exchange is absent, the process exhibits a "hard" GIM suppression, and therefore is more sensitive to the top quark mass; ii) due to the form of the hadronic current, the amplitude is free from QCD renormalization effects. As a consequence, the original electroweak dependence on the top mass is preserved. For $m_t > 80 \; GeV$, considering that three neutrino species are produced, one finds that $BR(B \to X, \sum \bar{\nu}\nu)$ is above 10^{-5}, as shown in fig. 3. Once again, exclusive final states such as $K\bar{\nu}\nu$ or $K^*\bar{\nu}\nu$, are expected to account for a $10-20\%$ of the total rate.

d) $b \to s$ "$glue$"

It has become customary to denote by $b \to s$"g" the set of contributions to hadronic B decays with net strangeness, induced by the effective $b - s - gluon$ vertex. Three different classes of diagrams can generally be distinguished and characterized by the kinematical regime of the momentum q of the gluon : i) $q^2 = 0$, fig. 4a ; ii) $q^2 < 0$, non-spectator diagrams, fig. 4b ; iii) $q^2 > 0$, fig. 4c. Within the SM framework it turns out that the contributions corresponding to fig. 4c are largely dominant. Once again, on the basis of the previous discussions, it is not difficult to guess why. The graph in fig. 4a is subleading because of the absence of the large $\log(m_W/m_b)$ terms present in *both* the contributions 4b and 4c (virtual gluon exchange), analogously to the electromagnetic penguin. On the other hand, the non-spectator graph in fig. 4b is suppressed because of the B-meson wavefunction factor $(f_B^2/m_B^2 \ll 1)$. The emission of two gluons (fig. 4c) has also been partially included in ref. [54], and shown to give a a contribution comparable to the four-quark operator.

The problem of QCD corrections for this class of transitions has been considered in ref. [55]. In fig. 5 a summary of the QCD corrected and uncorrected results for $BR(b \to s$"g") is presented, as reported in ref. [55]. Although the curves refer to the dimensional reduction result, which is incorrect, it turns out that, once the needed changes are made, only the region of $m_t < 60 \; GeV$ is sensibly affected [42].

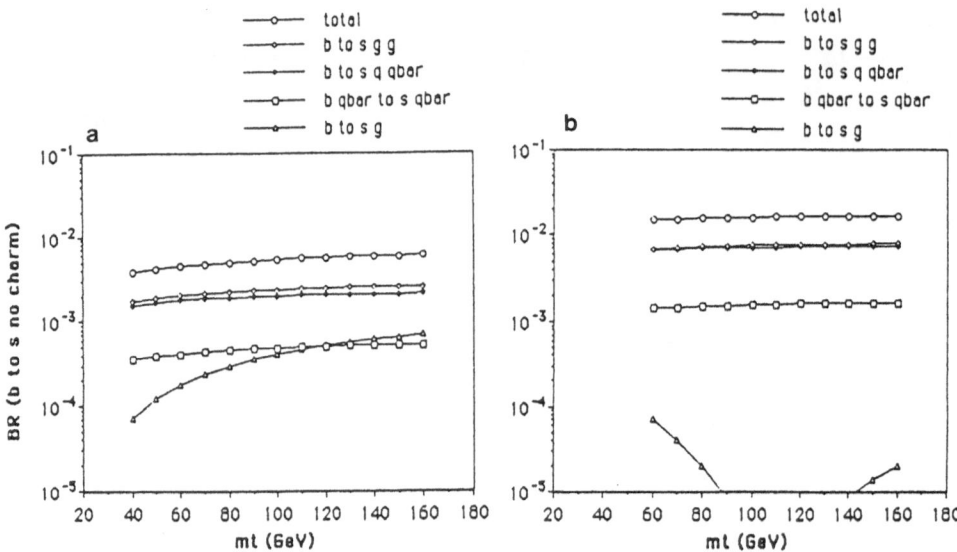

Figure 5. Predictions for the QCD uncorrected (a) and QCD corrected (b) rates for
$b \to s$ "*glue*" transitions, as a function of the top mass. The figures are taken from
ref. [55].

The different features (top dependence, size, ...) of the uncorrected amplitudes,
that we have emphasized in previous cases, are easily recognized in fig. 5a. Also
the pattern of QCD corrections closely follows the discussion in subsections *(a)* and
(b). The largest renormalization effect appears in the $q^2 = 0$ term, analogously to
$b \to s\gamma$, with a noteworthy difference : here, due to destructive interference in the
operator mixing, the corrected amplitude for $b \to s$ *gluon* is further, and dramatically,
suppressed.

With $BR(B \to X_s$ "*no charm*") at the percent level in SM, we may reasonably
expect exclusive few-body final states, such as $K\pi$ or $K\varphi$, in the $10^{-4} - 10^{-5}$ range
[56]. This is not too far from the present experimental bounds [46,57].

e) $B^0 - \bar{B}^0$ *mixings*

In the SM the leading $\Delta B = 2$ transition, responsible for the $B_q^0 - \bar{B}_q^0$ mixings
($q = d, s$), is induced by box-diagrams with W and top exchange, giving rise to the
local operator $(\bar{q}_L \gamma^\mu b_L)^2$. For $m_t > m_W$, the SM prediction fits naturally the experimental result for $B_d^0 - \bar{B}_d^0$ mixing (for a recent and detailed review on the subject see
ref. [58]). QCD corrections are multiplicative (we recall here that the multiplicative
scaling does not depend on the nature of the "heavy" particles exchanged in the loop),
and affect by a factor $\eta^{-6/23} \simeq 0.86$ the electroweak transition [59]. In ref. [60] it is
however argued that the effective low energy scale at which the RG equations are to
be integrated should be given by $\sqrt{m_b m_q}$ instead of m_b. In this case the correction
factor reduces to $\simeq 0.6$. Since in the remainder of this talk we will be interested only
in the relative size of the additional contributions with respect to the standard ones,
this multiplicative uncertainty will not affect our conclusions.

In this section we have briefly reviewed some effective FCNC transitions which can provide (with the caveat of the still large uncertainties in the hadronic matrix elements) further tests of the minimal $SU(3)_C \times SU(2)_L \times U(1)_Y$ structure of the strong and electroweak interactions. We have particularly emphasized the features related to the presence of a heavy top quark and the effects of QCD corrections. In the next two sections we will consider two typical (and pedagogical) extension of the SM which exhibit an enlarged Higgs sector, and we will see how this can affect the SM predictions. In order to incorporate QCD corrections in the extended framework it is convenient to observe that, if

i) no *new* effective operators are added to the basis relevant for the process considered (this is, for instance, the case if all the *additional* particles that can be exchanged in the loop are "heavy", with respect to the characteristic scale of the process).

ii) possible additional tree level contributions to effective four-fermion operators, like $O_{1,2}$, are suppressed (a typical case is when Yukawa couplings are involved),

iii) the newly exchanged particles can be integrated away together with the W gauge boson (we will assume the same range of applicability as for the case of the top quark),

then, as it should be clear from the previous discussions, it follows that

a) the new contributions are only multiplicatively renormalized,

b) the correction factors are the same as for the SM amplitudes.

Conditions i) and ii) are generally satisfied in the models considered here. In addition, we will be interested in mass ranges for the new particles which satisfy iii). In this case, the extension of QCD corrections to non-standard contributions becomes straightforward.

A TWO-HIGGS DOUBLET MODEL

A simple Higgs extension of the SM is obtained by duplicating the minimal scalar doublet. It is known that the addition of Higgs doublets or singlets (with non-vanishing vacuum expectation values) does not affect the tree level identity $m_W/m_Z \cos\theta_W = 1$ (Veltman's ρ parameter). which is experimentally well established. Phenomenological problems may however arise in the Yukawa sector, since the simultaneous coupling of different scalar multiplets to fermions, spoils, in general, the simultaneous diagonalization of fermion masses and Yukawa couplings. As a consequence, tree level FC currents mediated by neutral scalars arise, which induce potentially large contributions to a plethora of rare FCNC processes. This problem can be overcome by "adjusting" the masses of the relevant scalars at a scale high enough to suppress the "unwanted" contributions. A second approach is to impose, by the introduction of an *ad hoc* discrete symmetry, that at most one neutral Higgs field is coupled to a given isospin component of the fermion doublets [61]. This solution is considered more "natural". according to the "symmetry dogma" which guides present day model building (sometimes also interpreted as : " symmetries are always beautiful. no matter how awful they may look ").

For the sake of the present discussion, we will adopt a specific realization of the second alternative, in which, with the help of a "simple" Z_2 symmetry, one Higgs doublet, say H_1, couples to the right-handed component of down-type fermions, whereas H_2 couples to the right-handed up-type fermions. This scenario is naturally achieved in supersymmetric extensions of the SM, as we will discuss in the next section. A detailed description of the model and references can be found in ref. [62].

After spontaneous symmetry breaking, five scalar degrees of freedom are left in the physical spectrum: three are neutral (two CP even, $H_{1,2}^0$, and a CP odd, H_3^0), and two are charged ($H^{\pm}$). The various scalar couplings are described in terms of two mixing angles, usually denoted by α and β, of which one is simply related to the two vacuum expectation values v_1 and v_2, say $\tan \beta = v_2/v_1$ (as a reminder, v_2 is responsible for the mass of the up-type fermions). It is useful for the discussion that follows to report explicitely the form of the couplings of the physical charged scalar to quarks:

$$\frac{g}{\sqrt{2}} \left[\frac{m_{u_i}}{m_W} \tan^{-1} \beta \bar{u}_{Ri} K_{ij} d_{Lj} + \frac{m_{d_j}}{m_W} \tan \beta \bar{u}_{Li} K_{ij} d_{Rj} \right] H^+ + h.c. \qquad (10)$$

In eq. (10), K_{ij} are the elements of the usual Cabibbo-Kobayashi-Maskawa matrix, and a sum over $i, j = 1, 2, 3$ is understood.

a) $b \rightarrow s + H_i^0$

The decay $B \rightarrow X_s + H^0$ may now involve more than one scalar, and an additional contribution arises from charged Higgs exchange. In addition to m_t, the total amplitude depends therefore on m_{H^+} and the two mixing angles α and β (for the production of H_3^0, only β appears). The various possibilities have been discussed in the literature [63]. Because of the number of free parameters involved no definite conclusion can be inferred on limits on the light scalar masses. Sets of parameter values exist for which large cancellations occur between the W and H^+ induced contributions, thus leading to negligeably small decay rates for the process, although kinematically unsuppressed.

b) Other Loop-Induced $b \rightarrow s$ Transitions and Mixing

What are the effects of the additional scalar contributions to the rare B processes discussed in the previous section ? As it follows from eq. (10), in addition to the top mass, the total amplitudes depend on the charged Higgs mass and the ratio of the two vacuum expectation values. In particular, from the structure of the relevant quark currents (see the previous section) and the form of the couplings in eq. (10), it can be inferred that at least part of the amplitudes are proportional to $\tan^{-1} \beta = v_1/v_2$. This fact has been originally exploited in the literature in order to claim for potential large enhancements, for this class of processes. when a large "inverse hierarchy" of the two vacuum expectation values is present (remember that with our conventions the down- and up-type fermion masses are proportional to v_1 and v_2 respectively). This effect is in fact particularly relevant for the $B^0 - \bar{B}^0$ mixing: there the additional box

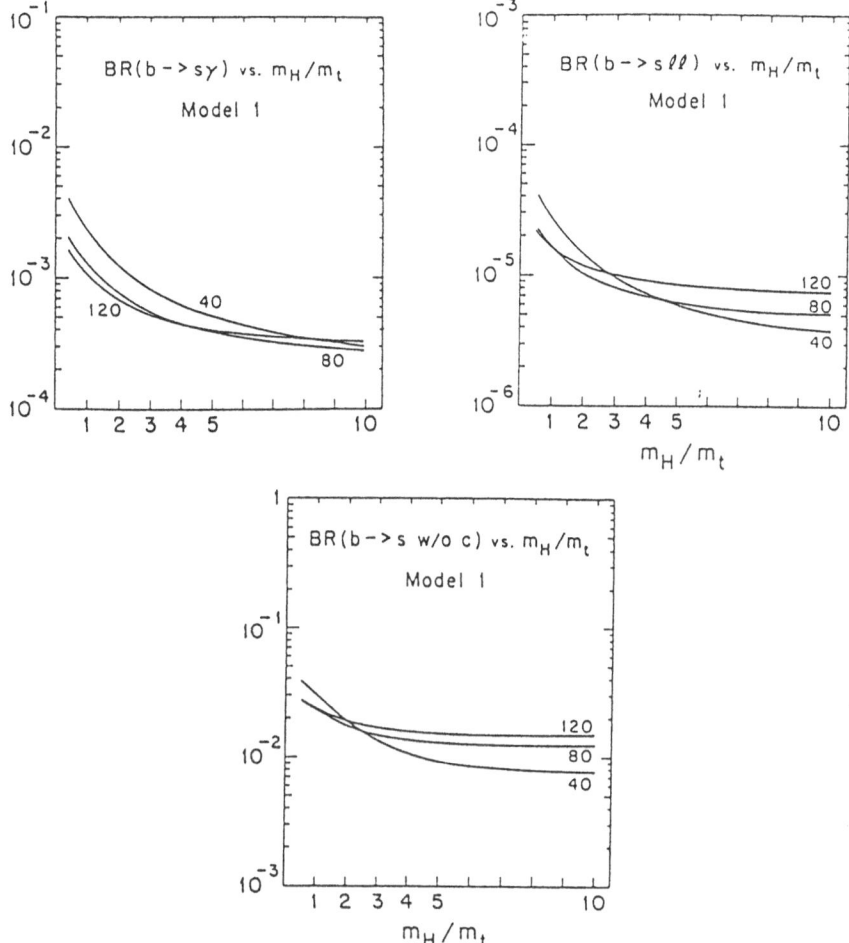

Figure 6. Contributions to various $b \to s$ transitions in the two Higgs doublet model as a function of the charged Higgs boson mass, for $m_t = 40, 80, 120\ GeV$. The figures are taken from ref. [54]. See the text for comments.

diagrams with $W^+ H^+$ and $H^+ H^+$ exchange are proportional to $(v_1/v_2)^2$ and $(v_1/v_2)^4$ respectively (and interfere constructively with the WW contribution). Given therefore a range of values for the SM parameters, a combined constraint can be extracted on $\tan^{-1} \beta$ and m_{H^+} from the comparison with the present measured value for the $B_d - \bar{B}_d$ mixing [52,53]. This comparative analysis has been presented in ref. [54], from which figs. 6a,b,c are borrowed. By comparison with the SM results shown in figs. 3 and 5, it is clear that, when the constraint from the measured mixing is included, only an enhancement of at most a factor 3-4 is allowed for the various decays in the presence of a light charged Higgs boson.

In fig. 6b, QCD corrections were only partially included: the correction to A_1 in eq. (8) is missing. However, from the discussion following eq. (8) and the results of fig. 3, it follows that this omission is of negligeable numerical consequence.

A more complete analysis, which includes also some contraints coming from Kaon physics (Δm_K, ϵ, ϵ'/ϵ, $K^+ \to \pi^+ \bar{\nu} \nu$), is now available [64], while others are in progress [65]. From ref. [64], I like in particular to mention that the inclusion of the non-zero NA31 result on ϵ'/ϵ induces a constraint on v_1/v_2 2-3 times more severe, than when considering the mixing alone. This is enough to exclude any sensible enhancement in the class or rare B decays considered, due to charged Higgs exchange.

THE MINIMAL SUPERSYMMETRIC STANDARD MODEL

In this section we will examine a particularly simple supersymmetric (SUSY) extension of the SM, derived from spontaneously broken minimal N=1 supergravity (for a review see refs. [66,67]). The field content of the model corresponds to the minimal supersymmetrization of the SM spectrum (two Higgs doublets are required by the chiral structure of the superfields). In addition, we will invoke a grand-unified scenario at a scale $M_X \simeq 10^{16}\ GeV$. R-parity is kept as a good symmetry in order to prevent a too fast proton decay. As just mentioned, the minimal Higgs strucure required reproduces the scenario discussed in the previous section. The presence of a larger symmetry limits the form of the terms present in the Higgs potential. Quartic couplings, for instance, are directly related to the usual gauge couplings (D-terms).

The spontaneous breaking of local supersymmetry (supergravity) in the "hidden" sector at a scale M_{SUSY} (in the following we will neglect any effect related to the difference between M_{SUSY} and M_X), produces an effective *globally* supersymmetric Lagrangian with the addition of *soft* breaking terms (which have dimension less than four, with the exception of the gaugino mass terms [68]). The crucial non renormalization properties of the supersymmetric theory are then preserved.

The presence of the soft breaking terms introduces, in the minimal form, three parameters: i) a mass m common to every scalar; ii) a gaugino mass term M; iii) a dimensionful trilinear coupling, usually parametrized as Am. In addition a fourth dimensionful parameter μ appears in the superpotential, in the $H_1 H_2$ coupling. For given values of the *four* abovementioned parameteres, and of the standard gauge and Yukawa couplings, the structure of the effective Lagrangian is completely determined.

Let us now examine the conditions needed in order to achieve a correct breaking of $SU(2)_L \times U(1)_Y$. At the M_X scale, the scalar potential along the neutral direction is given by:

$$V^{(0)} = m_1^2 |H_1^0|^2 + m_2^2 |H_2^0|^2 - m_3^2 (H_1^0 H_2^0 + h.c.) + \frac{g^2 + g'^2}{8}(|H_1^0|^2 - |H_2^0|^2)^2 \qquad (11)$$

where (at M_X)

$$m_1^2 = m_2^2 = m^2 + \mu^2, \quad m_3^2 = -(A-1)m\mu \qquad (12)$$

The existence of an absolute minimum and the request of spontaneous breaking of the electroweak symmetry are granted by the conditions

$$m_1^2 + m_2^2 > 2|m_3^2|, \quad m_1^2 m_2^2 < m_3^4 \qquad (13)$$

(a recent and detailed discussion on the problem of stable color- and electric- charge breaking vacua can be found in ref. [69]). It is clear from eq. (12) that the above inequalities cannot be simultaneously satisfied at M_X. In other words, no tree level breaking of $SU(2)_L \times U(1)_Y$ can be achieved in the present model.

It was observed that the addition of a singlet superfield is enough to trigger a correct electroweak breaking [70], and explicit implementations have been discussed in the literature. In this class of models at least one extra dimensionful parameter appears. On the other hand, when considering the renormalization of the Lagrangian down to the Fermi scale (more precisely to the scale of the processes considered), the relations in eq. (12) may be sufficiently modified so as to satisfy the conditions in eq. (13), thus avoiding any further extension of the model [71]. The driving terms in the RG equations for the relevant parameters in the scalar potential are the top mass and the common gaugino mass M. In particular the form of the equations suggest that m_2 is driven to values substantially smaller than m_1. Explicit numerical solutions show indeed that the correct breaking is achieved for a large range of initial boundary conditions at M_X. A stable vacuum then arises for $\langle H^0_{1,2} \rangle = v_{1,2} > 0$, while the following relations hold [72] :

$$\sin 2\beta \;=\; \frac{2m_3^2}{m_1^2 + m_2^2}, \tag{14}$$

$$\tan^2 \beta \;\equiv\; \left(\frac{v_2}{v_1}\right)^2 = \frac{m_1^2 + m_Z^2/2}{m_2^2 + m_Z^2/2} \;\; (> 1) \tag{15}$$

The inequality in eq. (15) just follows from the pattern of renormalization which leads to $m_2^2 < m_1^2$. This result is very important for our present discussion: no inverse hierarchy of the vacuum expectation values is allowed in the minimal supersymmetric standard model (MSSM), independent of experimental constraints. On the contrary, the contributions proportional to $\tan^{-1} \beta$ tend to be suppressed, as we will see below.

From eq. (15) a second constraint on v_2/v_1 follows. For a given value of m_t, $\tan \beta$ is indeed generally bounded to be smaller than $\sim m_t/m_b$. In fact, for such a large ratio of v_2/v_1 the Yukawa couplings responsible for top and bottom masses become comparable. This, in turn, entails a comparable renormalization of m_1 and m_2, thus jeopardizing the conditions for obtaining the correct vacuum (eq. (13)). A detailed numerical analysis of this bound is given in ref. [73].

It is useful to recall that the presence of supersymmetry, although softly broken, implies strict relations among the scalar masses. In particular one obtains

$$
\begin{aligned}
m_{H^+}^2 \;&=\; m_W^2 + m_{H_3^0}^2 \\
m_{H_2^0} \;&<\; m_Z < m_{H_1^0} \\
m_{H_3^0} \;&>\; m_{H_2^0}
\end{aligned}
\tag{16}
$$

where the notation follows the one introduced in the previous sections. In particular it is important to notice that: i) there exists at least one neutral scalar which is lighter than m_Z; ii) the charged Higgs boson is always heavier than m_W.

Before proceeding to the discussion of the numerical results, we have to recall that a novel feature, induced in this class of model by renormalization effects, is the presence of gluino-squark-quark (analogously. neutralino-squark-quark) FC vertices [74]. This may in fact be relevant for those FC processes which can be induced by gluino exchange, since one can take avantage of the replacement of the weak coupling by the strong one. We will come back to this in the following.

a) $b \rightarrow s + H_i^0$

The possibility of the emission of a light scalar in a minimal supersymmetric scenario has been considered in refs. [75,76]. In ref. [75] the role of gluino and chargino exchange were analized. In particular, it was found that there is a cancellation of leading terms in the gluino mediated diagrams, when the lightest CP even scalar (in our notation H_2^0) is produced, thus making gluino exchange negligeable. On the other hand, it was shown that this cancellation is not present when the CP odd scalar (H_3^0) is emitted, leading to phenomenologically excluded branching ratios. One must say, however, that the possibility of having such a light pseudoscalar is, in this class of models, highly disfavoured (for a detailed discussion see ref. [75]). The contributions of charginos were shown to be possibly relevant only for a light top quark.

For the case of H_2^0 emission, a complete analysis of the process is given in ref. [76]. There it is argued that, for the range of parameters that allow for a sufficiently light scalar, there is a contrived cancellation between the W and H^+ induced contributions (the only two "large" contributions present in the model), such that the total SUSY amplitude turns out to be always smaller than that of the SM. Nonetheless, there are region of parameters, for which a light SUSY Higgs boson can be excluded on the basis of the experimental bounds discussed in the second section.

b) Other Loop-Induced $b \rightarrow s$ Transitions and Mixings

It was pointed out already a few years ago that the presence of effective gluino-squark-quark FC couplings, may have a sizeable impact on a class of rare flavour violating B decays [77,78]. What was argued to be a necessary (but not sufficient) requirement in order to achieve some SUSY enhancement, is that the process under consideration exhibits, in the SM, a "hard" GIM suppression. This request singled out $b \rightarrow s\gamma$ and $b \rightarrow s\ gluon$ ($q^2 = 0$) as promising processes. It was in fact shown [77] that radiative B decays could reach, at the inclusive level, branching ratios above 10^{-3} (the QCD corrected SM prediction is at the level of a few times 10^{-4}), for squark and gluino masses at the W mass scale. Analogously, $b \rightarrow s\ gluon$ may induce non-charmed decays at the 10% level, which is within the experimental error on charm counting [2].

At present, however, limits on squark and gluinos masses have grown above $100\ GeV$, thus spoiling the leading role of gluino exchange., A more detailed and complete analyisis of all the contributions to rare processes, present in the model, is needed in order to ascertain possible SUSY effects. This has been done in ref. [79], where also a careful study of the process of renormalization is presented. For each process there are five classes of contributions, which can be characterized as follows : a) W exchange; b) charged Higgs exchange; c) chargino exchange; d) gluino exchange; e) neutralino exchange. The last two classes of diagrams are present because of the flavour-changing vertices induced by renormalization effects.

It turns out, that when the present experimental bounds on SUSY particles are implemented, the contributions (c)-(e) become generally small (with the exception of $b \to s$ $gluon$, where (d) still plays a relevant role due to the large gluino-gluino-gluon coupling). On the other hand, the effect of charged Higgs exchange is minimal, since, as we know, m_{H^+} is bound to be larger than m_W, and actually substantially heavier values are favoured in the parameter space.

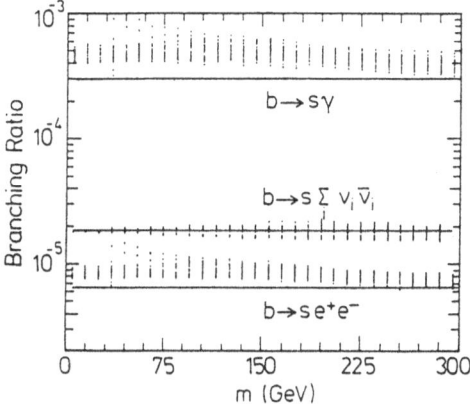

Figure 7. Contributions to various $b \to s$ transitions in the minimal supersymmetric standard model, compared to the SM predictions (solid horizontal lines). The figure is taken from ref. [79]. See the text for further comments.

Some of the results are summarized in figs. 7 and 8, for $m_t = 100 \ GeV$ and $\tan \beta = 2$ (a discussion on the influence of these two parameters is found in [79]: no substantially different features are found). The figures show the size of the various processes as a function of the soft breaking mass m. The other parameter needed to determine the structure of the model completely is taken here to be the gaugino mass M, which is varied between $\pm 200 \ GeV$. Its variation is responsible for the vertical width of the band which represents the SUSY result. The corresponding SM results are also reported (solid lines). For the case of the $B_d^0 - \bar{B}_d^0$ mixing, the ratio with the standard transition amplitude is shown, which is independent on the hadronic matrix element. Only deviations of at most a few tens of percent with respect to the SM result are possible. This result is not surprising, since a "direct" hierarchy of the vacuum expectation values suppresses the additional Higgs contributions (see the discussion in the previous section). The smallness of the gluino contributions in the case of the mixing had been already emphasized in ref. [80].

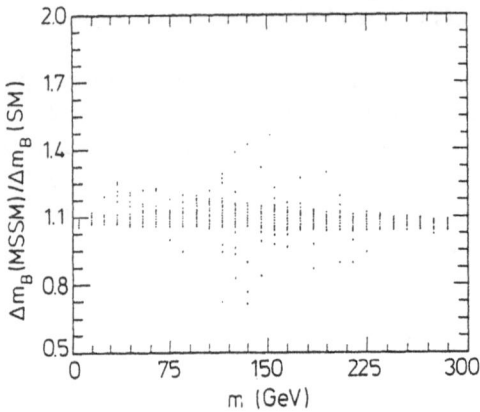

Figure 8. Ratio of the MSSM contribution to the $B_d^0 - \bar{B}_d^0$ mixing over the corresponding SM amplitude. An analogous result holds for the $B_s^0 - \bar{B}_s^0$ mixing. The figure is taken from ref. [79].

Analogous conclusions hold for the $B_s^0 - \bar{B}_s^0$ mixing. The MSSM contributions to hadronic non-charmed B decays are also confined to branching ratios below the few percent level. In general, no large effects are induced by supersymmetry, given the present experimental constraints.

It is important to recall that these results are obtained in the minimal framework, and therefore do not exclude in general the possibility of measurable supersymmetric effects in other contexts. At any rate, they can provide a useful "guidance" for further analysis.

CONCLUSIONS

We have examined various implications between rare B processes and the Higgs sector, focussing on some minimal (and popular) extensions of the standard scenario. As far as the SM Higgs boson is concerned, the recent experimental studies on the $B \to X_s + H^0$ decays exclude, for $m_t \simeq 80~GeV$, a light Higgs scalar with mass in the $2m_\mu - 2m_\tau$ range. Bounds from $\Upsilon \to H^0 + \gamma$ are at present weaker, due to the large uncertainties in the QCD corrections and an additional suppression due to relativistic bound state corrections.

Rare B processes, like loop induced FC decays and mixings, may be a good test for extensions of the standard Higgs scenario, where physical charged scalars are present. For this purpose we have attempted a brief summary of the theoretical predictions within the standard electroweak model, paying particular attention to the role of QCD effects and the top quark, and indicated how they may affect additional contributions.

Recent analysis of two-Higgs doublet models. show that no sensible enhancement over the SM predictions can be achieved when the constraints deriving from present experimental data on the $B_d - \bar{B}_d$ mixing and Kaon physics are taken into account. Of particular relevance for this result is the possibility of a non zero value of ϵ'/ϵ, as recently claimed by the NA31 collaboration.

In the minimal supersymmetric extension of the SM, aside from the leading W exchange, charged Higgs bosons play (or have a chance to play) the most relevant role, among the various additional contributions. However, only mild enhancements are possible.

In conclusion, the accumulation of precise experimental data, together with the direct available bounds on new particles, pushes the possibility of the discovery of exotic physics more and more forward in energy and time, and leaves seemingly space only for tiny indirect effects. This being the case, it becomes questionable whether rare B decays will play a relevant role in such precision tests, mainly because of the crucial theoretical uncertainties related to the evaluation of the hadronic matrix elements. Lattice results are in this respect critical. On the other hand, this should not discourage the experimental effort in this direction. After all, Nature might be far more "imaginative" than we are, and (why not ?) surprises may just be waiting "around the corner".

Acknowledgements

I would like to express my gratitude to A. Ali for inviting this contribution and for the opportunity of participating in a stimulating workshop. I am also indebted to F. Borzumati, A Masiero and G. Ridolfi for the collaboration on part of the results reported. I profited from discussions with A. Ali, H. Dreiner, G. Grigjanis, J.F. Gunion, H.E. Haber, J.L. Hewett, P. Krawczyk, J. Lee-Franzini, I. Montvay, H. Steger, M. Sutherland, J. Swaine, R.S. Willey, and D. Wyler.

REFERENCES

[1] NA31 Coll., H. Burkhardt et al., *Phys. Lett.* **B 206** (1988) 169.

[2] H. Albrecht, these Proceedings.

[3] G. Altarelli, *Theory of Precision Electroweak Experiments*, talk presented at the XIV International Symposium on Lepton and Photon Interactions, Stanford CA, August 6-12, 1989, to appear in the Proceedings.

[4] P.W. Higgs, *Phys. Rev. Lett.* **12** (1964) 132.

[5] D. Callaway, *Phys. Rep.* **167** (1988) 241.

[6] J. Jersak, these Proceedings.

[7] A.D. Linde, *JETP Lett.* **23** (1976) 64; S. Weinberg, *Phys. Rev. Lett.* **36** (1976) 294.

[8] H.D. Politzer and S. Woltram, *Phys. Lett.* **B 82** (1979) 242; P.Q. Hung, *Phys. Rev. Lett.* **42** (1979) 873.

[9] M.J. Duncan, R. Phylippe and M. Sher *Phys. Lett.* **B 153** (1985) 165; M. Sher and H.W. Zauglauer, *Phys. Lett.* **B 206** (1988) 537; M. Lindner, M. Sher and H.W. Zauglauer, *Phys. Lett.* **B 228** (1989) 139.

[10] M. Sher, *Phys. Rep.* **179** (1989) 273.

[11] M.E. Zeller, these Proceedings; H.L. Yu, these Proceedings.

[12] F. Wilczek, *Phys. Rev. Lett.* **39** (1977) 1304.

[13] J. Lee-Franzini, Proceedings of the XXIV Int. Conference on *High Energy Physics*, R. Kotthaus and J.H. Kühn (Eds.), Springer-Verlag, Berlin (1989); these Proceedings.

[14] M.I. Vysotsky, *Phys. Lett.* **B 97** (1980) 159; J. Ellis, K. Enquist, D.V. Nanopoulos and S. Ritz, *Phys. Lett.* **B 158** (1985) 417, (E) *ib.* **163** (1985) 408; P. Nason, *Phys. Lett.* **B 175** (1986) 223.

[15] R. Barbieri, R. Gatto, R. Kögerler and Z. Kunzt, *Phys. Lett.* **B 57** (1975) 455.

[16] H. Goldberg and Z. Ryzak, *Phys. Lett.* **B 218** (1989) 348.

[17] G. Fäldt, P. Osland and T.T. Wu, *Phys. Rev.* **D 38** (1988) 164.

[18] I.G. Aznauryan, S.S. Grigorian and S.G. Matinyan, *JETP Lett.* **43** (1986) 646.

[19] R.S. Willey and H.L. Yu, *Phys. Rev.* **D 26** (1982) 3086; *ib.* **26** (1982) 3287.

[20] B. Grzadkowski and P. Krawczyk, *Zeit. für Physik* **C 18** (1983) 43.

[21] T.N. Pham and D.G. Sutherland, *Phys. Lett.* **B 151** (1985) 444; F. Botella and C. Lim, *Phys. Rev. Lett.* **56** (1986) 161; R. Willey, *Phys. Lett.* **B 173** (1986) 480; R.M. Godbole, U. Türke and M. Wirbel. *Phys. Lett.* **B 194** (1987) 302.

[22] R.S. Chivukula and A.V. Manhoar, *Phys. Lett.* **B 207** (1988) 86; (E) *ib.* **217** (1989) 568.

[23] B. Grinstein, L. Hall and L. Randall, *Phys. Lett.* **B 211** (1988) 363.

[24] J. Ellis, M.K. Gaillard and D.V. Nanopoulos, *Nucl. Phys.* **B 106** (1976) 292.

[25] J. Ellis and P.J. Franzini, CERN report, CERN-TH 4952/88 (1988); J. Ellis and F. Pauss, in *Proton-Antiproton Collider Physics*, World Scientific, Singapore (1988); M.S. Chanowitz, *Ann. Rev. Nucl. Part. Sci.* **38** (1988) 323.

[26] H.E. Haber, these Proceedings.

[27] R.S. Willey, these Proceedings.

[28] J.F. Gunion, H.E. Haber, G.L. Kane and S. Dawson, *The Higgs Hunter's Guide*, UC Davis report, UCD-89-4 (1989), to be published in *Frontiers in Physics*, Addison-Wesley (1990).

[29] H.E. Haber, A.S. Schwarz and A.E. Snyder, *Nucl. Phys.* **B 294** (1987) 301.

[30] B. Grinstein, N. Isgur and M.B. Wise, *Phys. Rev. Lett.* **56** (1986) 298; B. Grinstein, N. Isgur, D. Scora and M.B. Wise, *Phys. Rev.* **D 39** (1989) 799.

[31] CLEO Coll., M.S. Alam et al., *Phys. Rev.* **D 40** (1989) 712.

[32] TASSO Coll., M. Althoff et al., *Zeit. für Physik* **C 22** (1984) 219.

[33] S. Dawson, J.F. Gunion and H.E. Haber, UC Davis report, UCD-89-12 (1989).

[34] T. Inami and C.S. Lim, *Prog. Theor. Phys.* **65** (1981) 297.

[35] S.L. Glashow, J. Iliopoulos and L. Maiani, *Phys. Rev.* **D 2** (1970) 1285.

[36] B.A. Campbell and P.J. O'Donnell, *Phys. Rev.* **D 25** (1982) 1989.

[37] S. Bertolini, F. Borzumati and A. Masiero, *Phys. Rev. Lett.* **59** (1987) 180.

[38] N.G. Deshpande, P. Lo, J. Trampetic, G. Eilam and P. Singer, *Phys. Rev. Lett.* **59** (1987) 183.

[39] B. Grinstein, R. Springer and M. Wise, *Phys. Lett.* **B 202** (1988) 138.

[40] R. Grigjanis, P.J. O'Donnel, M. Sutherland and H. Navelet, *Phys. Lett.* **B 213** (1988) 355.

[41] M.A. Shifman, A.I. Vainshtein and V.I. Zacharov, *Phys. Rev.* **D 18** (1978) 2583.

[42] M. Sutherland, talk presented at the Workshop on $B\bar{B}$ *Factories and Related Physics Issues*, Blois, France, June 26-July 1, 1989; R. Grigjanis, P.J. O'Donnel, M. Sutherland and H. Navelet, University of Toronto report, UTPT-89-21 (1989).

[43] T. Altomari, *Phys. Rev.* **D 37** (1988) 677.

[44] C.A. Dominguez, N. Paver and Riazuddin, *Phys. Lett.* **B 214** (1988) 459.

[45] P.J. O'Donnel, in *Quark Gluons and Hadronic Matter*, Eds. R. Viollier and N. Warner, World Scientific, Singapore (1987); *Phys. Lett.* **B 175** (1986) 369.

[46] CLEO Coll., P. Avery et al., *Phys. Lett.* **B 223** (1989) 470.

[47] ARGUS Coll., H. Albrecht et al., *Phys. Lett.* **B 229** (1989) 304.

[48] B. Grinstein, M.J. Savage and M.B. Wise, *Nucl. Phys.* **B 319** (1989) 271.

[49] R. Grigjanis, P.J. O'Donnel, M. Sutherland and H. Navelet, *Phys. Lett.* **B 223** (1989) 239.

[50] N.G. Deshpande and J. Trampetic, *Phys. Rev. Lett.* **60** (1988) 2583.

[51] D. Wyler, talk presented at the Workshop on $B\bar{B}$ *Factories and Related Physics Issues*, Blois, France, June 26-July 1, 1989;

[52] ARGUS Coll., H. Albrecht et al., *Phys. Lett.* **B 209** (1988) 119.

[53] CLEO Coll., M. Artuso et al., *Phys. Rev. Lett.* **62** (1989) 223.

[54] W.-S. Hou and R.S. Willey, *Nucl. Phys.* **B 326** (1989) 54.

[55] R. Grigjanis, P.J. O'Donnel, M. Sutherland and H. Navelet, *Phys. Lett.* **B 224** (1989) 209.

[56] L.-L. Chau and H.Y. Cheng, *Phys. Rev. Lett.* **59** (1987) 958; M.B. Gavela, A. Le Yaouanc, L. Oliver, O. Pene, J.C. Raynal, M. Jarfi and O. Lazrak, *Phys. Lett.* **B 154** (1985) 425.

[57] ARGUS Coll., H. Albrecht et al., DESY report, DESY 89/096 (1989).

[58] P.J. Franzini, *Phys. Rep.* **1** (1989) 1.

[59] A.J. Buras, W. Slominski and H. Steger, *Nucl. Phys.* **B 245** (1984) 369.

[60] W.A. Kaufman, H. Steger and Y.-P. Yao, *Mod. Phys. Lett.* **A3** (1988) 1479.

[61] S.L. Glashow and S. Weinberg, *Phys. Rev.* **D 15** (1977) 1958; E.A. Paschos, *Phys. Rev.* **D 15** (1977) 1966.

[62] S. Bertolini, *Nucl. Phys.* **B 272** (1986) 77.

[63] L.J. Hall and M.B. Wise, *Nucl. Phys.* **B 187** (1981) 397; J.M. Frère, M.B. Gavela and J.A.M. Vermaseren, *Phys. Lett.* **B 125** (1983) 275; R.M. Barnett, G. Senjanovic and D. Wyler, *Phys. Rev.* **D 30** (1984) 1529.

[64] A.J. Buras, P. Krawczyk, M.E. Lautenbacher and C. Salazar, Max-Planck Inst. report, MPI-PAE/PTh 52/89 (1989).

[65] J.F. Gunion, private communication; J.L. Hewett, private communication.

[66] H.P. Nilles, *Phys. Rep.* **110** (1984) 1.

[67] H.E. Haber and G.L. Kane, *Phys. Rep.* **117** (1985) 75.

[68] L. Girardello and M.T. Grisaru, *Nucl. Phys.* **B 194** (1982) 65.

[69] J.F. Gunion, H.E. Haber and M. Sher, *Nucl. Phys.* **B 306** (1988) 1.

[70] A.H. Chamseddine, R. Arnowitt and P. Nath, *Phys. Rev. Lett.* **49** (1982) 970; R. Barbieri, S. Ferrara and C.S. Savoy, *Phys. Lett.* **B 119** (1982) 343.

[71] K. Inoue, H. Komatsu and S. Takeshita, *Prog. Theor. Phys.* **70** (1983) 330; L.E. Ibanez, *Nucl. Phys.* **B 218** (1983) 514; J. Ellis, D.V. Nanopoulos and K. Tamvakis, *Phys. Lett.* **B 121** (1983) 123; L.E. Ibanez and C. Lopez, *Phys. Lett.* **B 126** (1983) 54; L. Alvarez-Gaumé, J. Polchinski and M. Wise, *Nucl. Phys.* **B 221** (1983) 495.

[72] K. Inoue, H. Komatsu and S. Takeshita, *Prog. Theor. Phys.* **68** (1982) 927.

[73] G.F. Giudice and G. Ridolfi, *Zeit. für Physik* **C 41** (1988) 447.

[74] M.J. Duncan, *Nucl. Phys.* **B 221** (1983) 285; J.F. Donoghue, H.P. Nilles and D. Wyler, *Phys. Lett.* **B 128** (1983) 55.

[75] S. Bertolini, F. Borzumati and A. Masiero, *Nucl. Phys.* **B 312** (1989) 281.

[76] A.A. Johansen, V.A. Khoze and N.G. Uraltsev. Leningrad Nucl. Phys. Inst. report (1988).

[77] S. Bertolini, F. Borzumati and A. Masiero, *Phys. Lett.* **B 192** (1987) 437.

[78] S. Bertolini, F. Borzumati and A. Masiero, *Nucl. Phys.* **B 294** (1987) 321.

[79] S. Bertolini, F. Borzumati, A. Masiero and G. Ridolfi, DESY report, in preparation.

[80] S. Bertolini, F. Borzumati and A. Masiero, *Phys. Lett.* **B 194** (1987) 545; (E) *ib.* **198** (1987) 590.

SEARCHES FOR HIGGS PARTICLES
IN THE DECAYS OF OPEN AND HIDDEN BEAUTY

Juliet Lee-Franzini

Physics Department
SUNY at Stony Brook
Stony Brook, N.Y. 11794, U.S.A.

1. INTRODUCTION

1.1 WHY HIGGS AND HOW MANY?

In the past decade we have been inundated by the numerous successes of the Standard Model (SM), the most spectacular of which are the observation of the weak gauge bosons at the predicted masses.[1] However, the mechanism by which masses are given to these originally massless gauge bosons implies the appearance of a physically observable "Higgs" scalar of unspecified mass, which so far has eluded experimental observation.

The number of such "Higgs" scalar bosons (hereafter referred to as Higgs') needed is also variable. In the minimal scenario, the spontaneous symmetry breaking is accomplished by introducing one complex, weak-I-spinor scalar field, to generate a potential which has a minimum for a non zero value of the field. The low energy phenomenology of the weak interactions specifies the scale of the symmetry breaking via the relation $v^2 = (\sqrt{2}G_F)^{-1} = 246^2$ GeV^2, where $v = \langle\phi_0\rangle$ is the vacuum expectation value of the neutral scalar Higgs and G_F is the Fermi coupling constant. In this construct it is also possible to introduce "by hand" in the Lagrangian, Yukawa couplings between Higgs' and fermions with strength m_f/v together with induced mass terms $(G_f v/\sqrt{2})f\bar{f}$ (one arbitrary constant, G_f, per fermion), which give masses to all fermions.

Minimal Supersymmetric Standard Model (MSSM)

In most supersymmetric models, two complex SU(2) Higgs doublets are required. In which case we have the relations:

$$\langle \phi_1 \rangle^2 + \langle \phi_2 \rangle^2 = 246^2 \text{ GeV}^2$$

$$\tan \beta = \langle \phi_2 \rangle / \langle \phi_1 \rangle.$$

In the only non minimal model that we consider in this paper, the minimal supersymmetric extension of the standard model (MSSM), the hypercharge Y=-1 (Y=1) doublet is coupled only to the down–type (up–type) quarks and leptons. In MSSM the physically observable states are: a pair of charged Higgs ($H^{\pm}$), two neutral scalars (H^0, h^0) where one (H^0) is likely to be very heavy, and one light (h^0), and one neutral pseudoscalar particle (A). The couplings for h^0 and A, relative to the minimal standard model, are modified by the multiplicative factors:

$$h u \bar{u} : \quad \frac{\cos \alpha}{\sin \beta} \qquad h d \bar{d}, \ h l \bar{l} : \quad -\frac{\sin \alpha}{\cos \beta}$$

$$A u \bar{u} : \quad \cot \beta \qquad A d \bar{d}, \ A l \bar{l} : \quad \tan \beta$$

where α is the mixing angle between H^0 and h^0.

1.2 THEORETICAL BOUNDS ON THE HIGGS' MASSES

Since there are no real constraints on the mass of the Higgs from fundamental principles, hypothetically masses from electron volts to TeV's are allowed, albeit in the latter case the Higgs would look like a strongly interacting particle.[2]

Linde-Weinberg Bound

In the present paper, since we are searching for Higgs' from open or bound beauty-states, we are interested only in whether there are lower bounds on the Higgs' mass, or whether a light Higgs is already excluded from other considerations. In fact, until recently, there was one such interesting theoretical lower bound. Based on arguments on the stability of the vacuum, a lower bound for the minimal Higgs' mass of ~ 7 GeV had been obtained by Linde and by Weinberg.[3]

However, the 7 GeV lower bound is valid only if one can neglect fermion mass terms relative to those due to W and Z bosons. Now that the mass of the top quark is most likely > 65 GeV,[4] we should evaluate its contribution. Specifically, assuming one heavy fermion, the top, and one Higgs doublet, the Linde-Weinberg bound is given by $M_H^2 > (3(2M_W^4 + M_Z^4 - 4m_t^4))/16\pi^2 v^2$. This bound is shown as the lower branch of the curve in fig. 8. For an m_t value of ~ 79 GeV, there is no lower bound. The branch for $m_t > 79$ GeV was computed following Cabibbo et al.[5] who for this region obtain the lower bound $M_H^2 > \sqrt{3(4m_t^4 - 2M_W^4 - M_Z^4)}$, also found by Gross and Duchovni.[6]

These curves are applicable only in the minimal Higgs case with three generations of fermions. If one replaces M_H^2 by the sum of the masses of n Higgs doublets, and/or adds on the right hand side of the inequality more generations of heavy fermions, the lower bound on the Higgs mass from the vacuum stability constraint evaporates.

As an amusing aside I call attention to a recent study by Langacker[7] to look at the implications of the new $M_{Z,W}$ and neutral current measurements for the top quark and Higgs masses. He found that while the M_{top} limits do exhibit some sensitivity to M_H, there is no constraint on the Higgs mass!

2. HIGGS FROM UPSILON DECAYS

2.1   THE WEINBERG–WILZCEK FORMULA

Ever since the discovery of the Υ, it was recognized that its radiative decay would be a good source for Higgs of mass less than that of the Υ. From the strength of the $q\bar{q}H$ vertex, $m_q/v = m_q\sqrt{\sqrt{2}\,G_F}$, we can compute the annihilation rate. Fig. 1 illustrates the Feynman diagrams involved. Uncertainties about the value of the wave function at the origin, $|\Psi(0)|^2$ are elegantly avoided here by comparing to the decay $\Upsilon \to \mu\mu$. This is the well known Weinberg–Wilczek result (WW):[8]

$$\frac{\Gamma(\Upsilon \to \gamma + H)}{\Gamma(\Upsilon \to \mu\mu)} = \frac{G_F m_b^2}{\sqrt{2}\pi\alpha}\left(1 - \frac{M_{H^0}^2}{M_\Upsilon^2}\right)X_\Upsilon^2,$$

which for Υ decaying to a very light Higgs gives $BR(\Upsilon \to \gamma + H) \sim 2.3 \times 10^{-4}X_\Upsilon^2$. $X_\Upsilon = -\sin\alpha/\cos\beta$ in the MSSM, and equals one for the SM.

a)

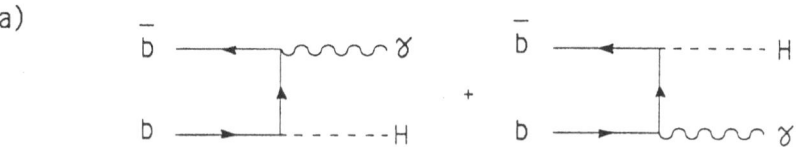

b)

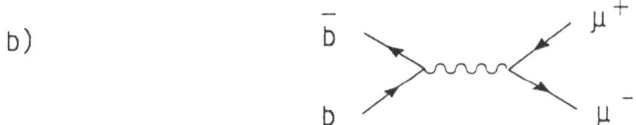

Fig. 1. Amplitudes for (a) the decay $\Upsilon \to H^0\gamma$ and (b) the decay $\Upsilon \to \mu^+\mu^-$.

Radiative Corrections to the WW Formula

In 1985 CUSB reported, from a study of 400,000 Υ's decays, a limit on $BR(\Upsilon \to \gamma + H)$ smaller than the WW value for Higgs masses less than ~ 4 GeV.[9] It was promptly pointed out by J. Ellis and others that our result was not very significant because of the large lowest order QCD radiative corrections which strongly suppress $\Upsilon \rightarrow \gamma + H$.[10] The single gluon diagrams involved are shown in fig. 2. Specifically, the multiplicative factor is $[1 - 4\alpha_s(\mu)a(\kappa)/3\pi]$ where $a(\kappa)$ is approximately 10 for scalar particle, and 6.5 for a pseudoscalar particle. Combining the effect of the lowest order QCD radiative corrections to both the $\mu\mu$, where the multiplicative factor is $[1 - 16\alpha_s(\mu)/3\pi]$, and to the $H^0\gamma$ rates result in a reduction of the above BR by a factor of about two.

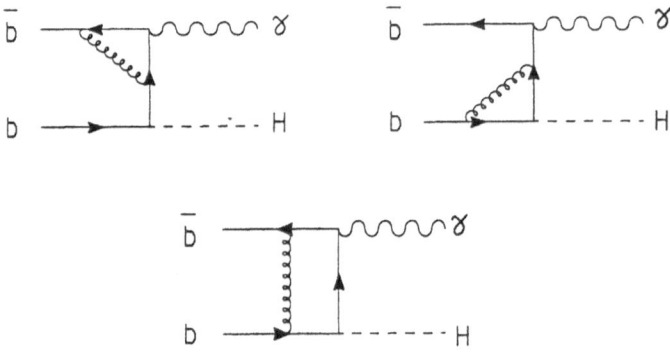

Fig. 2. QCD radiative corrections to lowest order.

A particularly unsatisfactory aspect of the calculations of the radiative correction to the $H^0\gamma$ rate is that, as usual, the energy scale, μ, at which the calculation might be reliable is not known. In addition the correction varies by over an order of magnitude for $m_b/2 < \mu < 2m_b$. Goldberg and Ryzak (GR) have reanalyzed the problem, performing the calculation in the $\overline{\text{MS}}$ scheme.[11] They argue that the scale should be $\mu > m_b$ and show that the result is stable for $m_b < \mu < 2m_b$. They obtain in this way a suppression considerably smaller than the ones obtained by Vysotsky, Ellis et al., or Nason, hereafter referred to as VEN.

Of course there are considerable arguments about a scheme dependent QCD correction. Obviously theorists should be calculating the next order QCD corrections so the question of the magnitude of the QCD correction can be *hopefully* settled in the future.

Bound State and Relativistic Corrections

In the wake of the birth and demise of a Higgs candidate, the ς particle, there were a number of papers seeking to understand the bound state effects on the WW formula which was derived assuming point $Hb\bar{b}$ interaction.[12] The consensus is that for $M_H \approx M_\Upsilon$ and $M_H \approx M_{\chi_b}$ large corrections would be present. In the first case, E1 transitions between the excited Υ's and a nearby χ_b modify the $\Upsilon \rightarrow H^0\gamma$ rate and in the latter case, very interesting mixing effects could appear between the Higgs and the $J = 0$ P-wave states.

272

Attempts were also made to include relativistic corrections to the nonrelativistic potential model calculations of $|\Psi(0)|^2$. Faldt et al. [13] were only able to solve the problem for $M_H \approx M_\Upsilon$ and obtained a reduction similar to that found by Pantaleone et al. [12] from studying bound state effects. Calculations by I. Aznauryan et al. [14] in an infinite-momentum system, obtain a curve which looks almost identical to that of Vysotsky [10] from considering lowest order QCD corrections. So, to this author at least, it seems as if some double counting has taken place.

Thus, because the mass of the Higgs which are excluded in the next two experiments is far from M_Υ and because I do not understand whether the Aznauryan et al. suppression factor is already included in the QCD calculations, the data will be compared only to predictions with and without lowest order QCD corrections.

2.2 LIMITS FROM CUSB

The CUSB-II Detector

Since Higgs searches do involve experimental apparatus, I take this opportunity to show a spectrometer which was optimized for monochromatic photon spectroscopy, namely CUSB-II. It is the first bismuth germanate (BGO) calorimeter to come into operation in high energy physics. It has achieved the best resolution for photons in the energy range from 30 MeV to 5 GeV. [15] Fig. 3 is a perspective drawing of the whole detector seen through partially cut-out counters used for the muon trigger. A detail of the BGO array is also shown. This consists of 36 azimuthal sectors, each covering ten degrees in ϕ. Each sector, twelve radiation lengths (λ_0) thick at normal incidence, is divided into two polar halves, covering the θ ranges $45° - 90°$ and $90° - 135°$ and five radial layers, for a total of 360 crystals in the whole array.

Between the beam pipe and the BGO cylinder is a mini drift chamber. The BGO cylinder is surrounded by a square array of 328 NaI crystals, 8.8 λ_0 thick at normal incidence, arranged in five radial layers, 32 azimuthal sectors, and two polar halves. Between NaI crystal layers are four proportional chambers with x and y cathode strip readout, used for tracking non interacting charged particles. The NaI array is surrounded in turn by one layer of 8 × 8 lead glass blocks, 7 λ_0 thick. Minimum ionizing non interacting particles are identified, by their energy loss, measured five times in BGO, five times in NaI and once in lead glass, this material corresponding to two and a half nuclear interaction lengths. The detector covers a solid angle of 66% of 4π. Plastic scintillation counters covering $\approx$ 29% of 4π provide a muon trigger. Calibration is done in real time during data taking with sources which are embedded between crystal layers. The fractional photon energy resolution obtained during running conditions for ^{137}Cs (0.66 MeV) and for ^{65}Zn (1.11 MeV) at FWHM (full width half max) are 24% and 18%, respectively in BGO. The necessary dynamic range is accomplished by having two signal paths, such that two different sensitivity channels measure collision events and source signals. CUSB has achieved $\leq$ 0.1% channel to channel calibration error over the whole detector, and thereby has obtained a photon energy resolution of 0.95% for $\approx$ 5GeV electrons from Bhabha scatterings.

Electromagnetic Showers in CUSB

Experimentally the signature for $\Upsilon \rightarrow H + \gamma$ is a monochromatic photon, i.e. a high energy neutral electromagnetic (em) shower, emitted opposite to the decay products of

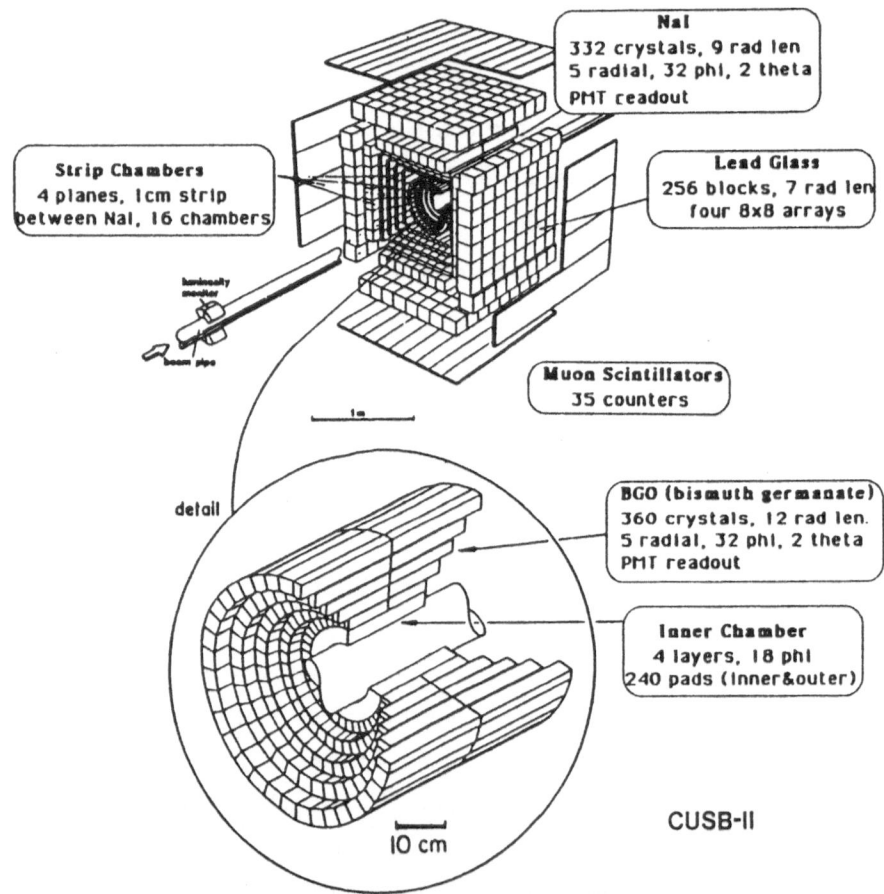

Fig. 3. The CUSB detector.

the Higgs. As an example of how such em showers look in CUSB we show in fig. 4 the energy deposition pattern of an $e^+e^- \rightarrow e^+e^-\gamma$ event in the two halves of the CUSB BGO cylinder.

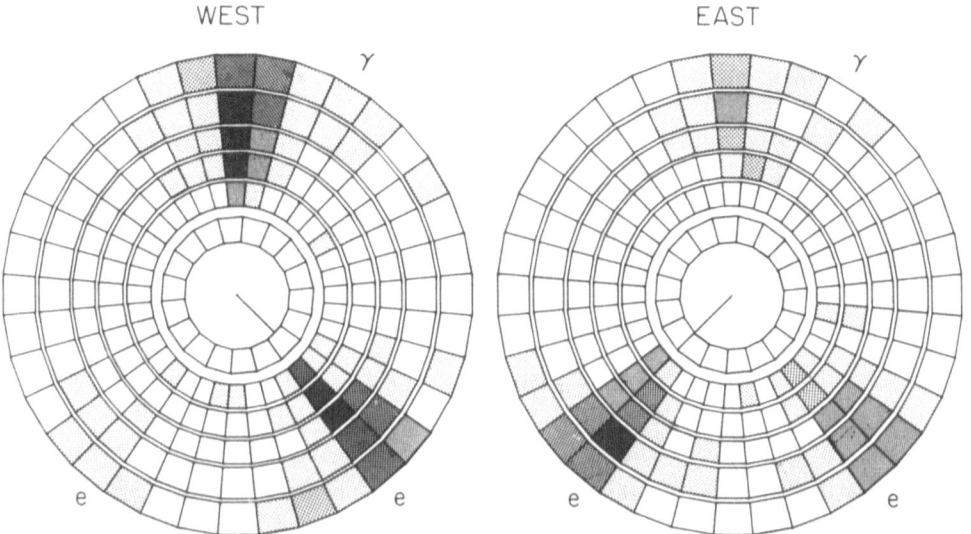

Fig. 4. Energy deposition of $e^+e^- \rightarrow e^+e^-\gamma$ event in the two halves of the CUSB BGO cylinder. Density of shading is proportional to the energy deposited.

A high energy em shower is characterized by distinctive longitudinal and transverse development. The CUSB-II photon energy resolution can be parametrized by $\sigma_E/E_\gamma \sim 1.8\%/\sqrt[4]{E_\gamma}$, roughly a factor of two improvement with respect to CUSB-I. The improved resolution also allowed us to improve our algorithm for rejection of π^0's, thereby improving the rejection of QED continuum background. Fig. 5 shows a typical continuum subtracted photon spectrum taken at the $\Upsilon(1S)$.

Mass Limits from CUSB

Since 1985 CUSB has also tripled their statistics. The new data consists of 400,000 Υ and 600,000 Υ'', taken with CUSB-II, in addition to the previously mentioned old sample. Recall that the sensitivity of Higgs search is proportional to $\sqrt{N}$ and $1/\sqrt{\sigma}$, where N, σ are the number of decays collected and the photon resolution respectively.

Fig. 6 shows the branching ratio limit resulting from all CUSB data combined. In the minimal scenario, at 90% CL, we do not observe monochromatic photons corresponding to Higgs production for a mass smaller then $\sim$5 GeV comparing to the (WW+VEN) prediction, and smaller than $\sim$5.75 GeV comparing to the (WW+GR) prediction.[16] In

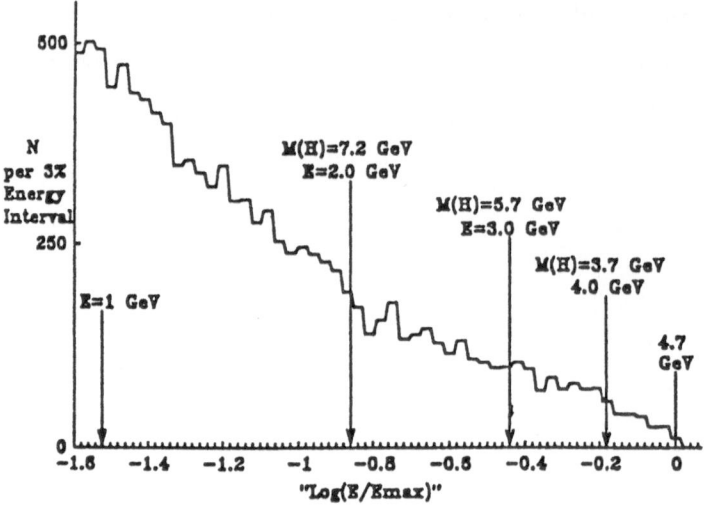

Fig. 5. Typical "photon" spectrum from Υ decay, continuum subtracted.

fig. 7 we show the region in the $\sin \alpha / \cos \beta$, M_H plane excluded by the CUSB data. In fig. 8 we show the CUSB limit superimposed over the theoretical lower bounds computed for the minimal SM. Our result excludes light Higgs in the region of $M_{top} \approx 79$ GeV where there are no theoretical lower bounds.

Search for Light gluinos

I hope I'm allowed a small digression. The same combined CUSB data can be used to exclude gluinos of $0.1 \leq M_{gluino} \leq 3.6$ GeV, improving our previous limits.[17] This is shown in fig. 9 where the CUSB data is compared to theoretical curves from Kühn and from Keung and Khare.[18] We note that this result, being independent of the squark mass, neatly complements previous light gluino searches.[19] See fig. 10.

2.3   ARGUS RESULTS ON $\Upsilon \rightarrow H + \gamma$

ARGUS has searched for $\Upsilon \rightarrow \gamma X$ where X decays into two oppositely charged pions, kaons or protons. In the $\pi\pi$ invariant mass region $0.3 - 0.57$ GeV the limits vary from 3 to 4.5×10^{-5}. Using the assumption that $BR(H^0 \rightarrow \pi\pi) \geq 45\%$ in this Higgs mass region, we see that the ARGUS limit lies below the (WW+VEN) formula, see fig. 11, hence excluding Higgs masses in the range 0.29 to 0.57 GeV.[20]

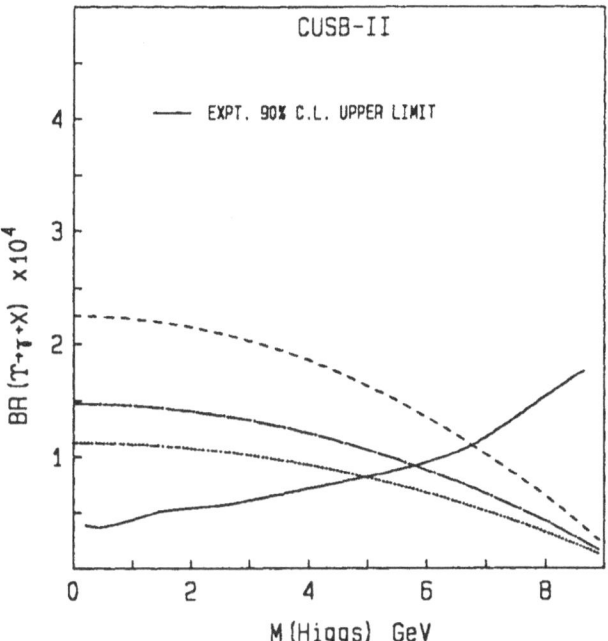

Fig. 6. 90% CL upper limit from CUSB for $BR(\Upsilon \rightarrow \gamma + X)$ (solid line), compared to the WW calculation (dashed line), the same with the correction of reference 10 (dotted line), and (dash–dotted line) with the correction of reference 11.

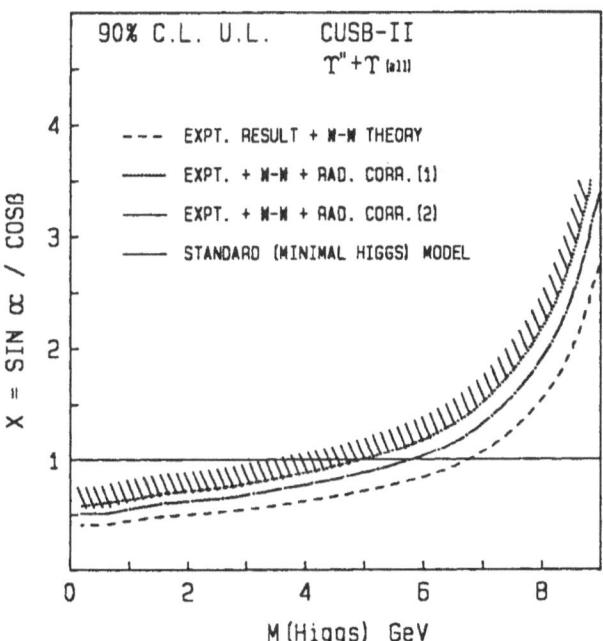

Fig. 7. 90% CL upper limit from CUSB on $\sin \alpha / \cos \beta$ (solid line), compared to the WW calculation (dashed line), the same with the correction of reference 10 (dotted line), and (dash–dotted line) with the correction of reference 11.

277

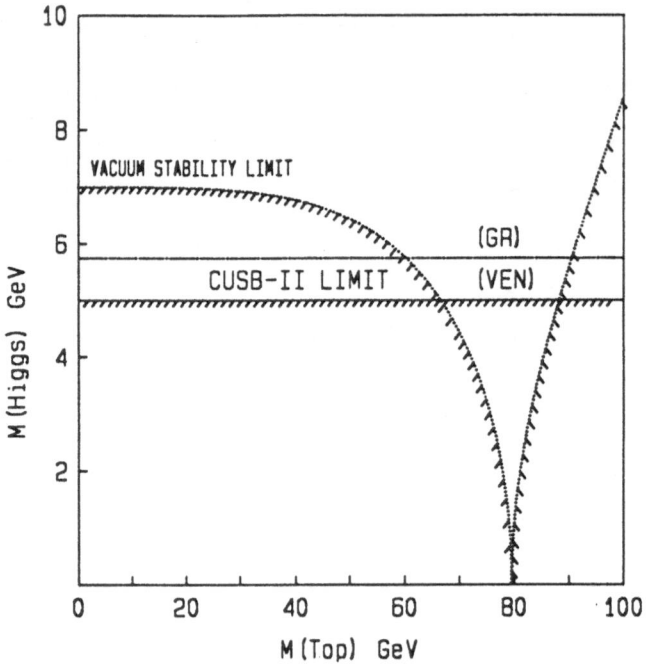

Fig. 8. CUSB Limit superimposed on the theoretical lower bounds.

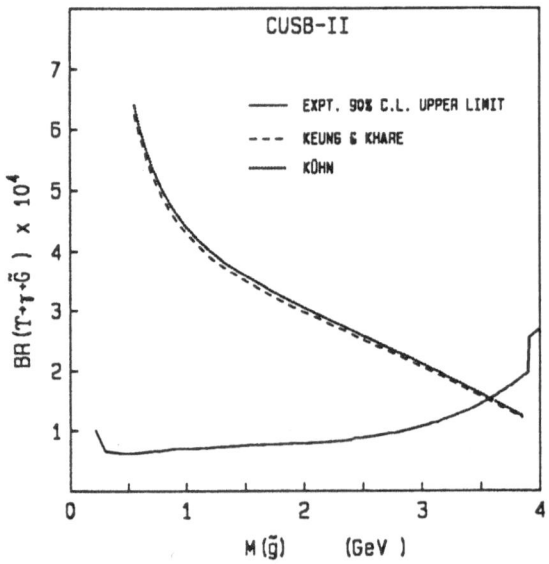

Fig. 9. The 90%CL upper limit for the $BR(\Upsilon \to \gamma \eta_{\tilde{g}})$ from all CUSB Υ data combined vs gluino mass. The two upper curves are theoretical model predictions for $\Upsilon \to \gamma \eta_{\tilde{g}}$.

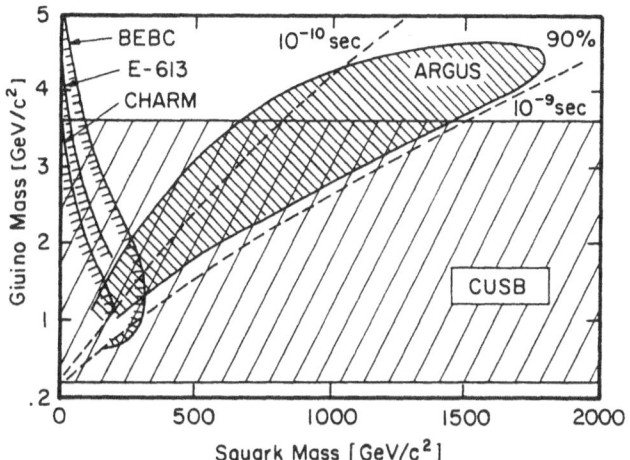

Fig. 10. The gluino mass region excluded at the 90%CL by this experiment (CUSB), superimposed on beam dump and ARGUS results.[19]

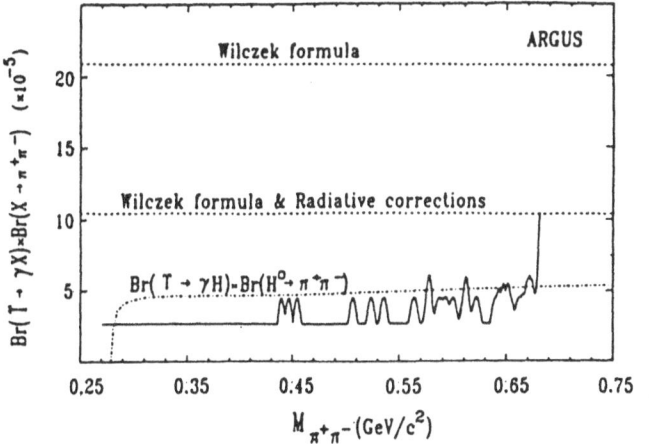

Fig. 11. 90% CL upper limit from ARGUS for $BR(\Upsilon \to H + \gamma) \times BR(H^0 \to \pi\pi)$.

2.4 LIMITS ON VERY LIGHT HIGGS

Limits from Atomic and Nuclear Physics

Bounds for a massless or very light Higgs usually come from atomic and nuclear physics experiments. In order to draw limits one needs to estimate the effective Higgs-Nucleon coupling. Traditionally one writes it as $g_{H^0 NN} = \frac{gM_N}{2M_W}\eta$ where M_N is the nucleus mass and η parametrizes the effect of treating the nucleon as a quark. Most authors think that $\eta \geq 0.3$. The limits quoted below use this reasonable (but not guaranteed) value. [21]

One type of experiment compares the measured and theoretically calculated muonic x-ray energies in transitions (e.g. $^3d_{5/2} - {}^2p_{3/2}$) among high Z elements (e.g. ^{24}Mg and ^{28}Si) to very high accuracy. [22] The agreement to within three parts per million indicates the absence of long range forces due to Higgs' of mass less than 8 MeV.

Experiments have also been performed to search for light Higgs production in the transitions of ^{16}O(6.05 MeV) [23] and ^{4}He(20.2 MeV) to the ground state, with the Higgs decaying into an electron–positron pair. [24] The null results exclude Higgs' in the mass range $2.8 \leq M_H \leq 11.5\ MeV$.

Limits from Combined J/ψ and Υ Decays for massless Higgs

Given the uncertainty in η it is very nice that there exists alternate methods to search for massless or very light Higgs. Both the Crystal Ball and CUSB have searched for events of the type:

$$J/\psi \text{ or } \Upsilon \rightarrow \gamma + NOTHING.$$

We use the results from CUSB, [25] where in 2.07×10^4 Υ decays we found no scalar candidates, and in 5.87×10^4 Υ'' decays we also found no scalar candidate. Therefore:

$$BR(\Upsilon \rightarrow \gamma + H^0) < 1.19 \times 10^{-4} \text{ at } 90\%\text{CL}$$
$$BR(\Upsilon \rightarrow \gamma + H^0) < 0.59 \times 10^{-4} \text{ at } 68.4\% \text{ CL}$$

and the X–Ball results, [26]

$$BR(J/\psi \rightarrow \gamma + H^0) < 1.4 \times 10^{-5} \qquad \text{at } 90\% \text{ CL}$$
$$BR(J/\psi \rightarrow \gamma + H^0) < 0.96 \times 10^{-5} \quad \text{at } 68.4\% \text{ CL}$$

Thus the 90 % CL limit on the product of the two branching ratios is:

$$BR(\Upsilon \rightarrow \gamma + H^0)BR(J/\psi \rightarrow \gamma + H^0) < 5.7 \times 10^{-10} \text{ at } 90\% \text{ CL},$$

To obtain the corresponding theoretical prediction for the product of the two branching ratios we used for $B_{\mu\mu}$ and α_s the values:[27]

$$B_{\mu\mu}(J/\psi) = 0.07 \pm 0.01$$

$$B_{\mu\mu}(\Upsilon) = 0.0257 \pm 0.0009$$

$$B_{\mu\mu}(\Upsilon'') = 0.0173 \pm 0.0016$$

$$\alpha_s(q = m_b = 4.9 \text{ GeV}) = 0.1742 \pm 0.0022$$

$$\Lambda_{\overline{MS}} = 158 \pm 7 \text{ MeV}$$

$$\alpha_s(q = m_c = 1.5 \text{ GeV}) = 0.25, \text{ from } \Lambda_{\overline{MS}} \text{ above.}$$

We thus obtain:

$$BR(\Upsilon \to \gamma + H^0)BR(J/\psi \to \gamma + H^0) = B_{\mu\mu}(\Upsilon)B_{\mu\mu}(J/\psi)\frac{G_F^2 m_c^2 m_b^2}{2\pi^2\alpha^2} = 115.1 \times 10^{-10}.$$

Next we compute the QCD corrections mentioned in section 2.1:

$$\frac{\Gamma(V \to H + \gamma)}{\Gamma(V \to \mu\mu)} = (WW)[1 - (4.25 - 1.70)\alpha_s)] = (WW)(1 - 2.55\alpha_s)$$

The radiative correction factor is:

$$(1 - 2.55\alpha_s^{\Upsilon})(1 - 2.55\alpha_s^{\psi}) = (1 - 2.55 \times 0.174)(1 - 2.55 \times 0.25) = 0.20$$

which yields for the predicted product branching ratio to be:

$$BR(\Upsilon \to \gamma + H^0)BR(J/\psi \to \gamma + H^0) = 23 \times 10^{-10},$$

a value about 4 times larger than the experimental limit, thus excluding a massless Higgs.

For MSSM models one has to multiply the theoretically predicted product BR's by $(\cos^2\alpha/\sin^2\beta) \times (\sin^2\alpha/\cos^2\beta)$ which in the SM equals 1 because then $\cos\alpha = \sin\beta$ and $\sin\alpha = -\cos\beta$.

<u>Quarkonia Limit on the Mass of Very Light Higgs</u> One can also obtain a limit on the mass of very light Higgs (i.e. mass below $2m_\mu$, so that it decays into an electron-positron pair), by noting that both Xtal ball and CUSB require not seeing anything opposite the high energy gamma up to a distance of 0.5 meters. This implies that both Υ and J/ψ BR's get multiplied by a term $\exp(l/l_0)$ where l is the detector length, l_0 is the mean free path of the Higgs decaying into an electron-positron pair, given by $l_0 = \gamma c\tau_{H^0}$, $\gamma = M_V/2M_{H^0}$ and

$$\Gamma(H^0 \to e\bar{e}) = \frac{\sqrt{2}G_F m_e^2 M_{H^0}}{8\pi}(1 - 4m_e^2/M_{H^0}^2)^{3/2}\frac{\sin^2\alpha}{\cos^2\beta}$$

which yields the 90% CL upper limit $M_{H^0} > 61$ MeV in the minimal SM.

In the MSSM model $M_{h^0} > \cos\beta/\sin\alpha \times 61$ MeV.

<u>Quarkonia Limit on the Mass of an Axion</u>

By a similar calculation one obtains that a massless pseudoscalar (axion) is excluded with a factor 12 safety. The 90% CL upper limit for M_{A^0} is $\cot\beta \times 82$ MeV.

Limit from $\pi \to e^+\nu_e H^0$ Decays

For completeness I should also mention that the M_{H^0} region just discussed has also recently been partially excluded by an experiment at PSI.[28] They study $\pi^+ \to e^+e^+e^-\nu$ to obtain limits on light Higgs. For a Higgs emitted by the virtual W, one finds:

$$\frac{\Gamma(\pi^+ \to e^+\nu_e H^0)}{\Gamma(\pi^+ \to \mu^+\nu_\mu)} = \frac{\sqrt{2}G_F m_\pi^4 f(x)}{48\pi^2 m_\mu^2 (1 - m_\mu^2/m_\pi^2)^2} = 6.5 \times 10^{-9} f(x)$$

where $f(x) \equiv (1 - 8x + x^2)(1 - x^2) - 12x^2 \log x$ with $x \equiv M_{H^0}^2/m_\pi^2$. The SINDRUM Collaboration at PSI obtains $\Gamma(\pi^+ \to e^+\nu_e e^+e^-)/\Gamma(\pi^+ \to \mu^+\nu_\mu) = (3.2 \pm 0.5) \times 10^{-9}$.

Comparison with the theoretical prediction excludes the region 10 MeV $\lesssim M_{H^0} \lesssim$ 110 MeV. Because of an error, found by H-L Yu,[29] which reduces the above theoretical prediction by a factor of two, the margin of safety is reduced correspondingly.

Limit from Higgs Bremsstrahlung

A new experiment was performed at Orsay by Davier and Ngoc to search for Higgs in the very low mass region. They use production of Higgs from bremsstrahlung off electrons in the Coulomb field of large Z nuclei and detect the Higgs decaying into an electron-positron pair.[30] The null result allows them to reject at 90% CL the existence of a SM Higgs from 1.2 to 52 MeV. Non-standard Higgs are also excluded in a large range of couplings, see fig. 12. It is very reassuring that this low mass region can be excluded using so many different methods.

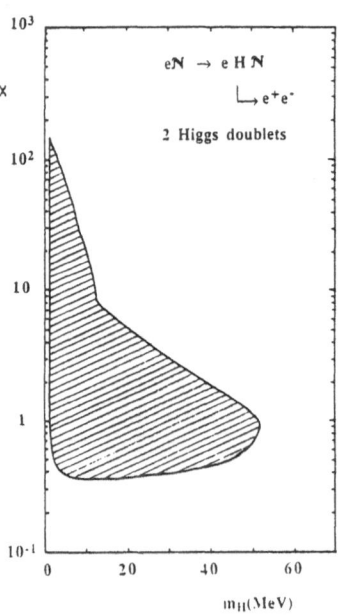

Fig. 12. The 90% CL excluded mass domain from the Davier and Ngoc experiment for scalar and pseudoscalar Higgs in the MSSM scheme.

<u>Limit from $K \to H + \pi$</u>

Finally, very low mass Higgs can also be excluded in principle by studying $K \to H + \pi$ reactions. I include some of such results here partially to contrast them relative to the $B \to H + K$ processes to be discussed in section 3.2. Assuming $m_t > 80$ GeV and limits on the elements of the KM matrix from the 1986 PDG, Dawson obtains $BR(K^+ \to H + \pi^+) > 5.7 \times 10^{-6} \beta_H$ where β_H is the Higgs velocity in the K rest frame.[21]

A KEK experiment[31] which searched for a peak in the pion momentum spectrum resulting from stopping K^+ decays gave $BR(K^+ \to X + \pi^+) < 5.5 \times 10^{-6}$. Comparing with the above values we can exclude 10 MeV $< M_H <$ 100 MeV. Background from the $K \to \pi^0 \pi^+$ decay limits the sensitive region to $M_H < 100$ MeV.

One could also extract mass bounds from experimental limits on $BR(K_L \to \pi \ell \ell)$ by using

$$BR(K_L \to \pi \ell \ell) = \frac{\Gamma(K_L \to \pi H)}{\Gamma(K_L \to all)} BR(H^0 \to \ell \ell).$$

Controversies in the interpretations of the experimental bounds stem from uncertainties in calculations of $\Gamma(K_L \to \pi H)$. For example, one by Chivukula and Manohar (CM),[32] incorporated the effect of the $\Delta I = 1/2$ enhancement hitherto ignored. Unfortunately the contribution from this term to the amplitude $A(K_L \to \pi^0 H) = A(K^+ \to \pi^+ H)$ interferes destructively with the one due to the flavor–changing two–quark interactions induced at one loop. Furthermore, the CM result depends on an unknown parameter B. Consequently, the above amplitudes are estimated to be 10^{-11} ($B = 1$) to 0.78×10^{-10} ($B = 0$) GeV which implies $BR(K^+ \to H^0 + \pi^+) \sim 6 \times 10^{-8}$ to 6×10^{-6}, and $BR(K_L \to H^0 + \pi^0) \sim 1.5 \times 10^{-8}$ to 1.5×10^{-6}.

Comparing these predictions with the recent results of NA31[33] which finds that $BR(\pi^0 ee) < 4 \times 10^{-8}$ would lead one to conclude that $2m_e < M_H < 2m_\mu$ is also excluded from the absence of $K \to H$ decays.

3. LIMITS FROM B DECAYS

3.1 LIMITS FROM $B \to H$+ANYTHING; $H^0 \to \mu^+ \mu^-$

Obtaining limits on the Higgs mass from B meson decay is in principle simpler. The major contribution to $\Gamma(B \to H^0 + X)$ is from the flavor changing vertex $b \to s$. See fig. 13.

Fig. 13. Flavor changing diagrams which contribute to $\Gamma(B \to H^0 + X)$.

Willey and Yu (WY) obtained[34]

$$\frac{BR(B \to HX)}{BR(B \to \ell\nu X)} = \frac{27\sqrt{2}}{64\pi^2} G_F m_b^2 \frac{|V_{tb}V_{ts}^*|^2}{|V_{cb}|^2} \left(\frac{m_t}{m_b}\right)^4 \frac{\left(1 - \frac{M_H^2}{m_b^2}\right)^2}{f(m_c^2/m_b^2)}.$$

Setting both V_{tb} and $|V_{ts}/V_{cb}|$ to one, $m_b = 4.9$ GeV, the $b \to c\ell\nu$ phase space factor $f(m_c^2/m_b^2) = 0.48$ and $BR(B \to \ell\nu X) = 0.11$ yields

$$BR(B \to H^0 X) = 0.042 \left(\frac{m_t}{50 \text{GeV}}\right)^4 \left(1 - \frac{M_H^2}{m_b^2}\right)^2$$

which is plotted for m_t=30 GeV and 50 GeV, in fig. 14.

In inclusive analyses, the Higgs is detected through its dimuon decay mode. That is, one measures $BR(B \to H + X) \times BR(H^0 \to \mu^+\mu^-)$ and has to assume values of $BR(H^0 \to \mu^+\mu^-)$ to be able to use the WY formula to obtain limits on M_H as a function of m_t. For example, using the fact that CLEO[35] has obtained, for $M_H > 1.0$ GeV, $BR(B \to H + X) \times BR(H^0 \to \mu^+\mu^-) < 0.008$, one would conclude that as long as $BR(H^0 \to \mu^+\mu^-) \geq 0.02$, for $80 \leq m_t \leq 100$ GeV, $1.0 \leq M_H \leq 3.5$ GeV is excluded.

The CLEO experiment is not able to detect the dimuon mode for $M_H < 1.0$ GeV because of cuts on the muon momentum. Experiments done at PETRA to search for a b' by searching for $B \to \mu\mu X$ do not have such cuts. Thus in principle we could infer from the TASSO[36] limit of $BR(B \to \mu^+\mu^-) < 0.02$ for $2m_\mu < M_H \leq 5$ GeV that the region $2m_\mu < M_H \leq 2m_\pi$ for $m_t \geq 40$ GeV is excluded. Similarly, one might use the JADE[37] limit of $BR(B \to \mu^+\mu^-) < 0.007$ for $0.3 < M_H \leq 5$ GeV to exclude a Higgs with a mass in that region.

Two caveats are in order here:

1. The efficiencies are computed for $B \to \mu^+\mu^- X$, i.e. for three body final states and not the sequential decays $B \to HX \to \mu^+\mu^- X$.

2. There are serious questions such as what is $BR(H^0 \to \mu^+\mu^-)$ and the effect of final state interactions and resonances on this BR.

<u>$BR(H^0 \rightarrow \mu^+\mu^-)$</u>

The uncertainty in the prediction of $BR(H^0 \rightarrow \mu^+\mu^-)$ is not due to ambiguities in the calculation of $\Gamma(H^0 \rightarrow \mu^+\mu^-)$ which is given simply by the following expression for Higgs decaying to quarks and leptons:

$$\Gamma(H^0 \rightarrow f\bar{f}) = \frac{C_f G_F m_f^2 M_{H^0}}{4\sqrt{2}\pi}(1 - 4m_f^2/M_{H^0}^2)^{3/2}$$

where the color factor C_f is 1 for leptons and 3 for quarks.

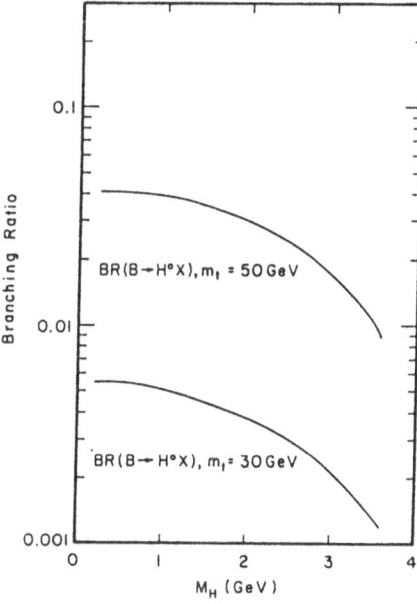

Fig. 14. $BR(B \rightarrow H^0 X)$.

The problem is to compute the hadronic decay rate which enters in the total Higgs' decay width. For example, we need to include the gluonic width, $\Gamma(H^0 \rightarrow gg)$. Fig. 15 shows the Feynman diagrams of interest. At the parton level the Higgs decay rate to gluons is given by

$$\Gamma(H^0 \rightarrow gg) = \frac{G_F M_{H^0}^3}{36\sqrt{2}\pi^3}\alpha_s^2 N_{eff}^2$$

where N_{eff} is the effective number of heavy quark types with mass above $M_{H^0}/2$.[38] The argument is how this relates to meson final states and the effect of final state interactions, etc.

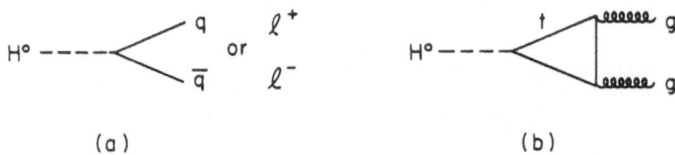

Fig. 15. Higgs decay diagrams.

Recently Raby and West,[39] RW, examined the effects of final state interactions on the decay of a light Higgs into two pions. They found, as expected, that even in the region $2m_\pi \leq M_H \leq 2m_K$, final state interactions strongly enhanced the two pion mode, leading to a much suppressed branching ratio of the Higgs to two muons. By including a broad $\pi\pi$ resonance at about 850 MeV and a narrow one at 975 MeV they compute $f \equiv \sum \Gamma(H^0 \to \pi^+\pi^-)/\Gamma(H^0 \to \mu^+\mu^-)$, from which we extract $BR(H^0 \to \mu^+\mu^-)$ as tabulated below.

M_{H^0} (GeV)	0.3	0.4	0.5	0.6	0.7	0.8	0.9
f	23	23	23	33	49	90	332
$BR(H^0 \to \mu^+\mu^-)(\%)$	4.2	4.2	4.2	2.9	2.0	1.1	0.3

Truong and Willey[40] recently carried out a coupled channel analysis of final state interactions and found that in the range 0.95 to 1.1 GeV, there is a strong S^* enhancement of the $\pi\pi$ and $K\bar{K}$ modes of the Higgs which depresses the branching ratio for $H^0 \to \mu^+\mu^-$ to practically nil. Their f as a function of M_H is shown in fig. 16.

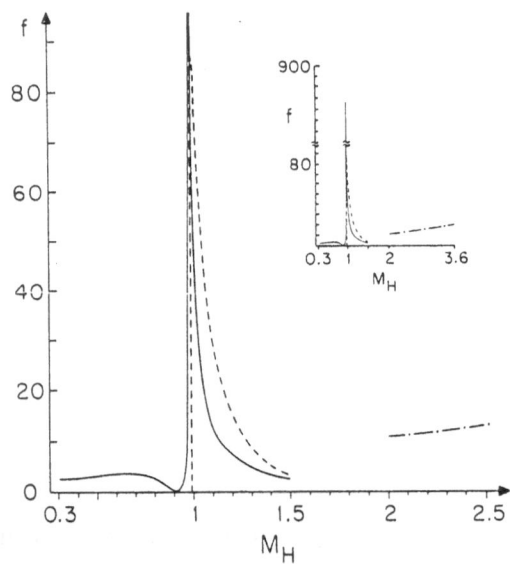

Fig. 16. Truong and Willey's f as a function of M_H.

3.2 LIMITS FROM $B \to H + K$; $H^0 \to \mu^+ \mu^-$

$\underline{B \to H + K}$

If we use exclusive channels such as $B \to H + K$ then, in addition to the disputes with regard to $BR(H^0 \to$ final state), we have the added controversy re the hadronic matrix elements: $\langle K | \cdots | B \rangle = \cdots \times F(q^2)$ ($\cdots$ stand for something well known, the form factor $F(q^2)$ is the problem) which affect estimates of $\Gamma(B \to HK)/\Gamma(B \to HX)$.

Haber et al.,[41] give

$$\frac{\Gamma(B \to HK)}{\Gamma(B \to HX)} = \frac{0.7}{(1 - M_H^2/m_b^2)(1 - M_H^2/\Lambda^2)}.$$

The dot-dash curve in fig. 17 is a plot of this equation using $m_b = 4.9$ GeV and $\Lambda = 6.1$ GeV.

Previously CLEO,[42] using the exclusive branching ratio given by Haber et al. and the decay branching ratios given by Voloshin, concluded that either the mass of the top quark is less than 46 GeV or that a light Higgs is excluded in the mass range between 0.3 and 3 GeV. Fig. 18 shows the effect of using RW's $BR(H^0 \to \mu^+ \mu^-)$ on the CLEO analysis.[43]

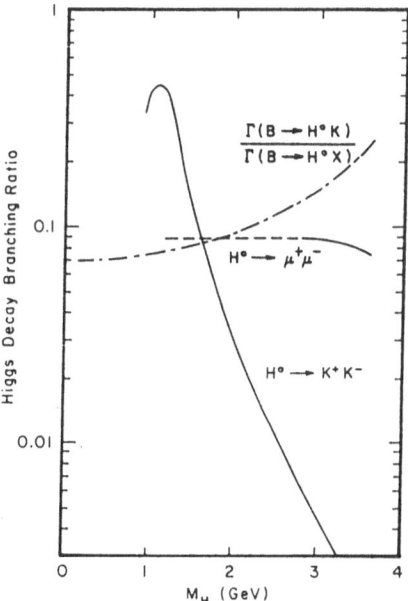

Fig. 17. Predicted fraction of $B \to H^0 K / B \to H^0 X$ (dashed-dot curve).[41] CLEO's assumptions for Higgs decay branching ratios to $\mu^+ \mu^-$ (dash-solid curve) and to $K^+ K^-$ (solid curve).

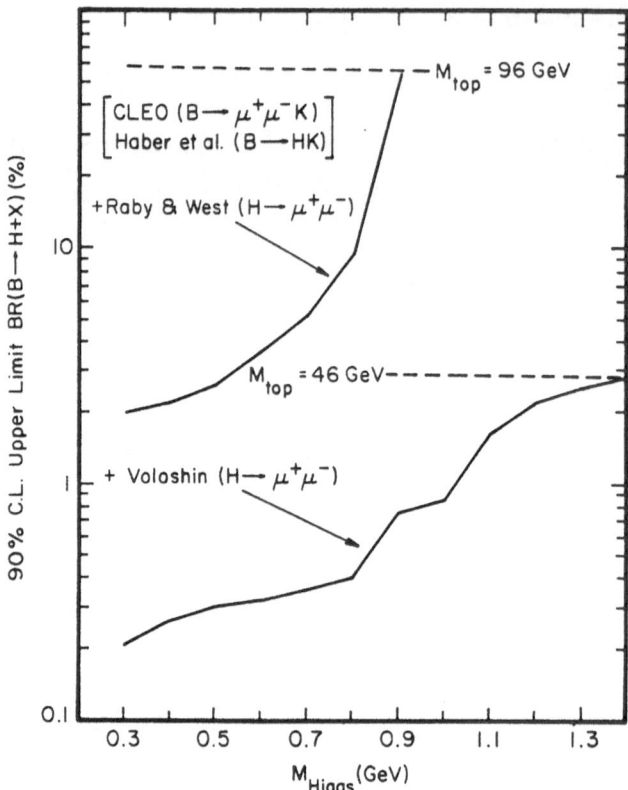

Fig. 18. 90% CL upper limit on $B \to H + X$ using CLEO data, together with the theoretical prediction for $BR(H^0 \to \mu^+\mu^-)$ from Raby and West (the upper curve) and Voloshin (the lower curve).

Limits from the Latest CLEO Results

CLEO[44] recently completed an exhaustive search for $B \to \mu^+\mu^- X$ in the mass region 1 to 3.5 GeV, see fig. 19, and $B \to (\mu^+\mu^- + \pi^+\pi^-)K^-$ for the mass region between $2m_u$ to 1 GeV, and $B \to \mu^+\mu^- K^-$, $B \to (K^+K^-)K^-$ in the mass region 1 to 3.5 GeV. Assuming the decay rates and branching ratios shown in fig. 17 they obtain the upper limits (90% CL) for $BR(B \to H^0 X)$ as a function of the Higgs' mass as shown in fig. 20.

Limit on Top Mass vs Higgs Mass from CLEO

Fig. 21 shows the CLEO limits in the $m_t - M_H$ plane using the WY formula. Note again that the limits for Higgs mass below 1 GeV depend on the Haber rate of ~ 0.09, and the results above depend on assuming $BR(H^0 \to \mu^+\mu^-)$ to be ~ 0.07.

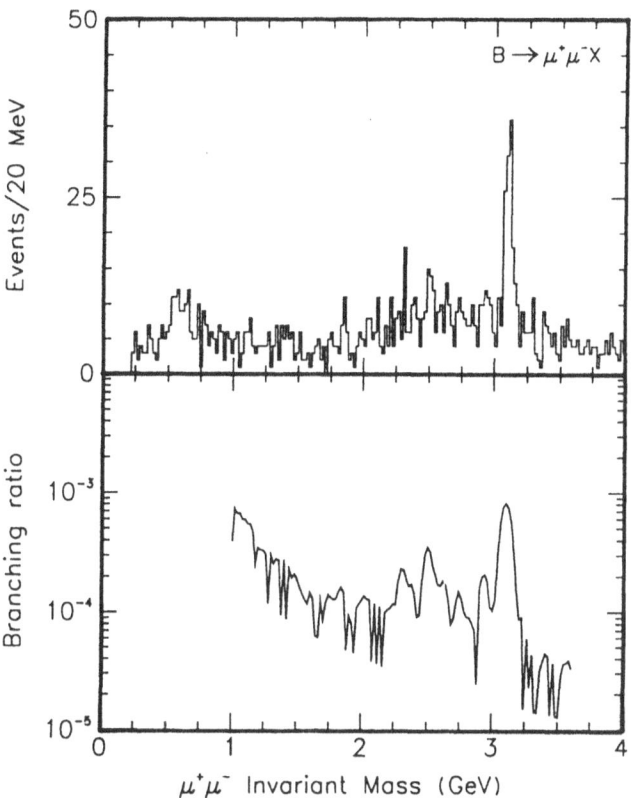

Fig. 19. Top: observed number of candidates for $B \rightarrow H^0 X$, $H^0 \rightarrow \mu^+\mu^-$ as a function of M_H. Bottom: upper limit (90% CL) for the product branching ratio.

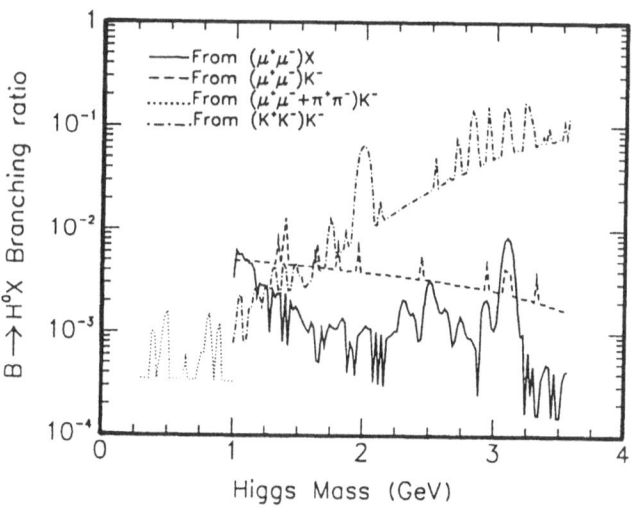

Fig. 20. Upper limits (90% CL) for $B \rightarrow H^0 X$ as a function of M_H.

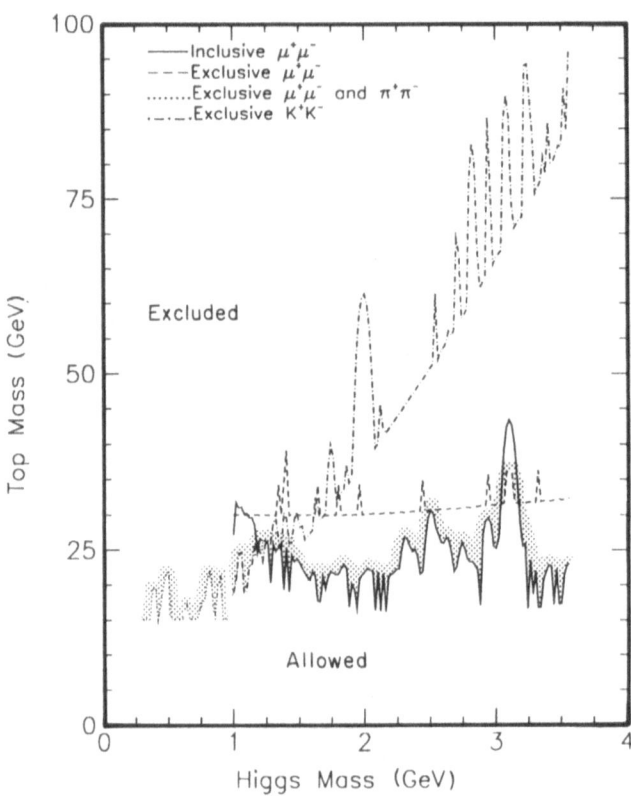

Fig. 21. Maximum value (90% confidence level) allowed by the CLEO experiment for m_t as a function of the assumed value of M_H, as implied by the WY prediction.

4. CONCLUSIONS

1. Light Higgs' have not been found in the decays of open or hidden beauty.

2. The SM Higgs' mass region excluded by the experimental searches, with possibly a few holes, is

$$0 \leq M_H \leq 5 \text{ GeV}$$

if the present "calculations" are correct.

3. New calculations which are desperately needed:

   (a) Higher order QCD radiative corrections.

   (b) $BR(H^0 \to \mu^+\mu^-)$ for Higgs' of mass in the range 1.1 to 3.5 GeV.

   (c) Exclusive channel branching ratios.

ACKNOWLEDGEMENTS

The author thanks Paolo and Paula Franzini, Meenakshi Narain and P. Michael Tuts for discussions and help in the preparation of this paper. She also thanks Ahmed Ali and Nino Zichichi for the warm hospitality extended to her at the Eloisatron Workshop of the CENTRO CULTURA SCIENTIFICA ETTORE MAJORANA.

REFERENCES

1. P. Langacker, in Proceedings of XXIV International Conference on High Energy Physics, eds. R. Kotthaus and J. H. Kuhn, Springer-Verlag (1988) p.190.

2. L. Hall, in Proceedings of XXIV International Conference on High Energy Physics, eds. R. Kotthaus and J. H. Kuhn, Springer-Verlag (1988) p.335.

3. A.D. Linde, JETP Lett. **23** (1976) 64; S. Weinberg, Phys. Rev. Lett. **36** (1976) 294.

4. K. Eggert, in Proc. of XIV Int. Symp. on Lepton and Photon Interactions, Stanford 1989; L. Di Lella, *ibid*; P. Sinervo, *ibid*.

5. N. Cabibbo, L. Maiani, G. Parisi and R. Petronzio, Nucl. Phys. **B158** (1979) 295.

6. E. Gross and E. Duchovni, Phys. Rev. **D38** (1988) 2308.

7. P. Langacker, Preprint UPR 0400T (1989).

8. F. Wilczek, Phys. Rev. Lett. **40** (1978) 220; S. Weinberg, *ibid*. 223.

9. P. Franzini et al., Phys. Rev. **D35** (1987) 2883.

10. M. I. Vystosky, Phys. Lett. **97B** (1980) 159; J. Ellis et al., *ibid.* **158B** (1985) 417; P. Nason, *ibid.* **175B** (1986) 223.

11. H. Goldberg and Z. Ryzak, Phys. Lett. **218B** (1989) 348.

12. See for. ex. J. Pantaleone, M.E. Peskin, and H. Tye, Phys. Lett. **1 49B** (1984) 225.

13. G. Faldt, P. Osland and T.T. Wu, Phys. Rev. **D38** (1988) 164.

14. I. Aznauryan, S. Grigoryan and S. Matinyan, JETP lett. **43** (1986) 646.

15. J. Lee-Franzini, Nucl. Inst. Meth. **A263** (1988) 35; P.M. Tuts, *ibid* **A265** (1988) 243; P. Franzini and J. Lee-Franzini, in Proc. of Oregon Meeting, ed. by R.C. Hwa, World Scientific (1986) p.1009.

16. M. Narain et al., included in J. Lee-Franzini, in Proc. of XXIV Int. Conf. on High Energy Physics, eds. R. Kotthaus and J. H. Kuhn, Springer-Verlag (1988) 1432.

17. P.M. Tuts et al., Phys. Lett. **B186** (1987) 233.

18. J.H. Kühn and S. Ono, Phys. Lett. **142B** (1984) 436; W.Y. Keung and A. Khare, Phys. Rev. **D29** (1984) 2657.

19. S. Komamiya, in Proc. of the 1985 Int. Symp. on Lepton and Photon Interactions at High Energies (ISLEPH85), eds. M. Konuma and K. Takahashi, Kyoto University, Kyoto (1985) p.612.

20. H. Albrecht et al., Z. Phys. **C42** (1989) 349.

21. S. Dawson, in Proceedings of Storrs Meeting, ed. K. Haller et al., World Scientific (1989) 711.

22. I. Beltrami et al., Nucl. Phys. **A451**, (1986) 679.

23. D. Kohler, B.A. Watson and J.A. Becker, Phys. Rev. Lett. **33** (1974) 1628.

24. S.J. Freedman et al., Phys. Rev. Lett. **52** (1984) 240.

25. M. Sivertz et al., Phys. Rev. D **26** (1982) 717.

26. C. Edwards et al., Phys. Rev. Lett. **48** (1982) 903.

27. T. Kaarsberg et al., Phys. Rev. Lett. **62** (1989) 2077.

28. S. Egli et al., PSI Preprint PSI-PR 89-02 (1989).

29. H-L Yu, Preprint IP-ASTP-02-89-R (1989).

30. M. Davier and H. Nguyen Ngoc, Preprint, Lab. de l'Accelerateur Lineaire, ORSAY (1989).

31. T. Yamazaki et al., Phys. Rev. Lett. **52** (1984) 1089.

32. R.S. Chivukula and A.V. Manohar, Phys. Lett. **B207** (1988) 86.

33. H. Nelson, NA31 Preprint in preparation.

34. R.S. Willey and H.L. Yu, Phys. Rev. **D26** (1982) 3086.

35. K. Chadwick et al., Phys. Rev. Lett. **46** (1981) 88.

36. M. Althoff et al., Zeit. Phys. **C22** (1984) 219.

37. W. Bartel et al., Phys. Lett. **132B** (1983) 241.

38. M.B. Voloshin, Sov. J. Nucl. Phys. **44** (1986) 478.

39. S. Raby and G.B. West, Phys. Rev. **D38** (1988) 3488.

40. T.N. Truong and R.S. Willey, Univ of Pittsburgh Report PITT-89-05 (1989).

41. H.E. Haber et al., Nucl. phys. **B294** (1987) 301.

42. CLEO Collaboration, P. Avery et al., contribution to the Lepton and Photon Symposium, Hamburg, 1987.

43. P. M. Tuts, private communication.

44. M.S. Alam et al., Phys. Rev. **D40** (1989) 712.

EXPERIMENTAL SEARCHES FOR THE HIGGS PARTICLE

IN HIGH ENERGY COLLISIONS

Abraham Seiden

Santa Cruz Institute for Particle Physics
University of California, Santa Cruz, CA 95064

ABSTRACT

We look at the potential for discovery of the Higgs Particle at the SSC. As a function of the Higgs mass, different search strategies will be necessary, leading to different experimental challenges. We look at some of these experimental issues.

INTRODUCTION

Since the Higgs mass, and indeed the Higgs sector particle content is unknown, preparation of experiments for the SSC requires that a broad search capability be built into these experiments. This implies that we need to map out a discovery strategy as a function of the Higgs mass and translate this into an appropriately general and powerful experiment or groups of experiments. A great deal of work has gone into this program covering the full energy range and we refer the reader to two general references[1,2] which contain details on the Higgs search strategy as well as references to the original literature. We will here review some of the recent progress, focusing specifically on SSC energies. Many of the studies are ongoing and require more work in order to yield definitive conclusions.

Most of the discussion below will center on the single Standard Model Higgs. We will occasionally discuss the case of models with several Higgs doublets as occurs, for example, in supersymmetric models.

We take, below, the Higgs mass region of interest for the SSC to be

$$m_{H^0} \geq 80 \text{ GeV}.$$

For masses below this value, LEP-I and LEP-II will have either found the Higgs or set a lower limit around this value. The region above 80 GeV breaks up into three mass regions distinguished by the allowable Higgs decay modes and the event rates at the SSC. These regions are as follows:

(1) $m_{H^0} < 2m_W$, where the Higgs decay into two real gauge bosons is kinematically forbidden.

(2) $2m_W < m_{H^0} \lesssim 600$ GeV, where the rates for $W \to Z^0 Z^0 \to 4$ charged leptons allow detection in the very clean all charged lepton final state. The upper limit of 600 GeV, determined by the mass at which only a handful of events per year are produced, is at present a soft number because the production rates are quite sensitive to the unknown top quark mass.

(3) $m_{H^0} \gtrsim 600$ GeV, where the more copius, and experimentally more difficult channels $H^0 \to ZZ \to l^+l^-\nu\bar\nu$ and $H^0 \to ZZ \to l^+l^- + 2$ hadronic jets, must be used.

We will discuss each of the regions, in turn, below.

To calculate rates we need to specify a detector. For some of the difficult, specialized measurements, we will explicitly discuss the detection characteristics needed. Otherwise we will assume detection in a generic solenoidal detector described in reference 3. This includes:

For transverse energy measurement, calorimetry in the rapidity range $|Y| < 5.5$.

Calorimetric jet reconstruction in the rapidity range $|Y| \leq 3$.

For the latter region calorimeter cells forming towers covering 0.05×0.05 in $\Delta\phi\Delta Y$.

Electromagnetic energy resolution: $15\%/\sqrt{E}$.

Hadronic energy resolution: $50\%/\sqrt{E}$.

Rates quoted will correspond to an integrated luminosity $L = 10^{40}$cm^{-2}(SSC year). Monte Carlo calculations are usually performed using the Pythia[4] simulation program. This is of course an approximation to the real QCD physics and implies an uncertainty in the analyses which cannot always be estimated, particularly for the tails of various distributions.

HIGGS PRODUCTION AND DECAY

The expected Higgs production cross section at the SSC[5] as a function of its mass is shown in Fig. 1. Per SSC year, the production rates vary from about 5×10^6 for a 100 GeV H^0 to about 10,000 for a 1 TeV H^0. The rates are dependent on the mass of the top quark as shown in Fig. 1.

The primary production diagrams of interest are shown in Fig. 2. Diagram (c) is only large at low mass, contributing 25,000 events for $m_H = 100$ GeV, as opposed to millions

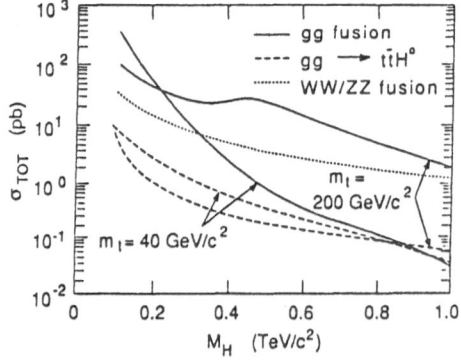

Fig. 1. The Higgs production cross sections for various processes as a function of the Higgs mass for two different t quark masses.

of events for diagram (a). It is of interest mainly if the presence of a $W^{\pm}$ in the final state helps in the rejection of backgrounds.

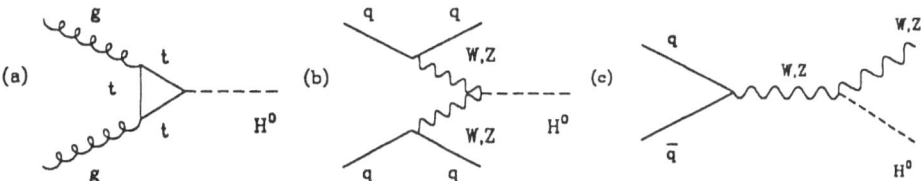

Fig. 2. Diagrams for Higgs Production at the SSC.

The decay modes of the H^0 are dependent on its mass. We will call the Higgs heavy if it has a mass above 180 GeV (*i.e.*, above the threshold for decay into $Z^0 Z^0$ gauge boson pairs) and light if its mass is below this value.

Well above the gauge boson pair decay threshold the H^0 decays 2/3 into $W^+ W^-$ and 1/3 into $Z^0 Z^0$. For a sufficiently light top quark, it will decay almost entirely into $t\bar{t}$ below this threshold. Clearly, the measurement of the t quark mass is critical for predicting the production cross section and determining a search strategy for the Higgs. Present limits on the top quark mass, however, indicate that the mass range over which $t\bar{t}$ decay dominates is very small or nonexistent.

For a heavy top quark, the light Higgs would decay mostly into $b\bar{b}$. Fig. 3 shows the expected branching ratios in this case.[6] The ZZ^* rate is for a final state with one real and one virtual Z. Not shown in the figure are the WW^* branching ratio, the gg branching ratio, and the small branching ratios into light fermions, since these do not yield useful signatures for discovery.

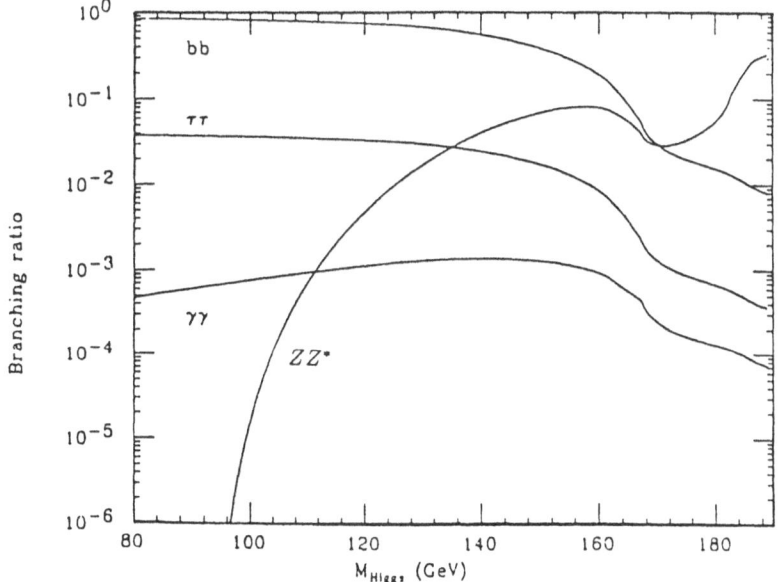

Fig. 3. The branching ratios for various decays of the Higgs boson. It has been assumed that $m_H < 2m_t$.

The decay into $b\bar{b}$ has enormous background from QCD processes, making diagram (c) in Fig. 2 the only useful production process from this decay mode. The decay into $\gamma\gamma$ has a small branching ratio but a very nice signature since it produces isolated photons with an invariant mass of m_H. We discuss these processes in the next section.

STUDY OF $H^0 \to \gamma\gamma$[7]

If the light Higgs lies above top threshold, its decays to top quarks will dominate to the extent that the reaction $H^0 \to \gamma\gamma$ will be undetectable at the SSC. For a Higgs lying below top threshold, the small branching ratio into $\gamma\gamma$ of about 10^{-3} may be detectable. This process has a clean signature, yielding two isolated photons in the final state, however it sits on a large continuum background. Hence the signal to noise will be directly proportional to the mass resolution achievable with the photon detector. This process has two kinds of backgrounds. The first is irreducible and comes from the processes $q\bar{q} \to \gamma\gamma$ and $gg \to \gamma\gamma$. The second comes from QCD jet–jet events where fluctuations yield jets indistinguishable in the detector from photons. The jet–jet rate is expected to be about 5×10^7 as large as the rate from the Higgs. These potential background events will require a rejection factor of $> 10^7$. Monte Carlo analyses indicate that, although difficult, this may be possible. This issue will require further study. Below, we discuss only the irreducible background of real $\gamma\gamma$ events.

The analysis of signal and background is based on the following cuts:

$$|Y_\gamma| < 3, \quad |\cos\theta^*| < 0.8, \quad E_T^\gamma > 10 \text{ GeV},$$

where θ^* is the photon angle in the $\gamma\gamma$ rest frame. For an 80 GeV top quark mass this would give about 500 signal events for a 100 GeV Higgs and about 800 for a 150 GeV Higgs. The background falls by about a factor of five over this mass range, so the signal to noise increases substantially with increasing Higgs mass up to about 160 GeV, where the expected H^0 to $\gamma\gamma$ branching ratio starts falling rapidly (see Fig. 3).

To compare the signal to background, a calorimeter resolution must be assumed. Two models have been looked at:

$\sigma_E/E = 10\%/\sqrt{E} + 1\%$, Excellent Resolution.

$\sigma_E/E = 3\%/\sqrt{E} + 0.5\%$, Extraordinary Resolution.

Fig. 4(a) and (b) show the expected $\gamma\gamma$ mass distribution for the signal from a 100 GeV Higgs superimposed on the irreducible background for the two detector models. Fig. 5(a) and (b) are the analogous distributions for an assumed 150 GeV Higgs. These plots are then fitted to a known background shape plus a signal of width given by the detector resolution. The results are given in the table below.

Higgs Mass	Detector Resolution	Mass Resolution	Statistical Significance
100 GeV	Excellent	1.44 GeV	None
100 GeV	Extraordinary	0.55 GeV	2.8σ
150 GeV	Excellent	1.91 GeV	7.6σ
150 GeV	Extraordinary	0.80 GeV	12.0σ

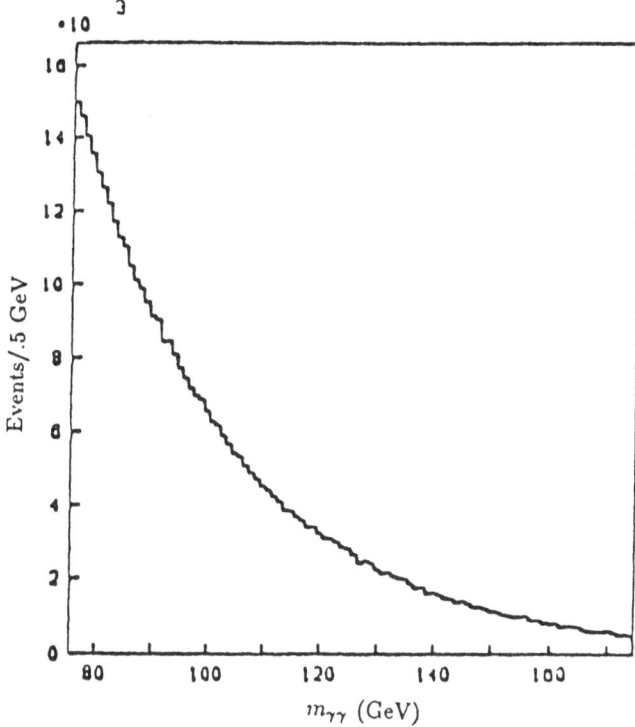

Fig. 4(a). Mass of $\gamma\gamma$ pairs in a detector of excellent resolution for a
100 GeV Higgs plus $\gamma\gamma$ background.

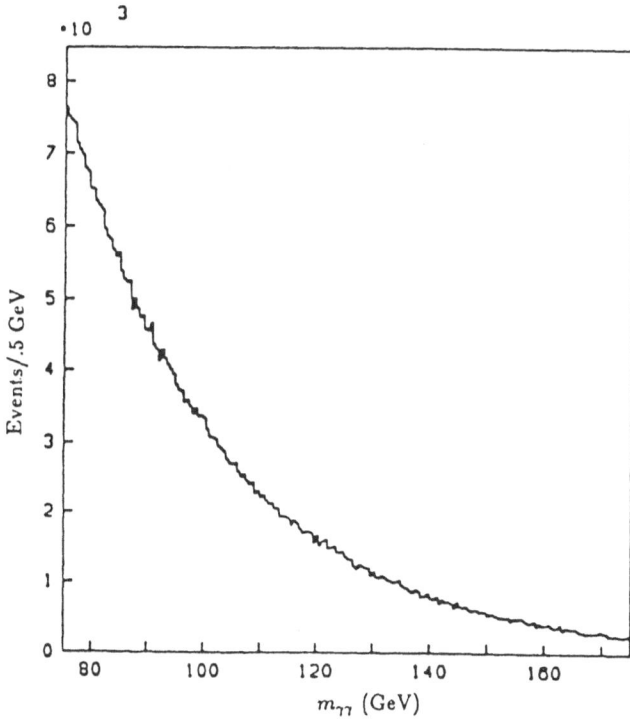

Fig. 4(b). Mass of $\gamma\gamma$ pairs in a detector of extraordinary resolution
for a 100 GeV Higgs plus $\gamma\gamma$ background.

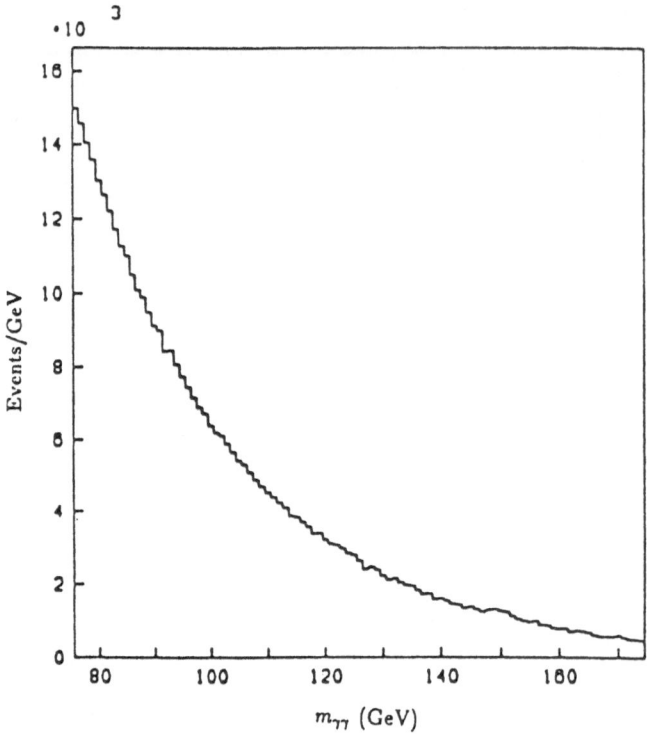

Fig. 5(a). Mass of $\gamma\gamma$ pairs in a detector of excellent resolution for a 150 GeV Higgs plus $\gamma\gamma$ background.

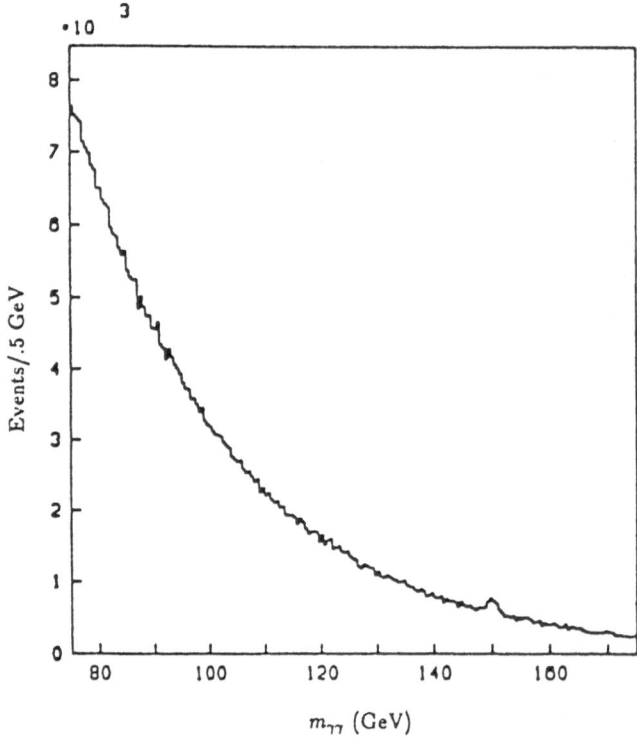

Fig. 5(b). Mass of $\gamma\gamma$ pairs in a detector of extraordinary resolution for a 150 GeV Higgs plus $\gamma\gamma$ background.

Unless the Higgs mass is a priori known, a signal of a few σ is not credible since the number of potential mass bins being searched in is large. It is therefore unlikely that the Higgs can be found in this mode alone for masses $\lesssim 100$ GeV. The region near 150 GeV seems possible provided the calorimeter is of very high quality and the jet–jet background can be rejected. Note, in addition, the detector must have good geometrical acceptance and be capable of triggering on isolated photons with transverse energy $\gtrsim 10$ GeV, without generating an unacceptably high trigger rate. These are all very stringent requirements and may require a dedicated experiment.

STUDY OF $H^0 \to b\bar{b}$ [8]

The process being searched for in this case is

$$q\bar{q} \to W^{\pm}H^0 \to b\bar{b}.$$
$$\qquad\quad \hookrightarrow l\nu$$

The presence of a $W^{\pm}$ in the final state is essential for background rejection, since the $H^0 \to b\bar{b}$ by itself would be swamped by QCD processes yielding $b\bar{b}$ pairs. We will assume that the top quark is sufficiently heavy that the dominant decay of the H^0 is into $b\bar{b}$ in the mass range being considered. The potential background processes are

$$qg \to W^{\pm}q$$
$$q\bar{q} \to W^{\pm}g$$

which are both large, and

$$q\bar{q} \to W^{\pm}Z^0$$

which is comparable to the signal and only a significant background for $m_H \simeq m_Z$.

The table below, based on a Monte Carlo study, shows the number of potential signal events for a 100 GeV H^0, as a function of the analysis cuts. The background starts off at a ratio 10^4 as large as the signal. The 20 GeV mass cut around m_H reduces this factor to about 10^3. To reduce the background substantially below this requires tagging the b decays, since the $b\bar{b}$ component of the QCD background is small in this mass range. With b tagging (including rejection of $c\bar{c}$ events) the background can be

Selection Criteria	Signal/Year
$pp \to W^{\pm}H^0$, $P_T > 50$ GeV	24,000
$W \to l\nu$	7,500
$P_T^l > 25$ GeV, $E_T^{\mathrm{miss}} > 40$ GeV	2,900
2 Highest E_T jets in $|Y| < 2.5$, mass in a 20 GeV mass bin around m_H	750

reduced by a factor of about 50 relative to the signal. To reach a signal to noise ratio of 1, requires additional background rejection of a factor of about 20. This must come from the event structure. The feasibility of finding the signal is thus very dependent on whether such differences exist and whether further cuts based on these differences leave enough signal to provide a convincing mass peak at m_H in the jet–jet mass plot.

A detailed Monte Carlo analysis has been done implementing b tagging criteria which involve requiring large impact parameter tracks for each potential b jet and at least one jet containing a lepton with large transverse momentum (> 1 GeV) to the jet direction. Requiring that each b jet is contained in a narrow cone and that the W and $b\bar{b}$ systems nearly balance in transverse momentum rejects most of the QCD background. These cuts leave about 40 signal and 80 background events for a 100 GeV Higgs. This result is encouraging. It is clear, however, that factors of two in signal or background rates strongly affect the possibility of discovery in this channel.

STUDY OF $H^0 \rightarrow Z^0 Z^0 \rightarrow 4$ CHARGED LEPTONS[9]

The process $H^0 \rightarrow Z^0 Z^0 \rightarrow 4$ charged leptons provides the cleanest channel for a Higgs search. In addition, it allows a search over a substantial region in mass, $180 \lesssim m_H \lesssim 600$ GeV. It is, however, rate limited, since the branching ratio for the Higgs into this final state is only 1.4×10^{-3}. Potential backgrounds from Z^0+ jets are large, yielding $Zc\bar{c}$, $Zb\bar{b}$, and $Zt\bar{t}$ final states which produce significant numbers of events with 4 leptons. The real $Z^0 Z^0$ events are distinguishable from background by using the Z^0 mass constraints, isolation criteria to eliminate leptons from light quark jets, and rapidity and transverse momentum cuts.

The processes $q\bar{q} \rightarrow Z^0 Z^0$ and $gg \rightarrow Z^0 Z^0$ provide a substantial continuum signal on which the H^0 resonance would sit. The table below indicates the rates expected per year, based on Monte Carlo studies. The detected rates assume the follows cuts: $P_T^l > 10$ GeV, $|Y^l| < 2.5$, $P_T^Z > 50$ GeV, $|\Delta m_Z| < 10$ GeV.

Mass Choices	Raw Rates	Detected Rates, After Cuts
$m_t = 40$ GeV		
$m_H = 200$ GeV	575	70
$m_H = 400$ GeV	144	89
$m_H = 600$ GeV	60	40
$m_t = 200$ GeV		
$m_H = 200$ GeV	687	93
$m_H = 400$ GeV	560	345
$m_H = 600$ GeV	225	146
Continuum Background from $q\bar{q}, gg \rightarrow Z^0 Z^0$	1500	280

The detector used for this process needs to be adequate for detection of the signal and elimination of background as discussed above. Since the event rates are small, further losses in efficiency from tracking or particle identification must be minimal. The detector has to have adequate mass resolution to avoid spreading the Higgs peak, since it sits on a continuum signal from which it has to be distinguished. Fig. 6 illustrates this point by showing the predicted distribution of $Z^0 Z^0$ reconstructed invariant mass in detectors with varying tracking or calorimetric resolutions. For this Monte Carlo study the Higgs mass has been chosen to be 400 GeV.

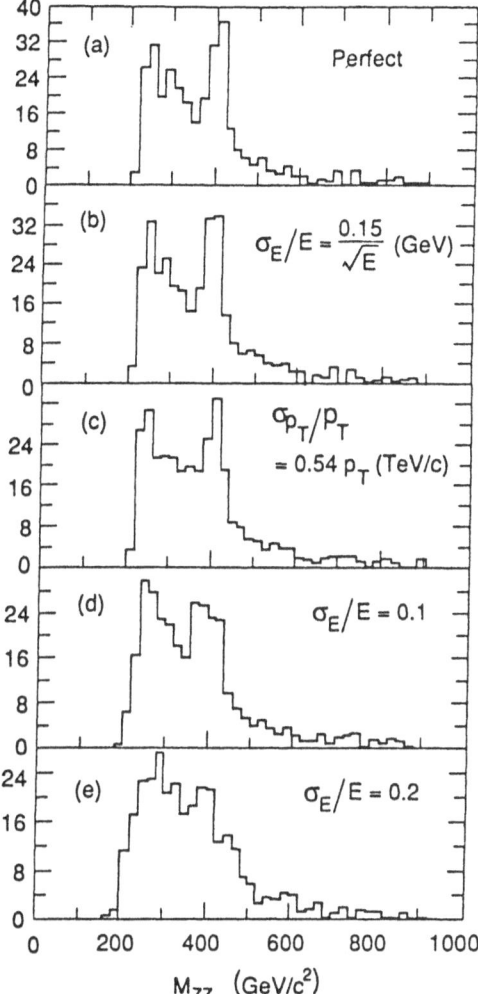

Fig. 6. $Z^0 Z^0$ mass distribution for events reconstructed with
detectors of varying resolution. $m_H = 400$ GeV.
Continuum background is included.

A possible strategy to increase the heavy Higgs event rate would be to run at very high luminosity ($> 10^{34} \text{cm}^{-2}\text{sec}^{-1}$) detecting only muons outside of a substantial absorber. To minimize background, a good magnetic tracking system should be used. Such a strategy would most probably not allow the lepton isolation cuts possible in a solenoidal detector. A detailed analysis, in particular of the background for such a detector configuration, would be of interest.

The $H^0 \to Z^0 Z^0 \to 4$ charged leptons allows a very large mass range to be searched. It could turn out to contain multiple neutral Higgs scalars or partners of lighter Higgs scalars. It is important that experiments maintain sensitivity to such possibilities. For such a case the individual Higgs couplings to gauge bosons are reduced and the coupling to fermions is uncertain. The gauge boson fusion diagram still, however, provides a lower limit for the production cross section, which is now reduced. In addition, the Higgs width is correspondingly narrower reflecting the reduced couplings.

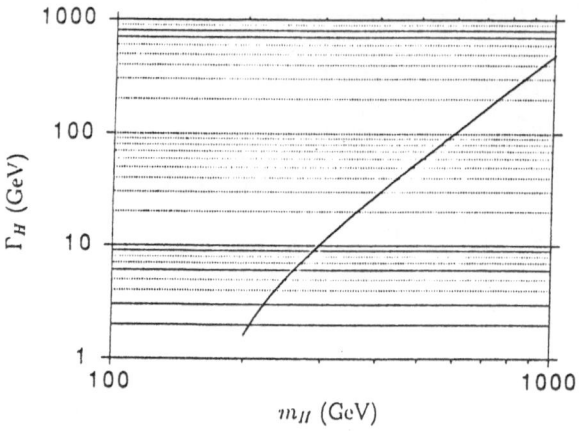

Fig. 7. Higgs width for simplest single Higgs. Masses that have been used are: $m_Z = 91$ GeV, $m_W = 83$ GeV, $m_t = 90$ GeV.

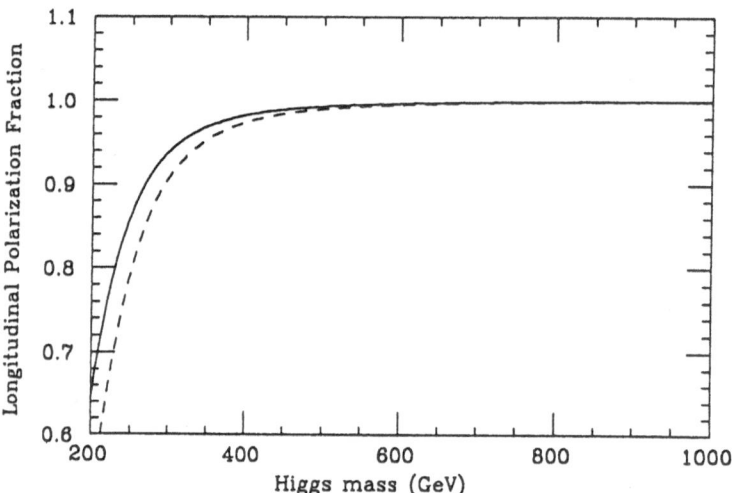

Fig. 8. Fraction of gauge bosons with longitudinal polarization for $H^0 \rightarrow ZZ$ (dashed) and $H^0 \rightarrow WW$ (solid).

To allow maximum sensitivity, it would be useful to have sufficiently good mass resolution that a signal is not spread out in mass more than necessary. Figure 7 shows the Standard Model Higgs width as a function of m_H. A practical mass resolution is about 1–2% of m_H. Thus for $m_H > 250$ GeV, good mass resolution will be useful in searching for a Higgs narrower than the simplest single Higgs. In addition, for a Higgs mass of 300 GeV the gauge bosons are already mostly longitudinally polarized as shown in Fig. 8. Thus in the mass region beyond 300 GeV, further background rejection using a polarization selection for each Z^0 is possible. With such capabilities, an SSC experiment should be sensitive to a Higgs with about 15% of the Standard Model gauge boson decay width.

DETECTION OF GAUGE BOSON PAIRS FROM HEAVY HIGGS DECAY[10]

For a Higgs mass above about 600 GeV the rate for $H^0 \rightarrow Z^0 Z^0 \rightarrow 4$ charged leptons is too small to allow unambiguous detection of the Higgs, unless the top quark mass is close to its maximum allowed value of 200 GeV. For a smaller top quark mass detection of one gauge boson through its hadronic decays allows the greatest reach in Higgs mass. For a 1 TeV Higgs about 1,000 events per year will be produced in the final state:

$$pp \rightarrow H^0 \rightarrow WW \rightarrow \text{jets}$$
$$ \hookrightarrow l\nu.$$

The expected background from the mixed electroweak–QCD process

$$pp \rightarrow W + \text{jets}$$
$$ \hookrightarrow l\nu$$

is expected to be about 100 times as large after loose selection criteria. Selecting events with jet systems of invariant mass close to the W, for example, $70 \leq m_{\text{jet system}} \leq 90$ GeV still leaves about 10 times as much background as signal. Thus it is necessary to find an additional factor of ten background suppression to allow the extraction of a clear signal. In particular this would then allow the observation of a mass peak at the W mass, indicating that we are indeed detecting WW final states.

The main diagrams for the signal and background are shown below in Fig. 9.

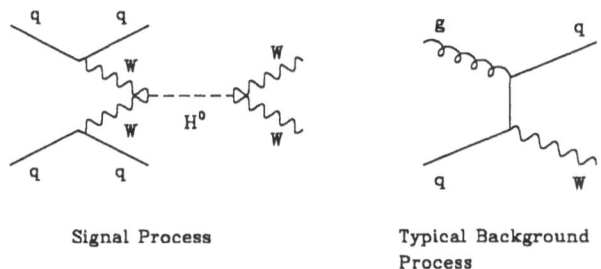

Signal Process   Typical Background Process

Fig. 9.   Primary production and background diagrams for very heavy Higgs.

Note, the signal process involves no exchange of color and a momentum transfer governed by the W mass scale. The background process involves an exchange of color between the protons and a momentum transfer scale given by the Higgs mass. Thus for the signal we are observing a boosted W which yields a multiplicity of hadrons like a W at rest, while the background contains a full 1 TeV scale hadronic jet. Simulating these processes using Pythia we find two dramatic differences:

1) The W jet [Fig. 10(a)] is typically collimated much more tightly than the QCD jet [Fig. 10(b)] which resembles a broad "fan".

2) The event multiplicity is lower by about a factor 2.5 in the Higgs case.

The charged particle multiplicity distribution from the simulation is shown in Fig. 11. A cut at 40 charged tracks leaves 78% of the signal and only 7.3% of the background. Thus

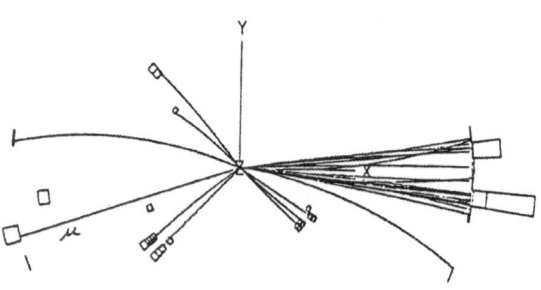

Fig. 10. Event display which shows charged tracks with $p_t \geq 0.5$ GeV
and calorimeter cells with $E_t > 0.5$ GeV in $|Y| < 2.5$ to-
gether with the isolated lepton from the W decay. a) Higgs
$\rightarrow W^+ W^+$ event with $M_{WW} = 1230$ GeV, $N_{ch} = 33$ and
$m_j = 73$ GeV. b) Mixed EW-QCD event with invariant mass
$M_{Wjet} = 1280$ GeV, $N_{ch} = 73$ and $m_j = 82$ GeV.

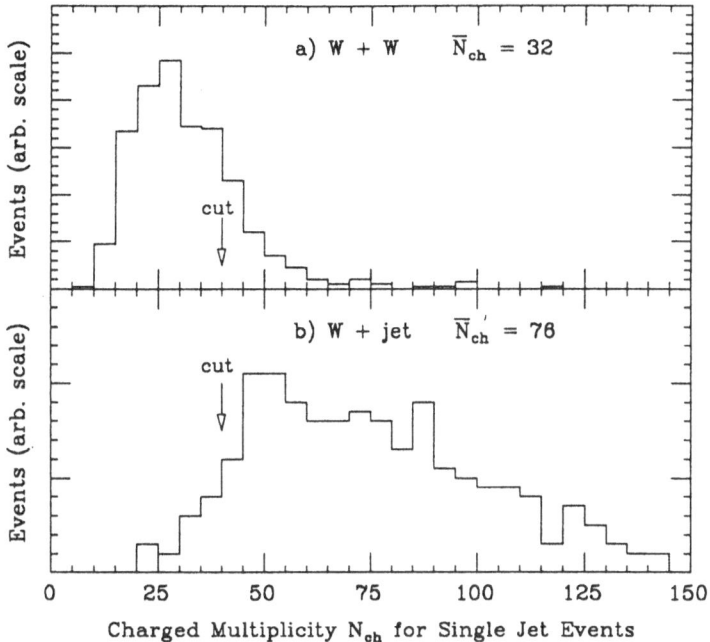

Fig. 11. Charged event multiplicity N_{ch} for tracks with $|Y| < 2.5$ and $p_t \geq 0.5$ GeV: *a)* Higgs $\rightarrow WW$, *b)* $W+$ jet.

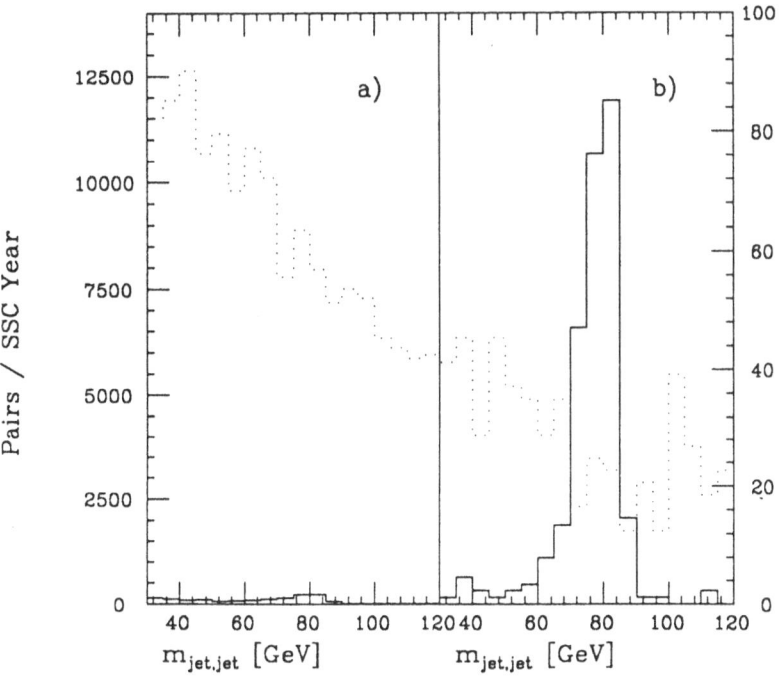

Fig. 12. Jet pair mass $M_{jet,jet}$ for the Higgs $\rightarrow WW$ decay (solid line) and mixed E.W. decay (dotted line). a) before cuts, b) after cuts.

this cut has the potential to provide the needed extra factor of 10 of background rejection. It also has the advantage that little of the signal is lost and it is independent of the W helicity. The latter is of interest since the very massive Higgs produces mainly longitudinal W's and it would be an important measurement if one could determine the W helicity. Fig. 12 shows the jet mass distribution from a calorimetric detector before and after the multiplicity cut. Note, in the latter case, the W mass peak should be clearly visible over the background, which is essential in order to demonstrate that two W's are present. In the analysis two jets in the central rapidity region are reconstructed in the calorimeter, each with large transverse momentum, and used to calculate the invariant mass. Depending on details of the analysis a 1 TeV Higgs should yield between 200 and 500 detected events in the decay channel discussed for 1000 initially produced.

We note, finally, that a similar analysis can be applied to

$$pp \rightarrow H^0 \rightarrow Z^0 Z^0 \rightarrow \text{jets.}$$
$$ \hookrightarrow l^+ l^-$$

The use of the hadronic decays yields a factor 20 increase in statistics, as compared to the all charged lepton decays. This channel typically results in about a factor 5 fewer events than in the WW channel because of the smaller $H^0 \rightarrow Z^0 Z^0$ and $Z^0 \rightarrow l^+ l^-$ branching ratios. It would, however, allow a corroboration of the signal seen in the WW channel and in addition allows a more direct measurement of the Higgs mass and Z^0 helicity.

The reconstruction of the invariant mass of hadronic decays of strongly boosted W or Z^0 bosons relies heavily on calorimetric energy measurement. The smearing of the measured energies still allows a reasonably good mass resolution of approximately 5 GeV. The incident energy is, however, rather narrowly confined in angle making shower spreading in the rear of the calorimeter a potentially important contributor to the mass resolution.[11] This issue deserves more attention in deciding on both an optimum algorithm for calculating the invariant mass from the shower measurements and an optimum calorimeter design.

For Higgs masses in the TeV range, the WW scattering channel becomes strongly interacting. A particular model for this is given by technicolor theories.[12] In these models, the Standard Model diagrams shown in Fig. 13 are supplemented by interactions through the bound states of the technicolor force as shown in Fig. 14. We choose for our example $Z^0 W^\pm$ scattering, with analogous diagrams for the other boson pairs. In addition, if we assume this interaction also generates the fermion masses (extended technicolor) there must be direct fermion couplings, as shown in Fig. 15. These latter couplings could produce very large rates since they do not require the radiation of gauge bosons from the incoming particles. These models can produce anywhere from small enhancements in rates as compared to the Standard Model, to large increases as much as a factor of 50. In addition, dramatic resonances in the gauge boson pair mass could appear. The technicolor theory presents an exciting possibility for SSC physics.

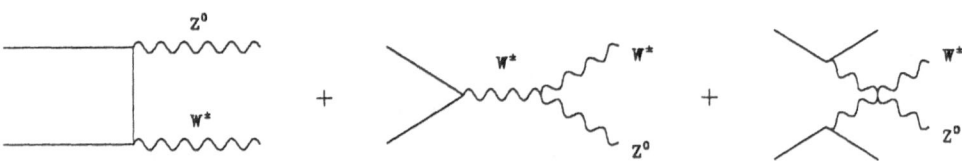

Fig. 13. Standard Model Diagrams.

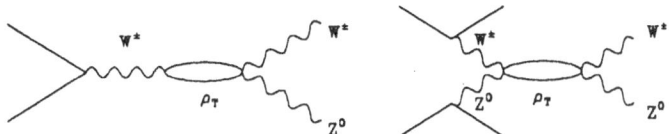

Fig. 14. Technicolor couplings.

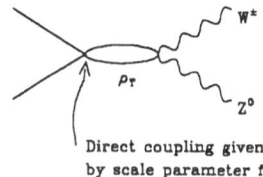

Direct coupling given
by scale parameter *f*

Fig. 15. Extended technicolor
coupling to fermions.

REFERENCES

1. J.F. Gunion, H.E. Haber, G.L. Kane and S. Dawson, *The Higgs Hunter's Guide* (Addison-Wesley, Reading, MA 1990); also UCD Preprint 89-4 (1989).

2. F.J. Gilman and S. Jensen, editors, *Proceedings of the 1988 Summer Study on High Energy Physics in the 1990's,* Snowmass, Colorado, June 27–July 155, 1988, to be published.

3. Donaldson, R. and Gilchriese, M.G.D., editors, *Proceedings of the Workshop on Experiments, Detectors, and Experimental Areas for the Supercollider,* Berkeley, CA (1987).

4. H. Bengtsson and T. Sjostrand, *Comput. Comm.* **46,** 43 (1987).

5. R.N. Cahn *et al.,* in Ref. 3; Gunion *et al.,* U.C. Davis Preprint UCD 86-15 (1986).

6. I. Hinchliffe, in ref. 3.

7. C. Barter *et al.,* in ref. 2.

8. J.E. Brau, K.T. Pitts and L.E. Price, in ref. 2.

9. E.M. Wang, G.G. Hanson *et al.,* in ref. 2.

10. H.F.-W. Sadrozinski, A. Seiden and A. Weinstein, in ref. 2.

11. Wang, E.M., in ref. 2.

12. R.S. Chivukula, Boston University Preprint BUHEP-88-16.

TESTING THE HIGGS MECHANISM AT HIGH ENERGY COLLIDERS

Michael S. Chanowitz

Theoretical Physics Group
Lawrence Berkeley Laboratory
Berkeley, California 94720

1. INTRODUCTION

In this talk I will review the implications of symmetry and unitarity for the physics of electroweak symmetry breaking and describe some of the signals of that physics that may emerge above 1 TeV at multi-TeV pp colliders. Though there is no direct experimental evidence, the Higgs mechanism[1] is universally regarded as the only viable agent of $SU(2)_L \times U(1)_Y$ symmetry breaking.[2] General considerations[3,4] based only on unitarity and gauge symmetry insure that decisive experiments can be done within the next ten years to test the Higgs mechanism. (This ten-year unitarity bound does require the cooperation of the Good Lord and the U.S. Congress. Caution is therefore advisable: while the Former has always honored unitarity, the latter is a known source of unitarity violations.) The outcome of these experiments is certain to be exciting. If the Higgs mechanism is not confirmed, it would mean either that the electroweak interactions are not described by a gauge theory or that a breaking mechanism exists which is unimagined today. If the Higgs mechanism is confirmed than there may or may not be Higgs bosons, but there is necessarily a new force (the real #5) and associated new quanta.

This talk assumes that the Higgs mechanism is correct. I refer to the new force and quanta generically as $\mathcal{L}_{SB}$, the lagrangian of the $SU(2) \times U(1)$ symmetry breaking sector. The $\$64 \times 10^8$ question (in then-year dollars, more or less) is "*What is $\mathcal{L}_{SB}$?*" It could be the Higgs sector of the Weinberg-Salam model, the Higgs sector of a multi-Higgs boson model as in supersymmetric theories, a dynamical model with no elementary Higgs boson like technicolor, or something else that no one has yet imagined. Our ignorance of $\mathcal{L}_{SB}$ is vast: in particular we know neither the coupling strength of the new force, λ_{SB}, nor the mass scale of the new quanta, M_{SB}.

One purpose of this talk is to present the model independent consequences for $\mathcal{L}_{SB}$ that follow from the assumption of an $SU(2)_L \times U(1)_Y$ gauge theory that is spontaneously broken via the Higgs mechanism. The two ingredients are symmetry and unitarity.

Higgs Particle(s)
Edited by A. Ali
Plenum Press, New York, 1990

Symmetry implies low energy theorems for $W_L W_L$ scattering,[3,4] the scattering of L = longitudinally polarized gauge bosons, which are essentially the Goldstone bosons of $\mathcal{L}_{SB}$, much like the pions of $\mathcal{L}_{QCD}$. There are two possibilities: $\mathcal{L}_{SB}$ may be weak or strong. If $\mathcal{L}_{SB}$ is weak then $W_L W_L$ scattering is never strong (except at narrow resonances), the low energy theorems show that the associated quanta lie below 1 TeV, $M_{SB} < 1$ TeV, and there are Higgs bosons among the quanta of $\mathcal{L}_{SB}$. In this case the symmetry breaking sector can be analyzed with perturbation theory. If $\mathcal{L}_{SB}$ is strong the low energy theorems show that $W_L W_L$ scattering is strong for $\sqrt{s_{WW}} = $ 1-2 TeV and that the new quanta of $\mathcal{L}_{SB}$ are above 1 TeV, $M_{SB} > 1$ TeV. Such theories cannot be analyzed perturbatively.

For a strongly coupled $\mathcal{L}_{SB}$ the low energy theorems together with unitarity imply an upper bound on the energy, which I call Λ_{SB}, at which the dynamics of $\mathcal{L}_{SB}$ must begin to influence $W_L W_L$ scattering. That limit is[3]

$$\Lambda_{SB} \lesssim 4\sqrt{\frac{\pi}{\sqrt{2}G_F}} \cong 1.8 \ TeV. \tag{1.1}$$

The most plausible dynamics is that the typical mass scale of the quanta is of order Λ_{SB}, $M_{SB} = O(\Lambda_{SB})$, in which case (1.1) implies that the lowest lying resonances will not be much heavier than ~ 2 TeV.

The experimental strategy is to measure $W_L W_L$ scattering above 1 TeV using the WW fusion mechanism[5] shown in figure (1.1). The initial W_L pair are each off mass-shell and must rescatter to become on-shell and appear in the final state. For a strong $\mathcal{L}_{SB}$, the leading contribution to the amplitude above 1 TeV is from strong rescattering by $\mathcal{L}_{SB}$, of order $g_{weak}^2 \cdot \lambda_{SB}$. The leading background for W pair production, $\bar{q}q \to WW$, is of order g_{weak}^2. Therefore we expect an observable signal from WW fusion if and only if λ_{SB} is strong, and we can learn from this measurement by observing either the signal or its absence. That is, if we are sure we can see the signal if it is present, then by observing its absence we learn that $\mathcal{L}_{SB}$ is weakly coupled, with Higgs bosons waiting to be found below 1 TeV.

The low energy theorems are the basis for an estimate of the magnitude of the strong WW scattering cross section, since they specify the energy scale, eq. 1.1, at which the leading partial wave amplitudes approach unity. This information in turn allows us to estimate the collider energy and luminosity that is needed to be able to detect the strong scattering signal. The parameters of the SSC, $\sqrt{s} = 40$ TeV and $\mathcal{L}_{SB} = 10^{33} cm^{-2}s^{-1}$, are a near minimal configuration to accomplish this goal.

While in theory, substantially lower collider energy (say a factor two or more) could be compensated by higher luminosity, in practice this appears very difficult if not impossible. In addition to the practical problems of accelerator and, especially, detector technology, there are too important problems of principle. First, not only the signal cross section but also the signal : background ratio is degraded at lower energy. One consequence is that the systematic theoretical uncertainty is a fixed fraction of the background (perhaps $\sim 30\%$ – see reference 6a) independent of luminosity, tending to obscure the signal. This problem remains for any luminosity. Second, the channels with the largest branching ratios that are detectable at $10^{33} cm^{-2}s^{-1}$ are not detectable at higher luminosity because of additional backgrounds from the increased number of events per beam crossing. An example of this problem emerged in the High Luminosity Option Study[6b] for the LHC, in the case of the heavy Higgs decay $H \to ZZ \to \bar{\ell}\ell + \bar{\nu}\nu$. This issue is discussed at more length in reference 6.

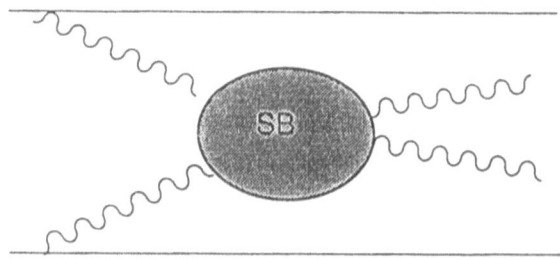

Figure 1.1 Generic $W_L W_L$ fusion via interactions of the symmetry breaking sector $\mathcal{L}_{SB}$.

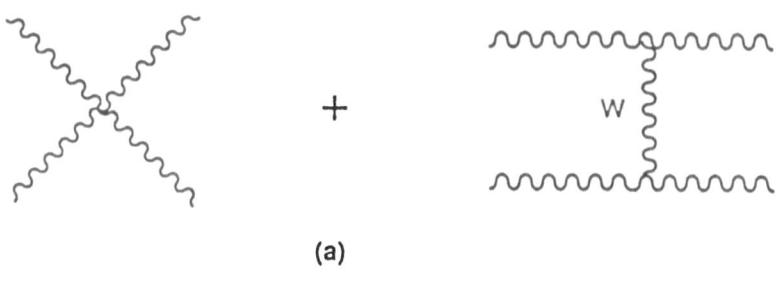

(a)

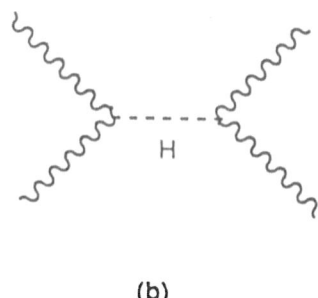

(b)

Figure 2.1 Leading diagrams for $W^+ W^- \rightarrow Z Z$, including interactions from the gauge sector (a) and the s–channel Higgs boson exchange (b) — see eq. (2.11).

Strong WW scattering could also be studied at multi-TeV e^+e^- colliders, if we knew how to build them. Though less intensively studied, it seems that a collider of $\sqrt{s} \cong$ 2-3 TeV and $\mathcal{L} \gtrsim 10^{33} cm^{-2}s^{-1}$ is a minimal configuration for the task.[7]

We are entering an exciting decade since the presently approved colliders, including LEP, LEP 200, and the SSC, offer a nearly complete capability to discover the $SU(2) \times U(1)$ breaking mechanism (with the exception of the "intermediate mass" Higgs boson discussed below). However it is equally clear that these facilities will be insufficient to provide the detailed studies that will be needed after the initial discovery has been made. The detectable event yields are in every case simply too small, never exceeding more than a few hundred in the most optimistic cases and typically much smaller. Eventually higher energy and/or higher luminosity will certainly be needed, for high statistics studies of the detailed structure of the fifth force and its associated quanta. For instance, discovery of a 30 GeV Higgs boson would undoubtedly motivate an upgrade of LEP to a $10^{33}cm^{-2}s^{-1}$ Z factory, with a factor 100 increase in Higgs boson statistics. Discovery of strong WW scattering, indicating a strongly interacting $\mathcal{L}_{SB}$ with quanta above 1 TeV, would require even more dramatic measures, providing compelling motivation for pp collisions at $\sqrt{s} \geq$ 200 TeV and/or e^+e^- collisions at $\sqrt{s} \geq 5 - 10$ TeV!

The outline of the talk is as follows:

- Section 2 is a brief review of the general implications of symmetry and unitarity for the mass scale and signal cross sections of the $SU(2) \times U(1)$ breaking physics.

- Section 3 discusses the signals of symmetry breaking above 1 TeV as they can be studied at pp colliders. I focus on the purely leptonic signals; the more ambitious "mixed" modes will be discussed by other speakers at the meeting.[8]

- Section 4 is a brief conclusion.

- The appendix contains a discussion of the relationship between the so-called "fine-tuning" problem and the importance of the search for the intermediate mass Higgs boson.

A general review of electroweak symmetry breaking, including theoretical and experimental prospects, may be found in reference 9.

2. SYMMETRY AND UNITARITY

In this section we review the most general implications of the Higgs mechanism, without committing to any particular model or class of models. The discussion is based on work done with Gaillard, Golden, and Georgi, cited in references 3 and 4. The presentation is brief since the material has been reviewed on other occasions, including the INFN Eloistron workshop held here last year.[6]

The skeleton of the argument is as follows:

- The equivalence theorem (first proved in tree approximation in reference 10 and to all orders in reference 3) establishes the equivalence of longitudinal gauge boson scattering, e.g., $W_L^+ W_L^- \to Z_L Z_L$, with the scattering of the underlying Goldstone bosons, e.g., $w^+w^- \to zz$, at energies $E_i >> M_W$, i.e.,

$$\mathcal{M}(W^+W_L^- \to Z_L Z_L) = \mathcal{M}(w^+w^- \to zz)_R + O\left(\frac{M_W}{E_i}\right). \qquad (2.1)$$

The subscript R indicates that the right side is evaluated in a renormalizable gauge, in which the unphysical Goldstone bosons $w^\pm, z$ still appear in the lagrangian. It is crucial that (2.1) hold to all orders in the interaction λ_{SB} of the symmetry breaking sector $\mathcal{L}_{SB}$, since we apply it when $\mathcal{L}_{SB}$ is strongly interacting.

- Goldstone bosons obey low energy theorems that are exact at threshold; even today they are among the few exact results known in strong interaction theories. The Weinberg $\pi\pi$ low energy theorems, e.g.,

$$\mathcal{M}(\pi^+\pi^- \to \pi^0\pi^0) \cong \frac{s}{F_\pi^2}, \quad s << min.\{m_\rho^2, (4\pi F_\pi)^2\} \qquad (2.2)$$

when $F_\pi = 93$ MeV are the prototypes for the theorems of interest to us,[4] e.g.,

$$\mathcal{M}(w^+w^- \to zz) \cong \frac{1}{\rho}\frac{s}{v^2}, \quad s << min\{M_{SB}^2, (4\pi v)^2\} \qquad (2.3)$$

where $v = (\sqrt{2}G_F)^{-1/2} = 246$ GeV. The rho parameter

$$\rho = \left(\frac{M_W}{M_Z \cos\theta_W}\right)^2 \cong 1 \qquad (2.4)$$

appears in (2.3), with no counterpart in (2.2), because we do not know if $\mathcal{L}_{SB}$ has a global $SU(2)$ symmetry analogous to hadronic isospin symmetry. The experimental fact that $\rho \cong 1$ is however sufficient to ensure that the low energy interactions of the Goldstone bosons of $\mathcal{L}_{SB}$ are symmetric under this "custodial" isospin.[4]

- Combining the equivalence theorem (2.1) and the Goldstone boson low energy theorem (2.3), we obtain the low energy theorem for longitudinal gauge boson scattering,

$$\mathcal{M}(W_L^+ W_L^- \to Z_L Z_L) \cong \frac{1}{\rho}\frac{s}{v^2}, \qquad (2.5)$$

actually valid in an intermediate energy domain (which may or may not occur in nature, depending on the value of M_{SB}),

$$M_W^2 << s << min.\{M_{SB}^2, (4\pi v)^2\}. \qquad (2.6)$$

The low energy theorems for the other $2 \to 2$ scattering processes are given in reference 4.

- Partial wave unitarity implies an upper limit on the energy at which eq. (2.5) can apply. Since the $J = 0$ partial wave cannot be larger than 1,

$$\begin{aligned} a_0(W_L^+ W_L^- \to Z_L Z_L) &= \frac{1}{32\pi}\int_{-1}^1 d\cos\theta \mathcal{M}(W_L^+ W_L^- \to Z_L Z_L) \\ &= s/16\pi v^2 \end{aligned} \qquad (2.7)$$

we see that the linear growth of (2.5) must be modified at a scale Λ_{SB} that is bounded by the requirement that $|a_0| \leq 1$, that is[3]

$$\Lambda_{SB} \lesssim 4\sqrt{\pi}v \cong 1.8 \; TeV. \qquad (2.8)$$

Equation (2.8) is a model independent upper bound on the energy at which the symmetry breaking sector $\mathcal{L}_{SB}$ must begin to have a measureable effect on WW scattering. It should not be confused with the 1.0 TeV "unitarity" bound on the Weinberg-Salam Higgs boson,[11] which does not forbid larger Higgs boson masses but just indicates the value at which perturbation theory breaks down because the Higgs sector is strongly interacting. Rather (2.8) is more like the bound on the scale of the weak interactions[12,9] obtained by applying partial wave unitarity to Fermi's four-fermion lagrangian, the difference being that in the present application a low energy theorem replaces a phenomenological lagrangian as the starting point. The analogous bound on the weak interaction scale is just a factor 2 smaller then (2.8) (at least in the accounting of reference 9),

$$\Lambda_{Fermi} \lesssim \sqrt{2\pi} G_F^{-\frac{1}{2}} = 2\sqrt{\pi} v = 0.9 \; TeV. \tag{2.9}$$

- $\mathcal{L}_{SB}$ is weakly or strongly interacting depending on whether Λ_{SB} approaches the 1.8 TeV upper bound. Above the cutoff Λ_{SB} we have typically

$$a_0(\sqrt{s} \cong \Lambda_{SB}) \cong \frac{\Lambda_{SB}^2}{16\pi v^2} = \left(\frac{\Lambda_{SB}}{1.8 \; TeV}\right)^2 \tag{2.10}$$

so that if $\Lambda_{SB} \ll 1.8$ TeV then $a_0 \ll 1$ and $\mathcal{L}_{SB}$ is weakly interacting. For $\Lambda_{SB} \gtrsim 1$ TeV we find $a_0 \gtrsim 1/3$, in which case unitarity is nearly saturated, perturbation theory is unreliable, and $\mathcal{L}_{SB}$ is strongly coupled.

- $\mathcal{L}_{SB}$ cuts off the growing amplitude (2.7) by exchange of quanta from $\mathcal{L}_{SB}$. Consequently the bound on Λ_{SB} is also an approximate upper limit on the mass scale M_{SB}. A simple example is the Weinberg-Salam model evaluated in tree approximation. We have

$$\mathcal{M}(W_L^+ W_L^- \to Z_L Z_L) = \mathcal{M}_{gauge} + \mathcal{M}_{SB} \tag{2.11}$$

where $\mathcal{M}_{gauge}$ is the result of the gauge sector diagrams of Figure (2.1a)

$$\mathcal{M}_{gauge} \cong \frac{g^2 s}{4\rho M_W^2} = \frac{s}{\rho v^2} \tag{2.12}$$

and $\mathcal{M}_{SB}$ is the Higgs boson exchange contribution, Figure (2.1b)

$$\mathcal{M}_{SB} \cong -\frac{g^2 s}{4 M_w^2} \frac{s}{s - m_H^2}. \tag{2.13}$$

Both (2.12) and (2.13) are given to leading order for $s \gg M_W^2$. For $s \ll m_H^2$ we have $\mathcal{M} \cong \mathcal{M}_{gauge}$, indicating the first step of an alternate (U-gauge !) derivation[4] of the low energy theorem (2.5), while for $s \gg m_H^2$ there is a cancellation (since $\rho = 1$ in this model) of (2.12) and (2.13) resulting in

$$\mathcal{M}|_{s \gg m_H^2} \cong -\frac{m_H^2}{v^2} \tag{2.14}$$

or

$$|a_0|_{s \gg m_H^2} \cong \frac{m_H^2}{16\pi v^2}. \tag{2.15}$$

Comparing (2.15) with (2.10) we see that m_H indeed plays the role of Λ_{SB} in the Weinberg-Salam model. That is, the mass scale of $\mathcal{L}_{SB}$, in this case $M_{SB} = m_H$, is roughly the energy scale Λ_{SB} at which it cuts off WW scattering, $\Lambda_{SB} \cong m_H$.

In a strongly coupled theory we expect $M_{SB} \cong \Lambda_{SB} \cong O(1.8 \text{ TeV})$. The analogous estimate from $\pi\pi$ scattering would be

$$m_{HADRON} \cong 4\sqrt{\pi} F_\pi \cong 700 \; MeV, \tag{2.16}$$

i.e., the scale at which the extrapolation of (2.2) saturates $J = 0$ partial wave unitarity is a good order of magnitude estimate of the hadron mass scale. This is no surprise: resonances occur when scattering becomes strong.

The two generic possibilities are shown in Figure (2.2): weakly coupled $\mathcal{L}_{SB}$ with narrow resonances $\equiv$ Higgs boson(s) at $M_{SB} << 1.8$ TeV or strongly coupled $\mathcal{L}_{SB}$ with broad resonances at $M_{SB} \cong O(2)$ TeV which may not be recognizable as Higgs bosons.

Notice that analogous considerations also applied to the bound (2.9) on the weak interaction scale. In that case the exchanged quanta that cut off the growth of Fermi's phenomenological four fermion scattering amplitude turned out to be the W and Z bosons. The fact that $M_W < 0.9$ TeV verifies unitarity, while the fact that $M_W << 0.9$ TeV means that the weak interactions are indeed weak. If instead $M_W \cong 0.9$ TeV were true, the so-called "weak interactions" would actually have been revealed as a strongly interaction theory above the W threshold!

A similar bound applies to the scale of quark and lepton mass generation,[13] which could in general be quite different than the scale Λ_{SB} associated with the W and Z masses. For instance, the chirality-flip amplitude for lepton annihilation to longitudinally polarized W's has a gauge sector contribution, analogous to eq. (2.12), given by

$$\mathcal{M}(L^+L^- \rightarrow W_L^+ W_L^-)_{gauge} \cong \frac{m_L \sqrt{s}}{v^2}. \tag{2.17}$$

Partial wave unitarity then requires a cutoff at

$$\Lambda_{Lepton} \lesssim \frac{16\pi v^2}{m_L} \cong \frac{(1.8 \; TeV)^2}{m_L}. \tag{2.18}$$

The corresponding bound for quarks is smaller by a factor $\sqrt{3}$ due to color. In Higgs boson models the growth, proportional to $\sqrt{s}$, of (2.17) is cancelled by Higgs exchange, just as in eqs. (2.12) and (2.13). Λ_{Lepton} and Λ_{Quark} may be much larger than Λ_{SB}, because (2.17) grows like $\sqrt{s}$ rather than s and is suppressed by the factor m_L. For instance, the bound on $\Lambda_{Electron}$ is $6 \cdot 10^6$ TeV! It would be interesting to establish whether there are sensible dynamical models which saturate these bounds for light ($< O$ (TeV)) fermions.

3. OVERVIEW OF STRONG WW SCATTERING SIGNALS AT pp COLLIDERS

In Section 2 we reviewed the low energy theorems for $W_L W_L$ scattering and showed that together with unitarity they require the dynamics of $\mathcal{L}_{SB}$ to affect the scattering at an energy scale $\Lambda_{SB} \lesssim 1.75$ TeV. The most probable mechanism is the exchange of particles from $\mathcal{L}_{SB}$, so that $\Lambda_{SB} \cong M_{SB}$, as shown in two examples in Section 2. In general just above the cutoff scale the $J = 0$ partial wave amplitude for scattering of the longitudinal modes $W_L^+ W_L^- \rightarrow Z_L Z_L$ is

$$a_0(W_L^+ W_L^- \rightarrow Z_L Z_L) \cong \frac{\Lambda_{SB}^2}{16\pi v^2} \tag{3.1}$$

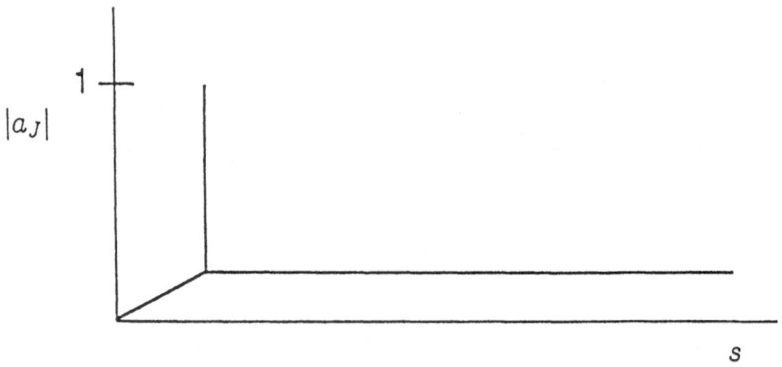

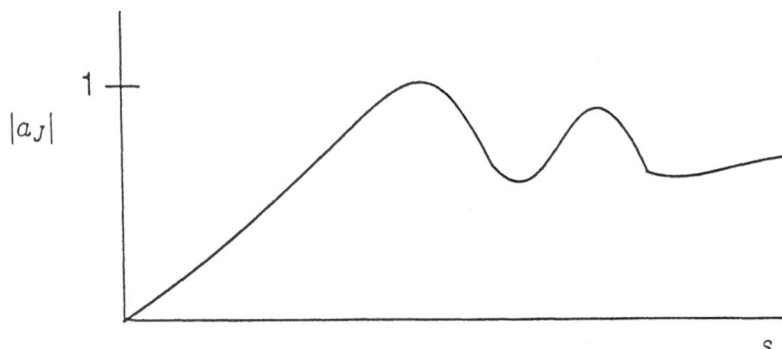

Figure 2.2 Typical behavior of partial wave amplitudes for $W_L W_L$ scattering for a weakly coupled model with narrow (Higgs) resonances (top figure) or a strongly coupled model with broad resonances in the 1-2 TeV region (bottom figure).

so that the scattering is strong if $\Lambda_{SB} << 1$ TeV and weak if $\Lambda_{SB} << 1$ TeV.

In fact there are three independent reaction channels, which can be chosen as $a_{IJ} = a_{00}, a_{11}, a_{20}$ where I is the index of the low energy effective $SU(2)$ discussed below eq. (2.4). In addition to (3.1) the complete list of $2 \to 2$ reactions is

$$W_L^+ W_L^- \to W_L^+ W_L^- \tag{3.2}$$

$$W_L^\pm Z_L \to W_L^\pm Z_L \tag{3.3}$$

$$W_L^+ W_L^+ \to W_L^+ W_L^+ \tag{3.4a}$$

$$W_L^- W_L^- \to W_L^- W_L^-. \tag{3.4b}$$

The corresponding low energy amplitudes are given in reference 4. All these channels will exhibit strong scattering for $\sqrt{s} > 1$ TeV if $\mathcal{L}_{SB} > 1$ TeV, and some will probably have s–channel resonances with masses M_{SB} of order $\mathcal{L}_{SB}$.

Therefore by measuring the $W_L W_L$ scattering amplitudes at high energy, $\sqrt{s} > 1$ TeV, we will learn whether $\mathcal{L}_{SB}$ is a strongly or weakly interacting theory and whether the mass scale of its quanta is in the TeV region or below. If strong we will probably also begin to observe the quanta directly as resonance effects in some of the $2 \to 2$ channels. A general strategy to accomplish this is based on the $W_L W_L$ fusion reaction, figure (1.1), that can be studied at a pp or $e^+ e^-$ collider. The initial state W_L's are off-mass-shell and must rescatter to appear on-shell in the final state. The contribution from rescattering by $\mathcal{L}_{SB}$ is $O(g^2 \lambda_{SB})$ where g is the $SU(2)_L$ gauge coupling constant and λ_{SB} the generic interaction strength of $\mathcal{L}_{SB}$. The dominant background from $\bar{q}q \to WW$ is $O(g^2)$. Therefore WW fusion contributes an observable increment if and only if the rescattering is strong, i.e., if and only if $\lambda_{SB}/4\pi = O(1)$ or equivalently $\Lambda_{SB} \gtrsim 1$ TeV.

Other backgrounds are $\mathcal{M}(gg \to W^+ W^-, ZZ) \sim \alpha_s g^2$ via heavy quark loops[14] (e.g., top), WW bremsstrahlung with gluon exchange between the quarks,[15] $\sim \alpha_s g^2$, and WW fusion by $\mathcal{L}_{SU(2) \times U(1)}$[24] which is $\sim g^4$. These backgrounds are illustrated in Figure (3.1). Though the backgrounds (except gg fusion) are dominated by transverse polarizations, this is not sufficient to separate them from the longitudinally polarized signal, though it can provide corroboration of a possible signal as discussed below.

The SSC is a minimal pp collider for this strategy. A collider of half the energy or less is not adequate, even with realistically likely higher luminosity. Because both the signal and the signal : background decrease at lower energy[6] and because the most important final states are inaccessible at high luminosity,[6b] an upgrade in $\mathcal{L}$ of two to three orders of magnitude would be needed to offset a factor three loss in energy.[6] An $e^+ e^-$ collider of $\sqrt{s} \cong 2 - 3$ TeV is probably minimal for the strong WW scattering signal,[7] though more study is needed. See Figure 3.2 for 1 TeV Higgs boson production cross sections at $e^+ e^-$ and pp colliders of various energies.[16]

In this section I consider three examples of signals for strong symmetry breaking:

1. The 1 TeV Weinberg-Salam Higgs boson

2. Strong $W^+ W^+$ and $W^- W^-$ scattering

3. Techni-rho production

I will consider purely leptonic final states, since they are experimentally cleanest. Larger yields will be possible if detection of $WW \to \ell v + \bar{q}q$ proves feasible.[8]

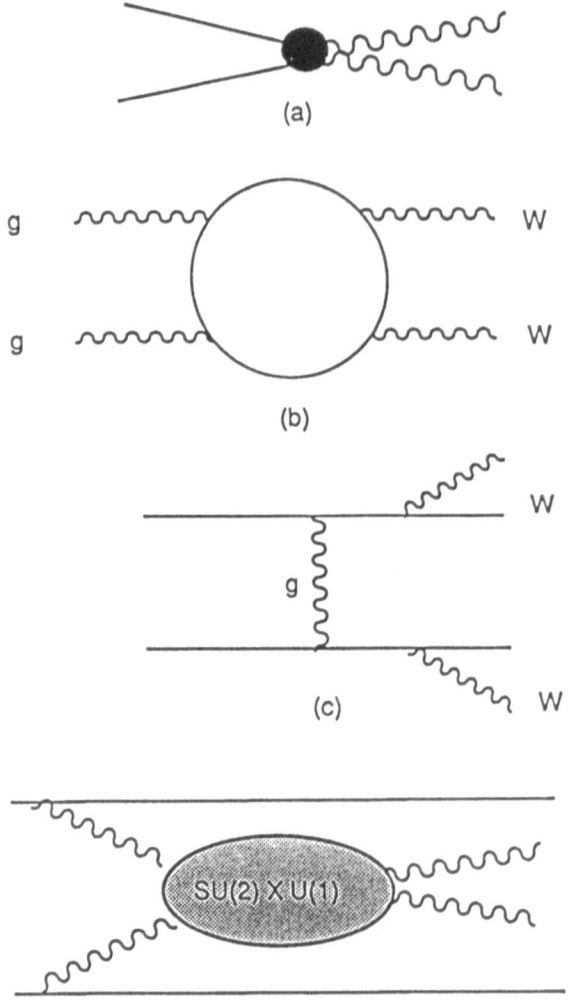

(a)

(b)

(c)

(d)

Figure 3.1 Backgrounds to $H \to WW$ signal from (a) $\bar{q}q \to WW$, (b) $gg \to WW$ via $\overline{Q}Q$ loops, (c) gluon exchange, and (d) higher order electroweak interactions including WW fusion as shown.

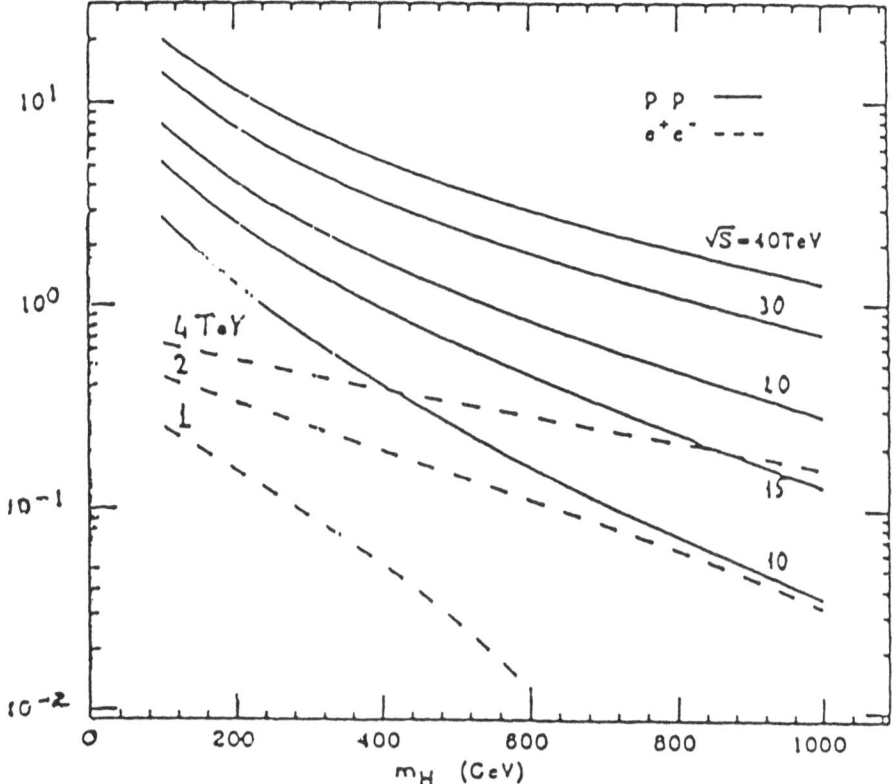

Figure 3.2 Higgs boson production cross sections in picobarns at e^+e^- and pp colliders with center of mass energies indicated (from ref. 16).

Table 3.1 Signal and background for the 1 TeV Higgs boson, $H \rightarrow ZZ \rightarrow e^+e^-/\mu^+\mu^- + e^+e^-/\mu^+\mu^-/\nu^+\nu^-$, with cuts as given in the text. In each entry the first number is the number of events for $10^4 pb^{-1}$ for Higgs decay, while the second is the background from $\bar{q}q$ and $gg \rightarrow ZZ$.

	$\sqrt{s} = 16$ TeV	$\sqrt{s} = 40$ TeV	$\sqrt{s} = 200$ Tev
$m_t = 50$ GeV	3/4	34/16	370/90
$m_t = 200$ GeV	9/5	95/22	920/130

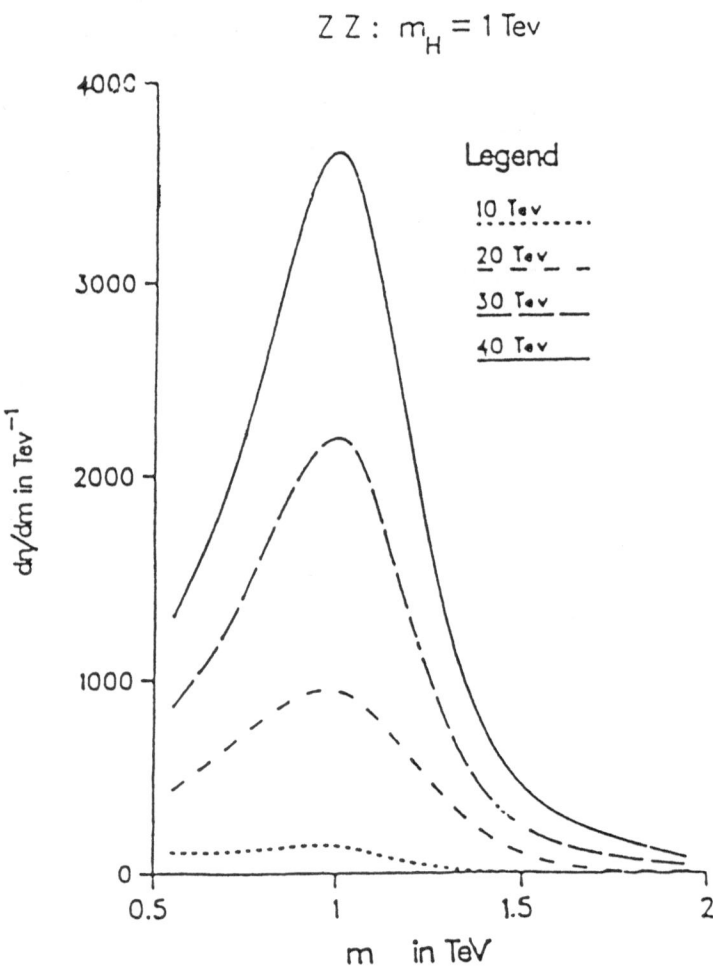

Figure 3.3 Yield dn/dm_{ZZ} in TeV^{-1} for $H \to ZZ$ at 10, 20, 30, and 40 TeV pp colliders, in events per $10^4 pb^{-1}$ with $|y_Z| < 1.5$ (from ref. 3).

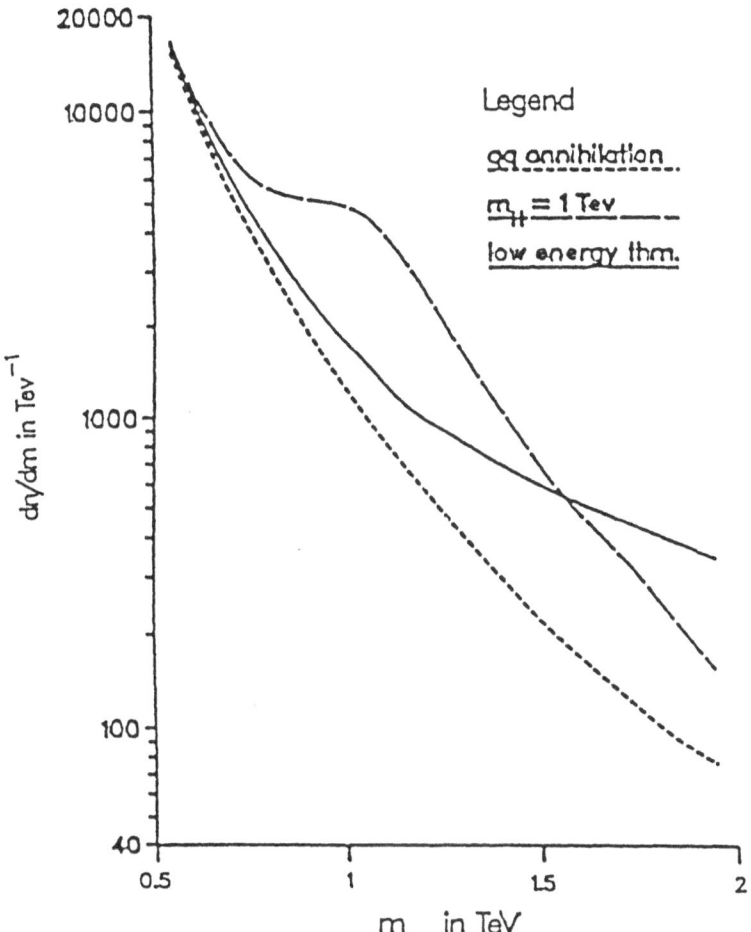

ZZ : signals + background
40 Tev

Legend

qq annihilation ....

$m_H = 1$ Tev ____

low energy thm. — —

Figure 3.4 Yields defined as in figure 3.3 for a 40 TeV pp collider. The short dashed line is the $\bar{q}q \rightarrow ZZ$ background while the long dashed line is the sum of the background and the $H \rightarrow ZZ$ signal. The solid line represents the sum of signal plus background for an extrapolation of the low energy theorem as discussed in Section 3.2 (from ref. 3).

323

The signals for examples 1) and 2) are excesses of events with no discernible structure. To detect them we must understand the background to ±30%, a goal consistent with the level at which we can expect to understand the nucleon structure functions and perturbative QCD.[6b] Realization of this goal requires an extensive program of "calibration" studies[17] at the SSC, to measure a variety of jet, lepton, and gauge boson final states in order to tune the structure functions and confirm our understanding of the backgrounds.

3.1 The 1 TeV Weinberg-Salam Higgs Boson

In the Weinberg-Salam model the generic figure 1.1 is replaced by s-channel Higgs boson exchange, figure 2.1b. Consider the leptonic final state,

$$H \rightarrow ZZ \rightarrow e^+e^-/\mu^+\mu^- + e^+e^-/\mu^+\mu^-/\bar{\nu}\nu \qquad (3.5)$$

for which the branching ratio is 1.1%, of which 6/7 of the events have one Z decay to $\bar{\nu}\nu$.[3,18] I require the observed Z's to be central, $|y_Z| < 1.5$, and in addition require either $m_{ZZ} > 0.9$ TeV or $(m_{ZZ})_T > 0.9$ TeV, where $(m_{ZZ})_T$ is the transverse mass, $2 \cdot \sqrt{m_Z^2 + p_T^2}$, computed from the p_T of the observed Z when the second Z decays to $\bar{\nu}\nu$. The cuts are needed in order to see the signal above $\bar{q}q \rightarrow ZZ$ background. For large m_H they are essentially equivalent to alternative cuts that have been suggested.[19]

An idea of the dependence of the signal on collider energy can be gotten from Figure 3.3, which shows the signal alone. Figure 3.4, showing the signal over the background, illustrates the need for the cut on m_{ZZ} or equivalently on $p_T(Z)$.

The results are shown in Table 3.1, for pp collider energies of $\sqrt{s} = 16, 40$, and 200 TeV, and for top quark mass values $m_t = 50$ and 200 GeV. The signals are increased by a factor ~ 3 for the larger value of m_t reflecting the added contribution of $gg \rightarrow \bar{t}t \rightarrow H$ via the $\bar{t}t$ loop; fortunately the background is not correspondingly augmented. The signal increases by a factor 10 as $\sqrt{s}$ is increased from 16 to 40 TeV and again from 40 to 200 TeV, while the background increases by smaller factors between 4 and 6. This affects not only the statistical significance of the signals but also the severity of the anticipated $\sim 30\%$ systematic uncertainty in the magnitude of the backgrounds. Note that the yields quoted for the signals are for the signals alone and do not include the background yields. In addition the quoted numbers do not include the cuts that might be needed to control the background due to QCD production of $Z+$ jets where the jets are somehow not observed and provide a fake $\not{E}_T$ signal – see reference 6a.

Except for $gg \rightarrow ZZ$, the backgrounds are predominantly transversely polarized Z's while the signal is purely longitudinal, resulting in different angular distributions for the decays $Z \rightarrow \bar{f}f$ where f is a lepton or quark. Define θ^* as the angle in the Z center of mass system between the fermion momentum $\vec{p}_f$ and the boost axis to the laboratory frame. Then the angular distributions for longitudinal and transverse polarizations are

$$P_L(\cos\theta^*) = \frac{3}{4}\sin^2\theta^* \qquad (3.6)$$

$$P_T(\cos\theta^*) = \frac{3}{8}(1 + \cos^2\theta^*) \qquad (3.7)$$

A strong cut against P_T throws out most of the P_L baby with the bath, and cannot be afforded given the small number of events. On the other hand, there are enough events to check that the signal is longitudinal as expected. For instance, a cut at $|\cos\theta^*| < 1/3$ reduces N_L by about 1/2 while reducing N_T by about 1/4 (see e.g. reference 20).

The like-charge $W_L W_L$ channel is controlled by the $I_{\text{Custodial}} = 2$ low energy theorem,[3,4]

$$a_{02} = -\frac{s}{32\pi v^2} \qquad (3.8)$$

where I have put $\rho = 1$. This is analogous to the exotic $I = 2$ channel in QCD, in which no resonance structure is observed. A simple model[3] for the continuum scattering in this channel is obtained by extrapolating the low energy theorem (3.8) to the unitarity limit at $\sqrt{32\pi v^2} \cong 2.5$ TeV,

$$|a_{02}| = \frac{s}{32\pi v^2}\theta(32\pi v^2 - s) + 1 \cdot \theta(s - 32\pi v^2) \qquad (3.9)$$

as shown in figure 3.5. We then use the effective W approximation[21] to compute the yield from WW fusion.

The model (3.9) can be thought of as a kind of "insurance policy" against the possibility that that the mass scale M_{SB} is much larger than the unitarity limit Λ_{SB}. As discussed in Section 3 this is physically implausible though not rigorously impossible. To see how this works, compare the analogous $\pi\pi$ scattering models with experimental data. For the three channels, $(I,J) = (0,0),(1,1),(2,0)$, the models analogous to eq.(3.9) are labeled by the curves a in figures (3.6), compared there with experimental data.[22] The model for $|a_{00}|$ describes the trend of the data well. For $|a_{11}|$ it underestimates the data because it fails to account for the ρ meson peak. For $|a_{02}|$ the model overestimates the data (note that since this is an exotic channel, $Im\ a_{02} \cong 0$ and $|a_{02}| \cong |Re\ a_{02}|$ to a good approximation), because it fails to include the effects of ρ exchange in the t and u channels. The model (3.9) is then a kind of worst case scenario: it works best if the resonances are much heavier than the unitarity bound for Λ_{SB} and therefore not directly observable at the SSC. For instance, if the ρ were heavier, say ≥ 1 GeV, then curve (a) in figure (3.8) would give a better fit (to larger s) than it now does. On the other hand, if the resonances are where we naively expect, $M_{SB} \cong \Lambda_{SB}$, then at least some channels will be dramatically enhanced relative to the model. We consider a resonant (technicolor) example below. First we consider strong WW scattering with no structure as in figure (3.5) and eq.(3.9).

The signal is defined by two isolated like-charge leptons,

$$W^+W^+ \to e^+\nu/\mu^+\nu + e^+\nu/\mu^+\nu. \qquad (3.10)$$

(Assuming $m_t > M_W$, the branching ratio is $(2/9)^2$.) Cuts imposed are $|y_\ell| < 2$ and $p_{T\ell} > 50$ GeV where $\ell = e, \mu$. In addition a "theorist's" cut of $M_{WW} > 800$ GeV is imposed to reduce background from $qq \to qqWW$ by gluon exchange, $O(\alpha_s g^2)$, and by higher order electroweak interactions, $O(g^4)$. This is a "theorist's" cut since the two ν's prevent it from being implemented experimentally. It can eventually be replaced by a set of cuts on observables, such as the charged dilepton mass and the transverse mass formed from the charged dilepton momentum.

The corresponding signals[23] based on eq. (3.9) for the three pp collider energies are shown in Table 3.2, where yields for the sum of W^+W^+ and W^-W^- are given per $10^4 pb^{-1}$. The backgrounds are $\sim 40\%$ from gluon exchange[15] and $\sim 60\%$ from $O(g^4)$ processes.[24] In estimating the backgrounds I have scaled the results of reference 24 to reflect the cuts imposed here. An independent calculation of the $O(g^4)$ processes (directly incorporating the above cuts) is currently under way.[25]

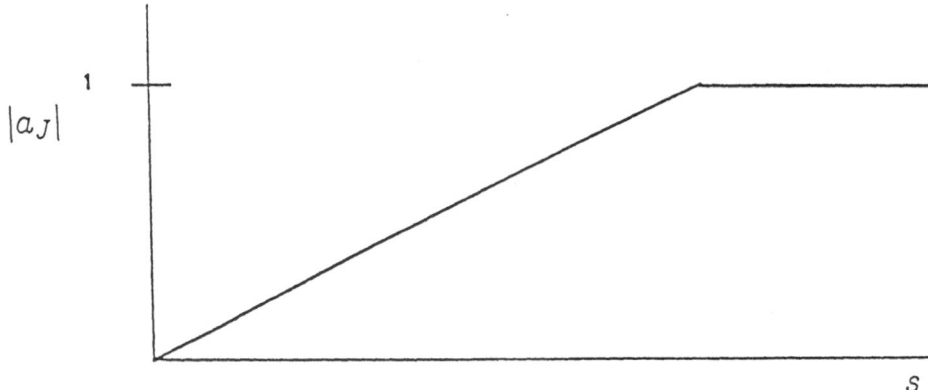

Figure 3.5 Extrapolated low energy theorem for strong W^+W^+ scattering, eq. (3.9).

Table 3.2 Signals and backgrounds for the model of strong W^+W^+ and W^-W^- scattering discussed in the text. The yields are for the leptonic final state, $e\nu/\mu\nu + e\nu/\mu\nu$, with cuts as defined in the text. In each entry the first number is events per $10^4 pb^{-1}$ for the strong scattering model and the second number corresponds to background events as discussed in the text. (The leptonic branching ratios assume $m_t \geq M_W$.)

$\sqrt{s} = 16$ TeV	$\sqrt{s} = 40$ TeV	$\sqrt{s} = 200$ TeV
5/7	53/34	800/170

Table 3.3 Signals and backgrounds for charged techni-rho production with subsequent decay $\rho_T^{\pm} \rightarrow W^{\pm}Z \rightarrow e^+\nu/\mu^+\nu + e^+e^-/\mu^+\mu^-$. Production mechanisms are $\bar{q}q$ annihilation and WW fusion. Cuts are defined in the text. Branching ratios asume $m_t \geq M_W$. In each case the first entry is the number of signal events per $10^4 pb^{-1}$ while the second is the background from $\bar{q}q \rightarrow WZ$.

$\sqrt{s} = 16$ TeV	$\sqrt{s} = 40$ TeV	$\sqrt{s} = 200$ TeV
1.8/.4	13/1.7	130/8

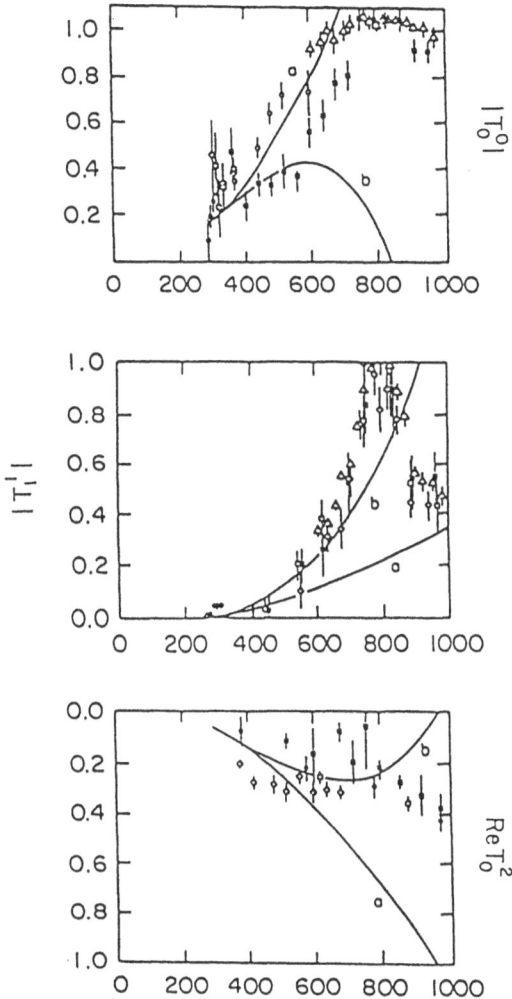

Figure 3.6 Data for $\pi\pi$ partial wave amplitudes compared with extrapolated low energy theorems (e.g., eq. (3.9)) for the three channels
$I, J = (0,0), (1,1), (2,0)$. The curves labeled a correspond to the naive
extrapolation as in eq. (3.9) and figure 3.5. The figures are from ref.
22.

3.3 Techni-rho Meson

As an example of the resonance production that may occur in strong interaction models, we consider the yields for the ρ_T meson of $SU(4)$ technicolor. Using the $3 \to \infty$ and $4 \to \infty$ limits (a popular theoretical way of planting one's head in the sand – lattice calculations would be very useful to check the resulting predictions given below) the ρ_T mass and width may be estimated from the mass and width of the $\rho(770)$:

$$m_{\rho_T} = \sqrt{\frac{3}{N_{TC}}} \cdot \frac{v}{F_\pi} m_\rho \tag{3.11}$$

$$\Gamma_{\rho_T} = \frac{3}{N_{TC}} \cdot \frac{v}{F_\pi} \Gamma_\rho. \tag{3.12}$$

For $N_{TC} = 4$ we then find

$$m_{\rho_T} \cong 1.8 \; TeV \tag{3.13}$$

$$\Gamma_{\rho_T} \cong 0.3 \; TeV. \tag{3.14}$$

There are two important production mechanisms: $W_L W_L \to \rho_T$ (reference 3) and $\bar{q}q \to \rho_T$ (reference 26). The dominant decay, analogous to $\rho \to \pi\pi$, is $\rho_T \to W_L W_L$. I consider just the clean leptonic decay channel,

$$\rho_T^{\pm} \to W_L^{\pm} Z_L \to e^{\pm}\nu/\mu^{\pm}\nu + e^+e^-/\mu^+\mu^-, \tag{3.15}$$

with branching ratio 0.015 for $m_t > M_W$. The signal and dominant background (from $\bar{q}q \to WZ$) are shown in Table 3.3, corresponding to $10^4 pb^{-1}$ and cuts $|y_{W,Z}| < 1.5$ and and $M_{WZ} > 1.6$ TeV. At these enormous WZ invariant masses the backgrounds are very small, so that the modest SSC signal is very clean. If $W \to \tau\nu$ is also reconstructible the yield increases by another 50%.

CONCLUSION

If we find that the Higgs mechanism is not the basis for electroweak symmetry breaking, it will clearly be an astonishing and very important discovery. Confirmation would be no less important since it entails discovery of nature's fifth force and an associated set of new quanta. In either case there is a guaranteed scientific treasure.

If we are able to measure strong WW scattering, between 1 and 2 TeV in the WW energy, we are sure to learn from the presence or absence of the signal. If absent it means that $\mathcal{L}_{SB}$, the symmetry breaking sector, lies below 1 TeV and is weakly coupled. If present, $\mathcal{L}_{SB}$ is strongly coupled and we expect to see resonances above 1 TeV but not much heavier than ~ 2 TeV. These statements depend on the low energy theorems for WW scattering, which assume the Higgs mechanism.

The yields quoted in Section 3 show that the SSC (just) has the no-lose capability to observe strong WW scattering. Except for the $O(2)$ TeV resonnances, this ability presupposes an intensive program of "calibration" studies to check the physics and distribution functions underlying the various backgrounds so that they are known to at least ± 30 %.

It is also clear from the SSC yields quoted in Section 3 and from the modest yields to be expected at LEP or LEP 200 for light Higgs bosons, that initial discovery of $\mathcal{L}_{SB}$ can only point the way to the detailed studies that will be needed to understand the new dynamics. It is clear that another generation of facilities will be needed, with much greater luminosity and/or energy. If strong WW scattering is observed above 1 TeV, then higher energy facilities will be imperative. The yields quoted for $\sqrt{s} = 200$ TeV in Section 3 show that a step to a pp collider of at least the 200 TeV considered for the ELOISATRON will be needed. The actual yields at a 200 TeV pp collider are likely to be much larger than given by the models considered in Section 3; higher partial waves and multiparticle production will also contribute – see for instance section 4.3 of reference 6 and Chanowitz and Gaillard, reference 21. If we have learned to detect the mixed decay modes, $WW \rightarrow \overline{q}q + \ell\overline{\nu}$, in the pp collider environment, they will provide another order of magnitude increase in signal yield. High energy e^+e^- colliders will also be desired, though with even higher energy than is presently being imagined for the TLC and CLIC.

APPENDIX: NATURALNESS AND THE INTERMEDIATE MASS HIGGS BOSON

There is one exception to my claim that the presently existing and planned complement of e^+e^- and pp colliders is certain to discover the symmetry breaking mechanism. If the Weinberg-Salam Higgs boson happens to fall in the so-called "intermediate" mass range, from about 70 to $\sim 140 - 160$ GeV, then it will be too heavy to discover at LEP 180 and may also be too light to observe a LHC/SSC. At this meeting Gunion will discuss some possible strategies for extending the search at the pp colliders to masses below ZZ threshold.[27] For now the only sure way to close this window would be to build an e^+e^- collider with $\sqrt{s} \gtrsim 300$ GeV and $\mathcal{L} \gtrsim 10^{32} cm^{-2} s^{-1}$, a facility not yet on the menu, though design studies are under way at SLAC.

The problem of the intermediate mass window is most severe in the Weinberg-Salam model because in that model a single Higgs boson is the only quantum of the symmetry breaking sector (apart from the longitudinal W and Z). In this sense it is the minimal model of $SU(2) \times U(1)$ breaking. Other models have either several Higgs particles or supersymmetric particles or a strongly interacting spectrum above 1 TeV.

The urgency of the problem is mitigated by our present understanding of a problem of theory, the "naturalness" or "fine-tuning" problem, which uniquely afflicts Higgs boson models. It may be thought of as an instability of the Higgs boson mass and vacuum expectation value against quantum corrections that tend naturally to drive them to violently larger values, say M_{Planck} or M_{GUT}. The problem arises because in $3 + 1$ dimensions, renormalizable field theories with elementary scalars are unique in having quadratic divergences. In the usual remoralization program these divergences are simply subtracted away. But that program renounces any attempt to understand the physical origin of the subtracted quantities, which are simply fit to experiment. The problem arises when we go beyond this limited perspective and ask questions about the origin of the subtracted quantities, assuming they will eventually be understood and calculable in the context of another theory formulated at a deeper level. It is in that framework that the necessary subtraction is artificial and implausible, requiring one loop integrals to be fine tuned to a precision of order $(M_W/\Lambda)^2$ where Λ could be M_{GUT} or M_{Planck}. Without a principle to assure the needed precision, there is little chance that such theories could describe nature.

There are two known solutions. Dynamical models like technicolor contain fermions and massless gauge bosons (techni-gluons) but no elementary scalars. Therefore, like QED and QCD, their divergences are at worst logarithmic and require no fine-tuning. Supersymmetric models contain fermionic partners that cancel the scalar-induced quadratic divergences, again leaving just logarithmic divergences. However the resulting integrals do grow quadratically with the mass differences of the fermion and boson partners, i.e., the supersymmetry breaking scale Λ_{SUSY}. If that scale is too large the fine-tuning problem reappears, roughly for $\Lambda_{SUSY} \gtrsim 1$ TeV. So if supersymmetry is the solution to fine-tuning, the superpartners cannot be much heavier than 1 TeV.

This discussion clearly has a bearing on the urgency of closing the intermediate mass window. Dynamical models of $SU(2) \times U(1)$ breaking give rise to strong WW scattering and probably also to resonances above 1 TeV, that can be observed at the SSC. Supersymmetric models contain gluinos and squarks that are produced by QCD mechanisms and can be hunted at the SSC to the 1 TeV mass range and beyond. In both of these classes of models there are sure to be observable signals at LEP and/or the SSC. If supersymmetry were discovered but no Higgs bosons, the discovery of the Higgs sector would be a well-defined problem for second generation experiments. The search would be constrained by the knowledge that we would be looking for the Higgs sector of s supersymmetric theory, and an e^+e^- collider to close the window would be very well motivated.

The fine-tuning problem occurs not only in the Weinberg-Salam model construed as a field theory with elementary scalars but also when it is realized as an effective low energy theory. For instance, a recent paper which obtains the Weinberg-Salam model as the effective low energy theory of a high energy four-fermion (alla Nambu Jona-Lasinio) interaction does not escape the problem. The construction assumes a strong top quark interaction at a very high scale $\Lambda >> M_W$ and the Higgs scalar emerges as a $\bar{t}t$ bound state. But the theory also requires fine-tuning of the gap equation to order $(M_W/\Lambda)^2$.

It should be emphasized that the line of reasoning presented in this section is not assured to be correct, since it invokes the *known* solutions of the fine-tuning problem. There is no theorem proving that other natural solutions will not be found which leave an intermediate mass Higgs boson as their only low energy manifestation. If neither SUSY nor strong WW scattering were discovered at the SSC, we would be forced to consider the possibility that nature knows a solution to the fine-tuning problem that is unknown to us.[28]

REFERENCES

1. P.W. Higgs, *Phys. Rev. Lett.* **12:132**, 1964; F. Englert and R. Brout, *ibid* **13:321**, 1964; P. W. Higgs, *Phys. Rev.* **145:1156**, 1966.

2. S. Weinberg, *Phys. Rev. Lett.* **19:1264**, 1967; A. Salam, *Proc. 8'th Nobel Symp.*, ed. N. Svartholm, p. 367, (Almqvist & Wiksells, Stockholm, 1968).

3. M. Chanowitz and M.K. Gaillard, *Nucl. Phys.* **B261:379**, 1985.

4. M. Chanowitz, M. Golden, and H. Georgi, *Phys. Rev.* **D36:1490**, 1987; *Phys. Rev. Lett.* **57:2344**, (1986).

5. R. Cahn and S. Dawson, *Phys. Lett.* **136B:196**(1984).

6. M. Chanowitz, LBL-26613, 1989 (to be published in Proc. of the INFN Eloisatron Workshop, Erice, 1988).

6a. R. Cahn et al., page 20, *Experiments, Detectors and Experimental Areas for the SSC*, eds. R. Donaldson and M. Gilchriese (World Scientific, Singapore, 1988).

6b High Luminosity Option for the LHC, ed. J. Mulvey, CERN 88-02, 1988.

7. M. Bento and C.H. Llewellyn Smith, *Nucl. Phys.* **B289:36**, (1987); G. Altarelli, B. Mele, F. Pitolli, *Nucl. Phys.* **B287:205**, (1987); J. Gunion, A. Tofighi-Niaki, *Phys. Rev.* **D36:2671**, (1987).

8. Contributions by A. Seiden and J. Gunion to these proceedings.

9. M. Chanowitz, *Ann. Rev. Nucl. Part. Sci.* **38:323**, 1988.

10. J.M. Cornwall, D. Levin, and G. Tiktopoulos, *Phys. Rev.* **D10:1145**, (1974).

11. B. Lee, C. Quigg, and H. Thacker, *Phys. Rev.* **D16: 1519**, 1977.

12. T.D. Lee, C.N. Yang, *Phys. Rev. Lett.* **4:307** (1960); B.L.. Ioffe, L. B. Okun, L. B. Rudik, *Sov. Phys.* JETP Lett. **20:1281** (1965).

13. M. Chanowitz and T. Appelquist, *Phys. Rev. Lett.* **59:2405** (1987).

14. E. Glover and J. van der Bij, *Nucl. Phys.* **B321:561**, 1989.

15. M. Chanowitz and M. Golden, *Phys. Rev. Lett.* **61:1053**, 1988, E **63:466**, 1989; D. Dicus and R. Vega, *Phys. Lett.* **B217:194**, 1989.

16. G. Altarelli, p.36, Proc. Workshop on Future Accelerators, La Thuile, 1987, ed J. Mulvey, CERN 87-07, Vol.I.

17. M. Chanowitz, p. 183, *Observable Standard Model Physics at the SSC*, eds. H-U Bengtsson et al. (World Scientific, Singapore, 1986).

18. R. Cahn and M. Chanowitz, *Phys. Rev. Lett.* **56:1327**, 1986.

19. V. Barger, T. Han, and R. Phillips, *Phys. Rev.* **D37:2005**, 1988.

20. G. Kane and C. Yuan, ANL-HEP-PR-89-43, 1989.

21. M. Chanowitz and M. Gaillard, *Phys. Lett.* **142B:85**, 1984; M. Chanowitz and M. Gaillard, ref. (3); S. Dawson, *Nucl. Phys.* **B29:42**, 1985; G. Kane, W. Repko, and W. Rolnick, *Phys. Lett.* **148B:367**, 1984.

22. J. Donoghue, C. Ramirez, G. Valencia, *Phys. Rev.* **D38:2195**, (1988).

23. M. Chanowitz and M. Golden, erratum to ref. 15, *Phys. Rev. Lett.* **E63:466**, (1986)

24. D. Dicus and R. Vega, UCD-89-9, 1989; *Phys. Rev.* **D37:2474** (1988).

25. M. Berger and M. Chanowitz, work in progress.

26. E. Eichten et al., *Rev. Mod. Phys.* **56:579**, 1984.

27. J. Gunion, these procedings.

28. In that event we would be strongly motivated to sharpen or replace the naturalness upper limit on Λ_{SUSY}, which is now a matter of taste.

PROBING HIGGS BOSONS/ELECTROWEAK SYMMETRY BREAKING IN PURELY LEPTONIC CHANNELS AT HADRON COLLIDERS*

J.F. Gunion

Department of Physics
University of California
Davis, CA 95616

ABSTRACT

I examine the ability of the LHC, SSC and Eloisatron hadron colliders to search for Higgs bosons in background-free purely leptonic modes. Such leptonic final state modes are surveyed for both the Standard Model Higgs scenario and a variety of non-minimal Higgs sectors (the features of which are reviewed along the way). In particular, I examine the ability of the SSC to probe exotic Higgs scenarios using purely leptonic channels, and discuss the extent to which a very high energy, high luminosity hadron collider such as the Eloisatron might be justified as a second generation SSC by virtue of its ability to employ leptonic channels to extend the SSC discovery reach or allow more detailed studies of already discovered Higgs bosons.

1. INTRODUCTION

It has long been appreciated that the most background-free mode in which to search for the Standard Model Higgs boson (ϕ^0) is the four-charged-lepton final state resulting from $\phi^0 \to ZZ \to l^+l^-l'^+l'^-$ where $l, l' = e, \mu$. The fact that the SSC can employ this search mode for the ϕ^0 over a substantial range of m_{ϕ^0} (reviewed below) has been an important litmus test in justifying the appropriateness of the SSC design energy and luminosity ($\sqrt{s} = 40 \ TeV$, $L_{year} = 10^4 \ pb^{-1}$). Clearly, it is of interest to extend these considerations by determining the extent to which such background-free purely leptonic modes can be used at the SSC to probe *non-minimal* Higgs sectors. This will be one focus of this report. My second focus will be to address the question of whether the design parameters of the Eloisatron ($\sqrt{s} = 200 \ TeV$, $L_{year} = 10^4 \ pb^{-1}$) would yield a sufficiently large extension of the purely-leptonic-channel Higgs boson discovery domains and/or allow use of leptonic channels to fill in enough crucial details concerning Higgs bosons and electroweak symmetry breaking (EWSB) that such a collider would be an appropriate second generation successor to the SSC and LHC ($\sqrt{s} \sim 17 \ TeV$, $L_{year} \gtrsim 10^4 \ pb^{-1}$). I will consider both the Standard Model and various non-minimal Higgs scenarios in addressing this latter issue. Not unexpectedly, the comparison between the Eloisatron and the SSC and LHC depends substantially upon the structure and complexity of the Higgs sector.

---

* Work supported, in part, by the Department of Energy.

More generally speaking, to understand the nature of electroweak symmetry breaking we must be able to study *all* $VV \rightarrow VV$ scattering processes in sufficient detail to either find or exclude Higgs boson (or other) resonances and/or broad enhancements over a very large mass range, up to and past the 1 TeV region. More precisely, we must determine at various m_{VV} values the size of $d\sigma/dm_{VV}$ and, hopefully, the polarization of the final V's. In particular, even if the source(s) of all EWSB are at mass scales below 1 TeV, a measurement of the m_{VV} spectrum above 1 TeV would be important; it would allow us to verify that the VV scattering amplitudes have the perturbative high energy behavior predicted in such a case. Since the cleanest final states in which to perform a complete study of the VV scattering processes are the purely leptonic decay channels of the VV pairs, for this first assessment of the appropriateness of the Eloisatron design parameters I will focus on these "gold-plated" leptonic modes. A more complete comparison between machines should, however, include consideration of other decay channels. Certainly, past studies of the SM Higgs scenario at the SSC and LHC have shown that alternative modes allow a large extension in exploratory reach over that possible using only the leptonic final states.

2. THE STANDARD MODEL HIGGS BOSON

The situation for the SM Higgs boson ϕ^0 was reviewed in an earlier INFN report.[1] I will briefly review the conclusions reached there, and update the background discussion. In ref. 1, we determined the range of m_{ϕ^0} over which one could employ the gold-plated four-charged lepton final state for Higgs detection ($l = e, \mu$). When $m_{\phi^0} < 2m_Z$, the four leptons are obtained via $\phi^0 \rightarrow ZZ^*$, followed by $Z \rightarrow l^+l^-$ and $Z^* \rightarrow l^{+\prime}l^{-\prime}$, where the $*$ indicates that one of the Z's is virtual. The $l^{+\prime}l^{-\prime}$ pair from the Z^* does not have a precisely determined mass, but the distribution in $M^*_{l^{+\prime}l^{-\prime}}$ is broadly peaked near the largest kinematically allowed value. When $m_{\phi^0} > 2m_Z$, $\phi^0 \rightarrow ZZ$ followed by double $Z \rightarrow l^+l^-$ yields the four-lepton final state; in this case both l^+l^- pairs have invariant mass narrowly peaked near m_Z. The advantages of the four-charged-lepton mode are obvious. The final state can be fully reconstructed with excellent mass resolution, and has minimal background.

Of course, a critical component in employing the four-charged-lepton ($4l$) decay mode of the Higgs is the branching ratio for this final state. This branching ratio is illustrated for the SM Higgs in fig. 1, for top quark masses of $m_t = 80$ GeV and $m_t = 200$ GeV. (Also appearing in this figure are branching ratios for several other Higgs boson models that will be discussed later.) In the computation of the $4l$ branching ratio, we have used tree-level computations for all the decay modes. In the ZZ^* region, where the dominant decay mode is $\phi^0 \rightarrow b\bar{b}$, this is a conservative procedure given that QCD corrections are expected to reduce $\Gamma(\phi^0 \rightarrow b\bar{b})$ by about a factor of two over the mass range of interest. [2]

Let us first focus on the region $m_{\phi^0} < 2m_Z$; there, the irreducible background derives from $q\bar{q} \rightarrow Zl^{+\prime}l^{-\prime}$ processes, where the final $l^{+\prime}l^{-\prime}$ comes from a virtual Z^* or γ^*. A priori, this background is large, but it can be severely reduced by two means: 1) demanding that the invariant mass of all four final leptons lie in a very narrow 2% mass bin centered on m_{ϕ^0}; and 2) demanding that $M^*_{l^{+\prime}l^{-\prime}}$ be close to the kinematical maximum. The first cut captures all of the Higgs signal (since for $m_{\phi^0} < 2m_Z$, $\Gamma_{\phi^0} < 0.02m_{\phi^0}$) while eliminating the bulk of the background. However, the background would still be larger than the signal without the second cut, that effectively eliminates the γ^*-mediated background (which peaks at small $M^*_{l^{+\prime}l^{-\prime}}$) and leaves a background-free signal. This procedure was studied in the case of the SSC in ref. 3 using the full $q\bar{q} \rightarrow Zl^{+\prime}l^{-\prime}$ cross sections obtained in ref. 4. Appropriate values for the minimum retained $M^*_{l^{+\prime}l^{-\prime}}$ as a function of m_{ϕ^0} were determined there. It is obvious that the ZZ^* detection mode is event rate limited because of the very small effective branching ratio for $\phi^0 \rightarrow Z(\rightarrow l^+l^-)Z^*(\rightarrow l^{+\prime}l^{-\prime})$

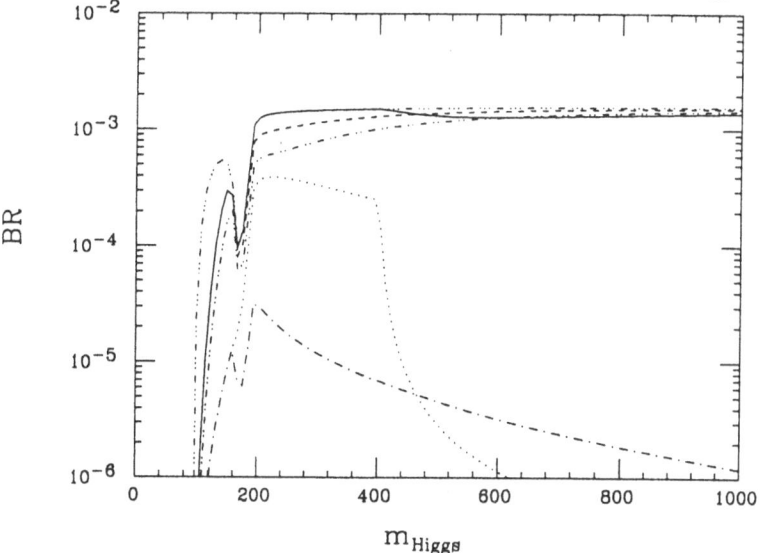

Fig. 1. The branching ratio for $h \rightarrow l^+l^-l^{+\prime}l^{-\prime}$ for a number of different Higgs models, and top quark masses. The Standard Model (SM) Higgs results are the solid and dashed lines. In the case of the minimal supersymmetric model (MSSM), the results (dots and dot-dash lines) apply to the heavier of the two scalar Higgs, the H^0, and we have chosen $\tan\beta = 1.5$. H^0 decays to neutralino and chargino pairs are assumed to be absent. The dash-doubledot curve is for a Higgs boson with SM quark couplings, but only $1/3$ of the SM Higgs coupling-squared to VV channels. Finally, the dash-tripledot curve is for a SM Higgs boson with no quark couplings.

Results for this branching ratio are given in fig. 1, for top quark mass cases where the top is sufficiently heavy that ϕ^0 decay to $t\bar{t}$ is forbidden for $m_{\phi''} < 2m_W$ (an increasingly probable situation as collider limits on the top quark mass increase). It is useful to quote the specific values:

$$BR(\phi^0 \rightarrow Z(\rightarrow l^+l^-)Z^*(\rightarrow l^{+\prime}l^{-\prime})) = \begin{cases} 1.4 \cdot 10^{-4}; & m_{\phi''} = 135 \; GeV \\ 5.5 \cdot 10^{-5}; & m_{\phi''} = 125 \; GeV \\ 1.3 \cdot 10^{-5}; & m_{\phi''} = 115 \; GeV . \end{cases} \qquad (2.1)$$

If we require 40 events before cuts (the ones considered above are 95% efficient for the signal assuming rapidity coverage of $|y_l| < 4$ and lepton momentum measurement down to $p_l \sim 5 \; GeV$) then at $L_{year} = 10^4 \; pb^{-1}$ we need cross sections of:

$$\sigma_{\phi''} = \begin{cases} 28 \; pb; & m_{\phi''} = 135 \; GeV \\ 73 \; pb; & m_{\phi''} = 125 \; GeV \\ 308 \; pb; & m_{\phi''} = 115 \; GeV . \end{cases} \qquad (2.2)$$

We shall return shortly to assess the ability of the various machines to achieve the required cross section levels.

335

For $m_{\phi^0} \gg 2m_Z$, the branching ratio of relevance is

$$BR(\phi^0 \to Z(\to l^+l^-)Z(\to l^{+'}l^{-'})) \simeq 1.57 \cdot 10^{-3}, \qquad (2.3)$$

where we have assumed that $Z \to t\bar{t}$ and $\phi^0 \to t\bar{t}$ are both forbidden. (Note, in particular, that the $ZZ \to l^+l^-l^{+'}l^{-'}$ branching ratio is $\simeq 0.0047$ for a heavy top, a considerable improvement over the $\simeq 0.0036$ appropriate for a light top used in past studies.) If $\phi^0 \to t\bar{t}$ is allowed, and the top is heavier than m_W, then this could be reduced by 10 to 20%. This is illustrated in fig. 1. Assuming that we have excellent mass resolution for the invariant masses of both the l^+l^- pairs, the main background derives from continuum $q\bar{q}, gg \to ZZ$ production of ZZ pairs. For m_{ϕ^0} near to $2m_Z$, the Higgs width is still relatively narrow and the Higgs production rate is always sufficiently large that there is no problem in finding the Higgs resonance enhancement on the ZZ continuum background. However, as we move to higher masses, σ_{ϕ^0} decreases and, in addition, Γ_{ϕ^0} increases, making it more and more difficult to detect the Higgs bump given the small four-charged-lepton Higgs branching ratio of eq. (2.3). We will estimate discovery reach at a given machine by requiring 60 events in this gold-plated mode before cuts. Using $L_{year} = 10^4 \; pb^{-1}$ and a conservative BR (to account for some $t\bar{t}$ component to ϕ^0 decay) of $BR \sim 1.3 \cdot 10^{-3}$, we find that $\sigma_{\phi^0} \simeq 5 \; pb$ is required. (Note that the 60 event criterion is roughly mass, and hence machine, independent since the ϕ^0 becomes increasingly narrow at lower m_{ϕ^0}, thereby compensating for the increased ZZ continuum background at lower mass.)

Having established our discovery criteria at low and high mass, the point at which one runs out of cross section at a given machine can be estimated. In ref. 1, we computed gg and VV fusion cross sections as a function of m_{ϕ^0} in the on-shell (s-channel pole imaginary part) approximation, with no cuts other than requiring $x \geq 10^{-4}$ for the distribution function arguments. This limitation on x was imposed simply because available distribution functions are not entirely reliable for smaller x values. It has relatively small impact on the LHC and SSC cross sections, but does yield Eloisatron cross sections that are $10 - 30\%$ (depending on m_{ϕ^0}) smaller than obtained without the cut. Cross sections were computed for $m_t = 80$ and $200 \; GeV$. Since at the Eloisatron the relevant m_{ϕ^0} values are at and above $1 \; TeV$ on the high mass end, and the ϕ^0 is becoming very broad, we also computed in this case the exact tree-level M_{ZZ} spectrum from $qq \to qqZZ$, both in the presence of a $m_{\phi^0} = 1 \; TeV$ Higgs boson and without (i.e. taking m_{ϕ^0} small). Integrating over the excess in $d\sigma/dM_{ZZ}$ for $|y_Z| < 3$ yields an effective Higgs cross section that is similar to, but somewhat larger than, that obtained from VV fusion computed in the on-shell approximation at $m_{\phi^0} = 1 \; TeV$. At high mass, for discovery reach limits the on-shell cross sections are generally a conservative estimate. For the reader's convenience, I have included below the graph of on-shell cross sections presented in the previous work, fig. 2. The discovery reach for an integrated luminosity of $L_{year} = 10^4 \; pb^{-1}$ at the three machines using the gold-plated mode is given in Table 1, for two different top quark masses. (The lower limit is independent of m_t to first approximation, since the gg fusion cross section that dominates at low masses does not vary much with m_t in the range of m_{ϕ^0} near 120 GeV.)

To understand how easy it is to find a very heavy Higgs in the gold-plated mode at the Eloisatron, let us note that at $m_{\phi^0} = 1500 \; GeV$ the ϕ^0 cross section from the on-shell computation of VV fusion alone is $\gtrsim 15 \; pb$. This yields some 50000 ZZ Higgs decays, implying somewhat more than 200 four-lepton events. Of the ZZ events, the exact matrix element computation referred to above finds that 75% will have both Z's longitudinally polarized. The exact computation (see ref. 1) also predicts a somewhat larger effective cross section, even though we require $|y_Z| < 3$, and shows that the (background) $q\bar{q} \to ZZ$ continuum (which would dominate the high

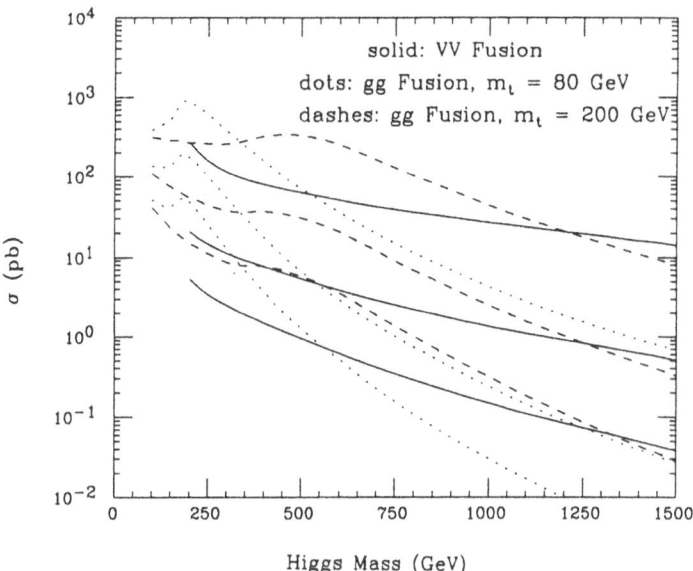

Fig. 2. Cross sections for SM Higgs production at the LHC, SSC, and Eloisatron, computed in the on-shell approximation as outlined in the text.

Table 1. $\phi^0 \rightarrow 4l$ Discovery limits (in GeV).

Machine	$m_{\phi^0}^{min}$	$m_{\phi^0}^{max}$ ($m_t = 80$)	$m_{\phi^0}^{max}$ ($m_t = 200$)
LHC	~ 135	~ 400	$500 - 550$
SSC	~ 125	~ 650	$800 - 850$
Eloisatron	~ 115	$\gtrsim 1500$	> 1500

mass M_{ZZ} spectrum if the Higgs were light) is much below the $d\sigma/dM_{ZZ}$ spectrum predicted by VV fusion in the presence of such a heavy Higgs. Of course, it should also be remembered that the Higgs cross section would be roughly doubled by gg fusion if $m_t = 200\ GeV$ (although the $gg \rightarrow ZZ$ background will also increase).

Another indication of the power of the Eloisatron is the fact that we could potentially study the ZZ spectrum above 1 TeV even if the Higgs boson has lower mass (say $< 800\ GeV$ so that the enhancement in the ZZ spectrum coming from it has entirely disappeared by 1 TeV). For instance, from the $q\bar{q} \rightarrow ZZ$ process we obtain the spectrum given in fig. 3 when we impose $|y_Z| < 3$. Integrating the spectrum of fig. 3 over ZZ masses from 1 to 1.5 TeV, we obtain 0.9 pb or about 45 events in the four-lepton gold-plated mode. This is enough to check that the cross section is behaving as expected over this interval, something that is completely impossible in the $4l$ mode at either of the lower energy machines. If $m_{\phi^0} < 500\ GeV$, $q\bar{q} \rightarrow ZZ$ integrated from $600 - 1000\ GeV$ yields a large enough cross section at the Eloisatron that one could even analyze the Z polarizations (to check that they were transverse) over this mass interval.

Of course, the $q\bar{q} \rightarrow ZZ$ reaction is simply the leading order tree-level source of ZZ pairs at high M_{ZZ}, if the Higgs boson is light. Higher order processes of many types will also contribute to the ZZ pair spectrum. The largest corrections to

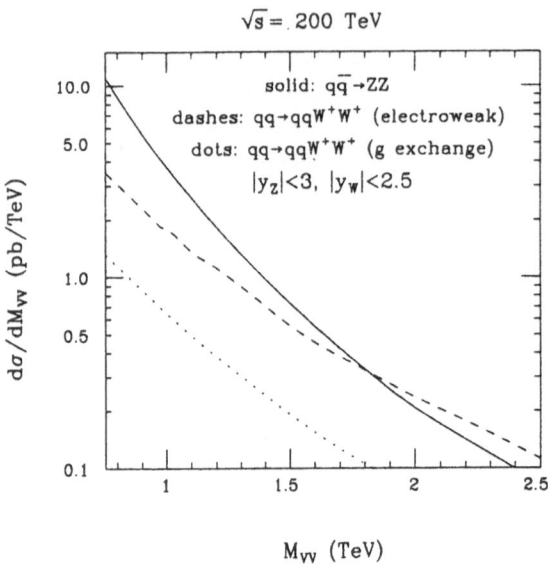

$\sqrt{s} = 200$ TeV

solid: $q\bar{q} \to ZZ$
dashes: $qq \to qqW^+W^+$ (electroweak)
dots: $qq \to qqW^+W^+$ (g exchange)
$|y_Z| < 3$, $|y_W| < 2.5$

Fig. 3. Results for vector boson pair continua at the Eloisatron: $q\bar{q} \to ZZ$ (solid); $qq \to qqW^+W^+$ electroweak (dashes); and $qq \to qqW^+W^+$ via gluon exchange (dots). We have required that the final state Z's have $|y_Z| < 3$ and that the final W^+'s have $|y_W| < 2.5$. For more details see text.

$q\bar{q} \to ZZ$ are clearly the QCD corrections of order α_s. Aside from virtual gluon and gluon emission corrections to the basic $q\bar{q} \to ZZ$ subprocess, there is also the $gq \to q\,ZZ$ subprocess involving the large gluon luminosity. Obviously, it is necessary to perform a complete calculation, including all order g_s graphs, to obtain a well-defined result at $\mathcal{O}(\alpha_s)$, free of collinear and infrared divergences. Electroweak corrections of $\mathcal{O}(\alpha)$ will also be present. Ultimately it may be important to measure the $2 \to 4$ electroweak $qq \to qqZZ$ amplitudes which include the interesting $W^+W^- \to ZZ$ and $ZZ \to ZZ$ subdiagrams. However, since these subprocesses contribute to the ZZ spectrum at order α^2 relative to $q\bar{q} \to ZZ$, special techniques are required to isolate these fundamental electroweak amplitudes when the Higgs boson is light and they behave perturbatively. Most probably, the ZZ spectrum measurement would have to be accompanied by double jet tagging so as to eliminate the larger $2 \to 2$ ($q\bar{q} \to ZZ$) and $2 \to 3$ (e.g. $gq \to q\,ZZ$) processes. Even with such tagging one must determine the level of background from $qq \to qq\,ZZ$ mediated via gluon exchange between the scattering quarks compared to electroweak contributions to $qq \to qq\,ZZ$. This comparison has not been made. But studies [5] of the analogous $qq \to qq\,WW$ electroweak and gluon exchange processes indicate that the latter can be suppressed below the level of the former by imposing a rapidity cut, e.g., $|y| < 3$, on the outgoing vector bosons. Such a cut is effective in isolating the electroweak signal of interest since the W's produced in the gluon exchange process are radiated from the quarks and emerge primarily in the forward and backward directions. As we shall discuss in a later section, this same rapidity cut is also effective in the case of $qq \to qq\,W^+W^+$, the spectra for which appear in fig. 3. A determination of whether enough event rate remains after jet tagging to detect $qq \to qq\,ZZ$ in the mode where both Z's decay leptonically must await further work.

Returning for a moment to the ZZ^* mode employed for $m_{\phi 0} < 2m_Z$, let me discuss a possible background that was pointed out in in ref. 6. The process

$pp \rightarrow Zb\bar{b}$ produces $Zl^{+'}l^{-'}+jets$ final states at a high rate. In ref. 6, the spectrum in $M_{Zl^{+'}l^{-'}}$ was computed and found to be larger than the Higgs signal. However, there are several improvements to their study that will, I believe, result in the elimination of this background as a source of difficulty. First, they employed a light top quark, $m_t = 40\ GeV$; larger m_t values consistent with current collider limits yield a much larger $\phi^0 \rightarrow ZZ^*$ branching ratio (and hence larger signal) than they employed. Second, they did not have time to perform several cuts that should make this background negligible. First, this background should be greatly reduced by requiring isolation for the l^+ and l^- to be associated with the Z^* — the background has soft jets accompanying these charged leptons. Second, it should be possible to impose an *anti-vertex* trigger to select against events containing b quarks. While a more complete study is needed for confirmation, I would anticipate that these cuts can be performed with at least 50% efficiency, leaving 20 background-free events at the $m_{\phi^0}^{min}$ discovery limits quoted earlier.

Before turning to non-minimal Higgs sector models, let me close this section by noting that the detector requirements that must be met in order to employ the gold-plated mode over the very large range of four-lepton invariant masses considered here are non-trivial. Aside from the need to impose lepton isolation requirements and (possibly) anti-vertex triggering mentioned just above, there are the more basic requirements emphasized in ref. 6 of very high tracking efficiency and energy resolution over a very large range of lepton momenta. To cover the full range of Higgs masses from $\sim 120\ GeV$ to $\sim 1500\ GeV$, we must deal with p_l values from $5\ GeV$ all the way up to $\sim 1\ TeV$. Given that the Higgs width increases rapidly for $m_{\phi^0} > 2m_Z$, ref. 6 shows that $\sigma_E/E \sim .15/\sqrt{E}$ (E in GeV) is adequate for a clear four-lepton invariant mass spectrum in the ZZ decay mode. However, they also demonstrate that this 'typical' SSC detector resolution is inadequate for the ZZ^* mode. A tracker with $\sigma_{p_T}/p_T \sim 0.26 p_T$ (p_T in TeV) (characteristic of a large magnetic detector when particles are constrained to come from the interaction region) begins to approach the required momentum resolution. And all of this must be done with high efficiency; if one is only 75% efficient for each track on average, the net efficiency is only $(3/4)^4 \sim 1/3$, a serious loss.

3. NON-MINIMAL HIGGS BOSONS

We have seen that the four-charged-lepton mode is extremely useful for detecting the SM neutral Higgs boson at a hadron collider, especially one with the very large energy of the Eloisatron. I now turn to an examination of the utility of purely leptonic modes, arising from decays to (possibly virtual) vector boson pair channels, for detection of the various Higgs bosons that arise in a number of non-minimal Higgs boson sectors. We shall see that their utility depends significantly on the model in question, and on the particular Higgs boson within that model. Along the way, we shall review the basic features of a number of different models. The reason for restricting consideration only to modes deriving from vector boson pairs, and the Higgs bosons that couple to such modes, is that the VV channels are the only ones directly related to the spontaneous generation of electroweak symmetry breaking (EWSB). To put it another way, since the strength of a Higgs boson's couplings to the appropriate VV channel(s) is directly proportional to its relevance to EWSB, it is clear that our ability to observe it using signals in the VV-mediated purely-leptonic final states also increases as its relevance to EWSB increases.

3.1 The Standard Model Higgs with no Quark Couplings

Despite the above introductory remarks, it is not obvious that every Higgs boson with strong VV couplings will be easily detected in purely leptonic modes. At a hadron collider, gg fusion production is a crucial component of the cross section for the neutral Higgs boson(s), especially when the mass is modest in size. It is not impossible that the mechanism for fermion mass generation is completely different from that for generating the vector boson masses. In this case, gg fusion contributions to production of a neutral Higgs boson involved in EWSB could be absent. The simplest example of such a case is that in which we adopt the Standard Model for the generation of W and Z masses, but assume that the SM ϕ^0 has no $f\bar{f}$ couplings. Fermion masses would then be generated by some other Higgs boson, or still other mechanism. We shall see explicitly that this is possible in the context of a two-doublet Higgs model in the following section. The discovery limits for this situation are easily extracted following the pattern of the discussion of the previous section. Only two alterations are required. First, we simply remove the gg cross section contributions, leaving only VV fusion. Second, we remove all fermion decay modes for the Higgs boson. The resulting $4l$ branching ratio was given in fig. 1. The VV fusion cross sections are precisely those given in fig. 2. The only difficulty is that, to obtain lower discovery limits in the ZZ^* region, the VV fusion cross sections must be extrapolated into a region where the effective-V technique employed in their computation becomes inaccurate. I adopt what I believe to be a conservative procedure of assuming that the VV fusion cross section is a constant in the region between $100\,GeV$ and $200\,GeV$ (the lowest point plotted in fig. 2). The resulting discovery limits are given in Table 2.

Table 2. $\phi^0 \rightarrow 4l$ Discovery limits (in GeV), no $\phi^0 q\bar{q}$ couplings.

Machine	$m_{\phi^0}^{min}$	$m_{\phi^0}^{max}$
LHC	~ 140	~ 200
SSC	~ 110	~ 600
Eloisatron	~ 100	$\gtrsim 1500$

Note that the ability of the LHC and SSC to probe Higgs masses below $2m_Z$ is not lost. The dramatic increase in the $\phi^0 \rightarrow ZZ^* \rightarrow 4l$ branching ratio found when $\phi^0 \rightarrow f\bar{f}$ modes are removed adequately compensates for the much smaller production cross section obtained when the gg fusion contribution is absent. In fact, the SSC lower-end reach actually improves. At the Eloisatron, the VV fusion and gg fusion cross sections for the SM ϕ^0 were already of similar size, and so the lower reach at this machine improves significantly. The most dramatic change from Table 1 is clearly the significant decrease in the maximum value of m_{ϕ^0} that can be probed at the LHC; gg cross sections are very important for the LHC to be a viable EWSB probe.

3.2 Higgs Sectors with Doublet and Singlet Fields

There is no particularly good reason for supposing that the Higgs sector should contain precisely one doublet field. A Higgs sector containing any number of doublet and singlet fields will continue to yield $\rho_{EW} = 1$ at tree level. Further, a Higgs potential such that the ground state of the theory breaks $SU(2) \times U(1)$ down to $U(1)_{E\&M}$ is easily constructed. Thus, we turn to an examination of several models of this type.

Two-Doublet Model. We imagine a Higgs sector containing precisely two doublet fields. We assume that the gauge boson sector is precisely that of the SM, i.e. with a charged W pair and a neutral Z boson. In this case, the physical Higgs bosons left over after electroweak symmetry breaking are five in number: two scalars (h^0 and H^0); a pseudoscalar (A^0); and a pair of charged Higgs bosons ($H^\pm$).[*]   The

---

[*] 'Scalar' and 'pseudoscalar' refer to 1 and γ_5 $f\bar{f}$ couplings, respectively.

parameters of the Higgs sector include the masses of these Higgs bosons, a mixing angle, α, having to do with diagonalizing the scalar sector mass matrix, and the ratio of the vacuum expectation values of the real neutral components of the two fields, $\tan\beta = v_2/v_1$. In the usual convention, $0 \leq \beta \leq \pi/2$ and $-\pi \leq \alpha \leq 0$. Of the Higgs bosons listed above, only the h^0 and H^0 have couplings to vector boson pair channels. Relative to the SM $\phi^0 VV$ couplings we write $f_{hVV} \equiv g_{hVV}/g_{\phi^0 VV}$. We then have:

$$f_{h^0 VV} = \sin(\beta - \alpha), \quad f_{H^0 VV} = \cos(\beta - \alpha), \qquad (3.1)$$

where $V = W$ or Z.

The important point to note with regard to detection of the h^0 and H^0 in the four-charged-lepton ZZ decay mode is the fact that the VV couplings of these two scalars are generally suppressed with respect to the SM Higgs value. This has three impacts: first, the vector boson fusion cross section for a given h is decreased by f_{hVV}^2; second, for $m_h > 2m_Z$, the width of $h \to ZZ$ (on-shell, two-body) decays is decreased proportionally to f_{hVV}^2; and, thirdly, for $m_h < 2m_Z$, $BR(h \to ZZ^*)$ is decreased by this same factor, divided by the factor which describes the change in $\Gamma(h \to b\bar{b}$ (or $t\bar{t}$)). This latter factor depends upon the model employed for $q\bar{q}$ couplings to the Higgs fields, and is a function of α and β. There are two models for $q\bar{q}$ couplings consistent with the absence of tree-level flavor-changing neutral currents (FCNC). In model I, both up and down quarks couple only to the second (by convention) of the Higgs doublet fields (ϕ_2). Defining $d_{hf\bar{f}} \equiv g_{hf\bar{f}}/g_{\phi^0 f\bar{f}}$, we have:

$$\begin{aligned} d_{h^0 t\bar{t}} &= d_{h^0 b\bar{b}} = \frac{\cos\alpha}{\sin\beta} \\ d_{H^0 t\bar{t}} &= d_{H^0 b\bar{b}} = \frac{\sin\alpha}{\sin\beta} . \end{aligned} \qquad (3.2)$$

In model II, up quarks couple to ϕ_2 while down quarks couple to ϕ_1, and we have the same $t\bar{t}$ coupling factors as above, but

$$d_{h^0 b\bar{b}} = -\frac{\sin\alpha}{\cos\beta}, \qquad d_{H^0 b\bar{b}} = \frac{\cos\alpha}{\cos\beta} . \qquad (3.3)$$

Either coupling pattern guarantees that good high energy behavior for $f\bar{f} \to VV$ scattering amplitudes is obtained.

Overall, there is clearly a lot of freedom. For instance, there is a limit in which we regain the SM for one of the scalar Higgs while the other decouples from the VV channels of interest; e.g. the couplings of the h^0 to VV and $f\bar{f}$ channels become the same as those of the SM Higgs when $\alpha \to \beta - \pi/2$. In this case, our previous SM discussion simply applies to the h^0, while the H^0 decouples from VV channels. In another extreme, the h^0 and H^0 could each have $f_{hVV}^2 = 1/2$, sharing equally the VV coupling strength squared. We will not pursue this case in detail, but wait to consider a generalization of it in a subsequent subsection.

However, before closing this section, let me demonsrate that this two-doublet model is capable of reproducing the scenario in which the Higgs boson responsible for electroweak symmetry breaking has negligible quark couplings. In order to obtain this limit we must employ Model I fermion couplings and take v_2 to be very small. In detail, if we take $\beta \to 0$ keeping $\alpha/\beta = \epsilon$ fixed at some arbitrarily small value, we find: the $h^0 VV$ couplings vanish; the $H^0 VV$ couplings become SM in strength; the $h^0 f\bar{f}$ couplings become of order $1/\epsilon$; and the $H^0 f\bar{f}$ couplings become of order ϵ.[†]

---

† Of course, if the $h^0 f\bar{f}$ couplings are too large, perturbation theory would break down.

Thus, the H^0 plays the role of the SM Higgs boson as far as electroweak symmetry breaking is concerned, but it is the large Yukawa couplings of the ϕ_2 doublet (of which the h^0 is the physical neutral scalar remnant) that are responsible for quark masses, despite the small vacuum expectation value associated with this doublet. While there are certainly more complicated realizations of the scenario in which the EWSB Higgs boson(s) have small or zero quark couplings, this is certainly the simplest. For another example see the triplet model discussion later in this report.

The Minimal Supersymmetric Model Higgs Sector. The minimal Higgs structure consistent with the requirements of supersymmetry is a two-doublet sector.[7,2] In addition, the supersymmetric structure imposes relations among the various parameters of the general two-doublet model, leaving only two free parameters. These can be conveniently chosen to be $\tan\beta$ and $m_{H^\pm}$. In terms of these we have:

$$m_{A^0}^2 = m_{H^\pm}^2 - m_W^2,$$
$$m_{h^0,H^0}^2 = \tfrac{1}{2}\left[m_{A^0}^2 + m_Z^2 \mp \sqrt{(m_{A^0}^2 + m_Z^2)^2 - 4m_Z^2 m_{A^0}^2 \cos^2 2\beta} \right]. \tag{3.4}$$

The mixing angle α is also easily computed in terms of the two basic parameters, from which we may compute the important factors $\cos^2(\beta - \alpha)$ and $\sin^2(\beta - \alpha)$ appearing in the VV couplings of eq. (3.1). There are several important phenomenological features of the results. First, we find that $m_{h^0} \leq min[m_Z, m_{A^0}]|\cos 2\beta|$, implying that the h^0 is always less massive than the Z; in particular, for $\tan\beta$ near 1 the h^0 can only have a small fraction of the Z mass. From eq. (3.4), we also see that $m_{H^\pm} \geq m_W$, that $m_{H^0} \geq m_Z$, and that $m_{H^\pm} \simeq m_{H^0} \simeq m_{A^0}$ for large $m_{H^\pm}$. It can also be shown that the mixing angle factor $\cos^2(\beta-\alpha)$ is $\leq \cos^2 2\beta$ and decreases very rapidly as $m_{H^\pm}$ increases; it is already severely suppressed by the time $m_{H^\pm} \gtrsim 2m_W$. Indeed, as $m_{H^\pm}$ increases, $\alpha \to \beta - \pi/2$ and the couplings of the h^0 all become identical to SM ϕ^0 couplings, while the H^0 decouples from VV channels.

Thus, it is highly likely that the h^0 will be found at either LEP or LEP-II. But, over what range of masses can we find the H^0 using the gold-plated mode? Let us temporarily assume that H^0 decays to neutralino and chargino pairs are not kinematically allowed. To analyze the $4l$ mode possibilities we adopt a moderate value of $\tan\beta = 1.5$ (in the range typically preferred by a grand unification scenario). Because of the reduced VV coupling, the dominant cross section for H^0 production of all masses is from gg fusion. Since the top quark loop dominates this reaction in the mass range of interest, we must modify the SM predictions for gg fusion by the factor $d_{H^0 t\bar{t}}^2$. In addition, the partial widths for the various H^0 decay channels are significantly modified by the various mixing angle factors. In particular, the VV decay channels are never dominant over the $b\bar{b}$ (or $t\bar{t}$) decays of the H^0. The H^0 remains quite narrow for all masses of interest. Thus, we will assess the discovery potential for the H^0 by requiring 40 gold-plated events before cuts, regardless of whether m_{H^0} is above or below the two-Z threshold. The $4l$ branching ratio for the H^0 obtained after incorporating all relevant modifications is presented in fig. 1, for $m_t = 80$ and $200\ GeV$. The results for discovery limits obtained after combining the appropriate $4l$ branching ratio with the gg production cross section computed for the H^0 are presented in Table 3.

As expected, the range of masses over which the four lepton decay mode can be employed for H^0 discovery is very restricted in comparison to that for the SM ϕ^0. Indeed, the LHC could not detect the H^0 in this mode, regardless of its mass. The SSC has a small detection window for $m_t = 80\ GeV$ and can reach up to $2m_t$ at the higher top quark mass. The Eloisatron has a somewhat larger range where detection would be possible. Substantially larger values of $\tan\beta$ ($\gtrsim 4$) would be required to

Table 3. $H^0 \rightarrow 4l$ discovery limits in the MSSM (in GeV).

	$m_t = 80$		$m_t = 200$	
Machine	$m_{H^0}^{min}$	$m_{H^0}^{max}$	$m_{H^0}^{min}$	$m_{H^0}^{max}$
LHC	−	−	−	−
SSC	~ 190	~ 220	~ 190	~ 400
Eloisatron	~ 155	~ 275	~ 170	~ 420

significantly alter the discovery ranges stated above. However, we must now return to the possibility that the H^0 can decay to pairs of neutralinos or charginos. Indeed, the simplest GUT treatment of the MSSM[8] suggests that, although the H^0 is likely to have mass in the SSC and Eloisatron windows of Table 3, the neutralinos and charginos could well be light enough that H^0 decays to ino channels are possible. In this case, as originally shown in ref. 7, the neutralino and chargino decay modes of the H^0 will dominate $H^0 \rightarrow b\bar{b}$ and $H^0 \rightarrow t\bar{t}$ and decrease the (rare) $4l$ decay mode branching fraction. Clearly, the detection windows of Table 3 would disappear, and it would be necessary to turn to detection of the H^0 directly in the dominant neutralino and chargino modes. The high H^0 production rates at the Eloisatron would, as always, provide a distinct advantage; however, a detailed analysis must await future work.

Let me end this subsection with a final remark. Although we have seen that it may be difficult to detect the H^0 using the purely leptonic ZZ decay modes, this is nothing more than a reflection of the fact that the H^0 is generally weakly coupled to VV channels and, therefore, to the issue of EWSB. If we regard unraveling the mechanism for electroweak symmetry breaking as the primary goal of the next generation of accelerators, then failure to find this or any other Higgs boson weakly coupled to VV channels will not be a crucial failure. Nonetheless, it would be nice to be able to explore the complete Higgs spectrum in as clean a fashion as possible while revealing the nature of EWSB.

A Higgs Sector with Two Doublets and One Singlet. As we expand the Higgs sector, retaining the SM gauge structure, more and more physical Higgs bosons emerge. In particular, the number of neutral scalars capable of VV couplings increases. In the two-doublet one-singlet model, there are three neutral scalars, h^0, H_1^0, and H_2^0. As in the two-doublet model, good high energy behavior in the VV scattering channels requires that the couplings of these neutral scalars provide the SM VV coupling strength squared. This leads to a sum rule of the general form:

$$\sum_h g_{hVV}^2 = g_{\phi''VV}^2 \qquad (3.5)$$

($i.e.$ $\sum_h f_{hVV}^2 = 1$), where the sum runs over all scalar Higgs bosons h. This sum rule holds in any model with only doublet and singlet Higgs representations, regardless of number; it guarantees good high energy behavior for the scattering of longitudinally polarized vector bosons, and is required in a renormalizable gauge theory of vector bosons whose mass is spontaneously generated by only doublet and singlet Higgs fields. If unitarity for VV scattering is to hold at tree-level, the sum rule (3.5) must be saturated by Higgs bosons with mass below 1 TeV. As we shall see, this type of sum rule does not hold if there are higher-dimensional Higgs representations ($e.g.$ triplet Higgs representations).

A sum rule similar to eq. (3.5) must be satisfied in order that $f\bar{f} \to VV$ scattering amplitudes all have good high energy behavior:

$$\sum_h g_{hVV}g_{hf\bar{f}} = g_{\phi^nVV}g_{\phi^nf\bar{f}}, \qquad (3.6)$$

(*i.e.* $\sum_h f_{hVV}d_{hf\bar{f}} = 1$). Because this sum rule does not involve a sum of squares, it does not imply a general suppression of $f\bar{f}$ couplings. Indeed, in the two-doublet model one can check that this sum rule is satisfied, and, yet, we saw that in the two-doublet model $d_{hf\bar{f}}$ could be larger than 1 in magnitude. Since the d's can be either positive or negative, in general, compensation between terms contributing to (3.6) can be arranged so as to obtain the SM result. Also, even if one f_{hVV} is near 1, this need not imply that $d_{hf\bar{f}}$ is near 1; some other h could have a very large d, so that its fd product could be of order 1 even though its f is small.

Aside from the sum rule constraints and possibly demanding perturbative unitarity, unless we impose supersymmetry or some other new physics model, the masses and hVV couplings of the neutral scalars of the Higgs sector are all free parameters. Of course, if one neutral scalar saturates the VV coupling strength, then, as we have discussed, good high energy behavior in $f\bar{f} \to VV$ amplitudes is easily arranged. (It is obtained most trivially if it has Standard Model couplings to fermions as well, in which case the SM discussion applies without alteration.) The other neutral scalars would not be easily detectable in the VV-mediated gold-plated channel, although by analogy with the MSSM discussion just completed, we would expect that, depending upon details, there could be small windows for their detection at the SSC and Eloisatron. However, it is a priori equally likely that the coupling strength to VV channels is shared rather equally among the various scalar Higgs bosons of the model.

Let us consider the case in which $g_{hVV}^2 = 1/3$ for $h = h^0, H_1^0, H_2^0$. In addition, for simplicity we shall assume that the fermion couplings of these h's are all of SM strength to first approximation. Thus, we must modify our SM Higgs discussion by accounting for the reduced VV fusion cross sections, and reduced ZZ decay widths of the neutral Higgs. Since in the above approximation all the Higgs have identical couplings, we need only discuss one, say the H_2^0. We use the same discovery criteria as in the SM discussion: 40 four-lepton events in the ZZ' region, and 60 four-lepton events in the ZZ region. The results for the discovery reach of each type of machine are presented in Table 4.

Table 4. $H_2^0 \to 4l$ discovery limits for $1/3\ VV$ strength (in GeV).

Machine	$m_t = 80$		$m_t = 200$	
	$m_{H_2^0}^{min}$	$m_{H_2^0}^{max}$	$m_{H_2^0}^{min}$	$m_{H_2^0}^{max}$
LHC	~ 140	~ 400	~ 145	~ 500
SSC	~ 130	~ 600	~ 135	~ 800
Eloisatron	~ 125	~ 1500	~ 125	> 1500

From this table, we see that an equal sharing of the VV coupling strength increases slightly the minimum Higgs mass accessible, due to the decreased branching ratio for the ZZ' decay mode of the Higgs. However, the maximum accessible Higgs mass is little changed, in the present case, from the SM results of Table 1. The reasons for this are simple. Except for the SSC and Eloisatron $m_t = 80\ GeV$ cases, the dominant contribution to the cross section in the mass region near the upper limits is from gg fusion, which is unaffected by a reduction in VV coupling strength. Further, since WW and ZZ channels are dominant in the decays, a decreased VV coupling does not affect their ratio. Except to the extent that the $t\bar{t}$ mode begins to become significant, this implies that the four-lepton branching ratio will be unaffected. In the case of $m_t = 80\ GeV$ at the SSC, the VV and gg fusion cross sections are roughly equal at the upper limit in the SM Higgs case, and decreasing the former by a factor of $1/3$ results in a $\sim 50\ GeV$ decrease in maximum accessible Higgs mass. In the case of the Eloisatron, even though at $m_t = 80\ GeV$ the VV fusion reaction dominates for Higgs mass near $1.5\ TeV$, a reduction by $1/3$ in the cross section does not make it impossible to probe Higgs masses out to $1.5\ TeV$, although there is less margin for inefficiencies. This is because in the case of the SM we had somewhat more events than required for discovery.

From the above, we see once again that the SSC has substantial discovery search, but that the Eloisatron exhibits a large advantage over the less energetic machines. As a further remark, let us note that in the shared VV coupling scenario, if two of the Higgs are relatively light, unitarity limits on the mass of the heaviest Higgs will be substantially weaker than the unitarity limit on the SM Higgs. Thus, there is a possibility for the H_2^0 to be heavier than $1\ TeV$ without violating the simplest forms of perturbative unitarity. Consequently, the extra reach of the Eloisatron might be important even if the Higgs sector is perturbative.

The generalization of the above type of scenario to the case of more and more Higgs scalars, sharing roughly equally the VV coupling strength squared, can be roughly outlined. So long as the VV strength is sufficient that VV modes dominate the decay of a given Higgs boson, the four-lepton final state branching ratio will be approximately independent of VV coupling strength. Thus, any upper limit that is determined by gg fusion cross sections will only be altered to the extent that $t\bar{t}$ coupling to the Higgs in question (which determines the magnitude of the gg-Higgs coupling) are not of SM strength. Any upper limit that is controlled by the magnitude of the VV fusion cross section will be reduced. The most notable example is the $m_t = 80\ GeV$ Eloisatron upper limit, which is VV fusion dominated. This will exhibit a progressive deterioration. Of course, as one continues to split up the VV coupling, the $t\bar{t}$ decay mode of the Higgs bosons will start to become important, if allowed. For $m_t = 200\ GeV$, this starts to happen at about $1/5$ SM strength VV coupling, depending of course on the Higgs mass. At this point, all upper limits for the $4l$ search mode will begin to deteriorate as the ZZ branching ratio starts to decline.

Theoretical Motivations for a Higgs Sector with Singlets. I have already noted that in the context of the SM there is no particular reason to suppose that the Higgs sector cannot contain an arbitrary number of singlet and doublet fields. Further, in models with a supersymmetric structure, the Higgs sector must contain *at least* two doublets. However, even the minimal version of the supersymmetric Higgs sector with exactly two doublets is not necessarily entirely satisfactory from the point of view of naturalness and hierarchy. In order to generate a non-zero mass for the A^0 in the MSSM, it is necessary to introduce a superpotential term of the form $\mu \hat{H}_1 \hat{H}_2$ ($\hat{H}_{1,2}$ are the doublet superfields of the model) and a soft SUSY-breaking potential term of the form $B\mu H_1 H_2$ which mixes the scalar fields themselves. Both B and μ are arbitrary mass parameters, and it is necessary to understand why they should not be of order m_{Planck} or m_{GUT} when the MSSM is placed in the context of supergravity or other grand-unified models. Were B and μ large, then all Higgs masses would be large. (For instance, the A^0 mass is determined by $m_{A^0}^2 \propto B\mu$.) This is sometimes referred to as the 'mixing hierarchy problem'. One of the most attractive ways in which a small mass scale for B and μ can be a natural result of the unified model is if they are not fundamental scales of the theory, but rather derive from the symmetry breaking of a Higgs potential containing at least one singlet field in addition to the two doublet fields. Consider the case of exactly one singlet field N in addition to the two doublet fields H_1 and H_2. A term in the superpotential of the form

$$W = \lambda \hat{N} \hat{H}_1 \hat{H}_2, \qquad (3.7)$$

combined with $\langle N \rangle = n$ yields an effective term of the MSSM form with $\mu = \lambda n$. As we shall discuss, n is expected to be of order $v_{1,2}$ in a typical unified theory, and so μ should be quite reasonable in size when λ is in a perturbative domain. In fact, a trilinear superpotential term, eq. (3.7), emerges automatically in string theories. Indeed, those superstring theories explored to date have *only* trilinear terms in the superpotential. Further, once a term of the form (3.7) is present in the superpotential, evolution from m_{GUT} to m_W will, in general, automatically generate a contribution to the soft SUSY-breaking potential of the form

$$V_{soft} \ni -\lambda A_\lambda N H_1 H_2. \qquad (3.8)$$

With $\langle N \rangle = n$ we obtain an effective $H_1 - H_2$ mixing term with strength specified by $B\mu = \lambda A_\lambda n$.

A number of models of this general type have been analyzed in the literature. For a review see ref. 2. Here we focus on the minimal non-minimal supersymmetric model (MNMSSM) in which there is just one singlet in addition to the two-doublets, and in which the low-energy gauge group is precisely that of the SM. [9] This model can be viewed as a stand-alone theory in its own right, but has the additional motivation of emerging from a very attractive four-dimensional superstring theory based on the 'flipped' $SU(5) \times U(1)$ grand unification group.[10] The full superpotential of the model is specified by

$$W = \lambda \hat{N} \hat{H}_1 \hat{H}_2 - \frac{k}{3} \hat{N}^3. \qquad (3.9)$$

No bilinear terms are included; this avoids the mixing hierarchy problem discussed above and, in addition, is the structure predicted in the superstring context. Aside from the V_{soft} term given in eq. (3.8), there are the usual $m_1^2|H_1|^2 + m_2^2|H_2|^2 + m_N^2|N|^2$ quadratic terms and a term of the form $V_{soft} \ni -(kA_k/3)N^3$. The $m_{1,2,N}$ parameters can be replaced by the vacuum expectation values obtained after spontaneous symmetry breaking, $v_{1,2}$ and n, using the potential minimization conditions at the global minimum. Further, $m_W^2 = (g^2/2)(v_1^2 + v_2^2)$ constrains $v_{1,2}$ so that only the ratio defined by $\tan\beta = v_2/v_1$ is a free parameter. Thus, in the end we have six parameters specifying the Higgs sector. They may be chosen as: λ, k, $r = n/\sqrt{v_1^2 + v_2^2}$, $\tan\beta$, and two mass parameters which can be taken to be $m_{H^\pm}$ and A_k. Clearly, if we took all these parameters as varying independently of one another, there would be a great deal of freedom in the Higgs sector. However, the primary motivation for the model is in the grand-unification context. Thus, in ref. 9 the predictions for the Higgs sector in the GUT context were analyzed. There, the most attractive scenario, in which supersymmetry breaking derives entirely from a universal non-zero value for the gaugino mass at m_{GUT}, was analyzed. By combining this SUSY-breaking scenario with the requirements of correct symmetry breaking at scale m_W (*e.g.* no charge or color breaking) and of no conflicts with experimental limits on SUSY particles, remarkably definitive results for the Higgs boson masses and properties were obtained. We sketch those appropriate for the case where the top mass is moderate in size (of order 90 to 100 GeV).

First, we must note that with two doublet Higgs fields and one singlet field, and with no additional low-energy gauge structure, the physical Higgs spectrum will contain three scalars (h^0, H_1^0, H_2^0), two pseudoscalars (A_1^0, A_2^0) and a charged

Higgs pair $(H^\pm)$. In particular, the scalar Higgs spectrum (on which I shall focus) is precisely that discussed in the previous phenomenological subsection. Thus, it is appropriate to analyze the impact of the specific predictions of the MNMSSM model in the context examined there. The most notable result of the model is the prediction that all the Higgs masses are extremely modest in size. Typically, only $m_{H_2^0}$ can be larger than $100\ GeV$, and usually not by much. The lightest scalar h^0 is often very light, e.g. $m_{h^0} \lesssim 20\ GeV$, and the H_1^0 is often below $50\ GeV$ in mass. Thus, the h^0 and H_1^0 are likely to be accessible at LEP and LEP-II (provided the VV couplings, which I shortly discuss, are adequate) and it is only the H_2^0 that will certainly require a supercollider to detect. However, the favored H_2^0 masses are obviously at the lower end of the mass range that might be accessible to detection using the $4l$ decay mode. In fact, an analysis of the VV couplings shows[9] that the coupling strength squared is very evenly split among the three scalars, and we have precisely the $1/3$ SM strength scenario discussed earlier. Further, there are some new features. Most importantly, the $H_2^0 \to h^0 h^0$ decay mode dominates the $H_2^0 \to b\bar{b}$ decay mode, so that the rare $4l$ decay mode will be further suppressed (by about a factor of 5) compared to the assumptions made in the previous subsection. Thus, even the Eloisatron would have difficulty detecting the H_2^0 in the $4l$ mode. Other techniques, such as associated production, $pp \to W^\pm H_2^0$ followed by triggering on the $H_2^0 \to h^0 h^0 \to b\bar{b}b\bar{b}$ final decay mode (using impact parameter tagging), would probably have to be employed. The high energy and luminosity of the Eloisatron would be a distinct advantage for this detection mode, just as for the $4l$ mode. But, these alternative approaches are outside the scope of the present review.

Of course, the specific grand-unification scenario employed in the MNMSSM context need not be nature's choice. It does, however, illustrate the possibility of obtaining a set of scalar Higgs bosons with similar masses and with VV coupling strength quite evenly divided among them. If, in addition, a different GUT scheme led to an overall mass scale for the scalars in the region above $120\ GeV$, then the analysis of the preceding subsection could become completely relevant. However, corrections to the $h \to 4l$ event rates resulting from h decays to lower mass Higgs pairs (for $h = H_1^0$ and H_2^0) and also to supersymmetric neutralino and chargino pair states might be necessary. In this case, the results of Table 4 would have to be altered by keeping in mind the fact that the gg and VV fusion cross sections would remain more or less the same, while the $4l$ branching ratios could be significantly decreased. In such a situation, the advantage of the Eloisatron's high production rates are apparent. For instance, if the $4l$ branching ratio of the H_2^0 were reduced by a factor of 5, while keeping the production cross sections the same, at $m_t = 80\ GeV$ the Eloisatron might still be able to probe up to $m_{H_2^0} \sim 500\ GeV$ in the $4l$ mode, whereas H_2^0 detection at the LHC and SSC would require use of alternative detection techniques (such as focusing on the neutralino and chargino decay modes).

3.3   Higgs Sectors Containing Triplet Representations

Certainly, Higgs representations containing only singlet and doublet Higgs fields have a certain elegance and simplicity. Further, no phenomenologically viable superstring model has been constructed with sufficiently large low-energy matter representations that Higgs representations beyond doublets and singlets are present. However, there does not seem to be any fundamental reason as to why the Higgs sector cannot include triplet and higher representations, nor is there a theorem implying that a superstring model with higher Higgs representations is impossible. Thus, it is clearly of interest to evaluate the relative merits of different hadron colliders if higher Higgs representations are present. A number of models have been considered in the literature in which triplet Higgs fields are included in the Higgs sector. These are reviewed in ref. 2. Below we sketch the theoretical issues involved, and examine the phenomenology of one of the most interesting proposals. The most amusing aspect of this phenomenology is the expansion in the number of purely leptonic final states that, in general, must contain signals of Higgs bosons and electroweak symmetry breaking. Not surprisingly, the high event rates of the Eloisatron would provide a much broader range of access to such signals.

Theoretical Issues. Any Higgs sector that contains Higgs fields belonging to triplet or higher representations must be very carefully constructed in order to avoid violating the experimental fact that $\rho_{EW} \simeq 1$. The general expression for $\rho_{EW} \equiv m_W^2/(m_Z^2 \cos^2 \theta_W)$ computed at tree-level is:

$$\rho_{EW} = \frac{\sum_{T,Y} \left[4T(T+1) - Y^2 \right] |v_{T,Y}|^2 c_{T,Y}}{\sum_{T,Y} 2Y^2 |v_{T,Y}|^2}, \tag{3.10}$$

where the neutral member of a given representation has $\langle \phi(T,Y) \rangle = v_{T,Y}$ and $c_{T,Y} = 1(1/2)$ for complex (real) representations, respectively. A single representation can only satisfy this if $(2T+1)^2 - 3Y^2 = 1$: the first two such solutions are $T = 1/2, Y = \pm 1$ and $T = 3, Y = \pm 4$. The former is, of course, the standard doublet choice. The latter is clearly a very large representation. Here we pursue the possibility of including Higgs triplet representations in the Higgs sector. Since only doublets have the correct quantum numbers for coupling to fermion-antifermion channels, we must always include in our Higgs sector one or more doublets if we are to give mass to the quarks and leptons via the Higgs boson sector. Thus, we consider the possibility of a Higgs sector with one doublet and one or more triplet Higgs fields. If we arbitrarily add triplets, then eq. (3.10) makes it apparent that in general we will have $\rho_{EW} \neq 1$ at tree level. The magnitude of the deviation from 1 will depend on the size of the triplet vacuum expectation values. For instance, if we have a doublet field and a single $Y = 0$ or $Y = 2$ triplet field, then experimental limits on $\rho_{EW} - 1$ require that

$$|v_{1,0}|/|v_{1/2,1}| \leq 0.047, \quad |v_{1,\pm 2}|/|v_{1/2,1}| \leq 0.081. \tag{3.11}$$

Thus, a single triplet field can be added to the standard doublet field only if it has very little to do with EWSB. The simplest Higgs sector with triplet and doublet fields that has $\rho_{EW} = 1$ at tree level, independent of the size of the triplet vacuum expectation values, is one in which there is 1 doublet, 1 real ($Y = 0$) triplet, and 1 complex ($Y = 2$) triplet. By choosing $v_{1,0} = v_{1,2} \equiv b$ (we define $v_{1/2,1} = a$) we compute $\rho_{EW} = (2a^2 + 4b^2 + 4b^2)/(2a^2 + 0 + 8b^2) = 1$, where the 1st, 2nd, and 3rd terms in the numerator and denominator come from the doublet, real triplet and complex triplet, respectively. The model so obtained   was first considered in refs. 11 and 12.   The Higgs sector of this model is said to have a custodial $SU(2)$ symmetry at tree level, due to the equality of the vacuum expectation values of the neutral triplet fields. In ref. 13 it was shown that a Higgs potential for the model could be constructed in such a way that it preserves the tree-level custodial $SU(2)$ symmetry. This has the important implication that the custodial $SU(2)$ is maintained after higher-order loop corrections from Higgs self-interactions. Thus, the model provides an attractive example of an extension of the SM Higgs sector which contains Higgs triplets but no other new physics. I shall shortly return to the phenomenology of this type of Higgs sector following the discussion of ref. 14.

First, I wish to discuss a few more theoretical points. One very characteristic feature of a Higgs sector with triplets is a non-zero $H^- W^+ Z$ vertex. In general, in order to avoid such a vertex there are only two possibilities; (a) one must have no representations with $Y \neq 0$ in the Higgs sector; or (b) one must have *only* complex representations with the same Y and $T = Y/2$. Of course, any arbitrary number of singlets can be added to either case, and representations with $v_{T,Y} = 0$ (and, therefore, no connection to EWSB) do not count. Clearly (a) is not a viable choice since it leads to $\rho_{EW} = \infty$ at tree level. Choice (b) only yields $\rho_{EW} = 1$ if we choose doublets for our complex representations. In a little more detail, the conditions that $\rho_{EW} = 1$ and that there be no $H^- W^+ Z$ vertex can only be met simultaneously if we have:

$$\sum_{T,Y} Y^2 \left[4T(T+1) - Y^2 \right] |v_{T,Y}|^2 = \sum_{T,Y} 2Y^2 |v_{T,Y}|^2 . \qquad (3.12)$$

While the the condition for $\rho_{EW} = 1$ (see eq. (3.10)) can be satisfied in a variety of ways for arbitrarily large representations, the above condition, which guarantees the absence of any $H^- W^+ Z$ vertex (when $\rho_{EW} = 1$!) can never be satisfied for any combination of representations beyond doublets. This is because $4T(T+1) - Y^2 \geq 4T > T$ for $T > 1/2$ if $|Y| \leq 2T$, as required for a neutral member with $v_{T,Y} \neq 0$. Thus, the direction of violation of the equality of eq. (3.12) is the same for all representations beyond doublets, and no compensation between different representations is possible. Consequently, triplet models with $\rho_{EW} = 1$ always have a non-zero $H^- W^+ Z$ vertex. As a result, the H^+ will appear as an s-channel resonance in the $W^+ Z$ scattering amplitude, and will often influence other VV scattering processes (e.g. $W^+ W^- \to ZZ$) by making an exchange contribution in the t or u channel.

A second very characteristic feature of a triplet Higgs model with $\rho_{EW} = 1$ is the existence of a physical doubly-charged Higgs boson. Clearly, to have a triplet with a neutral member that can acquire a non-zero vacuum expectation value, we must have either $Y = 0$ or $Y = 2$. We have already argued above that if we had only $Y = 0$ representations, then $\rho_{EW} \neq 1$. The only possibility, without going to still higher T representations, is to include a $Y = 2$ triplet representation. Such a representation certainly contains a doubly-charged Higgs boson. Further, so long as $v_{T,Y} \neq 0$ (as required to fix up ρ_{EW} and in order that the triplet representation be relevant for EWSB), this doubly-charged Higgs boson will couple to the $W^+ W^+$ channel. Thus, it will appear as a s-channel resonance in the $W^+ W^+$ scattering amplitude, and will yield t-channel and u-channel exchange contributions to $W^+ W^-$ scattering.

Because of the diversity of possible signals, it is clearly of great interest to investigate the phenomenology that would be associated with a Higgs sector containing doublets and triplets. We focus on the simplest case specified earlier of one doublet, one real triplet, and one complex triplet with custodial $SU(2)$ at tree-level.

Phenomenology of a Higgs Triplet Sector with Custodial $SU(2)$. In the model of ref. 11, the Higgs fields take the form

$$\phi = \begin{pmatrix} \phi^{0*} & \phi^+ \\ \phi^- & \phi^0 \end{pmatrix} \qquad \chi = \begin{pmatrix} \chi^0 & \xi^+ & \chi^{++} \\ \chi^- & \xi^0 & \chi^+ \\ \chi^{--} & \xi^- & \chi^{0*} \end{pmatrix} , \qquad (3.13)$$

i.e. one $Y = 1$ complex doublet, one real ($Y = 0$) triplet, and one $Y = 2$ complex triplet. As already noted, tree-level invariance for the gauge boson mass terms under the custodial $SU(2)$ is arranged by giving the χ^0 and ξ^0 the same vacuum expectation value. (However, since the hypercharge interaction with the B field breaks the custodial $SU(2)$, there are potentially infinite contributions to $\rho - 1$ at one-loop.) We define $\langle \chi^0 \rangle = \langle \xi^0 \rangle = b$, and also take $\langle \phi^0 \rangle = a/\sqrt{2}$. It will be convenient to use the notation:

$$v^2 \equiv a^2 + 8b^2, \quad c_H \equiv \frac{a}{\sqrt{a^2 + 8b^2}}, \quad s_H \equiv \sqrt{\frac{8b^2}{a^2 + 8b^2}}, \qquad (3.14)$$

where c_H and s_H are the cosine and sine of a doublet-triplet mixing angle. The gauge boson masses obtained after absorbing appropriate Goldstone bosons are:

$$m_W^2 = m_Z^2 \cos^2 \theta_W = \tfrac{1}{4} g^2 v^2. \tag{3.15}$$

The remaining spin-0 physical states can be classified according to their transformation properties under the custodial $SU(2)$. One finds a five-plet $H_5^{++,+,0,-,--}$, a three-plet $H_3^{+,0,-}$ and two singlets, H_1^0 and $H_1^{0\prime}$. However, not all these states need be mass eigenstates. Only the doubly-charged $H_5^{++,--}$ and, for appropriately chosen phases, the H_3^0 cannot mix at tree level. In general, the remaining neutral Higgs can mix with one another, as can the singly-charged Higgs, depending upon the precise structure of the Higgs potential. The masses and compositions of the mass eigenstates are determined by the quartic interactions among the Higgs fields ϕ and χ. However, as we have already mentioned, it is desirable to choose the Higgs potential in such a way that it preserves the custodial $SU(2)$ symmetry, as done in ref. 13. In this case, the 5-plet and 3-plet states cannot mix at tree level with one another or with the singlets; the only possible mixing is between H_1^0 and $H_1^{0\prime}$. This latter mixing depends upon the parameters of the Higgs potential, and can range from zero to maximal. We shall adopt the language of zero mixing. One should note that when the Higgs potential preserves the custodial $SU(2)$, the 5-plet Higgs bosons are all degenerate in mass with one another, as are the 3-plet Higgs bosons. We call these common masses m_{H_5} and m_{H_3}. The masses $m_{H_1^0}$ and $m_{H_1^{0\prime}}$ can also be chosen freely. Once θ_H and these four masses are specified, all but one of the parameters of the most general $SU(2)$ custodial Higgs potential are determined.

From the Higgs boson couplings to fermions and vector bosons we can determine the basic phenomenological features of the Higgs sector of the model. The fermion couplings have been studied in ref. 14. There are two possible types. First, there are the standard Yukawa couplings of the doublet Higgs field to fermion-antifermion channels. We shall analyze these couplings in detail shortly. The only other possible couplings are ones closely analogous to those required in order to produce a "see-saw" mechanism for generating neutrino masses in left-right symmetric models; namely, couplings of the triplet Higgs fields (with $Y = 2$) to the lepton-lepton channels. However, in the present context, where we envision expanding only the Higgs sector of the Standard Model, there are no right-handed partners for the neutrinos, and the introduction of such couplings leads directly to Majorana masses for the neutrinos. Limits on such Majorana masses for the neutrinos are quite restrictive, and are reviewed in ref. 14. In the case of the electron neutrino, they are sufficiently strong that even if the coupling in question assumes its upper limit value, it will have no phenomenological impact. We shall ignore such couplings in the following discussion.

Returning to the standard doublet fermion-antifermion interactions, we see that all Higgs boson couplings to fermion-antifermion channels are determined by the overlap of the mass eigenstate Higgs fields with the doublet field. One finds that the $H_5^{++,--}$, $H_5^{+,-}$, H_5^0, and $H_1^{0\prime}$ states have no such overlap, and that only the $H_3^{+,-}$, H_3^0 and H_1^0 will have fermion-antifermion couplings. The Feynman rules for the various couplings are given below (to be multiplied by an overall factor of i):

$$
\begin{aligned}
g_{H_1^0 q\bar{q}} &= -\frac{g m_q}{2 m_W c_H} \quad (q = t, b), \\
g_{H_3^0 t\bar{t}} &= +\frac{g m_t s_H}{2 m_W c_H} \gamma_5, \\
g_{H_3^0 b\bar{b}} &= -\frac{g m_b s_H}{2 m_W c_H} \gamma_5, \\
g_{H_3^- t\bar{b}} &= \frac{g s_H}{2\sqrt{2} m_W c_H} \Big[m_t(1 + \gamma_5) - m_b(1 - \gamma_5) \Big],
\end{aligned}
\tag{3.16}
$$

where third-generation notation is employed for the quarks. Analogous expressions hold for the couplings to leptons. As pointed out in ref. 13 it is possible that $b \gtrsim a$, so that most of the mass of the W and Z comes from the triplet vacuum expectation values. In this case, the doublet vacuum expectation value $a/\sqrt{2}$ is much smaller than in the SM, and the Yukawa couplings of the doublet to the fermions must be much larger than in the SM in order to obtain the experimentally determined quark masses. Then, the Higgs bosons that do couple to fermions have much larger fermion-antifermion pair couplings and decay widths than in the SM.

Most interesting, however, are the tree-level couplings to vector bosons. The Feynman rules for these are specified as follows (we drop an overall factor of $ig_{\mu\nu}$):

$$
\begin{aligned}
H_5^{++}W^-W^- : &\quad \sqrt{2}gm_W s_H \\
H_5^+W^-Z : &\quad -gm_W s_H/c_W \\
H_5^+W^-\gamma : &\quad 0 \\
H_5^0W^-W^+ : &\quad (1/\sqrt{3})gm_W s_H \\
H_5^0ZZ : &\quad -(2/\sqrt{3})gm_W s_H c_W^{-2}
\end{aligned}
\qquad
\begin{aligned}
H_1^0W^-W^+ : &\quad gm_W c_H \\
H_1^0ZZ : &\quad gm_W c_H c_W^{-2} \\
H_1^{0\prime}W^-W^+ : &\quad (2\sqrt{2}/\sqrt{3})gm_W s_H \\
H_1^{0\prime}ZZ : &\quad (2\sqrt{2}/\sqrt{3})gm_W s_H c_W^{-2}
\end{aligned}
$$

$$(3.17)$$

where s_W and c_W are the sine and cosine of the standard electroweak angle, respectively. There are no non-zero tree-level couplings of the H_3 Higgs multiplet members to vector bosons. Obviously, the SM is regained in the limit where $s_H \to 0$, in which case the H_1^0 plays the role of the SM Higgs and has SM couplings. However, in this model with custodial $SU(2)$ symmetry, there is no intrinsic need for s_H to be small. Indeed, in the opposite limit, where $c_H \to 0$, the H_1^0 decouples from VV channels, and only those Higgs with no $f\bar{f}$ couplings are involved in EWSB. This is a more complicated example of the type of scenario discussed earlier, where we assumed that the SM Higgs was responsible for EWSB but had no $f\bar{f}$ couplings. Also note, in accord with our earlier discussion, the presence of a $H_5^+W^-Z$ coupling, in contrast to the absence of such a coupling of the charged Higgs in any model containing only Higgs doublets (and singlets). Overall, we have a remarkable dichotomy between the H_5 and the H_3 multiplets: ignoring for the moment the HV and HH type channels, the former couple and decay only to vector boson pairs, while the latter couple and decay only to fermion-antifermion pairs.[*] However, HV and HH decay channels can become important — we shall return to these as required in our later discussion. In this report, I wish to focus on those Higgs bosons most directly connected to EWSB, i.e. those with large couplings to the vector boson channels — the H_1^0, the $H_1^{0\prime}$, and the H_5 multiplet members.

It is both amusing and useful to examine the manner in which high-energy unitarity is preserved for longitudinal vector boson scattering processes in this model. For Higgs sector extensions involving only doublets and singlets, good high-energy behavior for longitudinal vector boson scattering is guaranteed by eq. (3.5). However, the manner in which good high-energy behavior is obtained in Higgs sector extensions containing triplets and higher representations is much more complicated. We give two examples in the context of the model being discussed. Consider $ZW^- \to ZW^-$. In the SM there is one t-channel graph involving the exchange of the ϕ^0, with effective strength proportional to $g^2 m_Z^2$. In our triplet model the couplings of eq. (3.17) make it clear that we have three t-channel graphs for the neutral Higgs bosons, and an s-channel and a u-channel graph for the singly-charged H_5^-. The latter s- and u-channel graphs combine together to give the same result as a t-channel graph except

---

[*] As a technical aside, we note that at one loop this dichotomy breaks down. For instance, one-loop corrections induce $f\bar{f}$ couplings for the $H_1^{0\prime}$, H_5^0 and H_5^+. In the case of the H_5^+ the induced couplings could be large, but need not be so. Thus, for purposes of the present discussion, we may neglect the one-loop-induced couplings and focus on the phenomenology implied for the H_5 multiplet members by the tree-level couplings. Further discussion appears in ref. 14.

for an overall sign difference. Thus, the four contributions have effective strength proportional to:

$$
\begin{aligned}
H_1^0 : & \quad g^2 c_H^2 m_Z^2 \\
H_1^{0\prime} : & \quad (8/3)g^2 s_H^2 m_Z^2 \\
H_5^0 : & \quad -(2/3)g^2 s_H^2 m_Z^2 \\
H_5^- : & \quad -g^2 s_H^2 m_Z^2,
\end{aligned}
\tag{3.18}
$$

where the minus sign in the H_5^- case is introduced to account for the sign difference alluded to above. Clearly the sum of all four terms gives back the original $g^2 m_Z^2$ of the SM Higgs t-channel exchange graph. However, a non-zero vertex for $W^- H_5^+ Z$ was crucial. It is because such a vertex cannot appear in a multi-doublet model that the unitarity sum rule takes a much simpler form in such models. It is also amusing to consider the case of $W^+ W^+ \to W^+ W^+$ scattering. In the SM there are two ϕ^0-exchange graphs: one is a t-channel and the other a u-channel graph. They can be thought of as combining together and having effective strength $g^2 m_W^2$. In our triplet model we have three t-channel and three u-channel neutral Higgs graphs, and an s-channel graph, the latter involving the H_5^{++}. An s-channel graph is equivalent to the sum of a t- and u-channel graph except for an overall sign. Thus the effective strengths of the various contributions are:

$$
\begin{aligned}
H_1^0 : & \quad g^2 c_H^2 m_W^2 \\
H_1^{0\prime} : & \quad (8/3)g^2 s_H^2 m_W^2 \\
H_5^0 : & \quad (1/3)g^2 s_H^2 m_W^2 \\
H_5^{++} : & \quad -2g^2 s_H^2 m_W^2,
\end{aligned}
\tag{3.19}
$$

and again these sum to give the SM result.

In the above, it should be particularly noted that good high energy behavior for VV scattering processes can result by *cancellation* between contributions coming from different Higgs bosons. This is to be contrasted with models containing only doublet and singlet Higgs representations, where all the (neutral) Higgs boson contributions enter with the same sign thereby yielding a restrictive sum rule (see eq. (3.5)) for the absolute squares of the Higgs VV couplings. For general $\tan\theta_H = s_H/c_H$, all Higgs bosons in the custodial $SU(2)$ triplet model can have large (SM-type strength times Clebsch-Gordon coefficient) coupling to the appropriate VV channel. In fact, for general $\tan\theta_H$, *all* VV scattering channels should be strongly influenced by the presence or absence of one or more of the Higgs bosons of the model.

Having understood the manner in which good high-energy behavior for amplitudes involving longitudinally polarized vector bosons is guaranteed, we can turn to approximate numerical constraints imposed upon Higgs boson masses by requiring that they be small enough that the various tree-level scattering amplitudes never numerically exceed their unitarity limits. The weakest form of this constraint in the case of the SM requires that the mass of the ϕ^0 be below roughly $1\,TeV$ in order that WW, ZZ, $\phi^0 Z$ and $\phi^0\phi^0$ coupled channel tree-level scattering amplitude matrix not violate the S-wave unitarity bound in the $s \to \infty$ limit.[†] If we combine this fact with the pattern of Higgs boson masses that emerges from the allowed form of the Higgs potential (see refs. 13 and 14), and with the amplitude contributions outlined in eqs.

[†] While the complete theory cannot violate unitarity even if m_{ϕ^0} is larger than this value, perturbative calculations are certainly inadequate for higher Higgs masses, since the Higgs sector becomes strongly interacting.

(3.18) and (3.19), it is easy to see[14] that m_{H_5}, $m_{H_1^0}$ and $m_{H_1^{0'}}$ must all lie below about 1 TeV in order that unitarity be obeyed perturbatively. Only if unnaturally large ratios among the coupling constants in the Higgs potential are allowed, could this rough perturbative unitarity limit be evaded. Thus, there is reason to hope that all of these Higgs bosons connected to EWSB could be accessible at machines such as the LHC, SSC and Eloisatron through a careful and complete study of all VV scattering processes.

<u>Experimental Probes Using Purely Leptonic Final States.</u> In this subsection, I focus on the H_1^0, $H_1^{0'}$, and H_5 Higgs bosons that are intimately connected to EWSB. As emphasized above, good high energy behavior for VV scattering processes is accomplished without any intrinsic sum-rule-type limitation on the strength of the couplings of these Higgs bosons to the appropriate VV channels — all such couplings can be of order gm_W. Thus, in general, *all VV* scattering processes will be influenced strongly by the Higgs bosons listed above. However, as apparent from eq. (3.17), their VV couplings do depend greatly upon the relative size of the doublet and triplet vacuum expectation values, as parameterized by $\tan \theta_H = s_H/c_H$ (see eq. (3.14)). In the limit of $\tan \theta_H \to 0$, the Standard Model is regained; the H_1^0 has SM couplings to ZZ, W^+W^- and $f\bar{f}$ channels, but the VV couplings of the $H_1^{0'}$ and the H_5's all vanish. However, as we have emphasized, there is no reason for $\tan \theta_H$ to be small. $\rho_{EW} = 1$ at tree-level regardless of the choice for $\tan \theta_H$. A very small, or very large, value for $\tan \theta_H$ requires large ratios between closely related parameters in the Higgs potential and thus can be regarded as somewhat unnatural. Most naturally, the vacuum expectation values a and b are similar in size, and $\tan \theta_H$ is somewhat larger than 1. For definiteness, I will assume $\tan \theta_H = 1$ in the following analysis of discovery limits. As in previous scenarios, my goal will be to estimate the mass range over which the above Higgs bosons can be discovered in purely leptonic final states at each of the three machines, the LHC, SSC and Eloisatron. Of course, the four-charged-lepton final state is only appropriate for the neutral Higgs bosons; for the H_5^+ and H_5^{++}, I employ the $l^+l^-l^+{}'\nu$ (3l for short) and the $l^+\nu l^+{}'\nu$ (2l for short) final states, respectively. These should be almost as background free as is the 4l final state. For instance, although mass resolution in the 2l final state will not be good, the isolated like-sign lepton signal should allow elimination of all but W^+W^+ continuum backgrounds if sufficient signal rate is available. And the W^+W^+ continuum background to a H_5^{++} resonance signal should itself be small given that there are no $q\bar{q}$, gq or gg induced processes contributing to like-sign W-pair production. In fact, it was shown in ref. 15 that the irreducible background from $qq \to qqW^+W^+$ via gluon exchange is small compared to a Higgs signal. I will say a bit more about this later. In estimating the largest accessible masses for the various Higgs bosons at each machine, I will require 60 events in the 4l channel for a neutral Higgs boson, but adopt a more conservative 100 event requirement in the 3l channel for the $H_5^{+,-}$ and demand 150 events in the 2l channel for the $H_5^{++,--}$.

As always, there are two basic ingredients in determining discovery reach in purely leptonic modes: 1) the magnitude of the production cross section; and 2) the decay branching ratio to the leptonic final state. Let us consider the relevant branching ratios first. Since I am interested in the upper mass limits in this section, the branching ratio is the product of the Higgs to VV branching ratio times the VV to leptonic state branching ratio. The relevant VV branching ratios are:

$$BR(ZZ \to 4l) \simeq 0.0047,$$
$$BR(W^\pm Z \to 3l) \simeq 0.015, \qquad (3.20)$$
$$BR(W^\pm W^\pm \to 2l) \simeq 0.05,$$

where I have assumed that top quark channels are forbidden in both W and Z decays. The Higgs to VV branching ratios require some discussion. Alternative decay channels are possibly important for all the Higgs bosons under consideration. The important alternative channels for the H_5^0, H_5^+ and H_5^{++} at high mass are $H_3 V$ and $H_3 H_3$ two-body final states; because of the degeneracy among the H_5's, final states involving any H_5 are kinematically forbidden. But, for the $H_1^{0\prime}$ and H_1^0 we must allow, as well, for the possibility of $H_5 H_5$ channels (there are no tree-level couplings to $H_5 V$ channels). Finally, in the case of the H_1^0 its fermion-antifermion couplings can yield a small contribution from the $t\bar{t}$ decay mode if m_t is large. Obviously, if the final state Higgs bosons are sufficiently massive, the HH and HV channels will be forbidden; but to be conservative, I shall allow for their presence. Since the H_3's couple to fermion-antifermion channels proportionally to fermion mass (as in the SM), see eq. (3.16), $H_3 V$ and $H_3 H_3$ modes will contribute a negligible amount to purely leptonic final states, while $H_5 H_5$ final states are generally too complex to allow for a clean signature. Thus, only the VV pair final states contribute significantly to the purely leptonic signals on which I focus. The relative importance of the VV channels in the decays of the various Higgs bosons is a function of $\tan\theta_H$. To illustrate, I present in fig. 4 (taken from ref. 14) the relevant branching fractions in the case of the H_5^+, for several $\tan\theta_H$ values.

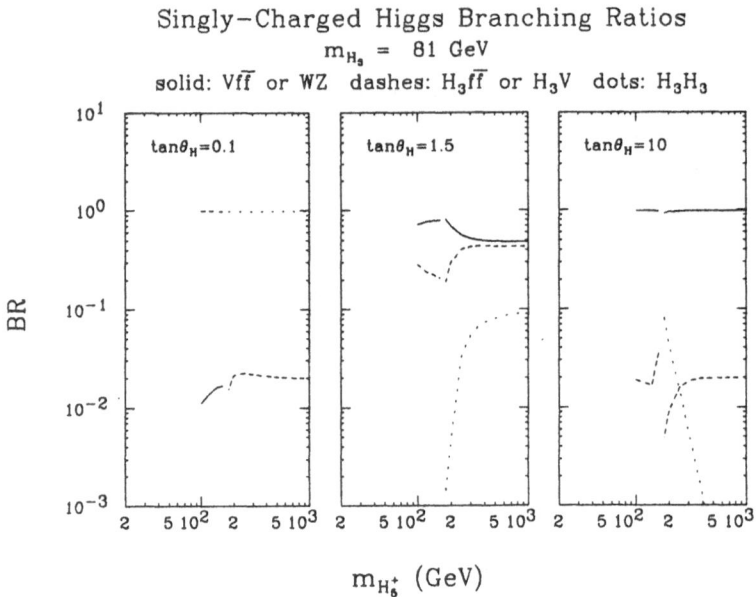

Singly–Charged Higgs Branching Ratios

m_{H_3} = 81 GeV

solid: $V f \bar{f}$ or WZ dashes: $H_3 f \bar{f}$ or $H_3 V$ dots: $H_3 H_3$

Fig. 4. Branching ratios for H_5^+ decay for $\tan\theta_H = 0.1$, 1.5, and 10. We have taken $m_{H_3} = 81\ GeV$. The WZ branching ratio of interest is given by the solid line. For further details see ref. 14.

From this figure, we see that at moderate $\tan\theta_H$ near 1 the $H_5^+ \to W^+ Z$ branching ratio is of order 0.5 at high m_{H_5}. Similary, for $\tan\theta_H$ near 1 and Higgs mass above about 300 GeV, we find $BR(H_5^0 \to ZZ + W^+ W^-) \simeq 0.5$, $BR(H_5^{++} \to W^+ W^+) \sim 0.5$, and $BR(H_1^{0\prime} \to ZZ + W^+ W^-) \sim 0.2$. $BR(H_1^0 \to ZZ + W^+ W^-)$ varies significantly for $m_{H_1^0}$ in the region above 200 GeV, rising from about 0.02 at $m_{H_1^0} \simeq 200\ GeV$ to about 0.1 at $m_{H_1^0} \simeq 1\ TeV$; let us for the moment adopt a rough typical value of 0.08. The small size of these latter two branching ratios results from allowing for the large number of alternative decay channels outlined above. For the neutral Higgs, we also use eq. (3.17) to determine the relative importance of the ZZ vs. $W^+ W^-$ mode. For H_5^0, $ZZ = (2/3) VV$, while for the $H_1^{0\prime}$ and H_1^0,

$ZZ = (1/3)VV$. Combining all the above factors we obtain the net branching ratios to the purely leptonic final state of interest given in Table 5. Also given is the cross section required for the target number of events (given earlier) for the different modes at $L_{year} = 10^4 \ pb^{-1}$. In the case of the charged Higgs, these are the cross sections required after summing over both negatively and positively charged Higgs.

Table 5. $H_2^0 \rightarrow 4l$ discovery limits for $1/3 \ VV$ strength (in GeV).

Channel	BR	σ req. at $L_{year} = 10^4 \ pb^{-1}$
$H_5^{++,--} \rightarrow 2l$	$\simeq 0.025$	0.6 pb
$H_5^{+,-} \rightarrow 3l$	$\simeq 0.0075$	1.3 pb
$H_5^0 \rightarrow 4l$	$\simeq 0.0016$	3.8 pb
$H_1^{0\prime} \rightarrow 4l$	$\simeq 0.0003$	20 pb
$H_1^0 \rightarrow 4l$	$\simeq 0.00013$	48 pb

Let us now turn to production cross sections for the various Higgs bosons of interest. For all but the H_1^0, there are no $q\bar{q}$ couplings, and hence the gg fusion mechanism is absent. For H_1^0, at $\tan \theta_H = 1$ the $t\bar{t}$ coupling squared is twice SM strength, and we employ the gg fusion results of Fig. 2 rescaled by a factor of 2. The VV fusion cross sections have been obtained in ref. 14 for the various Higgs bosons, and are presented in Fig. 5 for the energy of the SSC. As a function of machine energy they scale very much like the SM Higgs VV fusion cross sections given in Fig. 2.

Using the above cross section results, it is now straightforward to determine the Higgs mass values at which the production cross sections fall below those listed in Table 5 that are required in order to produce the bench mark event rates in the appropriate purely leptonic channels. In this way, we may determine the maximum mass reach for each of the Higgs bosons, for all three of the hadron colliders. The results are presented in Table 6. In the case of the H_1^0, I have adjusted for the varying VV branching ratio as a function of $m_{H_1^0}$. This is especially important at the LHC where the declining $H_1^0 \rightarrow ZZ$ branching ratio for decreasing $m_{H_1^0}$ makes it impossible to find the H_1^0, regardless of m_Z. Further, since gg fusion only contributes to the production of the H_1^0, only H_1^0 discovery limits depend on m_t. The non-parenthetical values are those for $m_t = 80 \ GeV$, and those in parentheses are for $m_t = 200 \ GeV$. The extended reach of the Eloisatron is especially striking for the neutral Higgs bosons. One should also note the much larger discovery reach of the SSC compared to the LHC in the case of the charged Higgs bosons of the triplet model.

Finally, we should not forget that even if we discover a Higgs boson in some VV channel at relatively low mass, it will then be important to verify that the VV channel cross section is well-behaved at higher energies, i.e. that all the Higgs bosons contributing in the various channels (s, t, and u channels) all lie below some energy scale. For this purpose, we must be able to measure the VV continuum in the absence of Higgs boson enhancements. We have already discussed this issue in the SM section for the neutral ZZ channel, with the conclusion that only the Eloisatron really had the capability to do this in the $4l$ mode. The predicted cross section levels in the W^+W^+ channel have also been explored.[16] The W^+Z cross section is currently under investigation.[16] The results for the ZZ and W^+W^+ continua at the Eloisatron appeared in Fig. 3. Recall that in the case of the ZZ final state we plotted $q\bar{q} \rightarrow ZZ$. Contributions from gluon-mediated diagrams of various kinds yield higher-order corrections beginning at $\mathcal{O}(\alpha_s)$. The interesting

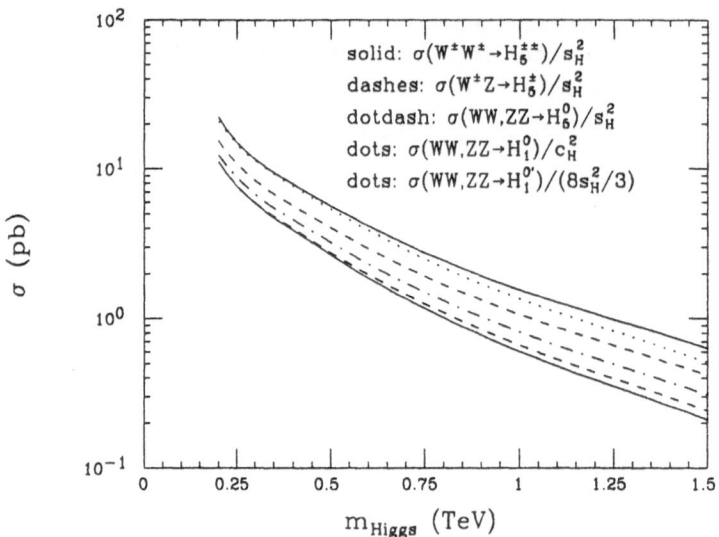

Fig. 5. Cross sections at the SSC for VV fusion production of the triplet model Higgs bosons. At $\tan\theta_H = 1$, these must be corrected by the indicated factors using $s_H^2 = c_H^2 = 1/2$. The upper (lower) solid curve is for H_5^{++} (H_5^{--}). The upper (lower) dashed curve is for H_5^+ (H_5^-).

Table 6. Discovery limits for triplet Higgs (GeV)

Higgs Boson	LHC	SSC	Eloisatron
$H_5^{++,--}$	~ 600	~ 1300	> 1500
$H_5^{+,-}$	~ 350	~ 800	> 1500
H_5^0	$-$	~ 300	~ 1500
$H_1^{0\prime}$	$-$	~ 250	~ 1500
H_1^0	$- \ (-)$	$\sim 350 \ (\sim 550)$	$\sim 750 \ (\sim 1250)$

$qq \rightarrow qq\, ZZ$ processes can only be isolated by double spectator jet tagging. As a result, the W^+W^+ continuum is a much more direct probe of the VV scattering subprocesses than the ZZ continuum. A W^+W^+ pair can only be produced beginning with the $2 \rightarrow 4$ subprocesses of the type $qq \rightarrow qq\, W^+W^+$. In this case, one has the possibility of probing the truly fundamental $W^+W^+ \rightarrow W^+W^+$ electroweak scattering subprocess which contributes to the $2 \rightarrow 4$ reaction without using jet tagging. However, one must now assess the magnitude of the $qq \rightarrow qq\, W^+W^+$ processes in which a gluon is exchanged between the scattering quarks and the W^+'s are emitted from the initial or final quarks. We have calculated this background as well as the electroweak process cross section (they don't interfere). In Fig. 3. I have plotted results for the $qq \rightarrow qq\, W^+W^+$ continuum, showing separately the contributions from the g exchange and electroweak processes. (To reemphasize, these are the appropriate W^+-pair mass spectra if all Higgs bosons are light.) Since the g exchange background yields W^+'s preferentially towards the forward or backward directions, the indicated rapidity cuts have adequately suppressed this background. Thus, the measured W^+W^+ spectrum will (regardless of Higgs mass) truly reflect the interesting $2 \to 4$ electroweak processes; this was not the case for the ZZ spectrum without simultaneously using jet tagging.

In any case, it is the $VV = ZZ$ and W^+W^+ spectra that must be determined above some m_{VV}^{min} value if one is to verify that there are no remaining Higgs boson resonances with mass near to or larger than m_{VV}^{min}. The results given in this figure are integrated over an appropriate mass region and multiplied by the appropriate branching ratio from eq. (3.20) in order to determine event rates in the $4l$ or $2l$ mode. In the case of the ZZ channel, we obtain the results discussed in the Standard Model section — some 45 $4l$ events for m_{ZZ} between 1 and 1.5 TeV. In the case of the W^+W^+ channel, we obtain (for the indicated rapidity cut) about 0.5 pb after integrating over $m_{W^+W^+}$ from 1 to 1.5 TeV. Using the $2l$ branching ratio of eq. (3.20) we obtain some 250 l^+l^+ events. This should be enough to infer the level of the W^+W^+ continuum, despite the fact that a given like-sign-lepton-pair mass bin will receive contributions from a large range of W-pair masses. Of course, a somewhat smaller signal will also be present in the W^-W^- channel. These results can be compared with the SSC results of ref. 15. The W^+W^+ spectrum level is roughly a factor of 10 lower, leaving an extremely marginal l^+l^+ signal given the indirect relationship between the l^+l^+ and W^+W^+ spectra.

Thus, when all Higgs boson enhancements are at low or moderate VV pair masses, we conclude that the Eloisatron rates are just sufficient to actually investigate the high mass VV continuum in the ZZ and W^+W^+ channels (and most probably the W^+Z channel as well), using purely leptonic final states. Further, it seems apparent that the lower-energy SSC event rates in these purely leptonic channels will not be adequate for this purpose.

4. CONCLUSION

Two important conclusions can be drawn from the "gold-plated" purely-leptonic mode evaluations performed above. First, it is clear that the SSC has substantial purely-leptonic-mode discovery reach in Higgs mass for many of the non-minimal Higgs sector models that we have considered. One can hope that other final state modes can be used to provide fairly complete coverage of all VV channels in the mass range below 1 TeV. However, further work is required to explicitly demonstrate that this is so. Nonetheless, it is also apparent that the Eloisatron, aside from giving essentially complete mass coverage for the SM Higgs boson, would also allow us to use purely leptonic modes to probe for Higgs bosons of a non-minimal Higgs sector over much broader VV mass ranges than the SSC (or LHC). Presumably other channels could be employed to verify signals found in the gold-plated channels. Further, the Eloisatron appears to allow use of the gold-plated modes for investigation of the high mass VV continuum cross sections in the case where all Higgs boson enhancements are at lower VV pair masses, and the continuum cross sections are perturbative in magnitude. Thus, a case can reasonably be made that the parameters of the Eloisatron are indeed appropriate for a successor to the SSC.

ACKNOWLEDGEMENTS

I would like to thank the Erice Center for hospitality during the time that much of this work was initiated. I am grateful to B. Grzadkowski, H. Haber, R. Vega, and J. Wudka for helpful conversations, and collaboration on projects closely connected to the present work.

REFERENCES

1. J.F. Gunion and R. Vega, INFN Eloisatron Report, CCSEM-EL-88-1, p. 42.

2. J.F. Gunion, H.E. Haber, G.L. Kane, and S. Dawson, "The Higgs Hunters Guide", Addison-Weseley Frontiers in Physics Series, in press.

3. J.F. Gunion, G.L. Kane, and Wudka, *Nucl. Phys.* **B299** (1988) 231.

4. J.F. Gunion, P. Kalyniak, M. Soldate, and P. Galison, *Phys. Rev.* **D34** (1986) 101.

5. See for example, J.F. Gunion, J. Kalinowski, and A. Tofighi-Niaki, *Phys. Rev. Lett.* **57** (1986) 2351.

6. E. Wang, G. Hanson, *et al.*, Proceedings of the 1988 Snowmass Workshop on "High Energy Physics in the 1990's", ed. by S. Jensen, World Scientific (1989).

7. See J.F. Gunion and H.E. Haber, *Nucl. Phys.* **B272** (1986) 1; *Nucl. Phys.* **B278** (1986) 449; *Nucl. Phys.* **B307** 445.

8. J. Ellis and F. Zwirner, preprint LBL-27310 (1989).

9. A detailed analysis of the Higgs sector appears in J. Ellis, J.F. Gunion, H.E. Haber, L. Roszkowski, and F. Zwirner, *Phys. Rev.* **D39** (1989) 844; a general analysis of the model appears in J. Ellis, J.S. Hagelin, S. Kelley and D.V. Nanopoulos, *Nucl. Phys.* **B311** (1988) 1.

10. I. Antoniadis, J. Ellis, J.S. Hagelin, and D.V. Nanopoulos, *Phys. Lett.* **205B** (1988) 459; *Phys. Lett.* **208B** (1988) 209.

11. H. Georgi and M. Machacek, *Nucl. Phys.* **B262** (1985) 463.

12. R.S. Chivukula and H. Georgi, *Phys. Lett.* **182B** (1986) 181.

13. M.S. Chanowitz and M. Golden, *Phys. Lett.* **165B** (1985) 105.

14. J.F. Gunion, R. Vega, and J. Wudka, preprint UCD-89-13 (1989).

15. R. Vega and D. Dicus, preprint UCD-89-9 (1989).

16. J.F. Gunion and R. Vega, work in progress, see also ref. 15.

VECTOR BOSON SCATTERING AT FUTURE COLLIDERS

Roberto Vega

Department of Physics
University of California at Davis
Davis, CA 95616

ABSTRACT

Vector boson scattering of like-charged W and W-Z pairs in pp collisions is discussed. Results are presented for the standard model and for extended models which allow for unsuppressed couplings of the type W^-ZH^+ and $W^+W^+H^{--}$. The W^+W^+ and W-Z fusion processes are found to be of prime importance in the search for new physics at the SSC and the Eloisatron.

INTRODUCTION

The scattering of vector bosons at high energy pp colliders plays an important role in the study of spontaneous breaking of the electro-weak symmetry (SSB). It is well that if the if the Higgs mass is large ($m_H \gtrsim 1$ TeV), or if the symmetry breaking sector is strong, vector boson scattering processes will provide the main tools in our attempts to understand SSB in the standard model[1]. With the advent of next generation high energy colliders capable of producing sufficient rates for these processes, their use in the study of SSB in gauge theories is close at hand. For this reason it has become increasingly urgent to understand in detail the standard model predictions for the scattering of vector bosons in pp collisions at planned future colliders[2]. Any deviations from these predictions will point to new physics.

At the cornerstone of non-abelian gauge theories, such as the standard model, is the appearance of self interacting gauge bosons. Since the W-W scattering processes involve these self interactions, they also provide the means of determining the nature of these vertices and verifying the high energy gauge cancellations which should occur. The study of how these gauge cancellations occur is in essence an indirect way of studying the mechanism of spontaneous symmetry breaking.

However, gauge boson scattering processes will be difficult to measure experimentally. Cross sections for these scattering processes are of order α^4. In general, they are swamped by lower order QCD backgrounds[2]. Stringent cuts imposed to reduce the QCD backgrounds also reduce the W scattering signal considerably. In the case of W^+W^- scattering it is not completely clear whether it will be possible to extract the scattering signal from the overwhelming backgrounds present at high luminosity colliders such as the SSC. It is important, to identify and study in detail those W scattering processes which have the highest rates and least backgrounds, and to determine the feasibility of measuring them at future colliders.

---

[1] This is one of the motivations for building a high luminosity, high energy pp collider.

[2] For an overview of this active area of research see ref. [1] and references therein.

Higgs Particle(s)
Edited by A. Ali
Plenum Press, New York, 1990

In this paper two possibly favorable scattering processes are considered, the $W^+W^+ \to W^+W^+$ and $WZ \to WZ$ scattering. Though these processes are not of primary importance in the search for the standard neutral Higgs, they may provide the most suitable means for studying gauge boson self interactions, and high energy gauge cancellations.

The W^+W^+ fusion channel is particularly attractive because the possible backgrounds are quite limited, the main background coming from the mixed strong-electro-weak gluon exchange process shown in fig. 1(a). This background turns out to be manageable without the stringent cuts required in other fusion processes.

It will be shown that the rate for like-charge vector pair production at the SSC is very promising. At the Eloisatron the outlook is even brighter, like-charge W fusion will become comparable to the rate of W^+W^- production. These facts, coupled with the limited backgrounds, put this vector fusion channel at the center stage in the search for departures from the standard model physics at future colliders. Furthermore, it will be argued that, if the W polarizations can be determined, this fusion channel will provide a means of differentiating between a weak and strong symmetry breaking sector[4].

The W-Z fusion channel, in turn, is interesting because its final state signature may also be easier to extract from the background. One would trigger on two nearly collinear charged leptons of opposite charge, and two jets with an invariant mass of order m_W. For this process, however, it is also necessary to use the same "quark tagging" techniques suggested in the search for Higgs signals in W^+W^- fusion channels[3]. Similarly to the W^+W^- fusion case, the QCD background to $WZ \to \bar{l}l + jj$ is very large and "quark tagging" is needed to reduce it. Nevertheless the rates for WZ scattering still turn out to be non-negligible, even after applying several types of signal enhancing cuts.

The W^+W^+ and W-Z fusion channels are also of interest because they are involved when considering the production and decay of a heavy, singly or doubly charged Higgs. These fusion channels provide the means of studying couplings of the type W^+ZH^- and $H^{++}W^-W^-$. These couplings may appear, unsuppressed, in extensions of the standard model which contain Higgs representations higher than a doublet representation[5,6]. If the charged Higgs are heavy, that is if $m_H > 2m_W$, then these couplings may dominate their interactions. The details and motivations for these models will not be discussed here[3]. This paper only attempts to answer the question, "what if" such couplings exist and are not suppressed. In particular, the prospects for detection of a heavy ($M_H > 2m_W$) doubly or singly charged scalar are discussed. This discussion must necessarily be within the context of the W^+W^+ and W-Z fusion processes.

Below the process $W^+W^+ \to W^+W^+$ is considered first, followed by a discussion of the process $W^+Z \to W^+Z$.

W-W FUSION AND H^{++} DECAY[4]

The Feynman diagrams for the process $W^+W^+ \to W^+W^+$ are depicted in Fig. 1. In addition to the graphs present in the standard model, charged Higgs 's' and 't' channel graphs have been added. These additional Higgs graphs come in to play when considering extended models which contain higher scalar representations of $SU(2) \times U(1)$ (e.g. triplet representations). The calculation is carried out with and without the additional Higgs graphs. For definiteness, in the extended model case, the results are obtained within the framework of a specific model suggested by Georgi and Machacek[5]. In this model the strength of the couplings $W^+W^+H^{--}$ and W^+ZH^- are parameterized by $sin\theta_H$, where $sin\theta_H = 0$ would correspond to the standard model. One can think of this parameter as a measure of the extent to which triplet scalar representations participate in the spontaneous symmetry breaking of the $SU(2) \times U_Y(1)$ symmetry. In the Georgi-Machacel model the scalar masses and $sin\theta_H$ remain unconstrained.

---

[3]For detail treatment of these models see Refs. [5,7,8,9].

[4]For more details on this fusion process see ref. [8].

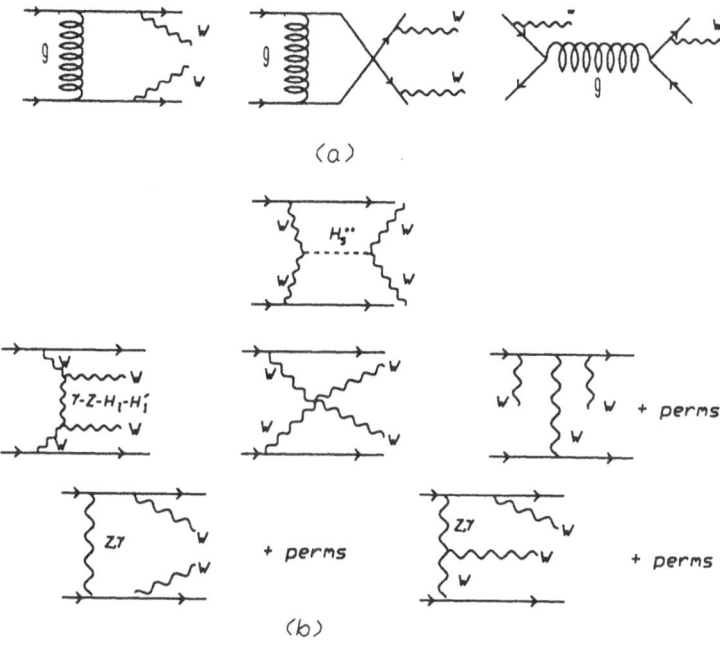

Figure 1. Feynman diagrams for like-charge production: (a) via gluon exchange and (b) via W fusion.

The details of the calculation will not be discussed here, the techniques used were developed in ref. [10]. It is important to stress, however, that in order to obtain correct results one has to carefully include the complete gauge invariant set of diagrams. It was checked that the diagrams shown fig. 1(b) do indeed form a complete set. The diagrams for the gluon exchange process (fig. 1(a)) form an independent gauge invariant set and do not interfere with the fusion process diagrams. They can be considered separately and were discussed previously in Ref. [11].

Results for the W^+W^+ process are displayed in Figs. 2-5. Results are shown for various combinations of the unconstrained parameters $\sin\theta_H$ and M_H, where all the scalars are taken to have equal mass, m_H. These results include rapidity cuts (Y_0) on the outgoing W's in the hadron CM frame. The EHLQ[1] (set 2) quark distribution functions were used throughout.

In Fig. 2 is shown the invariant mass distribution for the process

$$pp \rightarrow W^+W^+ + X$$

for various values of $\sin\theta_H$. The SM result corresponds to $\sin\theta_H = 0$. For comparison the contributions for the gluon exchange diagrams of fig. 1(a) are also shown in fig. 2. Notice that even for a mild rapidity cut of 2.5 the gluon exchange background is considerably below the fusion signal. The situation improves when a rapidity cut of 1.0 is imposed[8].

A doubly charged Higgs would produce a detectable signature if its couplings to the gauge bosons are not suppressed. For example, for $\sin\theta_H = .5$, the peak in the invariant mass is about three times bigger than the standard model signal. This can be seen in fig. 2 were it is noticed that the width of the doubly charged Higgs decreases with $\sin\theta_H$, but the height of the peak is practically independent of $sin\theta_H$. The peak becomes more defined, while the area below the peak decreases as $\sin^2\theta_H$ decreases, i.e. the cross section due to just the s-channel Higgs decreases as $\sin^2\theta_H$.

The total cross section as a function of m_H for various values of $\sin\theta_H$ is presented in fig. 3. For comparison, the contributions coming from only the H^{++} s-channel diagram is also shown. For SSC design parameters each picobarn corresponds to about 334 events per year where the W's decay into either $e\text{-}\bar{\nu}_e$ or $\mu\text{-}\bar{\nu}_\mu$.

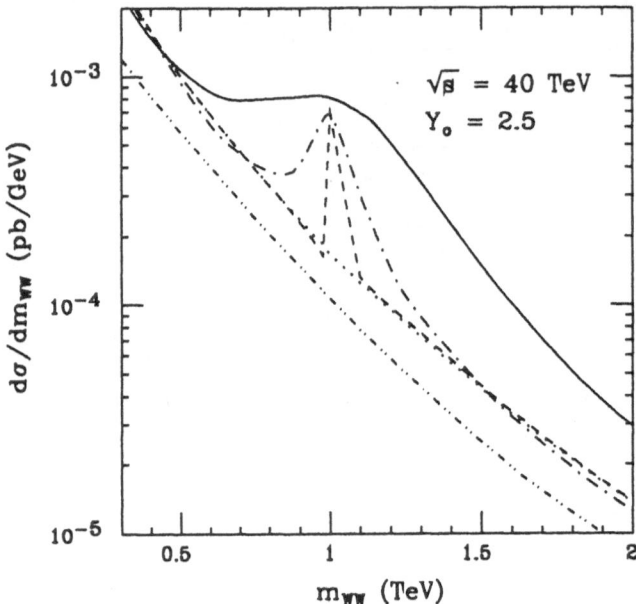

Figure 2. The invariant mass distribution for $pp \rightarrow W^+W^+ + X$ for $M_H = 1$ TeV. Y_0 is the rapidity cut, while $\sqrt{s}$ is the CM energy. The solid, dash-dot, dashed and dotted lines correspond to $\sin\theta_H = 1, 0.5, 0.1$ and 0 respectively. The dash-dot-dot line corresponds to like-charge W pair production via gluon exchange.

In the case of the extended model it is noted that the signal from just the H^{++} s-channel graph represents less than 5% of the total signal. Most of the like-charge W-pairs come from the SM W-W fusion mechanism. However, if one compares the total cross sections around the Higgs peak only, then most of the like-charge W's will come from the s-channel diagram. For example, for $M_H = 1$ TeV and $\sin\theta_H = 1$ one finds that the s-channel Higgs diagram contributes about 70% of the total of like-charge W pairs in a bin 450 GeV wide centered around the peak. This percentage decreases as $\sin^2\theta_H$ and increases with decreasing $M_{H^{++}}$.

For Eloisatron energies the situation improves dramatically, for several reasons. First the signal (W-fusion) to background (gluon exchange) ratio increases by a factor of about three from that of the SSC. When increasing the CM energy from 40 TeV to 200 TeV the fusion processes are enhanced over the gluon exchange processes, at least in the rapidity intervals considered here. This can be seen by comparing Fig. 4 with the appropriate lines in Fig. 2. For SSC energies the rate of like-charge pair production, from just W^+W^+ fusion, is about 220 W^+W^+-pair events per year whereas for the Eloisatron the rate is about 2000 W^+W^+-pair events per year. In essence, with moderate rapidity cuts almost all of the like charge W pairs will come from the W-fusion process, regardless of whether there is a Higgs pole or not. If there is a Higgs pole the signal to noise ratios will be even larger. These results compare favorably with those obtained for the W^+W^- fusion channel.

As mentioned in the introduction, at the Eloisatron, the production rate for like-charge W pairs is comparable with the production rate for ZZ and W^+W^- pairs, this can be seen by comparing, for example, Fig. 4 with the results of reference [12]. Furthermore, the background in the W^+W^- case is overwhelming[3] and stringent cuts have to be applied to reduce it. These cuts reduce the signal of interest considerably. It should be clear from this that the W^+W^+ fusion process stands out in clear advantage over other fusion processes in the study of the

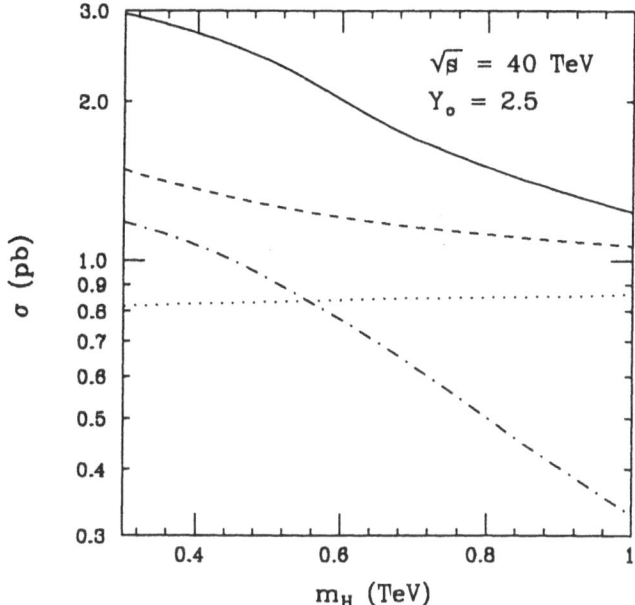

Figure 3. The total cross section for $pp \to W^+W^+ + X$ as a function of $M_{H^{++}}$. The solid, dashed and dotted lines correspond to $\sin\theta_H$ values of 1,0.5 and 0.0 respectively. The dashed-dot line is the contribution from just the s-channel diagram in Fig. 1b with $\sin\theta_H = 1$.

three gauge boson vertex and the gauge cancellations at high energies[5].

Polarization measurements for this process would also yield valuable information. The signal for a doubly charged Higgs is considerably enhanced over the background if polarization measurements of the final W's were made. This is clear from looking at Fig. 5. The same was found to be true for a heavy neutral Higgs in W^+W^- fusion[14,10]. However, in the present case, polarization measurements yield valuable information even in the absence of a Higgs signal.

In the absence of a heavy doubly charged Higgs, the like-charge W fusion process produces predominantly transversely polarized like-charge W pairs (fig. 5). This information is very useful because a strongly interacting symmetry sector will yield predominantly longitudinally polarized like-charge W pairs [4]. Thus, using polarization measurements for the W^+W^+ fusion process one would be able to distinguish between a weakly and a strongly interacting symmetry breaking sector[15] at future colliders such as the Eloisatron. Off course this presumes that the polarization of the W pairs could be measured, but with the high rates expected at the Eloisatron polarization measurements seem very plausible[16].

W-Z FUSION AND H^+ DECAY[6]

The analysis of the W-Z fusion process turns out to be quite a bit more complicated than for like-charge fusion case. The total number of gauge invariant Feynman diagrams for this process is over sixty two (depending on how you count them) for the case in which there are two initial u-quarks. The number and type of diagrams changing with the flavors of the incoming quarks. Fig. 6 shows only a representative set of the kind of diagrams that have to be included when there are two initial u-quarks. The complete set was used in carrying

---

[5] For further discussions on these points see Ref. [13].

[6] Most of the results presented in this section are from Ref. [17]

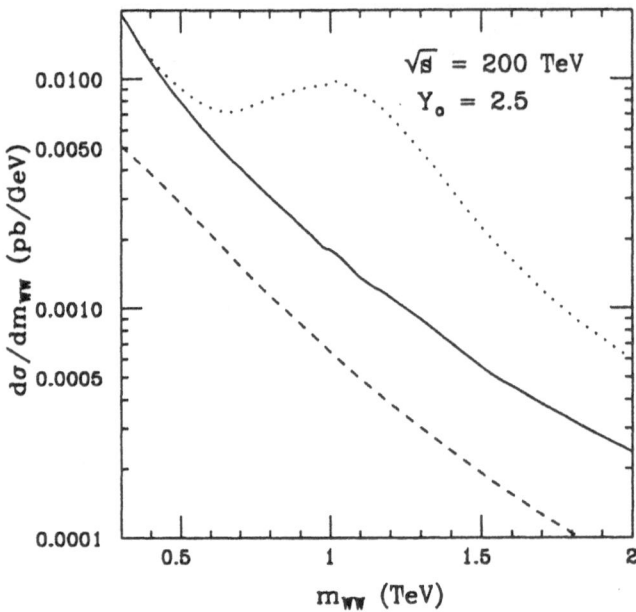

Figure 4. W^+ pair production at the Eloisatron. The dash line is the contribution from the gluon exchange graphs, the solid lines is for pair production in the standard model without a doubly charged scalar, the dotted line corresponds to the case where there is a doubly charged scalar with $M_H = 1$ TeV.

out the calculation. Again it is essential to have a complete set in order to preserve gauge invariance.

A further complication in this fusion channel is the appearance of the "photon exchange" diagrams. Particularly troublesome are those diagrams that have a photon connected to an on-shell quark line. For vanishing quark masses these diagrams diverge if the momenta of the incoming and outgoing quark are parallel.

For the purpose of this work this mass singularity issue is avoided by tagging the final quark lines. The "quark tagging"' is accomplished by requiring the rapidities of the final quarks to be between 3 and 5 and by imposing a cut ($P_T > 10$ GeV) on the transverse momenta of the final quarks. These cuts are sufficient to avoid the "mass singularity".

The selection of these cuts is guided by more practical reasons. The cross sections for W fusion processes decrease rapidly with energy and so it is important to find detectable final states with the largest possible branching ratios. The most promising channel for the W-Z fusion case seems to be $WZ \to \bar{l}ljj$, which has a branching ratio of about 10%. However, the QCD backgrounds from $Z + jj$ production, where the two jets fake a W can be overwhelming. Just as in the case of W^+W^- fusion[3] this background may be overcome by "quark tagging". Without it the $WZ \to WZ$ signal will be swamped by the QCD background which is roughly two orders of magnitude bigger. There is also a contribution to the background from $qq \to WZ$. This background is also greatly reduced using "quark tagging". For W fusion processes the bulk of the "tagging" jets will be in the rapidity interval between 3 and 5[3], hence the cuts are designed to reduce the backgrounds without excessivly deteriorating the signal.

An important question is how the above cuts affect a charged Higgs signal. It turns out that the above cuts will reduce the Higgs signal by a factor of about five from the case in which there are no cuts. Nevertheless, the signal is still quite sizeable even after the cuts are imposed.

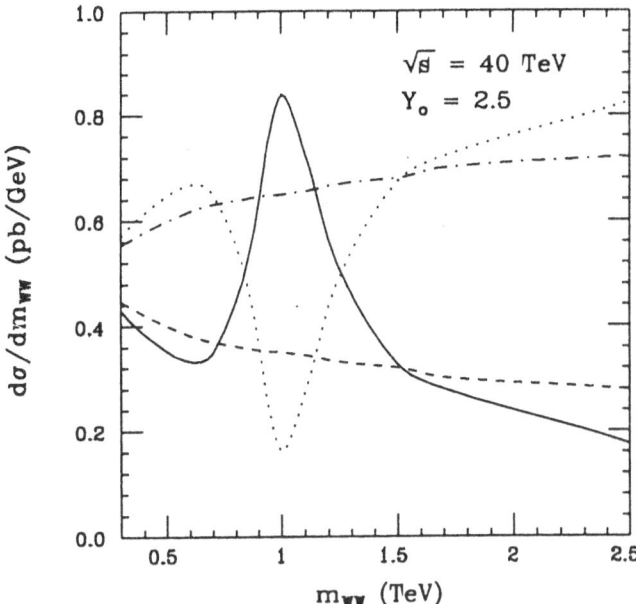

Figure 5. Fraction of longitudinal and transverse like-charge W-pairs for $M_H = 1.$ TeV. The solid and dotted lines correspond to $\sin\theta_H = 0.5$, while the dashed and dash-dot lines correspond to the $\sin\theta_H = 0.0$ case.

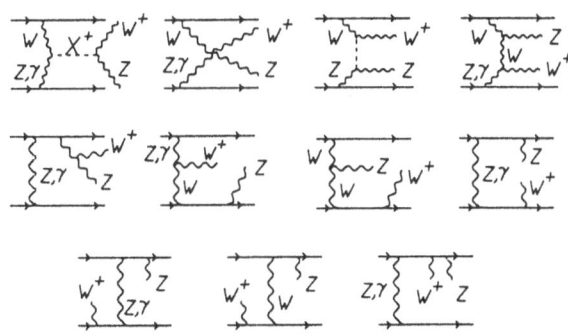

Figure 6. Representative set of feynman diagrams for the process $WZ \rightarrow WZ$ in pp collisions

The results for the W-Z fusion process are shown in Fig. 7. The cuts used are somewhat arbitrary, a more complete monte carlo study, that takes into account detector characteristics, is needed before determining what would be the most appropriate cuts. However, the results presented here will give an idea on what is to be expected. With that in mind one can see from Fig. 7 that even after imposing all the cuts, the rates for this fusion channel are better than one would expect.

The prospects for this fusion channel are not bad. Concentrating on the channel in which W decays hadronically and the Z decays to lepton pairs, the standard model W^+Z fusion process will yield around 80 events per year at the SSC[7]. If the W^-Z channel is taken into account than this number goes up to about 136 events per year[17]. By optimizing the cuts these numbers could be improved.

---

[7]This is for a bin, in the invariant mass distribution, from 500 to 2000 GeV.

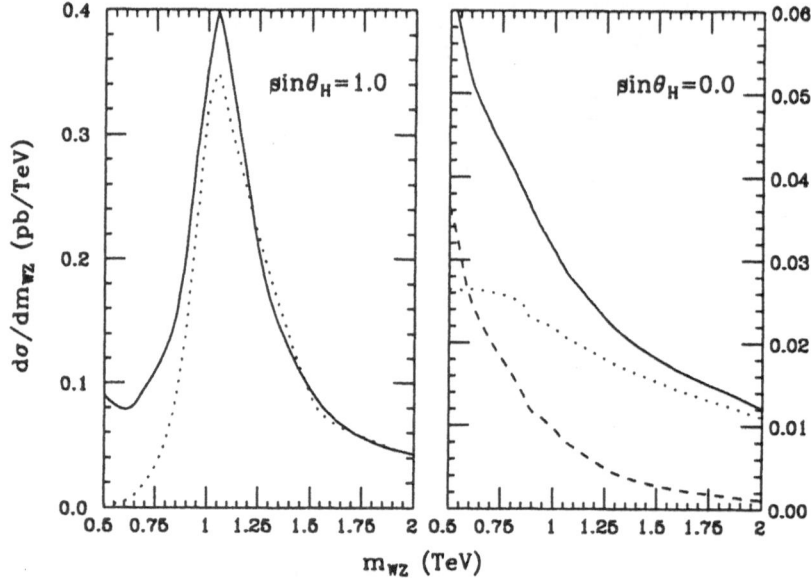

Figure 7. Production of WZ pairs at the SSC ($\sqrt{s} = 40$ TeV). The results presented here were obtained using rapidity cuts of $3 < |Y_q| < 5$ and a transverse momenta cut of $P_T > 10$ GeV on the final state quarks. In addition, a rapidity cut $Y_o < 2.5$ was imposed on each of the final W-Z gauge bosons. In the $\sin\theta_H = 1$ case, the solid line is the total sum over all final polarization states, the dotted line is the contribution from only the Higgs s-channel graph. In the $\sin\theta_H = 0$ case, the dotted (dashed) line is the contribution from the longitudinal (transverse) polarization modes of the final state gauge bosons, the solid line is the total sum over all final state polarizations.

If there exists an H^+ with unsuppressed couplings to the gauge bosons, then using the same final state as above, the number of events in a 300 GeV wide bin centered around the peak would be about 60 per SSC year. In the same bin the standard model ($\sin\theta_H = 0$) would yield only about 4 events per SSC year. This difference may be sufficient to see a charged Higgs "bump" for Higgs with masses at least up to 1 TeV.

Though results for WZ fusion at the Eloisatron are not yet available, rates are expected to go up by a factor of about ten[17].

CONCLUSION

It can be concluded that the signal for like charged W-pairs should be large at planned future colliders. The contributions to like-charged W pair production from both the W-W fusion mechanism and the gluon exchange mechanism are found to be quite sizeable. For example, from W-W fusion alone, using SSC design parameters, there would be around[8] 267 "like-sign" leptonic events per year where the W's both decay into either e-ν_e or μ-ν_μ. For a machine such as the Eloisatron the situation is found to improve quite drastically. The rates go up by a factor of about ten and the signal to noise ratio improves by a factor of about three over the SSC. It is also found that the like charged W pairs produced via standard

---

[8]The results presented here only include the contributions from like-sign positive lepton pairs. The situation improves somewhat when the negative like-sign lepton events are added in. Compared to W^+W^+ channel the W^-W^- contributions are smaller by a factor of about three[7].

model mechanisms are predominantly transversely polarized. Polarization measurements of the like-charge W pairs would be able to distinguish whether the $SU(2) \times U(1)$ symmetry breaking is weak or strong.

Although the results for H^{++} and H^+ production presented here were obtained within the context of a specific model[4] the general features should remain valid for any theory which has unsuppressed couplings between doubly charged scalars and the W vector bosons.

The signal for a doubly charged Higgs is found to rise clearly above the background. From the s-channel diagram alone, for a Higgs mass of 1 TeV, there will be tens and perhaps hundreds of "like-sign" leptonic events per year where the W's from the H^{++} both decay into either e-ν_e or μ-ν_μ. The exact number depends on the strength of the coupling. For a fixed coupling the number of these events increases with decreasing Higgs mass. Furthermore, if polarization measurements were possible then the signal for a doubly charged heavy Higgs could be considerably enhanced over the background and discovery is possible for a wide range $(0.2 < M_H < 1$ TeV$)$ of Higgs masses.

Thus the W^+W^+ fusion channel provides a useful tool for the study of spontaneous symmetry breaking, the three gauge boson vertex, and gauge cancellations at high energies at planned future colliders.

Much the same conclusions can be reached for the W-Z fusion channel. Event rates at the SSC are plentiful even after rapidity cuts for "quark tagging" and rapidity cuts on the final W-Z pairs.

References

[1] J.F. Gunion, H.E. Haber, G.L. Kane and S. Dawson, "The Higgs Hunters Guide", to be published by Addison Wesley.

[2] R. Kleiss and W.J. Stirling, Phys. Lett. **B200** (1988) 193.

[3] M.S. Chanowitz and M. Golden, Phys. Rev. Lett. **61** (1988) 1053.

[4] H. Georgi and M. Machacek, Nucl.Phys. **B262** (1985) 463.

[5] J.A. Grifols and A. Mendez, Phys. Rev., **D22**, (1980), 1275. A.A. Iogansen, N.G. Ural'tesev and V.A. Khoze, Sov. J. of Nucl. Phys., **46(5)**, (1987), 890.

[6] M.S. Chanowitz and M. Golden, Phys. Lett., **165B** (1985) 105.

[7] R.Vega and D.A. Dicus, to be published in Nucl.Phys. B.

[8] J.F. Gunion, R. Vega and J. Wudka, UCD–89-13 (1989).

[9] D.A. Dicus and R. Vega, Phys.Rev.Lett., **57** (1986) 1110; D.A. Dicus and R. Vega, Phys.Rev., **D37** (1988) 2474.

[10] D.A. Dicus and R. Vega, Phys. Lett. B, **217** (1989) 194.

[11] E. Eichten, I. Hinchliffe, K. Lane, and C. Quigg, Rev. of Mod. Phy.,**56** (1984) 579.

[12] J.F. Gunion and R. Vega, INFN Eloisatron, REPORT, CCSEM-EL-88-1, P. 42.

[13] J.F. Gunion, Probing the Higgs Bosons, these proceedings.

[14] M.J. Duncan, G.L. Kane and W.W. Repko, Nucl. Phys., **B272** (1986) 517.

[15] M.S. Chanowitz, On strong electroweak symmetry breaking, these proceedings.

[16] R. Vega, in preparation.

[17] R. Vega and J.F. Gunion, in preparation.

[18] M.S. Chanowitz and M.K. Gaillard, Nucl. Phys. **B261** (1985) 379.

[19] A. Abbasabadi, W.W. Repko, D.A. Dicus and R. Vega, Phys.Lett., **B213** (1988) 386.

TESTING THE HIGGS SECTOR AND THE THREE VECTOR BOSON COUPLING WITH ELECTROWEAK BOSON PAIRS

U. Baur

Physics Department, University of Wisconsin
Madison, WI 53706, USA

ABSTRACT

Electroweak vector boson pairs can be used to test the Higgs sector and the vector boson self-interactions of the Standard Model. They also constitute a background to new physics signals at future colliders. This is illustrated by vector boson pair production at large transverse momentum in hadronic collisions which may be a background which is relevant for Higgs tagging. The usefulness of electroweak boson pairs to probe the structure of the three vector boson vertex is demonstrated by $W\gamma$ production at the Tevatron. For an integrated luminosity of 100 pb^{-1} the $WW\gamma$ vertex can be measured with 25-40% accuracy in $p\bar{p}$ collisons at $\sqrt{s} = 1.8$ TeV.

1. INTRODUCTION

Although many aspects of the standard electroweak gauge theory (SM) have now been verified, crucial parts of the model are not yet tested experimentally. Most notably the Higgs sector and the vector-boson self-interactions which are uniquely given by the gauge theory structure of the SM still await experimental confirmation. Vector boson pairs play an important role in testing both sectors. In the following I shall discuss two examples which illustrate this.

Since the Higgs sector is responsible for the spontaneous symmetry breaking in the SM, the search for the Higgs particle is one of the most important tasks for future colliders. If its mass M_H is below 80 GeV the Higgs boson can be detected via the process $e^+e^- \rightarrow ZH$ at LEP200 [1]. For larger values of M_H the Higgs must be looked for at the LHC (pp collisions at $\sqrt{s} = 16$ TeV) or SSC (pp collisions at $\sqrt{s} = 40$ TeV). Studying the production of electroweak vector boson pairs is the most promising way to search for the Higgs boson at these colliders. If M_H is greater than twice the W-mass M_W, the Higgs will decay almost exclusively into a W or Z pair [2]. For $M_H < 2M_W$ it will mainly decay into $q\bar{q}$ states where the QCD background is overwhelming. However in this case one can consider rare decay modes of the Higgs boson proceeding via quark loops like $H \rightarrow \gamma\gamma$ or $H \rightarrow Z\gamma$ [3].

In all these cases the continuum production of vector boson pairs is an important background. In the first approximation vector boson pairs are produced via $q\bar{q}$ annihilation. For a more complete knowledge, higher order QCD corrections involving real or virtual gluons have to be considered. Virtual and soft real gluon corrections mainly contribute to the so-called K-factor which for some vector boson pair production processes has been calculated in ref. [4].

Higgs Particle(s)
Edited by A. Ali
Plenum Press, New York, 1990

Hard QCD corrections on the other hand lead to the topologically distinct signature of a vector boson pair accompanied by one or more jets with large transverse momentum. Processes such as $q\bar{q} \rightarrow ZZg$ or $qg \rightarrow ZZq$ will therefore be a potentially important background *e.g.* for Higgs production at large p_T [5,6]. The transverse momentum spectrum of the vector boson pair can also be considered as a test of perturbative QCD.

In the first part of my talk I discuss the production of vector boson pairs accompanied by a large p_T jet [7]. The calculation of the matrix elements, including the decay of W- and Z-bosons into massless fermion-antifermion final states, is briefly described. We then consider the situation when the jet is emitted in the forward direction while the boson pair is produced centrally. In this configuration vector boson pair plus jet production may be a background which is relevant for Higgs tagging [5,8].

In the second part I study the capability of future experiments at the Tevatron to probe the $WW\gamma$ vertex via $W\gamma$ production [9]. In the past many authors have considered this process [10-16] usually with emphasis on the anomalous magnetic moment of the W as a non gauge theory contribution. Here we go a step further and use the most general $WW\gamma$ coupling which is accessible in the annihilation processes $q\bar{q}' \rightarrow W\gamma$ of effectively massless quarks. Four different anomalous couplings are allowed by electromagnetic gauge invariance and Lorentz invariance [17,18], and we briefly review their properties. Apart from anomalies in the $WW\gamma$ vertex we assume the SM to be valid. In particular we assume the coupling of W and Z bosons to quarks and leptons to be given by the SM.

Our analysis is based on the calculation of helicity amplitudes for the complete process

$$q\bar{q}' \rightarrow W^{\pm}\gamma , \qquad W^{\pm} \rightarrow e^{\pm}\nu . \tag{1.1}$$

If finite W width effects are included, (1.1) and $q\bar{q}' \rightarrow W^{\pm} \rightarrow e^{\pm}\nu\gamma$ are described by the same gauge invariant set of Feynman diagrams and, in principle, these processes can no longer be distinguished. By imposing suitable kinematic cuts it is possible, however, to isolate regions in phase space where the major part of the cross section results from W-photon production. We describe these cuts in detail and discuss also the QCD corrections [19] to (1.1) and the background from W jet production, with the jet misidentified as a photon. We then discuss the signatures for anomalous $WW\gamma$ couplings. Constraints on anomalous contributions to the $WW\gamma$ vertex exist already, derived either from low energy experiments [20-27] or from unitarity considerations [21,28]. To avoid confusion we shall first discuss the expected signal without taking into account the low energy bounds. Only afterwards we compare the sensitivity to anomalous $WW\gamma$ couplings expected at the Tevatron with the low energy bounds and with the sensitivity expected from $ep \rightarrow eW^{\pm}X$ at HERA and $e^+e^- \rightarrow W^+W^-$ at LEP200. Our conclusions are given in Section 4.

2. PRODUCTION OF ELECTROWEAK BOSON PAIRS AT LARGE TRANSVERSE MOMENTUM IN HADRONIC COLLISIONS AND HIGGS TAGGING

2.1 Matrix Elements

In order to compute the production cross section for electroweak vector boson pairs plus a jet at hadron colliders, the matrix elements of the process

$$q\bar{q} \rightarrow V_1 V_2 g \tag{2.1}$$

and the crossed processes

$$qg \rightarrow V_1 V_2 q , \quad \bar{q}g \rightarrow V_1 V_2 \bar{q} \tag{2.2}$$

have to be calculated. Here $V_1 V_2$ generically stands for all possible vector boson pair combinations

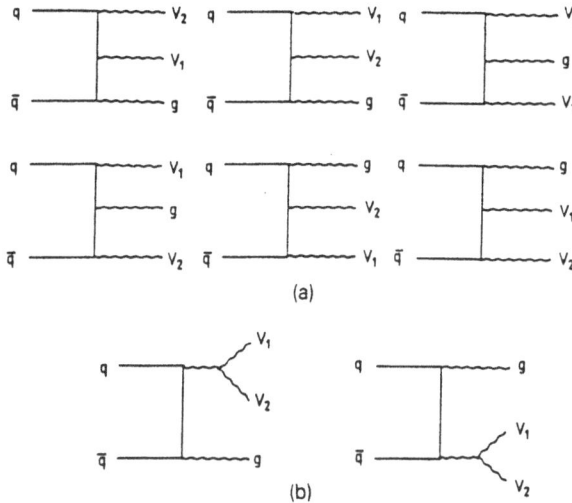

Fig. 1 . Feynman graphs contributing to $q\bar{q} \to V_1 V_2 g$, $qg \to V_1 V_2 q$ and $\bar{q}g \to V_1 V_2 \bar{q}$. The diagrams given in b) contribute only to processes with a $W^{\pm}$ in the final state. V_1 and V_2 generically denote the electroweak vector bosons.

$$V_1 V_2 = \gamma\gamma, \ Z\gamma, \ ZZ, \ W^{\pm}\gamma, \ W^+W^-, \ W^{\pm}Z \ . \tag{2.3}$$

The Feynman diagrams which contribute to (2.1) and (2.2) are shown in Fig. 1. Since the incoming quarks are effectively massless, the spinor technique of ref. [29] may be used to calculate the helicity amplitudes for (2.1) and (2.2). In this method massive vector bosons decay automatically into massless fermion-antifermion pairs so that one obtains the matrix elements for the full process including the decay of the W- and Z-bosons at no extra cost. Thus they are particularly convenient when properties of the final state particles are studied. Moreover, since amplitudes are evaluated directly the resulting expressions are, especially for complicated Feynman diagrams, much simpler than with the conventional method where the squared matrix element is calculated. The spinor technique of ref. [29] is of course not the only one that can be used to evaluate the helicity amplitudes for $V_1 V_2 j$ production. Another frequently used method is the one developed by Hagiwara and Zeppenfeld [30] which is particularly convenient for massive fermions.

The spin and color averaged matrix element square for (2.1) and (2.2) is given by

$$|\overline{\mathcal{M}}|^2 = \frac{1}{4} C \mathcal{G}^2 \sum_{\lambda_1 ... \lambda_{V_2} = \pm} |\mathcal{M}(\lambda_1, \lambda_3, \lambda_{V_1}, \lambda_{V_2})|^2 \ . \tag{2.4}$$

Here the factor 1/4 arises from initial state spin averaging,

$$C = \begin{cases} \frac{4}{9} & \text{for } q\bar{q} \text{ annihilation} \\ \frac{1}{6} & \text{for } qg \ (\bar{q}g) \text{ fusion} \end{cases} \tag{2.5}$$

is the color averaging factor and $\mathcal{G}$ is an overall coupling constant factor which depends on the process under consideration. Finally, $\mathcal{M}(\lambda_1, \lambda_3, \lambda_{V_1}, \lambda_{V_2})$ is the helicity amplitude, where λ_1 and λ_3 denote the helicity of the incoming quark and of the gluon, respectively. $\lambda_{V_1}(\lambda_{V_2})$ either stands for the helicity of $V_1(V_2)$ if it is a photon or for the helicity of one of the decay fermions if $V_1(V_2) = W^{\pm}, Z$. Since we only deal with massless fermions the helicity of any given antifermion is determined by the helicity of its fermion partner, e.g. $\lambda_1 = \lambda_2$ and need not to be specified in $\mathcal{M}$.

The calculation is simplified considerably by breaking up the helicity amplitude into a product of helicity coupling factors and subamplitudes,

$$\mathcal{M}(\lambda_1, \lambda_3, \lambda_{V_1}, \lambda_{V_2}) = \sum_i \mathcal{C}_i(\lambda_1, \lambda_3, \lambda_{V_1}, \lambda_{V_2}) \, \mathcal{M}_i(\lambda_1, \lambda_3, \lambda_{V_1}, \lambda_{V_2}) \qquad (2.6)$$

where the sum is over the contributing Feynman diagrams. The helicity coupling factors, $\mathcal{C}_i(\lambda_1, \lambda_3, \lambda_{V_1}, \lambda_{V_2})$, are diagram and process dependent, however the subamplitudes, $\mathcal{M}_i(\lambda_1, \lambda_3, \lambda_{V_1}, \lambda_{V_2})$, are common to several processes. For example, the subamplitudes for $q\bar{q} \rightarrow W\gamma g$ corresponding to the diagrams of Fig. 1a are simply given by the $q\bar{q} \rightarrow Z\gamma g$ subamplitudes $\mathcal{M}_i$ for lefthanded fermions. Eqs. (2.4) and (2.6) display the general structure of the amplitudes for all $V_1 V_2 j$ production processes. Explicit expressions for the matrix elements $\mathcal{M}_i$ can be found in ref. [7].

2.2 Higgs Tagging and non-central $V_1 V_2 j$ Production

Recently it has been suggested that, using forward detectors covering the rapidity range $3 < |y| < 5$, it may be possible to detect a heavy SM Higgs boson produced via WW-fusion in the $H \rightarrow WW \rightarrow \ell\nu jj$ decay channel [5,8] ($\ell = e, \mu$) at the LHC or SSC by tagging the two W-radiating quarks emerging with $p_T \sim M_W$. Without quark tagging the $H \rightarrow WW$ signal would be swamped by the Wjj background [31].

Since the Higgs boson decays isotropically, the signal for $H \rightarrow WW \rightarrow \ell\nu jj$ consists of the $\ell\nu$ and a jet pair all in the central region, and of two tagging jets in the forward region. Similarly, the decay $H \rightarrow ZZ \rightarrow \ell^+\ell^- jj$ results in a pair of tagging jets, two centrally produced jets, and in a charged lepton pair. To avoid misidentification between central and tagging jets a rapidity gap has to be required between the two jet types which we have chosen to be 0.5 (between 2.5 for the central and 3.0 for the forward region). Roughly 30% of the time, however, one of the tagging jets happens to have a rapidity between 2.5 and 3.0, or bigger than 5.0 and thus is lost [8]. In this case, where only one of the tagging jets is observed, W^+W^-j, $W^{\pm}Zj$ and ZZj produced with the jet in the rapidity range $3 < |y_j| < 5$ and the boson pair decaying into two leptons plus two jets in the central region are, together with $Wjjj$ and $Zjjj$ production, potentially dangerous background processes for Higgs tagging. If the Higgs signal would be large one simply could afford to ignore the portion of the cross section where one of the tagging jets is lost. Since this is not the case one would like to use such events as well and it is important to know how large the background from $V_1 V_2 j$ production is.

In Figs. 2 and 3 we compare the invariant mass and the p_T distribution of the vector boson pair in $pp \rightarrow W^+W^-j$ and $pp \rightarrow ZZj$ for jets produced in the central and in the forward direction at $\sqrt{s} = 40$ TeV. The curves for $W^{\pm}Zj$ are almost degenerate with the W^+W^-j lines and are not displayed to avoid overburdening the figures. In all cases we applied a p_{Tj} cut of 100 GeV and a rapidity cut of $|y| < 2.5$ on the vector bosons. Furthermore we imposed a p_T cut of

$$p_{TVV} > \varepsilon \cdot m_{VV} \,, \qquad (2.7)$$

where ε is a parameter to be discussed below [32]. Without the ε cut the differential cross sections for the lowest order process $pp \rightarrow ZZ$ and the first order process $pp \rightarrow ZZj$ with the jet in the central region become comparable at large m_{ZZ} [7]. Since $d\sigma/dm_{ZZ}$ for a fixed value of m_{ZZ} is dominated by the low p_{TZZ} region, one

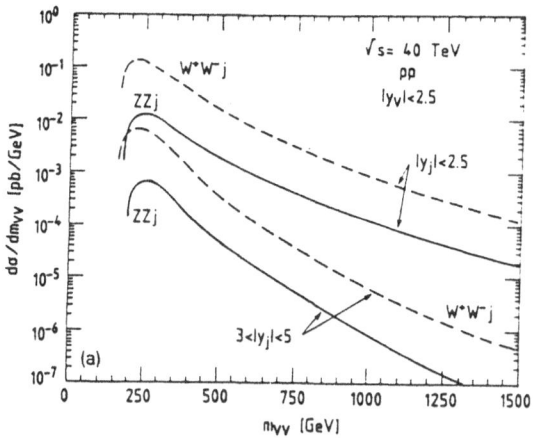

Fig. 2. Comparison of the invariant mass spectrum of the electroweak boson
pair for $|y_j| < 2.5$ and $3 < |y_j| < 5$ in $pp \to W^+W^-j$ (dashed line)
and $pp \to ZZj$ (solid line) at $\sqrt{s} = 40$ TeV. m_{VV} generically denotes
the invariant mass of the vector boson pair. In addition to eq. (2.7)
with $\varepsilon = 0.3$ a cut of 100 GeV is imposed on p_{TWW} and p_{TZZ}. W's
and Z's are required to have a rapidity $|y| < 2.5$.

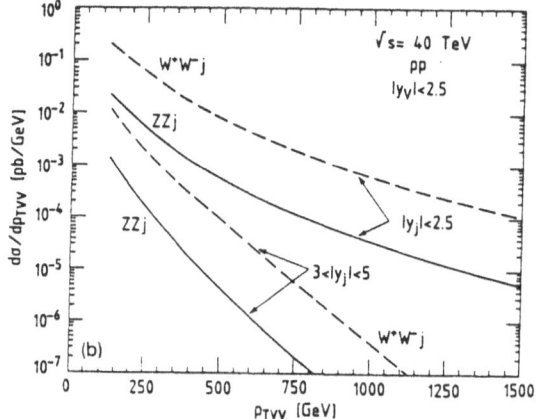

Fig. 3. Comparison of the p_T spectrum of the electroweak boson pair for $|y_j| <$
2.5 and $3 < |y_j| < 5$ in $pp \to W^+W^-j$ (dashed line) and $pp \to ZZj$
(solid line) at $\sqrt{s} = 40$ TeV. p_{TVV} generically denotes the transverse
momentum of the vector boson pair. In addition to eq. (2.7) with
$\varepsilon = 0.3$ a cut of 100 GeV is imposed on p_{TVV}. W's and Z's are
required to have a rapidity $|y| < 2.5$.

always enters a region where the main contribution to $d\sigma/dm_{ZZ}$ originates from the
soft region $p_{TZZ} \ll m_{ZZ}$, where multiple gluon processes become important, for
sufficiently large m_{ZZ}. Hence, for a fixed p_{TVV} cut the results for the invariant mass
distribution of the vector boson pair from first order processes become unreliable.
The precise value of m_{VV} where this happens depends of course on the value of the
p_T cut.

In the soft region terms of order

$$\frac{1}{p_{TVV}^2} \, \alpha_s^n(p_{TVV}^2) \, \ln^m\left(\frac{m_{VV}^2}{p_{TVV}^2}\right) \,, \qquad m \leq 2n - 1, \qquad (2.8)$$

become large. A cut of the form (2.7) with a sufficiently large value of ε will thus guarantee that these terms are under control for all powers of α_s. The cut (2.7) mainly suppresses the cross section at large m_{ZZ} and the result for $d\sigma/dm_{ZZ}$ from the first order processes is of $O(\alpha_s)$. The precise value of ε is of course somewhat arbitrary and could only be fixed if the full calculation including multiple gluon processes is done. In Figs. 2 and 3 we used $\varepsilon = 0.3$.

The SM parameters used in Fig. 2 and all subsequent figures are $\alpha = \alpha(M_W) = 1/128$, $M_W = 82$ GeV and $\sin^2 \theta_W = 0.23$. Our results remain unchanged if M_W is varied by a few GeV. For the parton distribution functions we use set 1 of Duke and Owens [33] with the scale Q^2 in Figs. 2 and 3 given by $\hat{s}/4$, where $\hat{s}$ is the parton center of mass energy squared.

It is obvious that the total cross section for the jet in forward direction is much smaller than for a centrally produced jet, typically a factor 20-30. While qg fusion dominates for jets in the central rapidity region, $q\bar{q}$ annihilation and qg fusion contribute about equally to the cross section when the jet rapidity is in the range $3 < |y_j| < 5$. The m_{VV} as well as the p_{TVV} spectrum drops for $3 < |y_j| < 5$ considerably faster than for $|y_j| < 2.5$.

A very crude estimate of the WWj, WZj and ZZj background in the region $p_{Tj} > 100$ GeV to Higgs tagging where one of the tagging jets is lost is obtained by integrating $d\sigma/dm_{VV}$ at $m_{VV} = M_H$ over twice the Higgs width Γ_H. It turns out that for the range of M_H which we have explored (300 GeV $\leq M_H \leq$ 700 GeV) signal and background are roughly comparable. However, since we have imposed (2.7) which suppresses $d\sigma/dm_{VV}$ for large values of m_{VV} we certainly underestimate the background which in reality may well be a factor 2 or 3 higher. Nevertheless our estimate demonstrates that the V_1V_2j background cannot be neglected in a detailed study of Higgs tagging where only one of the tagging jets is observed.

3. PROBING THE $WW\gamma$ VERTEX AT THE TEVATRON COLLIDER

3.1 The $WW\gamma$ Vertex

The CDF collaboration collected data corresponding to an integrated luminosity of about 4.7 pb^{-1} during the last run of the Tevatron $p\bar{p}$ collider [34], operating at $\sqrt{s} = 1.8$ TeV. The inclusive cross sections of W and Z boson production are about 21 nb and 6 nb at this energy. Thus a large sample of weak bosons has been obtained which makes it possible to study the properties of W and Z production in $p\bar{p}$ collisions in detail. In particular the production of weak bosons associated with one or more hadronic jets will provide further tests of perturbative QCD at energies beyond the reach of the CERN $p\bar{p}$ collider. In future runs an integrated luminosity of up to 100 pb^{-1} can be expected [35]. It will then be possible to investigate also electroweak corrections to weak boson production, e.g. $p\bar{p} \rightarrow W\gamma$.

$W\gamma$ production is particularly interesting theoretically since it depends on the $WW\gamma$ coupling which, so far, has not been tested experimentally. Within the Standard Model (SM), at tree level, the $WW\gamma$ vertex is completely fixed by the gauge theory structure of the model. The observation of the $WW\gamma$ coupling thus is a crucial test of the SM.

At the parton level, if finite W width effects are ignored, the reaction $p\bar{p} \rightarrow W^{\pm}\gamma$ proceeds via the Feynman graphs shown in Fig. 4a-4c. The $WW\gamma$ vertex in which we are interested enters via diagram 4c. The virtual and the onshell W both couple to

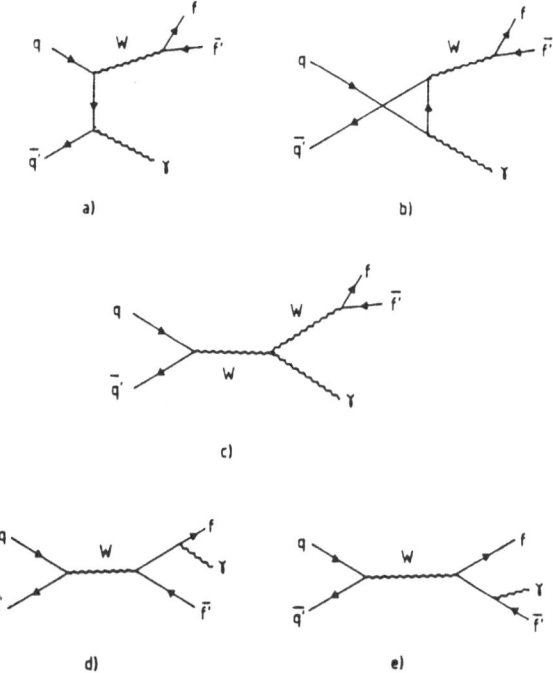

Fig. 4 . Feynman graphs for the parton level processes contributing to $p\bar{p} \rightarrow W\gamma;\ W \rightarrow f\bar{f}'$. If finite W width effects are taken into account, all diagrams have to be included in the calculation in order to preserve electromagnetic gauge invariance.

essentially massless fermions, which ensures that effectively $\partial_\mu W^\mu = 0$. This together with gauge invariance of the onshell photon restricts the tensor structure of the WW photon vertex sufficiently to allow just four free parameters, which are conveniently described by the effective Lagrangian [17,18,36]

$$\mathcal{L}_{WW\gamma} = -ie \left\{ \left(W^\dagger_{\mu\nu} W^\mu A^\nu - W^\dagger_\mu A_\nu W^{\mu\nu} \right) + \kappa W^\dagger_\mu W_\nu F^{\mu\nu} + \frac{\lambda}{M^2_W} W^\dagger_{\lambda\mu} W^\mu_{\ \nu} F^{\nu\lambda} \right.$$

$$\left. + \tilde{\kappa} W^\dagger_\mu W_\nu \tilde{F}^{\mu\nu} + \frac{\tilde{\lambda}}{M^2_W} W^\dagger_{\lambda\mu} W^\mu_{\ \nu} \tilde{F}^{\nu\lambda} \right\}. \tag{3.1}$$

Here A^μ and W^μ are the photon and W^- fields, respectively, $W_{\mu\nu} = \partial_\mu W_\nu - \partial_\nu W_\mu$, $F_{\mu\nu} = \partial_\mu A_\nu - \partial_\nu A_\mu$, and $\tilde{F}_{\mu\nu} = \frac{1}{2}\epsilon_{\mu\nu\rho\sigma} F^{\rho\sigma}$. e is the charge of the proton. and M_W represents the W boson mass.

The first term in eq. (3.1) arises from minimal coupling of the photon to the $W^\pm$ fields and is completely fixed by the charge of the W boson for onshell photons. While the κ and λ terms do not violate any discrete symmetries, the $\tilde{\kappa}$ and $\tilde{\lambda}$ terms are P odd and CP violating. Within the SM, at tree level,

$$\kappa = 1, \quad \lambda = 0,$$

$$\tilde{\kappa} = 0, \quad \tilde{\lambda} = 0. \tag{3.2}$$

375

The $\kappa(\tilde{\kappa})$ and $\lambda(\tilde{\lambda})$ terms are related to the magnetic (electric) dipole moment $\mu_W(d_W)$ and the electric (magnetic) quadrupole moment $Q_W(\tilde{Q}_W)$ of the W^+

$$\mu_W = \frac{e}{2M_W}(1 + \kappa + \lambda), \qquad (3.3a)$$

$$Q_W = -\frac{e}{M_W^2}(\kappa - \lambda), \qquad (3.3b)$$

$$d_W = \frac{e}{2M_W}(\tilde{\kappa} + \tilde{\lambda}), \qquad (3.3c)$$

$$\tilde{Q}_W = -\frac{e}{M_W^2}(\tilde{\kappa} - \tilde{\lambda}). \qquad (3.3d)$$

Tree level unitarity, e.g. for the process $e^+e^- \rightarrow W^+W^-$, uniquely restricts the $WW\gamma$ couplings to their (SM) gauge theory values at asymptotically high energies [37]. This implies that any deviation $a = \kappa - 1, \ldots, \tilde{\lambda}$ from the SM expectation has to be described by a form factor $a(q^2, \bar{q}^2, q_\gamma^2)$ which vanishes when one of the arguments, the square of the four momentum of one of the W bosons, q^2 or $\bar{q}^2$, or the square of the four momentum of the photon, q_γ^2, becomes large. For deviations of the three vector boson couplings from the gauge theory value, produced by some novel interactions operative at a scale Λ, one should expect that the form factors stay essentially constant for center of mass energies $\sqrt{\hat{s}} < \Lambda$ and start decreasing only when the scale Λ is reached or surpassed, very much like the well-known nucleon form factors. Present experimental data suggest that Λ is at least of the order of a few hundred GeV [38]. Since the energy region covered by the Tevatron is smaller than typically expected for Λ we may assume the form factors $a = \kappa - 1, \ldots, \tilde{\lambda}$ to be approximately constant in the following.

The effective Lagrangian (3.1) leads to cross section formulas for $q\bar{q}' \rightarrow W^\pm\gamma$ including the effect of anomalous $WW\gamma$ couplings. When the decay of the W into a massless fermion antifermion pair $f\bar{f}'$ and a finite width W-propagator are taken into account, all the Feynman graphs of Fig. 4 have to be included in the calculation in order to preserve electromagnetic gauge invariance. For our numerical simulations we have calculated the complete matrix elements corresponding to the diagrams of Fig. 4 by making use of the helicity techniques described in ref. [30].

When the $W \rightarrow f\bar{f}'$ decay and finite W-width effects are taken into account, $W\gamma$ production results in the same final state as radiative W decays, $q\bar{q}' \rightarrow W \rightarrow f\bar{f}'\gamma$, and both reactions are described by the same set of diagrams. As a result the two processes interfere and, in principle, can no longer be distinguished. Since anomalous $WW\gamma$ couplings influence different quantities in the two processes, it is necessary to isolate those regions in phase space where the dominant part of the cross section results from $W\gamma$ production. As we shall see in the next Section this can be achieved rather easily so that in practice it is legitimate to distinguish $p\bar{p} \rightarrow W\gamma; W \rightarrow f\bar{f}'$ from $p\bar{p} \rightarrow W \rightarrow f\bar{f}'\gamma$.

For our later phenomenological discussion we find it convenient to display the contributions of anomalous couplings to the $q\bar{q}' \rightarrow W\gamma$ helicity amplitudes, neglecting the W decay. The helicity of the incoming, effectively massless, quark (antiquark) is fixed to be $-\frac{1}{2}(+\frac{1}{2})$ by the $V - A$ structure of the $Wq\bar{q}'$ coupling. This means that the anomalous contributions to the $W\gamma$ production amplitudes depend only on the W and photon helicities, λ_W and λ_γ. Denoting these contributions by $\Delta\mathcal{M}_{\lambda_\gamma \lambda_W}$, one finds

$$\Delta \mathcal{M}_{\pm 0} = \frac{e^2}{\sin \theta_W} \frac{\sqrt{\hat{s}}}{2 M_W} \left[\kappa - 1 + \lambda \mp i(\tilde{\kappa} + \tilde{\lambda}) \right] \tfrac{1}{2} (1 \mp \cos \Theta)$$

$$\Delta \mathcal{M}_{\pm \pm} = \frac{e^2}{\sin \theta_W} \frac{1}{2} \left[\frac{\hat{s}}{M_W^2} \left(\lambda \mp i\tilde{\lambda} \right) + (\kappa - 1 \mp i\tilde{\kappa}) \right] \frac{1}{\sqrt{2}} \sin \Theta , \qquad (3.4)$$

where Θ denotes the scattering angle of the photon with respect to the quark direction, measured in the $W\gamma$ rest frame, and $\sqrt{\hat{s}}$ is the invariant mass of the W-photon system.

Since the structure of the $WW\gamma$ vertex enters $q\bar{q}' \to W\gamma$ via the s-channel exchange of a W boson (see Fig. 4c), only the four helicity combinations in eq. (3.4) are affected by anomalous couplings. The helicity combinations $(\lambda_\gamma, \lambda_W) = (+-)$ and $(-+)$ have the photon and W spins aligned along the photon momentum direction and hence have angular momentum $J \geq 2$: they cannot be reached by s-channel exchange of a vector boson. The fact that only the above four helicity combinations of the $J = 1$ partial wave can be reached by s-channel W exchange explains why four free parameters suffice to parametrize the effects of the most general $WW\gamma$ vertex in $W\gamma$ production.

A pronounced feature of $W\gamma$ production in $q\bar{q}'$ annihilation is the SM prediction of radiation zeros in all contributing helicity amplitudes at one value of the photon scattering angle Θ and hence in the differential cross section [11]. For $u\bar{d} \to W^+\gamma$ this radiation zero occurs at $\cos \Theta = -\frac{1}{3}$. In the presence of any anomalous contribution to the $WW\gamma$ vertex the radiation zero will be at least partially eliminated. This is obvious from eq. (3.4): none of the anomalous contributions to the scattering amplitudes vanishes at $\cos \Theta = -\frac{1}{3}$.

While the SM contribution to the $q\bar{q}' \to W\gamma$ scattering amplitudes is bounded from above for fixed scattering angle Θ, the anomalous contributions (3.4) rise without limit as $\hat{s}$ increases, eventually violating unitarity. This is the reason the anomalous couplings must show a form factor behavior at very high energies. Anomalous values of λ or $\tilde{\lambda}$ are enhanced by $\hat{s}/M_W^2$ in the amplitudes $\mathcal{M}_{\pm\pm}$, whereas terms containing κ and $\tilde{\kappa}$ mainly contribute to $\mathcal{M}_{\pm 0}$ and grow only with $\sqrt{\hat{s}}/M_W$. For large values of the $W\gamma$ invariant mass $\sqrt{\hat{s}}$, the amplitudes (3.4) will dominate the SM contributions and suffice to explain differential distributions of the photon and the W decay products.

3.2 Signal and Experimental Cuts

W-photon production leads to a final state consisting of a fermion antifermion pair and a photon when the W decay is taken into account. The hadronic decay modes of the W will be difficult to observe due to the QCD $jj\gamma$ background [39]. If only leptonic decay modes are considered, about 20% of all W decays are still observable (we neglect the $W \to \tau\nu$ decay in the following).

To be more specific we shall focus in our work on final states which contain either an electron or a positron. The signal we consider is

$$p\bar{p} \to e^\pm + \gamma + \not{p}_T , \qquad (3.5)$$

with the missing transverse momentum. $\not{p}_T$, resulting from the nonobservation of the neutrino arising from the $W \to e\nu$ decay. We include the leptonic branching fraction

$$B = \text{Br} \, (W \to e\nu) = 0.109 \qquad (3.6)$$

corresponding to a top quark mass $m_t > M_W$ in all subsequent figures.

377

A serious background to the signal (3.5) may be caused by W jet production with the jet misidentified as a photon. Such misidentifications originate mostly from jets hadronizing with a leading π^0, which carries away most of the jet energy. In the CDF detector the photons arising from the π^0 decay usually can be separated if the transverse momentum p_T of the π^0 is smaller than about 50 GeV [40]. Thus, the W jet background is not expected to pose problems in the region $p_{T\gamma} < 50$ GeV.

For photon transverse momenta $p_{T\gamma} > 50$ GeV the W jet background cannot be neglected. A precise value for the probability $P_{\gamma/j}$ that a jet of $p_T > 50$ GeV is misidentified as a single photon depends on details of the jet structure. Since π^0's and photons have rather different shower profiles in an electromagnetic calorimeter $P_{\gamma/j}$ will be small. In a preliminary study [41], values of $P_{\gamma/j} = (2 - 5) \cdot 10^{-3}$ were found for $p_{T\gamma} > 10$ GeV at Tevatron energies. Because $P_{\gamma/j}$ becomes rapidly smaller with increasing $p_{T\gamma}$ [42], one expects that the misidentification probability for $p_{T\gamma} > 50$ GeV will be significantly smaller than $5 \cdot 10^{-3}$. In the next Section we shall study in more detail the impact of the W jet background on the sensitivity of (3.5) to anomalies in the $WW\gamma$ vertex.

For our analysis of signatures of anomalous $WW\gamma$ couplings it is important that $W\gamma$ production can be distinguished efficiently from radiative W decays. In radiative W decays the $e^\pm \nu$ pair and the photon form a system with invariant mass $M(e\nu\gamma)$ close to M_W. For $W\gamma$ production, on the other hand, $M(e\nu\gamma)$ is always larger than M_W if finite W width effects are ignored. This difference suggests that $e^\pm \gamma p\!\!\!/_T$ events originating from radiative W decays can be separated by a $M(e\nu\gamma)$ cut from $W\gamma$ events which result in the same final state. However, due to the nonobservation of the neutrino, $M(e\nu\gamma)$ cannot be determined unambigously and the minimum invariant mass or the cluster transverse mass [43] is more useful:

$$M_T^2(e\gamma; p\!\!\!/_T) = \left[\left(M_{e\gamma}^2 + |\mathbf{p}_{T\gamma} + \mathbf{p}_{Te}|^2 \right)^{1/2} + p\!\!\!/_T \right]^2 - |\mathbf{p}_{T\gamma} + \mathbf{p}_{Te} + \mathbf{p}\!\!\!/_T|^2 \, , \qquad (3.7)$$

where $M_{e\gamma}$ denotes the invariant mass of the $e\gamma$ pair. For $W \rightarrow e\nu\gamma$ the cluster transverse mass sharply peaks at M_W [43] and drops rapidly above the W mass. Thus $e\gamma p\!\!\!/_T$ events originating from $W\gamma$ production and radiative W decays can be distinguished if $M_T(e\gamma; p\!\!\!/_T)$ is cut slightly above M_W [15]. In the following we shall identify events which satisfy

$$M_T(e\gamma; p\!\!\!/_T) > 90 \text{ GeV} \qquad (3.8)$$

with $W\gamma$ events. As demonstrated in detail in ref. [9] the M_T cut is quite efficient in distinguishing the two processes.

Effects of higher order QCD corrections are simulated in our analysis by a K-factor

$$K = 1 + \frac{8}{9}\pi\alpha_s(\hat{s}) \, , \qquad (3.9)$$

where α_s is the strong coupling constant. Recently the full $O(\alpha_s)$ QCD corrections to $W\gamma$ production have been computed [19]. It turns out that, for Tevatron energies, the QCD corrections to $q\bar{q}' \rightarrow W\gamma$ can be represented quite well by a K-factor. The radiation zero is affected insignificantly by the QCD corrections. By approximating QCD corrections by a K-factor one, however, ignores the transverse momentum of the produced $W\gamma$ system induced by the higher order terms.

Finite detector acceptance is simulated by cuts imposed on the final state particles. In the following we require a photon transverse momentum of $p_{T\gamma} > 10$ GeV, an electron photon separation in the pseudorapidity-azimuthal angle plane

$$\Delta R_{e\gamma} = \left[(\Delta\phi_{e\gamma})^2 + (\Delta\eta_{e\gamma})^2 \right]^{1/2} > 0.7 \,, \qquad (3.10)$$

and a missing p_T of $\not{p}_T > 20$ GeV. Uncertainties in the energy measurements of the $e^\pm$ and the photon in the detector are taken into account by Gaussian smearing of the particle four momenta with standard deviation $\sigma = (0.15 \text{ GeV}^{1/2}) \sqrt{E}$. In fact, without a finite $p_{T\gamma}$ and $\Delta R_{e\gamma}$ cut the cross section for (3.5) would diverge, due to the various collinear and infrared singularities which are present. The $\not{p}_T$ cut, on the other hand, has only a small effect on the cross section because the missing transverse momentum distribution shows the familiar Jacobian peak at $p_T \approx \frac{1}{2} M_W$.

Electrons can be identified by the CDF detector in the pseudorapidity range $|\eta_e| \lesssim 2$, but their charge can only be determined in the central region $|\eta_e| < 1.1$. The minimum electron p_T required in the CDF detector depends somewhat on η_e. For $|\eta_e| < 1.1$ a minimum transverse momentum of $p_{Te} > 15$ GeV is sufficient while for $|\eta_e| > 1.1$ a slightly higher threshold of ≈ 20 GeV is required [44]. For $q\bar{q}' \to W\gamma$; $W \to e\nu$ the p_{Te} distribution exhibits the familiar Jacobian peak at $p_{Te} \approx \frac{1}{2} M_W \approx 40$ GeV, smeared by the W's transverse momentum. Thus a transverse momentum cut of ≤ 30 GeV has almost no effect on the $W\gamma$ cross section. In the following we shall impose a pseudorapidity cut of $|\eta_e| < 2$ and an electron p_T cut of $p_{Te} > 20$ GeV in $p\bar{p} \to W\gamma$; $W \to e\nu$. Since the electron charge cannot be measured for $|\eta_e| > 1.1$ we shall sum over the W charges. Restriction to the central pseudorapidity region ($|\eta_e| < 1.1$) would result in only half the rate for $|\eta_e| < 2$ in the phase space region which is sensitive to anomalous $WW\gamma$ couplings.

The CDF detector can detect photons with good efficiency if their pseudorapidity is $|\eta_\gamma| < 1$. In events where one does not trigger on photons this region can be enlarged, perhaps to $|\eta_\gamma| < 3$. As we shall see in the next Section, anomalous $WW\gamma$ couplings affect mainly the small photon pseudorapidity region. In the following we thus require that $|\eta_\gamma| < 1$. Within the cuts described above and with an integrated luminosity of 100 pb^{-1} the Tevatron will provide about 50 clean $W^\pm\gamma$; $W^\pm \to e^\pm\nu$ events.

3.3 Signatures for Anomalous $WW\gamma$ Couplings

In the $q\bar{q}' \to W\gamma$ subprocess the effects of anomalies in the $WW\gamma$ vertex are enhanced at large energies, due to the $\hat{\gamma} = \sqrt{\hat{s}}/2M_W$ factors in the anomalous contributions to the amplitudes (3.4). This enhancement factor will particularly favor the observability of anomalous values of λ and $\tilde{\lambda}$, which are enhanced by $\hat{\gamma}^2$ in the amplitude whereas terms containing κ and $\tilde{\kappa}$ grow only linearly with $\hat{\gamma}$.

A typical signal for anomalous couplings will be a broad increase in the W-photon invariant mass spectrum at large values of $M_{W\gamma} = \sqrt{\hat{s}}$. The resulting effect on $B \cdot d\sigma/dM_{W\gamma}$ is shown in Fig. 5 for the illustrative values $\Delta\kappa = \kappa - 1 = 1$ and $\lambda = 0.5$. Only one $WW\gamma$ coupling at a time is chosen different from the SM prediction. For comparison the SM curve is included as a solid line. The scale Q^2 at which the structure functions are evaluated is chosen to be $\hat{s}$ in Fig. 5 and all subsequent figures.

At hadron colliders the $W\gamma$ invariant mass cannot be determined unambigously because the neutrino from W-decay is not observed. If the transverse momentum of the neutrino is identified with the missing transverse momentum of a given $W\gamma$ event, the unobservable longitudinal neutrino momentum $p_{L\nu}$ can be reconstructed, albeit with a twofold ambiguity, by imposing the constraint that the neutrino and the electron four-momenta combine to form the W rest mass [15,45]. Plotted in Fig. 5 is the "reconstructed" $W\gamma$ invariant mass spectrum obtained from the two solutions for $p_{L\nu}$ for each event.

In Fig. 5 the anomalous coupling curves are clearly distinguishable from the SM prediction. For the expected integrated luminosity of 100 pb^{-1} for future Tevatron

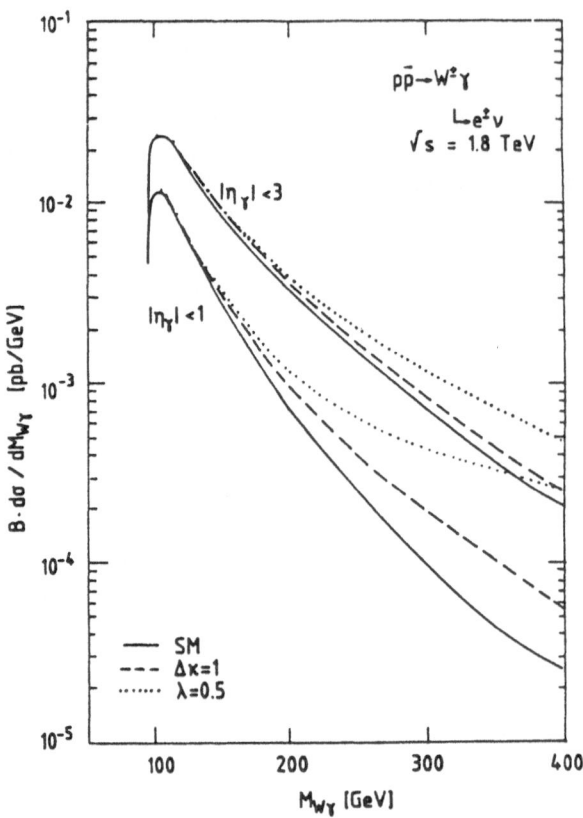

Fig. 5 . W photon invariant mass spectrum for the process $p\bar{p} \to W^{\pm}\gamma$; $W^{\pm} \to e^{\pm}\nu$ at the Tevatron. The curves are for the SM (solid curves), $\Delta\kappa = \kappa - 1 = 1$ (dashed curves) and $\lambda = 0.5$ (dotted curves). The lower solid, dashed and dotted curve give the results for the cuts as described in Section 3.2. The upper solid, dashed and dotted curve show the $M_{W\gamma}$ spectrum with the $|\eta_{\gamma}| < 1$ cut replaced by $|\eta_{\gamma}| < 3$, and all other cuts unchanged. Higher order QCD corrections are approximated by a K-factor.

runs, a differential cross section of 10^{-4} pb/GeV corresponds to one event per 100 GeV interval. Since terms in the helicity amplitudes which are proportional to λ grow much faster ($\sim \hat{\gamma}^2$) than κ terms ($\sim \hat{\gamma}$) for $\sqrt{\hat{s}} \gg M_W$, the $M_{W\gamma}$ spectrum for anomalous values of λ is much harder than the one for a nonstandard κ.

Besides the curves for the cuts described above, Fig. 5 also shows the $M_{W\gamma}$ distribution with the $|\eta_{\gamma}| < 1$ cut replaced by $|\eta_{\gamma}| < 3$. The increase of the cross section at large $M_{W\gamma}$ due to anomalous $WW\gamma$ couplings is much more pronounced for $|\eta_{\gamma}| < 1$ than for $|\eta_{\gamma}| < 3$. This is the reason we have chosen the rather stringent cut of $|\eta_{\gamma}| < 1$. For $|\eta_{\gamma}| > 1$ $W\gamma$ production at the Tevatron is very insensitive to anomalous $WW\gamma$ couplings. As far as a measurement of the $WW\gamma$ vertex is concerned, the only use of the $|\eta_{\gamma}| > 1$ data is to provide a check on the normalization of the $W\gamma$ production cross section.

At the Tevatron the sensitivity to anomalous couplings in $p\bar{p} \to W\gamma$ effectively stems from regions in phase space where the anomalous contributions to the cross section are considerably larger than the SM expectation. As a result, interference effects between the SM amplitude and the anomalous amplitude contributions (3.4) play a minor role, and an excess in counting rate, beyond the SM prediction, scales essentially like the square of the anomalous coupling.

As we shall see in Section 3.4, existing low energy bounds limit $|\tilde{\kappa}|$ to less than $O(10^{-3})$ and no visible effects of $\tilde{\kappa}$ are possible at the Tevatron. Because contributions to the helicity amplitudes containing λ and $\tilde{\lambda}$ differ only by a factor $\pm i$, cross sections for the same values of λ and $\tilde{\lambda}$ are almost identical. Results for anomalous $\tilde{\kappa}$ and $\tilde{\lambda}$ values are therefore not included in Fig. 5 and all subsequent figures.

As already mentioned, the $W\gamma$ differential cross section vanishes, in the SM, at one value of the photon scattering angle. For $u\bar{d} \to W^+\gamma$ ($d\bar{u} \to W^-\gamma$) the radiation zero occurs at $\cos\theta^* = -\frac{1}{3} (+\frac{1}{3})$, where θ^* is the scattering angle of the photon relative to the quark direction, in the $W\gamma$ center of mass frame. In practice, however, the zero is washed out considerably. In order to measure θ^* the $W\gamma$ rest frame has to be reconstructed. Since the unobservable longitudinal neutrino momentum can only be determined with a twofold ambiguity and, on an event to event basis, one does not know which solution is the correct one, both solutions have to be considered for each event and the zero is partially filled in. A similar effect is caused by the small fraction of radiative W decay events which passes the $W\gamma$ selection cuts [13] and by the fact that one does not know whether one has to associate the quark with the proton or the antiproton. Finally, when one sums over the W-charges, only $|\cos\theta^*|$ remains measureable, and the dip at $\cos\theta^* = \pm\frac{1}{3}$ is shifted to $\cos\theta^* = 0$.

In order to eliminate the strong peaking of the differential cross section at $\cos\theta^* = \pm 1$ which arises from the collinear singularity, it is convenient instead to study the rapidity distribution $B \cdot d\sigma/d|y^*|$ of the photon in the $W\gamma$ rest frame [18], with

$$y^* = \frac{1}{2}\ln\frac{1+\cos\theta^*}{1-\cos\theta^*}. \tag{3.11}$$

The resulting distribution is shown in Fig. 6 for the same anomalous couplings as in Fig. 5. The dip at $|y^*| = 0$, which is due to the radiation zero, is quite pronounced. In presence of any anomalous contribution to the $WW\gamma$ vertex the radiation zero is eliminated and the dip is filled at least partially.

Compared with the effects discussed above which obscure the radiation zero, the effect induced by higher order QCD corrections is small [19]. Soft QCD corrections which can be approximated by a K-factor exhibit the same radiation zero as the Born cross section. These effects are taken into account in our calculation. Hard QCD corrections which result in a finite p_T for the $W\gamma$ system are found to influence the radiation zero insignificantly at Tevatron energies [19].

As discussed in Section 3.2, the W jet background does not pose problems at the Tevatron for $p_{T\gamma} < 50$ GeV whereas for transverse momenta bigger than 50 GeV it can not be neglected a priori. The dash-dotted line in Fig. 6 shows the y^* distribution of the $p\bar{p} \to W^\pm$ jet background, calculated with $p_{Tj} > 50$ GeV and the cuts specified in Section 3.2. To represent the misidentification probability $P_{\gamma/j}$ of a jet as a photon the W jet y^* spectrum was multiplied by a factor $5 \cdot 10^{-3}$. In view of our discussion in Section 3.2 this value can be considered a safe upper limit for $P_{\gamma/j}$. Hence, our estimate of the W jet background is conservative. From Fig. 6 we observe that the background is considerably smaller than the SM $W\gamma$ cross section over the entire y^* range. Although the W jet cross section peaks at $y^* = 0$, and thus tends to fill in the dip caused by the radiation zero, we conclude that the W jet background for $p_{Tj} > 50$ GeV does not severely limit the sensitivity of the y^* distribution to anomalous couplings. A similar result is also obtained for other distributions which are sensitive to anomalous contributions to the $WW\gamma$ vertex.

Figure 6 demonstrates that anomalous couplings affect mainly the region of small center of mass rapidities. This is due to the fact that anomalous couplings only contribute via the s-channel W-exchange graph of Fig. 1, and hence only to the $J = 1$ partial wave, when fermion masses are neglected. The anomalous contributions are,

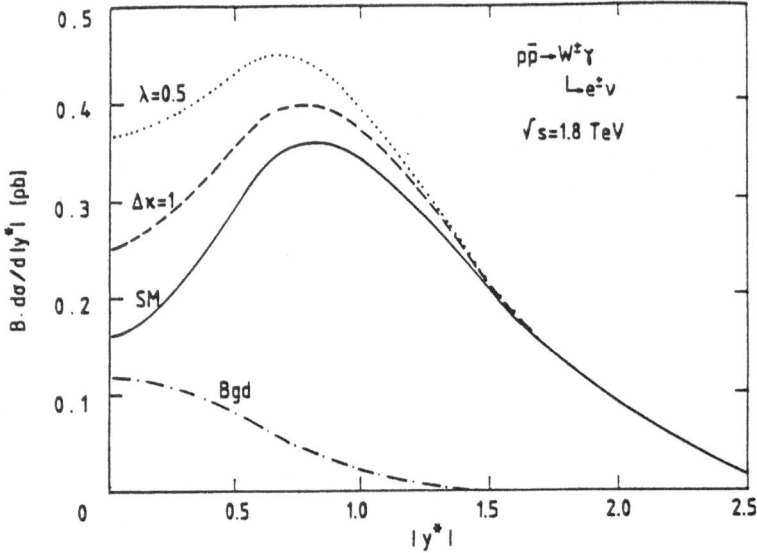

Fig. 6 . Rapidity spectrum of the photon in the $W\gamma$ rest frame for $p\bar{p} \rightarrow W^{\pm}\gamma$; $W^{\pm} \rightarrow e^{\pm}\nu$ at the Tevatron. The anomalous couplings are the same as in Fig. 5. Cuts are specified in Section 3.2. The dash-dotted line represents the background from $W^{\pm}$ jet production (see text for details). QCD corrections are simulated by a K-factor.

therefore, almost isotropic in the center of mass frame, while the u- and t-channel graphs of Fig. 4 result in a strong enhancement of the high rapidity region. Thus the finite acceptance cuts will largely eliminate the well known fermion exchange contributions to the cross-section and reduce by a much smaller amount possible signals of new physics, $i.e.$ the effects of anomalous $WW\gamma$ interactions.

The population of the small rapidity region, induced by anomalous couplings, considerably increases the average photon transverse momentum of events produced at a fixed value of the W photon invariant mass. The p_T distribution of the photon, $d\sigma/dp_{T\gamma}$, should be particularly sensitive to anomalous couplings. This fact is visible in Fig. 7 where the $p_{T\gamma}$ spectrum is plotted for the SM, and for $\Delta\kappa = 1$ and $\lambda = 0.5$. For an integrated luminosity of 100 pb^{-1} experiments at the Tevatron should be able to see the effect.

As we have demonstrated so far, the $W\gamma$ invariant mass spectrum, the rapidity distribution of the produced photon in the $W\gamma$ rest frame, and the photon transverse momentum spectrum are sensitive indicators of anomalous couplings. We now want to make these statements more quantitative by deriving those values of κ, λ and $\tilde{\lambda}$ which would give rise to a deviation from the SM at the 90% (69%) confidence level (CL) in either the $M_{W\gamma}$, the y^* or the $p_{T\gamma}$ spectrum. We assume an integrated luminosity of $\int \mathcal{L} dt = 100$ pb^{-1} at the Tevatron and the cuts described in Section 3.2. The parameter $\tilde{\kappa}$ is omitted since. as already mentioned, low energy data constrain this quantity to be less than $O(10^{-3})$. The confidence level is calculated by splitting the $M_{W\gamma}$ and y^* distributions into 6 bins each and the $p_{T\gamma}$ distribution into 5 bins with, typically, more than 5 events. In each bin the Poisson statistics is approximated by a Gaussian distribution. In order to achieve a sizable counting rate in each bin, all events with $M_{W\gamma} > 140$ GeV and $p_{T\gamma} > 30$ GeV are collected into a single bin. This procedure guarantees that in our calculation a high confidence level cannot arise from a single event at high $M_{W\gamma}$ or $p_{T\gamma}$ where the SM predicts, say, only 0.01 events. In order to derive realistic limits we include the W jet background for $p_{Tj} > 50$ GeV in our analysis and allow for a normalization uncertainty $\Delta\mathcal{N}$ of the SM $W\gamma$ production

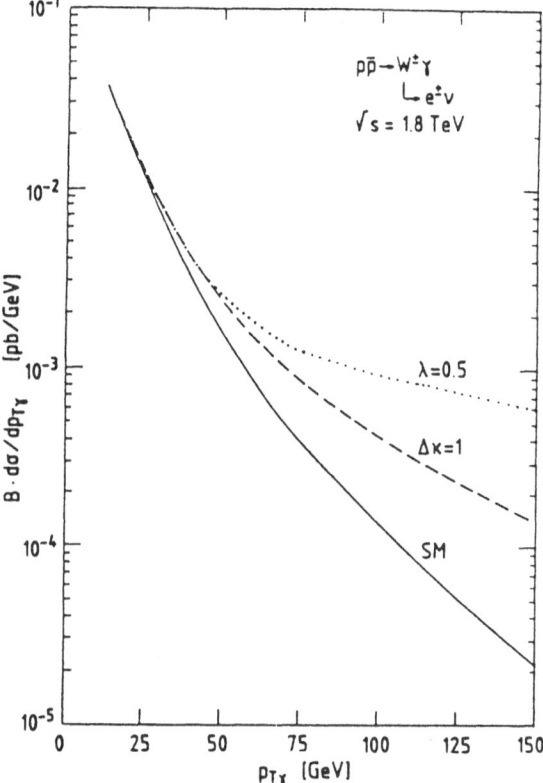

Fig. 7. Transverse momentum spectrum of the photon in $p\bar{p} \to W^{\pm}\gamma$; $W^{\pm} \to e^{\pm}\nu$ at the Tevatron. Parameters and cuts are chosen as in Fig. 6. A K-factor is used to simulate QCD corrections.

cross section of $\Delta\mathcal{N} = 50\%$. QCD corrections are simulated by a K-factor.

The resulting minimal anomalous couplings which would give rise to a 90% or 69% CL effect at the Tevatron are given in Table 1. Only one coupling at a time is assumed to be different from the SM value. The sensitivity bounds are given for two choices of the photon-jet misidentification probability, $P_{\gamma/j} = 5 \cdot 10^{-3}$ and a somewhat more optimistic value of $P_{\gamma/j} = 5 \cdot 10^{-4}$. These two values were chosen in order to display the effect of the W jet background on the limits that can be achieved.

In all cases the p_T spectrum of the photon is the most sensitive indicator of anomalous couplings. This can be easily understood by remembering that $d\sigma/dM_{W\gamma}$ as well as $d\sigma/d|y^*|$ both depend on the reconstructed longitudinal momentum $p_{L\nu}$ of the neutrino whereas this is not the case for the $p_{T\gamma}$ spectrum. Since $p_{L\nu}$ can only be determined with a twofold ambiguity, the effects of anomalous $WW\gamma$ couplings are less pronounced in the $M_{W\gamma}$ and y^* distributions than in the $p_{T\gamma}$ spectrum.

Table 1 shows that λ and $\tilde{\lambda}$ can be measured with 25-40% accuracy with an integrated luminosity of 100 pb^{-1} if $P_{\gamma/j} = 5 \cdot 10^{-4}$. These errors become larger by roughly 20% if $P_{\gamma/j} = 5 \cdot 10^{-3}$. $|\Delta\kappa|$, on the other hand, can only be constrained to be less than 0.8-1.5. The values presented in Table 1 thus directly reflect the different powers of $\hat{\gamma} = \sqrt{\hat{s}}/2M_W$ multiplying the various anomalous contributions to the amplitudes (3.4). At high energies ($\hat{s} \gg M_W^2$) terms proportional to λ and $\tilde{\lambda}$ grow much faster than κ terms and the sensitivity limits achievable for λ and $\tilde{\lambda}$ are thus

TABLE 1

Sensitivities achievable at the 90% and 69% CL for the anomalous $WW\gamma$ couplings $\Delta\kappa = \kappa - 1$, λ and $\tilde\lambda$ in $p\bar{p} \to W^{\pm}\gamma$; $W^{\pm} \to e^{\pm}\nu$ at the Tevatron ($\sqrt{s} = 1.8$ TeV) for an integrated luminosity of 100 pb^{-1}. The bounds are given for a photon-jet misidentification probability of $P_{\gamma/j} = 5 \cdot 10^{-3}$ and $P_{\gamma/j} = 5 \cdot 10^{-4}$. Only one coupling at a time is assumed to be different from the SM value.

$P_{\gamma/j}$	$\Delta\kappa$		λ		$\tilde\lambda$	
	$5 \cdot 10^{-3}$	$5 \cdot 10^{-4}$	$5 \cdot 10^{-3}$	$5 \cdot 10^{-4}$	$5 \cdot 10^{-3}$	$5 \cdot 10^{-4}$
90% CL	$+1.50$ -1.41	$+1.23$ -1.13	$+0.46$ -0.47	$+0.38$ -0.40	$+0.46$ -0.46	$+0.39$ -0.39
69% CL	$+0.99$ -0.90	$+0.82$ -0.72	$+0.30$ -0.31	$+0.25$ -0.26	$+0.31$ -0.31	$+0.26$ -0.26

considerably better than the ones for κ. Since interference effects between the SM amplitude and the anomalous contributions (3.4) to the amplitudes play only a minor role, the limits in Table 1 do not depend significantly on the sign of the anomalous coupling. Furthermore, the limits for λ and $\tilde\lambda$ are very similar because terms in the helicity amplitudes containing λ and $\tilde\lambda$ differ only by a factor $\pm i$.

The bounds presented in Table 1 have been derived by assuming that only one coupling at a time deviates from the SM. Since the leading contributions to the helicity amplitudes which are proportional to κ and λ grow with different powers of $\sqrt{\hat{s}}/M_W$ (see eq. (3.4)), effects of anomalous values of κ and λ cannot cancel at a significant level in the $p\bar{p} \to W\gamma$ distributions. Taking into account possible cancellations between terms proportional to $\Delta\kappa$ and λ in the helicity amplitudes, we find that the bounds given in Table 1 increase by at most 30%. Terms in (3.4) containing the CP violating coupling $\tilde\lambda$ contribute imaginary parts to the helicity amplitudes. Because the CP conserving parts are real, all interference effects between $\tilde\lambda$ and $\kappa\,(\lambda)$ terms vanish identically. Hence the sensitivities of Table 1 represent model independent upper bounds that can be set by experiments.

From the analysis we presented it is clear that a detailed investigation of W-photon production at Tevatron energies requires an integrated luminosity of at least 100 pb^{-1}. Nevertheless, a few events may already be found in the present data set. For $\int \mathcal{L}dt = 4.7$ pb^{-1} one expects 2 $W^{\pm}\gamma$; $W^{\pm} \to e^{\pm}\nu$ events within our standard set of cuts. Since the expected number of $W\gamma$ events is very small, the only meaningful observable which can be used to derive bounds on the anomalous couplings from present Tevatron data is the total $W\gamma$ cross section within cuts. Assuming that only one coupling at a time differs from the gauge theory value and that the SM $p\bar{p} \to W\gamma$ rate can be determined within 50%, sensitivities of

$$\Delta\kappa = \begin{smallmatrix} +3.1 \\ -2.9 \end{smallmatrix} \begin{pmatrix} +5.6 \\ -5.4 \end{pmatrix}, \tag{3.12a}$$

$$\lambda = \pm 1.1 \ (\pm 2.1), \tag{3.12b}$$

$$\tilde{\lambda} = \pm 1.1 \ (\pm 2.1) \qquad\qquad (3.12c)$$

can be reached at the 1σ (5σ) level from the total $p\bar{p} \to W\gamma$; $W \to e\nu$ cross section with an integrated luminosity of 4.7 pb^{-1}.

So far we have concentrated on $W\gamma$ production. As mentioned above radiative W decays are described by the same set of Feynman diagrams as $q\bar{q}' \to W\gamma$; $W \to e\nu$ and therefore are also sensitive to anomalous $WW\gamma$ couplings. However, in contrast to W-photon production, the available center of mass energy $\sqrt{\hat{s}}$ is fixed to $\sqrt{\hat{s}} \approx M_W$ in radiative W decays. Hence, the anomalous contributions to the $q\bar{q}' \to W \to e\nu\gamma$ helicity amplitudes do not dominate over the SM amplitudes, unless very large values are chosen for the anomalous couplings. Radiative W decays are thus much less sensitive to anomalous $WW\gamma$ couplings than $W\gamma$ production [9].

3.4 Low Energy Bounds and Expectations from HERA and LEP200

We have described the signatures which anomalies in the $WW\gamma$ vertex will produce in $W\gamma$ production at the Tevatron. In particular we have determined how large deviations from the SM must be in order to yield visible effects for an integrated luminosity of $\int \mathcal{L}dt = 100$ pb^{-1}. It is interesting to compare the sensitivity of the Tevatron with existing low energy limits on anomalous couplings and with the sensitivity to non gauge theory terms in the $WW\gamma$ vertex accessible via $e^+e^- \to W^+W^-$ at LEP200, and via single W production in ep collisions at HERA.

Low energy bounds on κ and λ are quite model dependent at present [26]. From loop contributions to $(g-2)_\mu$ [27] one estimates

$$\left| (\kappa - 1)\,\ln\frac{\Lambda^2}{M_W^2} + \frac{\lambda}{3} \right| < 3.7\,, \qquad\qquad (3.13)$$

where Λ is the scale at which weak bosons show novel strong interactions. Bounds derived from the photon propagator as measured at PETRA [22] and the W/Z mass ratio [21,22] are more stringent, but also more controversial because loop corrections are ill-defined in these cases. From Table 1 we may conclude that experiments at the Tevatron can significantly improve the present low energy bound on λ, derived from $(g-2)_\mu$. For κ the low energy level of accuracy can be reached. The situation is quite different for the CP violating coupling $\tilde{\kappa}$ which would contribute to the electric dipole moment of the neutron. From the present experimental limit on the neutron electric dipole moment one finds [24]

$$\left| \tilde{\kappa}\,\ln\frac{\Lambda^2}{M_W^2} \right| < 10^{-3} \qquad\qquad (3.14)$$

which clearly excludes any observable effect at the Tevatron. Curiously, as was observed in ref. [25], no such bound exists for the other CP violating coupling $\tilde{\lambda}$. Contributions of $\tilde{\lambda}$ to the neutron electric dipole moment are suppressed by at least another factor $(M_N^2/M_W^2) \approx 1.3 \cdot 10^{-4}$ (M_N being the neutron mass) compared to the $\tilde{\kappa}$ bound (3.14) and hence a constraint $|\tilde{\lambda}| < O(1)$ results at best. Comparison with Table 1 shows that the Tevatron would be able to explore almost one order of magnitude in $\tilde{\lambda}$. Of course, hadron supercolliders such as the LHC or the SSC would be even more sensitive to λ and $\tilde{\lambda}$. At these machines $\lambda\,(\tilde{\lambda})$ values as small as $O(10^{-3})$ can be probed [18].

Experiments at HERA, studying single W production via $ep \to eWX$, will also be able to probe the $WW\gamma$ vertex [46]. In ep collisions at $\sqrt{s} = 314$ GeV, for an integrated luminosity of 10^3 pb^{-1}, one expects about 90 $ep \to eW^{\pm}X$ events with $W \to e\nu, \mu\nu$. It turns out that HERA is considerably more sensitive to κ than to λ and $\tilde{\lambda}$. While κ can be measured with 30-50% accuracy, $|\lambda|$ and $|\tilde{\lambda}|$ can only be constrained to be less than 0.9-1.3 [46]. HERA will thus be able to put a better limit on $\Delta\kappa$ than the Tevatron. For λ and $\tilde{\lambda}$, on the other hand, the Tevatron with $\int \mathcal{L}dt = 100$ pb^{-1} will be able to give a limit which is about a factor 3-5 better than the bound which can be expected from HERA.

Combined, the reactions $p\bar{p} \to W\gamma$ at the Tevatron and $ep \to eWX$ at HERA will yield a precise direct measurement of the $WW\gamma$ vertex before W pair production can be studied at LEP200. Even in $e^+e^- \to W^+W^-$ at $\sqrt{s} = 190$ GeV only an accuracy of $|\Delta\kappa|, |\lambda|, |\tilde{\lambda}| \approx 0.2$ is expected [47], and in W^+W^- production it is actually a linear combination of $WW\gamma$ and WWZ couplings which will be measured. In view of our present poor knowledge of the values of κ, λ and $\tilde{\lambda}$, the direct measurement of the $WW\gamma$ couplings via $p\bar{p} \to W\gamma$ at the Tevatron and $ep \to eWX$ at HERA will constitute major progress. In the mid 1990's, these values will also be helpful in disentangling $WW\gamma$ and WWZ anomalous couplings in W pair production at LEP200.

4. CONCLUSIONS

Electroweak vector boson pairs are very useful in probing the Higgs sector and vector boson self-interactions at colliders. Vector boson pair production processes also constitute a background to the Higgs boson search at hadron colliders. We have illustrated this by the production of vector boson pairs accompanied by a high p_T jet where the jet is produced outside the central rapidity range $|y| < 2.5$. In this configuration W^+W^-j, $W^{\pm}Zj$ and ZZj production may contribute to the Higgs tagging background, notably when one of the tagging jets is not identified. A full analysis has not been done so far, but a crude estimate suggests that this background may be important.

W^+W^-, $W^{\pm}\gamma$ and $W^{\pm}Z$ production at hadron colliders are sensitive to the $WW\gamma$ and/or the WWZ coupling. We have demonstrated that future experiments at the Tevatron, where an integrated luminosity of order 100 pb^{-1} can be expected, can measure the $WW\gamma$ vertex with 25-40% accuracy in W-photon production. The WWZ coupling can be determined with similar precision from $W^{\pm}Z$ production at the Tevatron [48].

ACKNOWLEDGEMENTS

This research was supported in part by the University of Wisconsin Research Committee with funds granted by the Wisconsin Alumni Research Foundation, and in part by the U. S. Department of Energy under contract DE-AC02-76ER00881.

REFERENCES

1. See *e.g.* S. L. Wu *et al.* in *Proceedings of the ECFA Workshop on LEP200*, Volume II, Aachen, Germany, 29 September - 1 October 1986, CERN 87-08, p. 312 and references therein.

2. B. W. Lee, C. Quigg and H. B. Thacker, <u>Phys. Rev.</u> D16:1519 (1977).

3. J. F. Gunion, G. L. Kane and J. Wudka, <u>Nucl. Phys.</u> B299:231 (1988).

4. V. Barger, J. L. Lopez and W. Putikka, <u>Int. J. Mod. Phys.</u> A3:2181 (1988).

5. R. N. Cahn, S. D. Ellis, R. K. Kleiss and W. J. Stirling <u>Phys. Rev.</u> D35:1626 (1987).

6. R. K. Ellis, I. Hinchliffe, M. Soldate and J. J. van der Bij, <u>Nucl. Phys.</u> B297:221 (1988);
I. Hinchliffe and S. F. Novaes, <u>Phys. Rev.</u> D38:3475 (1988).

7. U. Baur, E. W. N. Glover and J. J. van der Bij, Nucl. Phys. B318:106 (1989).

8. R. K. Kleiss and W. J. Stirling, Phys. Lett. 200B:193 (1988).

9. U. Baur and E. L. Berger, CERN-TH.5517/89, ANL-HEP-PR-89-86, preprint (1989), to appear in Phys. Rev. D.

10. K. O. Mikaelian, Phys. Rev. D17:750 (1978);
    R. W. Brown, D. Sahdev and K. O. Mikaelian, Phys. Rev. D20:1164 (1979);
    K. O. Mikaelian, M. A. Samuel and D. Sahdev, Phys. Rev. Lett. 43:746 (1979).

11. Zhu Dongpei, Phys. Rev. D22:2266 (1980);
    C. J. Goebel, F. Halzen and J. P. Leveille, Phys. Rev. D23:2682 (1981);
    S. J. Brodsky and R. W. Brown, Phys. Rev. Lett. 49:966 (1982);
    R. W. Brown, K. L. Kowalski and S. J. Brodsky, Phys. Rev. D28:624 (1983);
    M. A. Samuel, Phys. Rev. D27:2724 (1983).

12. C. L. Bilchak, R. W. Brown and J. D. Stroughair, Phys. Rev. D29:375 (1984).

13. G. N. Valuenzuela and J. Smith, Phys. Rev. D31:2787 (1985).

14. J. C. Wallet, Z. Phys. C30:575 (1986).

15. J. Cortes, K. Hagiwara and F. Herzog, Nucl. Phys. B278:26 (1986).

16. S.-C. Lee and W. C. Su, Phys. Rev. D38:2305 (1988).

17. K. Hagiwara et al., Nucl. Phys. B282:253 (1987).

18. U. Baur and D. Zeppenfeld, Nucl. Phys. B308:127 (1988).

19. J. Smith, D. Thomas and W. L. van Neerven, CPT-89/P.2239, preprint (1989), to appear in Z. Phys. C.

20. F. Herzog, Phys. Lett. 148B:355 (1984);
    J. C. Wallet, Phys. Rev. D32:813 (1985);
    A. Grau and J. A. Grifols, Phys. Lett. 197B:437 (1987).

21. M. Suzuki, Phys. Lett. 153B:289 (1985).

22. J. J. van der Bij, Phys. Rev. D35:1088 (1987).

23. J. A. Grifols, S. Peris and J. Solà, Int. J. Mod. Phys. A3:255 (1988).

24. W. J. Marciano and A. Queijeiro, Phys. Rev. D33:3449 (1986).

25. F. Hoogeveen, preprint MPI-PAE/PTh 25/87 (1987).

26. G. L. Kane, J. Vidal and C. P. Yuan, Phys. Rev. D39:2617 (1989).

27. P. Méry, S. E. Moubarik, M. Perottet and F. M. Renard, CPT-89/P.2226, preprint (1989).

28. U. Baur and D. Zeppenfeld, Phys. Lett. 201B:383 (1988).

29. R. K. Kleiss and W. J. Stirling, Nucl. Phys. B262:235 (1985);
    J. F. Gunion and Z. Kunszt, Phys. Lett. 161B:333 (1985).

30. K. Hagiwara and D. Zeppenfeld, Nucl. Phys. B274:1 (1986).

31. W. J. Stirling, R. Kleiss and S. D. Ellis, Phys. Lett. 163B:261 (1985);
    J. F. Gunion, Z. Kunszt and M. Soldate, Phys. Lett. 163B:389 (1985).

32. E. W. N. Glover, K. Hagiwara and A. D. Martin, Phys. Lett. 168B:289 (1986).

33. D. Duke and J. Owens, Phys. Rev. D30:49 (1984).

34. S. Geer, these proceedings.

35. R. Johnson, FERMILAB-Conf-88/169, preprint (1988).

36. K. Gaemers and G. Gounaris, Z. Phys. C1:259 (1979).

37. J. M. Cornwall, D. N. Levin and G. Tiktopoulos, <u>Phys. Rev. Lett.</u> 30:1268 (1973); <u>Phys. Rev.</u> D10:1145 (1974);
C. H. Llewellyn Smith, <u>Phys. Lett.</u> 46B:233 (1973);
S. D. Joglekar, <u>Ann. Phys.</u> 83:427 (1974).

38. T. Kamae, in *"Proceedings of the XXIV Int. Conf. on High Energy Physics"*, Munich, August 4-10, 1988, R. Kotthaus and J. H. Kühn (Eds.), p. 156 and references therein.

39. F. A. Berends *et al.* , <u>Phys. Lett.</u> 103B:124 (1981);
P. Aurenche *et al.* , <u>Phys. Lett.</u> 140B:87 (1984); <u>Nucl. Phys.</u> B286:553 (1987);
V. Barger, T. Han, J. Ohnemus and D. Zeppenfeld, MAD/PH/515 preprint (1989).

40. R. Blair *et al.* (CDF Collaboration), preprint ANL-HEP-CP-89-07 and R. Blair, private communication.

41. S. A. Kahn *et al.* , BNL informal report 3/83 (1983).

42. Y. Morita, in *"Proceedings of the Summer Study on the Physics of the Superconducting Supercollider"*, Snowmass, Colorado, 1986, edited by R. Donaldson and J. Marx (Division of Particles and Fields of the APS, New York, 1987), p. 194.

43. V. Barger, A. D. Martin and R. J. N. Phillips, <u>Phys. Lett.</u> 125B:339 (1983);
E. L. Berger, D. DiBitonto, M. Jacob and W. J. Stirling, <u>Phys. Lett.</u> 140B:259 (1984).

44. F. Abe *et al.* (CDF Collaboration), <u>Phys. Rev. Lett.</u> 62:1005 (1989).

45. J. Stroughair and C. Bilchak, <u>Z. Phys.</u> C26:415 (1984);
J. Gunion, Z. Kunszt and M. Soldate, <u>Phys. Lett.</u> 163B:389 (1985);
J. Gunion and M. Soldate, <u>Phys. Rev.</u> D34:826 (1986);
W. Stirling *et al.*, <u>Phys. Lett.</u> 163B:261 (1985).

46. U. Baur and D. Zeppenfeld, <u>Nucl. Phys.</u> B325:253 (1989).

47. D. Zeppenfeld, <u>Phys. Lett.</u> 183B:380 (1987).

48. K. Hagiwara, J. Woodside and D. Zeppenfeld, MAD/PH/521 preprint, October 1989.

SEARCH STRATEGIES FOR HIGGS BOSONS AT HIGH ENERGY e^+e^- COLLIDERS[*]

Patricia R. Burchat

Santa Cruz Institute for Particle Physics

University of California, Santa Cruz, California, 95064

ABSTRACT

Search strategies for a minimal neutral Higgs boson at e^+e^- colliders with center-of-mass (c.m.) energies in the range 200 GeV to 2 TeV are reviewed. In addition, search stategies for a charged Higgs boson and nonminimal neutral Higgs bosons are discussed for c.m. energies near 1 TeV. With sufficient luminosity, searches for charged or minimal neutral Higgs bosons at e^+e^- colliders are sensitive to Higgs boson masses up to about 80% of the beam energy. However, there is a range of masses near the $W^\pm$ or Z^0 mass which can probably only be covered with an e^+e^- collider operating at a c.m. energy of about 300-400 GeV. For a limited range of masses, the nonminimal CP-even and CP-odd neutral Higgs bosons could be detected in e^+e^- collisions.

INTRODUCTION

A neutral Higgs boson mass has been excluded for most of the range below about twice the τ mass while a charged Higgs boson mass has been excluded below about 25 GeV.[1] Most of the former and all of the latter range has been excluded by experiments conducted at e^+e^- colliders. Existing hadron collider experiments are not sensitive to Higgs boson production at current luminosities. There is great hope that very high energy hadron colliders such as the SSC will provide some information about the Higgs sector. Yet studies show that this will be a very challenging task and that there will probably be a 'blind spot' for Higgs masses below about twice the $W^\pm$ mass where the backgrounds are too great to see a Higgs boson signal.[2] High energy e^+e^- colliders would provide a very attractive experimental environment for extending the range of Higgs masses to which we are sensitive and covering the range to which the SSC will probably be insensitive. However, e^+e^- colliders are technically extremely challenging. The SLAC Linear Collider has demonstrated this point; however, it has also demonstrated once again how much sensitivity one has to new physics with a very small number of events in e^+e^- collisions.[3]

---

[*] WORK SUPPORTED BY THE U. S. DEPARTMENT OF ENERGY.

In this paper, I will review search strategies for a Higgs boson at high energy e^+e^- colliders with a center-of-mass (c.m.) energy in the range 200 GeV to 2 TeV. I will show that experiments at e^+e^- colliders would be very sensitive to the Higgs sector with detectors which could feasibly be built today.

I will mainly review the following studies:

— the LEP 200 study[4] for a c.m. energy ($\sqrt{s}$) of 200 GeV;

— the SLAC TeV Linear Collider (TLC) study[5-7] for $\sqrt{s} \approx 0.5 - 1$ TeV;

— the 1988 Snowmass study[8] also at $\sqrt{s} \approx 0.5 - 1$ TeV;

— and the CERN Linear Collider (CLIC) study[9] for $\sqrt{s} = 2$ TeV.

In the first and last studies listed above, only the minimal neutral Higgs boson was considered, while in the SLAC and Snowmass studies, searches for a charged Higgs boson and nonminimal neutral Higgs bosons were also considered.

Sample Sizes

In these studies various sizes of data samples were considered. In Table 1, the data-sample sizes are listed along with a luminosity which would result in this sample in a reasonable amount of time ($\lesssim$ a few years). Generally, the size of the data sample increases with c.m. energy because annihilation cross sections generally decrease as $1/s$. However, as we will see below, the main production mechanism for the minimal Higgs boson in high energy e^+e^- colliders is not an annihilation process. In this situation, a larger data sample allows one to be sensitive to a higher Higgs boson mass – up to, or even greater than, the beam energy.

Table 1. Comparison of studies at different center-of-mass energies.

	LEP 200	TLC	CLIC
$\sqrt{s}$ (GeV)	200	500 - 1000	2000
Sample size (fb^{-1})	0.5	10 - 30	10 - 50
Example of running time and luminosity (cm^{-2}s^{-1})	5×10^7 s at 10^{31}	$1 - 3 \times 10^7$ s at 10^{33}	$1 - 5 \times 10^7$ s at 10^{33}
Major production mechanism	$e^+e^- \rightarrow Z^0 H^0$	$e^+e^- \rightarrow Z^0 H^0$ at 500 GeV $e^+e^- \rightarrow \nu_e \bar{\nu}_e H^0$ at 1 TeV	$e^+e^- \rightarrow \nu_e \bar{\nu}_e H^0$
Beamstrahlung effects	not included	included	not included
Detector simulation	Complete detector simulation	Parameterized at particle level	Parameterized at parton level

Higgs Boson Production Mechanisms

The primary Higgs production mechanisms are also listed in Table 1. For lower energy ($\sqrt{s} \approx 200 - 500$ GeV), the most important mechanism is e^+e^- annihilation to a virtual Z^0 which then couples to $Z^0 H^0$: $e^+e^- \rightarrow Z^0 H^0$. At higher c.m. energies ($\sqrt{s} \gtrsim 500$ GeV), the

"WW fusion" process becomes the most significant: $e^+e^- \to \nu_e\bar{\nu}_e H^0$. The Feynman diagrams for both of these processes are shown in Figure 1. For e^+e^- c.m. energies $\sqrt{s}$ which are large compared to the particle masses involved, the cross section[10] for $e^+e^- \to Z^0H^0$ decreases as s^{-1} while the cross section[11] for $e^+e^- \to \nu_e\bar{\nu}_e H^0$ increases logarithmically with s. The cross sections for the processes $e^+e^- \to Z^0H^0$ and $e^+e^- \to \nu_e\bar{\nu}_e H^0$ are shown in Figure 2 for the case $M_{H^0} = 100$ GeV. For this particular Higgs-boson mass, the cross sections are about equal at $\sqrt{s} \approx 400$ GeV.

Beamstrahlung

The extremely high density to which beam particles must be focused to produce sufficient luminosity for particle physics studies at a high-energy e^+e^- collider can result in a significant interaction rate between the particles of one beam and the collective electromagnetic fields produced by the particles in the opposite beam. A consequence of this is that the spectrum of c.m. energies at which e^+e^- interactions take place is not monochromatic. Radiation of photons during the beam-beam collision ("beamstrahlung") will result in a spectrum of particle interactions that depends in detail on the energy and bunch characteristics of each beam. In the TLC studies, the effects of beamstrahlung are included, while the LEP 200 and CLIC studies assume that all interactions take place at the nominal c.m. energy.

In the TLC studies, we have used a beamstrahlung spectrum that has been calculated[12] for a specific set of beam parameters, but which is typical of most machine designs that have been studied.[13] This calculation includes the effects of multiple radiation and beam-beam disruption for an e^+e^- machine operating at a nominal c.m. energy of 1 TeV and a luminosity of $3 \times 10^{33}\,\text{cm}^{-2}\text{s}^{-1}$. Approximately 30% of the luminosity lies within 2% of the nominal c.m.-energy-squared for the spectrum that we have chosen.

Included in the TLC Monte Carlo simulations are both the effective reduction in c.m. energy and the net longitudinal momentum created by the energy loss of each beam. The mean fractional energy loss that particles undergo during the beam-beam collision is 0.26 for the spectrum that we have used. We compute production cross sections by convoluting the luminosity spectrum with each cross section evaluated at the reduced c.m. energy of individual e^+e^- pairs.

Detector Resolution

Finally, the detail to which resolution effects in the detector are simulated varies widely between the studies and is summarized in Table 1. The LEP 200 study has the most complete and realistic detector simulation since they use the detector simulations for the ALEPH and OPAL detectors. The charged tracks are reconstructed through pattern recognition, for example, and the energy resolution and granularity effects are included for the calorimeters.

In the TLC study, the detector resolution is parameterized and the particle momentum and energy smeared accordingly. The analyses depend most heavily on energies measured with the hadron calorimetry. Photons and charged and neutral hadrons are treated indiscriminately as clusters of calorimetric energy. Tracks within 4° of one another are combined; then the direction of the combined track is smeared by a box function of size $\pm$ 2°. The energies of these tracks are smeared with

$$\frac{\sigma_E}{E} = \frac{50\%}{\sqrt{E}}\text{GeV}^{1/2} + 2\%.$$

To account for possible obstruction of the acceptance of the detector near the beam line by machine components and the hardware needed to support them, in the TLC study we simply

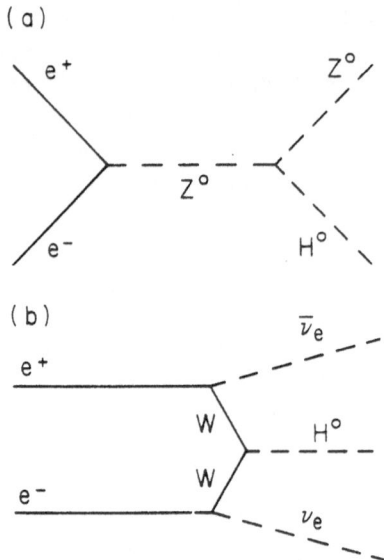

(a)

e^+

Z^0

Z^0

e^-

H^0

(b)

$\bar{\nu}_e$

e^+

W

H^0

W

e^-

ν_e

Figure 1. Feynman diagrams for Higgs boson production mechanisms at high energy e^+e^- colliders: (a) $e^+e^- \rightarrow Z^0 H^0$, (b) $e^+e^- \rightarrow \nu_e \bar{\nu}_e H^0$.

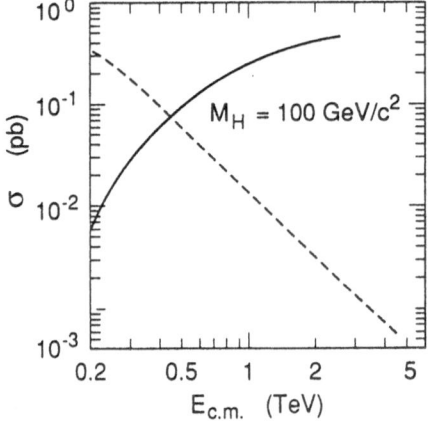

$M_H = 100$ GeV/c^2

Figure 2. Cross sections for the processes $e^+e^- \rightarrow Z^0 H^0$ (dashed curve) and $e^+e^- \rightarrow \nu_e \bar{\nu}_e H^0$ (solid curve) for a 100 GeV Higgs boson.

ignore particles within 10° of the beam axis. We assume, however, that the remainder of the detector is completely hermetic, and that the hardware performs well enough to avoid any further loss of sensitivity to the presence of particles in signal or background events.

In the CLIC study, calorimeter resolution is parametrized at the *parton* level. For example, it is assumed that jets from partons can be separated with a clustering algorithm if the angle between the initial partons is greater than 30°. Jets with energy less than 4 GeV are ignored as are events with a jet less than 15° from the beamline. The jet-energy resolution is assumed to be

$$\frac{\sigma_E}{E} = \frac{50\%}{\sqrt{E}} \text{GeV}^{1/2}$$

and $(10 \pm 5)\%$ of the jet energy is assumed to be lost.

Tagging of b Quarks

Because Higgs bosons are expected to decay preferentially through b quarks, the detection of the presence of b quarks in an event is an important tag which can be used to isolate the production of Higgs particles. The relatively long lifetime of the b quark gives rise to decay vertices displaced from the primary e^+e^- vertex or, equivalently, to tracks with large impact parameters. At a TeV linear collider, where the uncertainty in the position of the primary vertex is expected to be much smaller than the mean decay length of the b quark, especially in the plane perpendicular to the beam direction, precision vertex detectors may be used to identify such large-impact-parameter tracks which can then be exploited to enhance the signal-to-background ratio in searches for Higgs bosons. Only the TLC study considers the tagging of large-impact-parameter tracks. The exact simulation of the impact parameter measurement will be described in the discussion of selection criteria in the TLC study.

LEP 200 STUDY: $\sqrt{s} = 200$ GeV

The cross sections for Higgs boson production via $e^+e^- \to Z^0 H^0$ and for standard-model background processes are shown in Table 2 for $\sqrt{s} = 200$ GeV. Since the cross section for $e^+e^- \to Z^0 H^0$ is about 1 pb for a Higgs boson mass around $40 - 50$ GeV, about 500 events would be expected in 0.5 fb^{-1}, the sample size considered in the LEP 200 study. The cross sections for standard-model backgrounds are significantly larger, especially for $e^+e^- \to W^+W^-$ and $e^+e^- \to q\bar{q}$.

Table 2. Cross sections for Higgs-boson-production processes and for standard-model-background processes at $\sqrt{s} = 200$ GeV with no initial state radiation. (From Ref. 4.)

Process		Cross Section (pb)
$e^+e^- \to Z^0 H^0$,	$M_H = 40$ GeV	1.08
	$M_H = 60$ GeV	0.92
	$M_H = 80$ GeV	0.68
$e^+e^- \to W^+W^-$		16.1
$e^+e^- \to Z^0 Z^0$		2.2
$e^+e^- \to q\bar{q}$		74

There are three distinct signatures for $e^+e^- \to Z^0 H^0$, depending on how the Z^0 decays. All three were considered in the LEP 200 study. About 18% of the time, the Z^0 decays to $\nu\bar{\nu}$, about 6% to e^+e^- or $\mu^+\mu^-$, and about 73% to $q\bar{q}$. Although the case in which the Z^0 decays to a pair of charged leptons is the easiest to identify and generally the cleanest signature, it is also the one with the smallest branching fraction. The case in which the Z^0 decays to a quark - antiquark pair is the most copious but is the most difficult to identify. These three cases will now be discussed separately with the results summarized in Table 3. In all cases, it is assumed that the Higgs boson decays to $b\bar{b}$.

Table 3. Summary of results of LEP 200 study for $e^+e^- \to Z^0 H^0$. The data sample corresponds to 0.5 fb^{-1}. (From Ref. 4.)

Z^0 Decay Mode	M_H (GeV)	Number of Produced Events	Number of Reconstructed Events	Efficiency	Number of Background Events	Signal-to-Background Ratio
$Z^0 \to \nu\bar{\nu}$						
	40	107	49	46%	2	25
	60	83	34	41%	7	5
	80	56	12	21%	4	3
$Z^0 \to e^+e^-$ or $\mu^+\mu^-$						
	40	36	24	67%	0.2	Large
	60	28	17	61%	0.6	28
	80	18	11	61%	3	3.7
$Z^0 \to q\bar{q}$						
	40	430	54	13%	23	2.3
	60	340	60	18%	31	2

1. $e^+e^- \to Z^0 H^0$, $Z^0 \to \nu\bar{\nu}$

One would expect about 100 of these events in 0.5 fb^{-1} for a Higgs-boson mass of ≈ 50 GeV. The signature for this event is missing momentum transverse to the beamline,[*] missing mass equal to the Z^0 mass, and high sphericity in the rest frame of the visible particles. The exact selection criteria used in the LEP 200 study are listed here with the primary background which is removed by the cut indicated in parentheses:

— missing mass greater than 92 GeV ($e^+e^- \to W^+W^-$ and $e^+e^- \to q\bar{q}$);

— missing momentum transverse to the beam line greater than 30 GeV ($e^+e^- \to q\bar{q}$);

— missing momentum parallel to the beam line less than 40 GeV ($e^+e^- \to q\bar{q}\gamma$);

— sphericity in the rest frame of all observed particles greater than 0.02 ($e^+e^- \to W^+W^-$ and $e^+e^- \to Z^0 Z^0$);

— number of charged tracks greater than eight ($e^+e^- \to W^+W^-$);

— no charged lepton with energy greater than 25 GeV ($e^+e^- \to W^+W^-$);

— number of jets equal to two ($e^+e^- \to W^+W^-$ and $e^+e^- \to q\bar{q}$).

With the ALEPH simulation, the efficiency for the signal to pass the above selection criteria varies from 46% for a Higgs-boson mass of 40 GeV to 21% for a Higgs-boson mass of 80 GeV. This is shown in Table 3 along with the actual number of expected events for both the signal and background. The invariant mass of the observed particles in the event is calculated with the constraint that the missing mass in the event is equal to the mass of the Z^0. This sharpens the invariant mass peak of the Higgs boson considerably. The resulting invariant mass distribution is shown in Figure 3 for the signal (open region) and background (shaded region). These

---

[*] Missing momentum parallel to the beamline can be due to initial state radiation.

figures are from Ref. 4. For a Higgs-boson mass of 40 GeV, 49 signal events and 2 background events with an invariant mass within ± 10 GeV of the peak are expected to pass the selection criteria leading to a very clear signature for Higgs boson production. At a Higgs-boson mass of 60 GeV, the signal-to-background ratio is still high (≈ 5), but at 80 GeV, only 12 signal events are expected over a background of ≈ 4 events.

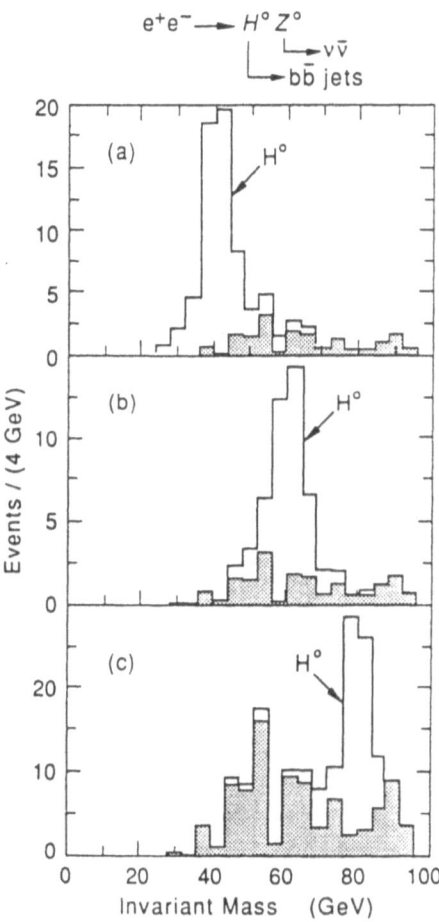

Figure 3. Invariant mass of all observed particles in the event with the constraint that the missing mass in the event equal the Z^0 mass, for signal (open region) and background (shaded region) for a Higgs-boson mass of (a) 40 GeV, (b) 60 GeV, and (c) 80 GeV. All distributions correspond to 0.5 fb^{-1} at $\sqrt{s} = 200$ GeV. (From Ref. 4.)

2. $e^+e^- \to Z^0 H^0$, $Z^0 \to e^+e^-$, $\mu^+\mu^-$

Only about 40 or 20 events of this type are expected for a Higgs-boson mass of 40 or 80 GeV, respectively. However, the selection criteria are simple and efficient:

— at least four charged tracks;

— a charged lepton pair with invariant mass greater than 60 GeV;

— missing mass in the event less than 15 GeV.

The recoil mass against the charged lepton pair is shown in Figure 4 for the OPAL detector simulation for a range of Higgs masses between 50 and 80 GeV. The selection criteria are about 60 - 70% efficient for the signal and the main background is, of course, from $e^+e^- \rightarrow Z^0 Z^0$. Table 3 lists the final signal-to-background ratios at the Higgs peak for a similar analysis using the ALEPH detector simulation. The signal is clear for a Higgs-boson mass up to about 80 GeV at which point the larger peak from $e^+e^- \rightarrow Z^0 Z^0$ dominates. In this analysis, the fact that the Higgs boson would decay almost completely to $b\bar{b}$ while the Z^0 background decays to $b\bar{b}$ only about 14% of the time was not used. The tagging of high-impact-parameter tracks would help to distinguish the Higgs-boson signal from the Z^0 background.

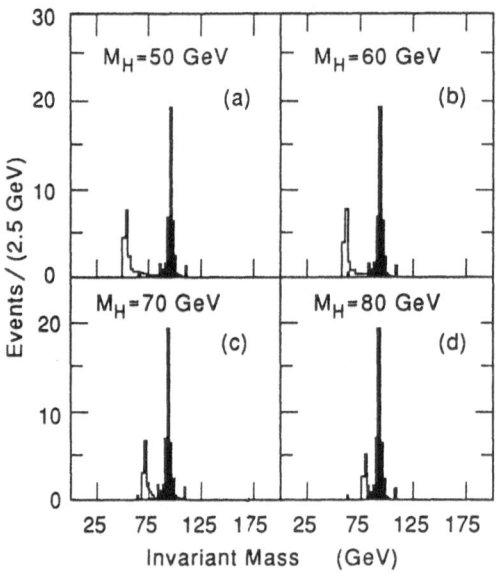

Figure 4. Recoil mass against the charged lepton pair for $e^+e^- \rightarrow Z^0 H^0$ where $Z^0 \rightarrow e^+e^-$ or $\mu^+\mu^-$ (open region) and for background (shaded region) for a Higgs-boson mass of (a) 50 GeV, (b) 60 GeV, (c) 70 GeV, and (d) 80 GeV. All distributions correspond to 0.5 fb^{-1} at $\sqrt{s} = 200$ GeV. (From Ref. 4.)

3. $e^+e^- \rightarrow Z^0 H^0$, $Z^0 \rightarrow q\bar{q}$

In a sample of 0.5 fb^{-1}, approximately 400 events of the type $e^+e^- \rightarrow Z^0 H^0$ where $Z^0 \rightarrow q\bar{q}$ would be produced if a Higgs boson with a mass in the range 40 - 60 GeV exists. The signature for such an event is four hadronic jets, two of which have an invariant mass equal to the Z^0 mass. In the LEP 200 study, with the ALEPH detector simulation, a cluster-finding algorithm is used to find four-jet events. The jet directions and energies are allowed to vary in a constrained fit to various hypotheses: that the jet-pair masses are equal (i.e. $e^+e^- \rightarrow W^+W^-$ or $e^+e^- \rightarrow Z^0 Z^0$), or that one jet-pair mass equals the Z^0 mass (i.e. $e^+e^- \rightarrow Z^0 H^0$). Then events are selected if they do not satisfy the hypothesis for $e^+e^- \to Z^0Z^0$ or $e^+e^- \to W^+W^-$ and do satisfy the hypothesis for $e^+e^- \to Z^0H^0$. Events are rejected if they contain a photon with energy greater than 40 GeV to eliminate $e^+e^- \to \gamma Z^0 \to \gamma q\bar{q}$. In addition, events are rejected if the sphericity is less than 0.15 or the missing momentum perpendicular to the beampipe is less than 2.5 GeV to eliminate $e^+e^- \to q\bar{q}$. The resulting constrained mass distribution for the jet pair opposite the Z^0 is shown in Figure 5 for a Higgs-boson mass of 40 and 60 GeV. The efficiency for identifying and reconstructing the signal events is only about 10-20% and the number of background events is high so that the final signal-to-background ratio is only about two. A Higgs boson with a mass greater than about 70 GeV is very difficult to distinguish from the $e^+e^- \to W^+W^-$ background.

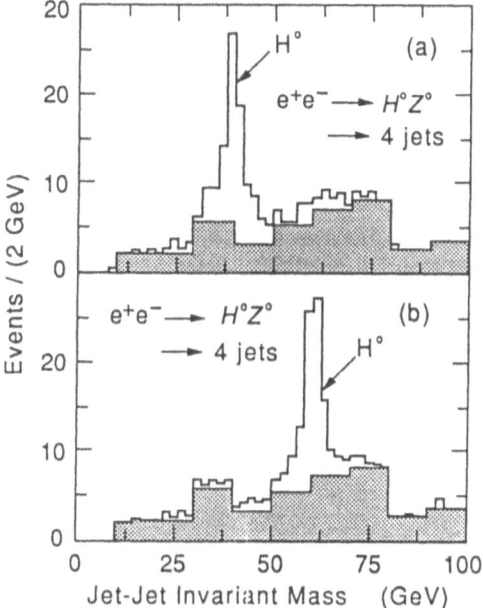

Figure 5. Jet-pair mass opposite a pair of jets consistent with the Z^0 mass from a constrained fit, for $e^+e^- \to Z^0H^0$ where $Z^0 \to q\bar{q}$ (open region) and for background (shaded region) for a Higgs-boson mass of (a) 40 GeV and (b) 60 GeV. All distributions correspond to 0.5 fb^{-1} at $\sqrt{s} = 200$ GeV. (From Ref. 4.)

4. SUMMARY FOR $\sqrt{s} = 200$ GeV

With 0.5 fb^{-1} of data at $\sqrt{s} = 200$ GeV, one is sensitive to Higgs boson production through $e^+e^- \to Z^0H^0$ for a Higgs-boson mass of less than about 80 GeV without tagging the b quarks from the decay $H^0 \to b\bar{b}$. For a mass greater than about 80 GeV, the backgrounds from $e^+e^- \to W^+W^-$ and $e^+e^- \to Z^0Z^0$ dominate.

The case in which the Z^0 decays to a charged lepton pair results in the smallest but cleanest signal whereas the case in which the Z^0 decays to a quark-antiquark pair results in the largest signal but the lowest signal-to-background ratio. Knowledge of the total energy and momentum of the event allows one to calculate constrained masses which significantly improve the signal-to-noise ratio.

TLC Study: $\sqrt{s} = (0.5 - 1)$ TeV

In this section, I will discuss searches for the minimal neutral Higgs boson, the charged Higgs boson and the nonminimal CP-even and CP-odd neutral Higgs bosons. With the exception of the Higgs particles that we are studying, we have assumed that the standard model with three generations of quarks and leptons is a correct description of nature. The LUND 6.3 model[14] with full showering of quark and gluon partons has been used to generate QCD background events and to fragment into hadrons final-state quarks and gluons in all signal and background processes. The mass of the top quark was assumed to be 40 GeV, unless otherwise specified. Numerical evaluations of the cross sections for signal and background processes at $\sqrt{s} = 1$ TeV are given in Table 4.

Table 4. Cross sections at a center-of-mass energy of 1 TeV with and without the effects of beamstrahlung.

Process		Cross Section (fb)	
		Without Beamstrahlung	With Beamstrahlung
$e^+e^- \to Z^0 H^0$	$m_H = 50$ GeV	12.7	43
	$m_H = 100$ GeV	12.4	35
	$m_H = 150$ GeV	12.0	25
	$m_H = 300$ GeV	9.8	13
	$m_H = 500$ GeV	5.6	4
$e^+e^- \to \nu_e \bar{\nu}_e H^0$	$m_H = 50$ GeV	234	196
	$m_H = 100$ GeV	216	163
	$m_H = 150$ GeV	147	124
	$m_H = 300$ GeV	70	53
	$m_H = 500$ GeV	23	13
$e^+e^- \to W^+W^-$		2310	3570
$e^+e^- \to Z^0 Z^0$		127	200
$e^+e^- \to q\bar{q}$		438	4050
$e^+e^- \to e^+\nu_e W^-$		9440	8770
$e^+e^- \to \nu_e \bar{\nu}_e W^+W^-$		800	
$e^+e^- \to e^+\nu_e W^- Z^0$		300	

1. Impact Parameter Tagging

We investigate the possibility of identifying b-quark decays through high-impact-parameter tracks by simulating the performance of a combined vertex detector and central tracking chamber in our Monte Carlo and using the impact parameter information in the analysis. The simulation is kept simple and generic in order that the success of the analysis not be dependent on specific details of detector design. Each track in an event is projected into the plane transverse to the beam line and the impact parameter is defined as the distance of closest approach of the track projection to the primary vertex. We compute the impact parameter for each track using the generated four-vectors and smear these calculated values with a Gaussian impact parameter resolution function with standard deviation $\sigma = \sqrt{A^2 + (B/p)^2}$ for a track with momentum p. The details of the vertex and tracking system are encompassed in the parameters A and B, which reflect in turn the intrinsic resolution of the tracking system and the effect of multiple scattering of particles as they traverse the beam pipe and detector. We choose values of A and B that are characteristic of combined vertex and tracking systems currently in use, namely $A = 5\,\mu$m and $B = 50\,\mu$m GeV. We assume that the uncertainty in the location of the collision point is small compared to the impact parameter resolution. For the analysis in this paper, a charged track is defined to be a 'high-impact-parameter track' if it lies within $|\cos\theta| \leq 0.9$ and has a (smeared) impact parameter less than 3 mm and greater than three times the resolution $\sigma = \sqrt{A^2 + (B/p)^2}$. The 3 mm upper limit suppresses contamination from K_S and Λ decays.

2. HEAVY HIGGS PARTICLES ($M_H \gtrsim 2M_W$)

If the mass of the Higgs boson exceeds twice the W mass, then the process $e^+e^- \to \nu_e\bar{\nu}_e H^0$ will be the most favorable source for its discovery. With our assumption that the top quark mass is less than the mass of the W, the Higgs boson will decay essentially 100% of the time to W^+W^- or Z^0Z^0. We note that if the mass of the top quark is larger than M_W, then the decay $t \to Wb$ will proceed immediately, and since the W will carry most of the momentum of the top quark, this case will be very similar to the direct decay of the Higgs boson to W-pairs.

The signature of the Higgs boson is a final state of two $W^\pm$ or Z^0 bosons which are typically not coplanar due to the transverse momentum generated by the boson propagators present in the production mechanism. (See Figure 1.) We define acoplanarity as the angle between the two planes that each contain the beam axis and one jet axis. Acoplanarity of zero degrees implies that the two jet axes are pointing in opposite directions in a projection perpendicular to the beam axis. It is very useful to work with momenta in the plane perpendicular to the beam axis in the presence of beamstrahlung, because the beamstrahlung provides a boost for the event along the beam axis but not perpendicular to it. The basic philosophy of our analysis is to search for events in which both W bosons decay to hadrons with little or no energy carried away by neutrinos or other particles that are not observed in the detector. The topology and kinematics of the signal are then used to eliminate backgrounds from QCD and two-photon processes and from $e^+e^- \to W^+W^-$ events in which a neutrino in the decay chain creates a significant acoplanarity angle.

In all of the analyses described in this section, we first boost all smeared four-vectors along the z-axis until the vectorial sum of all the z-components of momenta is zero. A thrust analysis is then performed in this boosted frame, and the event is rejected if the angle between the thrust axis and the beam, θ_{thr}, does not satisfy the condition $|\cos\theta_{thr}| < 0.8$. This procedure ensures that the event is well contained in the detector and preferentially rejects processes with differential cross sections which are sharply peaked along the beam axis (for example, $e^+e^- \to W^+W^-$, $e^+e^- \to Z^0Z^0$, and $e^+e^- \to q\bar{q}$).

A cluster analysis is performed with the detected particles in which the cluster-finding algorithm is constrained to divide the event into exactly two clusters. The invariant mass and the vectorial sum of the momenta for each cluster is calculated. We then select events in which the acoplanarity of these cluster momenta is greater than $10°$ and the net momentum in the event transverse to the beam line is greater than 50 GeV. These criteria remove two-photon events and $e^+e^- \rightarrow W^+W^-$ and $e^+e^- \rightarrow q\bar{q}$ events with one or more neutrinos in the final state. To isolate events with W^+W^- in the final state, the mass of the cluster with the smaller mass is required to lie between 66 and 94 GeV while that of the opposite cluster is required to lie between 75 and 110 GeV.

To further reduce backgrounds due to events with large beamstrahlung radiation and/or missing particles, we select events in which the direction of the missing momentum satisfies $|\cos\theta_{miss}| < 0.9$, where θ_{miss} is the angle between the beam axis and the missing momentum in the laboratory frame. We also reject events in which the total visible energy is greater than 600 GeV. This cut removes backgrounds but is extremely efficient for retaining the signal events.

The invariant mass of all particles that are detected in events that pass these requirements is calculated and shown in Figures 6(a) and 6(b) for Higgs-boson masses of 300 and 500 GeV, respectively. The figure gives the number of events that would be detected in a data sample of integrated luminosity 30 fb^{-1}. The background which is shown in the figure as a dashed line is dominated by the $e^+e^- \rightarrow W^+W^-$ process, but includes smaller contributions from $e^+e^- \rightarrow q\bar{q}$ and two-photon events.

In a CERN study[15] of search strategies for a heavy Higgs boson at CLIC ($\sqrt{s} = 2$ TeV), it was found that the process $e^+e^- \rightarrow e^+\nu_e W^- Z^0$ constitutes a significant background to signals from Higgs bosons with masses above 500 GeV. This background was not simulated in detail in our study but is not as significant[16] for searches for $M_{H^0} \lesssim 500$ GeV at $\sqrt{s} = 1$ TeV because the background cross section, compared to that of the signal, is lower and the Higgs boson resonance is considerably narrower. The cross section for this background depends strongly on the energy spectrum of beamstrahlung photons created during the collision.

The number of signal events surviving the cuts is 125 for a 300 GeV Higgs boson and 46 for a 500 GeV Higgs boson; either would easily be observed above the background. Higgs particles with masses above 500 GeV become more difficult to observe because the production cross section decreases with mass and because the width of the Higgs boson resonance increases as M_H^3.

3. INTERMEDIATE MASS HIGGS BOSON ($M_H \lesssim 2M_W$)

A Higgs boson with mass less than $2M_W$ (and greater than ≈ 10 GeV) will decay to either a top or bottom quark pair depending on the masses of the Higgs boson and the top quark. The philosophy of our analysis of the $e^+e^- \rightarrow \nu_e\bar{\nu}_e H^0$ reaction with $M_H < 2M_W$ is to search for events that contain two acoplanar jets each with mass below that of the W. We select events in which $|\cos\theta_{thr}| < 0.7$ and $|\cos\theta_{miss}| < 0.9$ where θ_{thr} and θ_{miss} are the polar angles of the thrust axis and the missing momentum in the event, respectively. We divide the event into two clusters using a clustering algorithm and select events in which the acoplanarity of the clusters is greater than 10° and in which each cluster has an invariant mass greater than 1 GeV (to reject leptonic decays of the $W^\pm$) and less than 50 GeV. Finally, events are accepted only if the missing momentum transverse to the beam is greater than 50 GeV and the number of charged particles outside the 10° hole around the beam axis is between 10 and 36. The probability that an event of the type $e^+e^- \rightarrow \nu_e\bar{\nu}_e H^0$ will pass the above selection criteria varies between about 35% and 50% depending on the mass and the decay mode of the Higgs boson. The background is dominated by the process $e^+e^- \rightarrow e^+\nu_e W^-$. When the invariant mass of the event is plotted in 4 GeV bins the peak bin for the background distribution contains about 25 to 50 times as many events as the peak bin for the signal! However, it should be possible to discover a Higgs boson with mass greater than about 150 GeV using just the above selection

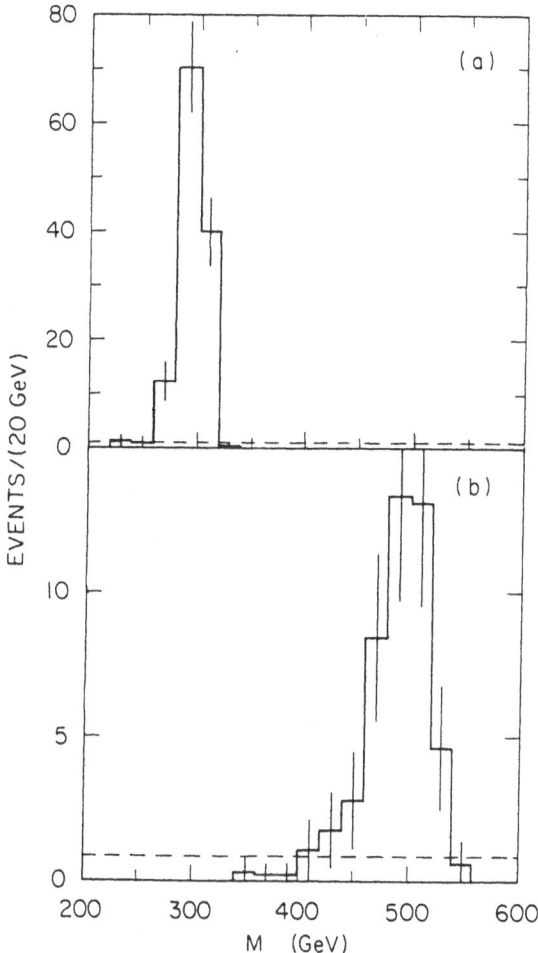

Figure 6. Reconstructed mass distribution, after selection criteria, for a Higgs-boson mass of (a) 300 GeV and (b) 500 GeV. The Higgs boson is produced in the process $e^+e^- \to \nu_e \bar{\nu}_e H^0$.

criteria.[6] The observability of Higgs bosons with masses closer to M_W depends critically on the resolution of the detector. For example, the Higgs boson signal remains significant down to a mass of about 130 GeV if the resolution of the hadronic calorimeters in the detector is improved from $\frac{50\%}{\sqrt{E}}\text{GeV}^{1/2}$ to $\frac{35\%}{\sqrt{E}}\text{GeV}^{1/2}$.

To further enhance the signal-to-background ratio, we count the number of high impact parameter (h.i.p.) tracks in each event. The definition of a h.i.p. track was given in Sec. 3.1. Events are accepted if the number of h.i.p. tracks is at least four. We first consider the case in which the H^0 decays to $b\bar{b}$. The efficiency for signal events to pass the cut on h.i.p. tracks, after all other cuts, is about 92% while for the background it is about 30%. The invariant mass distributions for a 120 GeV Higgs boson and a 150 GeV Higgs boson are shown in Figs. 7a and 7b, respectively. The figures include standard model backgrounds. The 150 GeV Higgs boson signal is clearly separated from the large W peak due to the $e^+e^- \to e^+\nu_e W^-$ background. However, the 120 GeV Higgs boson signal is just a shoulder on the side of the $W^\pm$ mass peak. The 120 GeV signal stands out clearly, though, if the resolution of the hadron calorimeter is improved from $\frac{50\%}{\sqrt{E}}\text{GeV}^{1/2}$ to $\frac{35\%}{\sqrt{E}}\text{GeV}^{1/2}$.

For Fig. 7, we assumed that the Higgs boson decays to a bottom quark pair. If we assume that the Higgs boson decays to a top quark pair, we can further increase the signal-to-background ratio by accepting events only if they have at least four clusters found by an algorithm which groups particles until further additions to the group would result in a cluster of particles with invariant mass greater than a value larger than the bottom quark mass but less than the top quark mass. The technique does not depend critically on the exact cutoff mass. In Fig. 8, we plot the invariant mass of events with at least four high-impact-parameter tracks and at least four clusters found by the clustering algorithm. The distribution includes a 150 GeV Higgs boson and backgrounds. The W peak from the $e^+e^- \to e^+\nu_e W^-$ background has been further reduced so that now the peak signal is over half of the peak background.

We conclude that impact parameter tagging is a useful tool for enhancing the signal of an intermediate mass Higgs boson relative to standard model backgrounds. In addition, if the Higgs boson decays to $t\bar{t}$, the number of low mass clusters provides a useful discriminator between signal and background.

4. Charged Higgs Boson

Charged Higgs bosons can be produced in pairs in e^+e^- annihilation via a virtual photon, or via a real or a virtual Z boson. The cross section is not very model dependent except near the Z peak since the photon couples to all charged particles with the same strength. The cross section relative to that for $e^+e^- \to q\bar{q}$, $e^+e^- \to W^+W^-$ and $e^+e^- \to Z^0Z^0$, at $\sqrt{s} \approx (0.5 - 1.0)$ TeV, is approximately given by

$$\frac{\sigma(e^+e^- \to H^+H^-)}{\sigma(e^+e^- \to q\bar{q}) + \sigma(e^+e^- \to Z^0Z^0) + \sigma(e^+e^- \to W^+W^-)} \approx \frac{0.30 \cdot \beta^3}{7 + 20 + 1} \approx 0.01 \cdot \beta^3$$

where β is the velocity of the charged Higgs boson. The large effects due to radiative corrections and beamstrahlung effects are not included here. Therefore, one can expect approximately a few hundred pairs of charged Higgs bosons to be produced at a high energy e^+e^- collider, if the $H^\pm$ exists with a mass below $\sqrt{s}/2$. If the top quark mass is greater than that of the charged Higgs boson plus the bottom quark, top quarks, which are produced in pairs at a rate roughly an order of magnitude greater than that for H^+H^-, can decay to a charged Higgs boson plus a bottom quark: $t \to H^+b$.

The decay process $H^- \to b\bar{t}$ is expected to be the dominant decay mode if it is kinematically allowed. Otherwise, the decay mode $H^- \to \tau^-\bar{\nu}_\tau$ can be significant. Another possibility is the

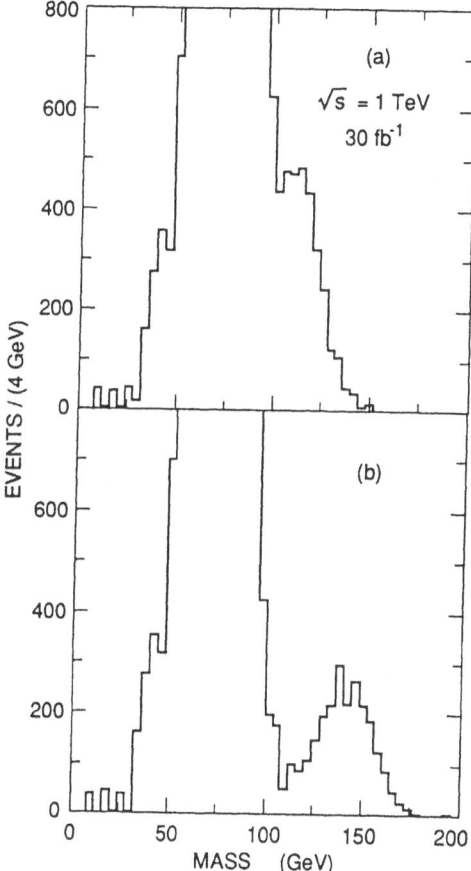

Figure 7. Reconstructed mass distribution, after selection criteria, for background events plus (a) a 120 GeV Higgs boson and (b) a 150 GeV Higgs boson. The Higgs boson is produced in the process $e^+e^- \rightarrow \nu_e \bar{\nu}_e H^0$ at $\sqrt{s} = 1$ TeV and decays to $b\bar{b}$. The selection criteria include a cut on the number of high-impact-parameter tracks. The number of events in the peak bin near the $W^\pm$ mass is about 3000.

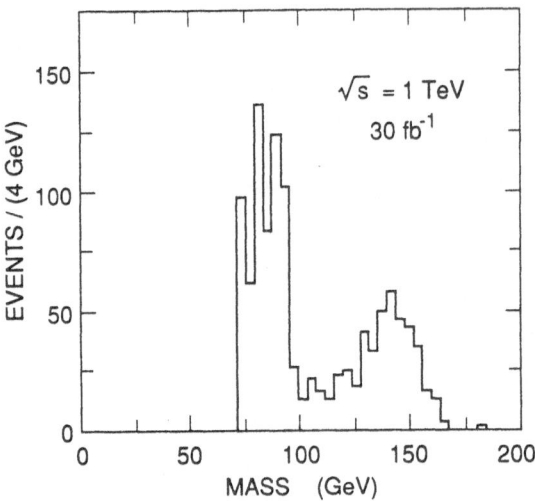

Figure 8. Reconstructed mass distribution, after selection criteria, for background events plus a 150 GeV Higgs boson. The Higgs boson is produced in the process $e^+e^- \rightarrow \nu_e \bar{\nu}_e H^0$ at $\sqrt{s} = 1$ TeV and decays to $t\bar{t}$. The selection criteria include a cut on the number of high-impact-parameter tracks and on the number of clusters found.

mode $H^- \rightarrow H_i^0 W^+$ where H_i^0 is one of the physical neutral Higgs bosons.

Because of the two possible production mechanisms for $H^\pm$, the large variety of possible decay modes, and the special problems of a charged Higgs boson with mass close to the $W^\pm$ mass, there are many cases to consider. Most of the possible cases have been considered in detail in Ref. 7. I will briefly outline the signature or search strategy for each case and the result of the study for a sample of 10 fb^{-1}. The details of the selection criteria can be found in Ref. 7. The main conclusion is that, with an integrated luminosity of 10 fb^{-1}, a charged Higgs boson with mass as large as 80% of the beam energy is not difficult to find at an e^+e^- collider with $\sqrt{s} = (0.5 - 1)$ TeV. The region around the $W^\pm$ mass should not present special difficulties. The number of detected events is small however, so that a detailed study of the charged Higgs boson would require a larger data sample.

$\underline{M_{H^\pm} > M_t + M_b}$ Here we assume that the charged Higgs bosons are produced in pairs and each one decays to $t\bar{b}$. We consider separately the cases when $M_{H^\pm}$ is near or above $M_{W^\pm}$.

- $M_{H^\pm}$ close to $M_{W^\pm}$ - The signature is two equal mass jets each with a large number of charged particles and a large number of high-impact-parameter tracks. By selecting events with these properties, one arrives at a significant signal of ≈ 50 events with little background.

- $M_{H^\pm} \gtrsim 120$ GeV - The signature is four jets, with two pairs having the same invariant mass, with many high-impact-parameter tracks. Over 50 events can be selected with negligible background for $M_{H^\pm}$ up to about 80% of the beam energy.

$\underline{M_{H^\pm} < M_t + M_b}$ Here we consider two cases correponding to different $H^\pm$ production mechanisms.

- $e^+e^- \rightarrow H^+H^-$, $H^+ \rightarrow c\bar{s}$, $H^- \rightarrow \tau^-\bar{\nu}_\tau$ - The signature for this case is a high-invariant-mass jet ($c\bar{s}$) versus a low-invariant-mass jet (τ^-). Very simple selection criteria yield a significant signal with neglible background for large $M_{H^\pm}$. For the special case when $M_{H^\pm} \approx M_{W^\pm}$, a significant signal can only be achieved if one selects a high-invariant-mass jet versus an isolated pion. The sign of the charge and the direction of the isolated pion are used to reject the large W^+W^- background which has a strong forward-backward charge asymmetry.

- $e^+e^- \to t\bar{t}, t \to H^+ b, \bar{t} \to H^- \bar{b}$ - If the charged Higgs boson decays to $\tau^- \bar{\nu}_\tau$, the signature is a ratio of isolated pions to isolated leptons which is large compared to that expected from $W\pm$ decay. For a top quark mass of 200 GeV and a charged Higgs boson mass of 150 GeV, the effect has a significance of five standard deviations. The significance increases for a lower top quark mass.

$M_{H\pm} > M_{W\pm} + m_{H_i^0}$ We now consider the case in which the two charged Higgs bosons decay to $W^\pm H_i^0$. The H_i^0 is assumed to decay to $b\bar{b}$. If one $W^\pm$ decays to a pair of quarks and the other to a charged lepton and neutrino, the whole event will contain one isolated lepton and a large number of high-impact-parameter tracks. It is possible to select events with these properties with a large signal-to-background ratio if the relevant Higgs boson branching fractions are large enough

Summary of Search Strategies for Charged Higgs Bosons   With 10 fb^{-1} of data at $\sqrt{s} = (0.5 - 1)$ TeV it is possible to select events with charged Higgs bosons with little background for $M_{H\pm}$ less than about 80% of the beam energy. The number of selected events is typically around 50 but depends on the mass, production mechanism and decay mode. In almost all cases, the mass of the charged Higgs boson could be determined to better than 5 GeV. The region around the $W^\pm$ mass must be treated separately but a charged Higgs boson signal would still be significant.

At LEP-II it may be difficult to detect a charged Higgs boson with mass around the W mass with a luminosity of 500 pb^{-1}. The R value for a 83 GeV $H^\pm$ at $\sqrt{s} = 200$ GeV is only 0.057, so the number of expected events after cuts is very small and the W background is larger. At LEP energies, the event topology is not two back-to-back jets and the particles from H^+ and H^- will be mixed with each other due to the small velocity of the $H^\pm$. Therefore, a more complicated analysis may be required. In any case, the b-tagging and the polar angle cuts discussed in this report are useful at any energy.

5. NONMINIMAL NEUTRAL HIGGS BOSONS

In the simplest extension of the standard model, two Higgs doublets yield five physical Higgs particles, two charged and three neutral. The neutral Higgs particles consist of two CP-even states (h^0 and H^0) and a CP-odd state (A^0). The masses of the charged and three neutral Higgs bosons, and two mixing angles are free parameters in the most general case. Minimal supersymmetry can be used[17] to impose constraints which reduce the number of free parameters from six to two. These may be chosen to be the charged Higgs boson mass, $M_{H\pm}$, and the ratio of the vacuum expectation values of the two Higgs doublets, expressed as $\tan\beta$. Once $M_{H\pm}$ and $\tan\beta$ are specified, production cross sections, decay branching ratios, and masses of the neutral Higgs particles can be calculated. The results of Ref. 17 may be summarized briefly. The light scalar h^0 behaves much like the standard model Higgs boson and may be produced by the familiar mechanism $e^+e^- \to Z^* \to Zh^0$ with some possibility for discovery at SLC or LEP. By contrast, these production mechanisms are forbidden for the A^0 and suppressed for the H^0. The only production mechanism available to these heavy Higgs particles that has adequate cross section is $e^+e^- \to Z^* \to H^0 A^0$. At 1 TeV, the largest predicted cross section is approximately 10 fb, or about one tenth the lowest order QED cross section for $e^+e^- \to \mu^+\mu^-$.

In Ref. 8, we investigate the sensitivity of searches for nonminimal neutral Higgs bosons at high energy e^+e^- colliders. For this study we assume that the charged Higgs mass is 150 GeV,

Table 5. Parameters used to generate $e^+e^- \to H^0A^0$.

$M_{H^\pm} = 150$ GeV	$BR(H^0 \to t\bar{t}) = 79.6\%$
$M_{H^0} = 153.4$	$BR(H^0 \to b\bar{b}) = 6.5\%$
$M_{A^0} = 125.7$	$BR(H^0 \to h^0h^0) = 13.6\%$
$M_{h^0} = 29.2$	$BR(A^0 \to t\bar{t}) = 95\%$
$M_t = 60$	$BR(A^0 \to b\bar{b}) = 5\%$
$\tan\beta = 1.5$	$BR(h^0 \to b\bar{b}) = 100\%$

the top quark mass is 60 GeV, and the ratio of the Higgs vacuum expectation values $\tan\beta$ is 1.5. The masses of the neutral Higgs bosons and their decay modes and branching ratios with these assumptions are given in Table 5. We note that $M_{H^0} = 153$ GeV, $M_{A^0} = 126$ GeV, and the branching ratios are dominated by $t\bar{t}$ and $b\bar{b}$. The choice of 150 GeV for the charged Higgs mass is motivated by two considerations. If $M_{H^\pm} \leq 135$ GeV then $M_{A^0} \leq 100$ GeV and there is little hope of seeing the signal over the W and Z background. As $M_{H^\pm}$ becomes larger than 150 GeV, the signal is further from the W and Z background, and becomes only easier to see. Moreover, for large $M_{H^\pm}$, $M_{A^0} = M_{H^0}$, giving one additional handles on the signal that may be exploited. The choice of $M_{H^\pm} = 150$ GeV was therefore felt to be the most challenging case with any prospect of success.

Events of the type $e^+e^- \to H^0A^0$ are characterized by two jets of unknown, and not necessarily equal, mass. To maintain the highest level of generality, the analysis exploits only two features of the events: the fact that the jets should be distributed as $\sin^2\theta$, and the fact that many b quarks will be produced, leading to events with high multiplicity and tracks with large impact parameters. The dominant standard model backgrounds are $e^+e^- \to q\bar{q}$, $e^+e^- \to W^+W^-$ and $e^+e^- \to Z^0Z^0$. In contrast to H^0A^0 events, these backgrounds are characterized by angular distributions peaked in the forward and backward direction, by lower multiplicity, and by few tracks with large impact parameters. The cuts on multiplicity and impact parameter prove to be the most powerful in enhancing the signal-to-background ratio.

In a run of 10 fb^{-1} integrated luminosity, one expects to produce approximately 112 H^0A^0 events, 40000 $q\bar{q}$ events, 35000 W^+W^- events, and 2000 Z^0Z^0 events. To simulate such a run, we generate the correct numbers of background events, and 1120 H^0A^0 events which are then weighted by 0.1. We apply the following cuts:

1. The total number of detected charged tracks in the detector volume ($|\cos\theta| \leq 0.9$) is at least 40. This favors events with b quarks which tend to have high multiplicities.

2. The number of detected charged tracks in the central region of the detector ($|\cos\theta| \leq 0.71$) is at least 20. This cut favors events with $\sin^2\theta$ distribution.

3. The total number of high impact parameter tracks (as defined previously) is at least 16. This strongly favors events with b quarks.

4. When the event is divided into two jets by a cluster finding algorithm,[18] then the cluster containing fewer high impact parameter tracks must still contain at least seven such tracks. This tends to reject background events in which one jet has a large number of high impact parameter tracks due to a statistical fluctuation.

5. The invariant mass of each jet is formed and the event is rejected if *both* jets have a mass in the range 63 GeV to 112 GeV. This helps reject remaining W^+W^- and Z^0Z^0 events. Naturally it also excludes any possibility of finding H^0A^0 events with masses in

Table 6. Selection criteria and number of events which satisfy each criteria for the nonminimal neutral Higgs boson search. The simulated samples correspond to 10 fb^{-1} at $\sqrt{s} = 1$ TeV.

cut	$H^0 A^0$	$q\bar{q}$	$W^+ W^-$	Z^0
generated sample	1120	40000	35000	2000
total charged multiplicity ≥ 40	774	5934	995	53
central multiplicity ≥ 20	753	5505	853	45
# high impact parameter tracks ≥ 16	426	128	65	6
# h.i.p tracks on side with fewer ≥ 7	305	71	32	2
not both in 63 $GeV < M < 112$ GeV	302	66	14	1

this range, but in doing so enhances the prospects for finding such events with masses *near* the W and Z^0.

The results of this analysis are not sensitive to the exact number of detected charged tracks required in the first two cuts. The number required would have to be tuned for a particular detector configuration.

Table 6 shows the impact of each of these cuts on the signal and background events. The net efficiency for the $H^0 A^0$ signal is 27% and the rejection power against standard model backgrounds is approximately 1000:1. For a 10 fb^{-1} run we expect 30 signal and 81 background events, giving an overall signal-to-background ratio of about 1:3. A nonstandard model background which we do not address here is the process $e^+ e^- \to \gamma^*, Z^* \to H^+ H^-$ which is expected to have a cross section approximately three times larger than the $H^0 A^0$ cross section. Such events will pass our selection criteria. In the minimal supersymmetric model, it is predicted that $M_{H^\pm} \sim M_{H^0} \sim M_{A^0}$ as the mass of the charged Higgs particle becomes heavier ($M_{H^\pm} \gtrsim 150$ GeV). For such a case, it would be very difficult to distinguish the Higgs particles from one another unless some new characteristic parametrization other than the invariant mass is found. Outside the context of supersymmetry, however, no relation among the Higgs masses is predicted and the mass peaks need not be close to one another at all.

The invariant masses of the clusters are calculated with a beam energy constraint included. The beam-constrained mass calculation balances the momentum of the two clusters and the (unobserved) beamstrahlung photon in both the beam direction and the plane transverse to the beam direction, and requires in addition that the three energies sum to 1 TeV. We make the approximation that all beamstrahlung and initial state radiation is carried by a single photon. This constrained fit proves to be essential for identification of mass peaks above background.

Figures 9a and 9b show the mass distributions for the higher mass cluster and the lower mass cluster, after all analysis cuts are applied. In Fig. 9a the higher mass cluster stands out clearly at the correct (*i.e.* generated) mass of ~ 150 GeV while in Fig. 9b the lower mass cluster (expected at ~ 125 GeV) is obscured by the presence of $W^\pm$ and Z^0 backgrounds. The lower mass cluster may be enhanced by applying a cut on the higher mass cluster. Fig. 10a shows that when the higher mass is required to lie inside the range 120 to 200 GeV, a choice motivated by Fig. 9a, the lower mass cluster stands out more clearly. By contrast, Fig. 10b shows the lower mass cluster distribution when the higher mass is required to lie *outside* the range 120 − 200 GeV; in this case there is no evidence for a mass peak.

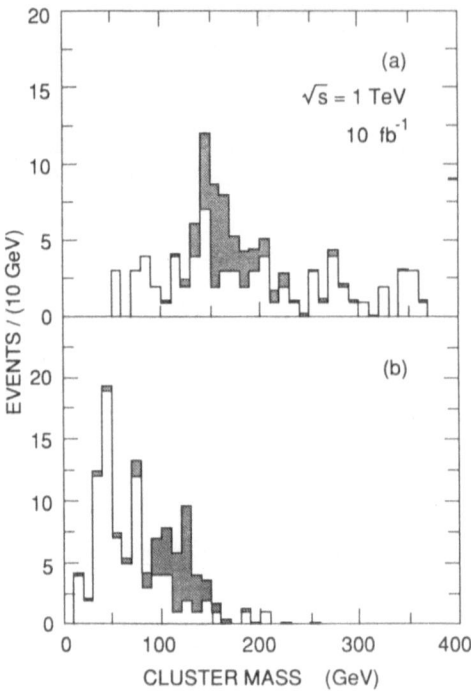

Figure 9. Reconstructed mass distribution for (a) the larger mass cluster and (b) the smaller mass cluster when all analysis cuts are applied. The shaded region is signal; the open region is background.

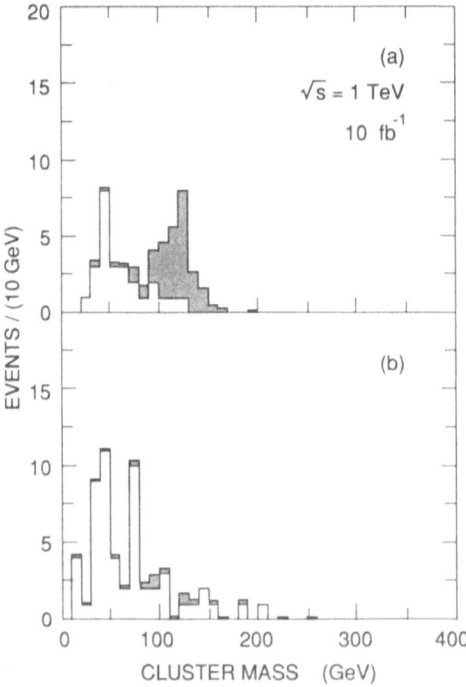

Figure 10. Reconstructed mass distribution of the smaller mass cluster for events in which (a) the larger mass lies *inside* the range 120 − 200 GeV and (b) the larger mass lies *outside* the range 120 − 200 GeV. The shaded region is signal; the open region is background.

We conclude that with modest impact parameter resolution and simple cuts designed to enhance events with b quarks and central distributions, a sample of 10 fb^{-1} integrated luminosity is sufficient to identify the presence of $H^0 A^0$ events with Higgs boson masses larger than $\sim$ 110 GeV and with decay modes dominated by heavy quarks.

CLIC STUDY: $\sqrt{s} = 2$ TeV

In reference 15, search strategies for a minimal neutral Higgs boson with mass in the range 500 GeV to 1 TeV at $\sqrt{s} = 2$ TeV were studied. After applying selection criteria (at the parton level) to simulate reconstruction of four jets from $e^+ e^- \to \nu_e \bar{\nu}_e H^0$ where $H^0 \to W^+ W^-$ and the $W^\pm$ bosons decay hadronically, three cuts are applied:

— the net momentum of the $W^+ W^-$ system perpendicular to the beam line is selected to be greater than 20 GeV;

— the angle θ_W^*, in the $W^+ W^-$ c.m. frame, between the W and the direction of the $W^+ W^-$ system in the laboratory frame is selected to satisfy $|\cos \theta_W^*| < 0.8$;

— events are rejected if there is a high energy electron or positron greater than 50 mr from the beam line.

The last criteria rejects $e^+ e^- \to e^+ e^- W^+ W^-$ events.

The results of applying these criteria to the signal process and to the two major background processes $e^+ e^- \to e^+ e^- W^+ W^-$ and $e^+ e^- \to e^+ \nu_e W^- Z^0$ are shown in Tables 7 and 8, and Figure 11 for Higgs boson masses in the range 500 GeV to 1 TeV. For the 500 GeV and 800 GeV masses, a sample size of 10 fb^{-1} was considered while for the 1 TeV mass, 50 fb^{-1} was considered. The efficiency for retaining the signal events with the above criteria is about 30%. In all cases, the number of signal events passing the criteria is at least a few hundred for the sample sizes considered. The dominant background is from $e^+ e^- \to e^+ \nu_e W^- Z^0$. In the region of the Higgs mass, the signal-to-background ratio is about 4, 2 and 1 for Higgs boson masses of 500, 800 and 1000 TeV, respectively.

The conclusion of the CLIC study is that an $e^+ e^-$ collider operating at a c.m. energy of 2 TeV would be sensitive to Higgs boson masses up to about 1 TeV.

CONCLUSIONS

Monte Carlo studies have been used to investigate the sensitivity of high-energy $e^+ e^-$ colliders to Higgs boson production. The general conclusion is that although large data samples are needed, clean signals can be extracted with fairly simple selection criteria over most of the kinematically-allowed mass range. The specific conclusions are the following:

- Minimal neutral Higgs boson - Experiments at $e^+ e^-$ colliders are sensitive to Higgs boson masses up to about $\sqrt{s}/2$ with data samples corresponding to integrated luminosities of $\approx$ 10fb$^{-1} \cdot \sqrt{s}$ TeV^{-1}. However, the LEP 200 study[4] concludes that the maximum Higgs boson mass accessible at $\sqrt{s} = 200$ GeV is about 80 GeV, while the TLC studies[6] conclude that a collider operating at $\sqrt{s} \approx 1$ TeV is not very sensitive to Higgs boson masses greater than the $W^\pm$ mass but less than about 120 GeV. It has been argued[19] that an $e^+ e^-$ collider with $\sqrt{s} \approx 300$ GeV would be sensitive to a Higgs boson with mass in the range 80 - 120 GeV. The strategy is to use the production mechanism $e^+ e^- \to Z^0 H^0$, to identify the Z^0 through its leptonic decays, to use b quark tagging to enhance the signal, and to identify the Higgs boson mass peak in the distribution of the recoil mass against the lepton pair. In Ref. 19, it is argued that one needs only about

Table 7. Selection criteria and number of events which satisfy each criteria for signal $(e^+e^- \rightarrow \nu_e \bar{\nu}_e H^0)$ and background $(e^+e^- \rightarrow e^+e^- W^+W^-$ and $e^+e^- \rightarrow e^+\nu_e W^- Z^0)$ processes for a Higgs boson mass of 500 GeV at $\sqrt{s} = 2$ TeV. The sample size corresponds to an integrated luminosity of 10 fb^{-1}. The letter V in the selection criteria refers to a vector boson (either $W^\pm$ or Z^0). The selection criteria are described in more detail in the text. (From Ref. 15.)

Selection Criteria	$e^+e^- \rightarrow \nu_e \bar{\nu}_e H^0$	$e^+e^- \rightarrow e^+e^- W^+W^-$	$e^+e^- \rightarrow e^+\nu_e W^- Z^0$
Produced events with $450 < M_{VV} < 550$	1050	3000	1200
Purely hadronic decays	550	1400	660
After acceptance and jet reconstruction	450	450	225
$p_T^{VV} > 20$ GeV	430	170	210
$|\cos\theta_V^*| < 0.8$	390	80	90
No $e^\pm$ above 50 mr	390	20	80

Table 8. Selection criteria and number of events which satisfy each criteria for signal $(e^+e^- \rightarrow \nu_e \bar{\nu}_e H^0)$ and background $(e^+e^- \rightarrow e^+e^- W^+W^-$ and $e^+e^- \rightarrow e^+\nu_e W^- Z^0)$ processes for a Higgs boson mass of 1 TeV at $\sqrt{s} = 2$ TeV. The sample size corresponds to an integrated luminosity of 50 fb^{-1}. The letter V in the selection criteria refers to a vector boson (either $W^\pm$ or Z^0). The selection criteria are described in more detail in the text. (From Ref. 15.)

Selection Criteria	$e^+e^- \rightarrow \nu_e \bar{\nu}_e H^0$	$e^+e^- \rightarrow e^+e^- W^+W^-$	$e^+e^- \rightarrow e^+\nu_e W^- Z^0$
Produced events with $500 < M_{VV} < 1500$	1200	42000	17000
Purely hadronic decays	640	19000	9400
After acceptance and jet reconstruction	530	6000	3000
$p_T^{VV} > 20$ GeV	500	2700	2800
No $e^\pm$ above 50 mr	500	600	2300
$|\cos\theta_V^*| < 0.8$	450	250	470

410

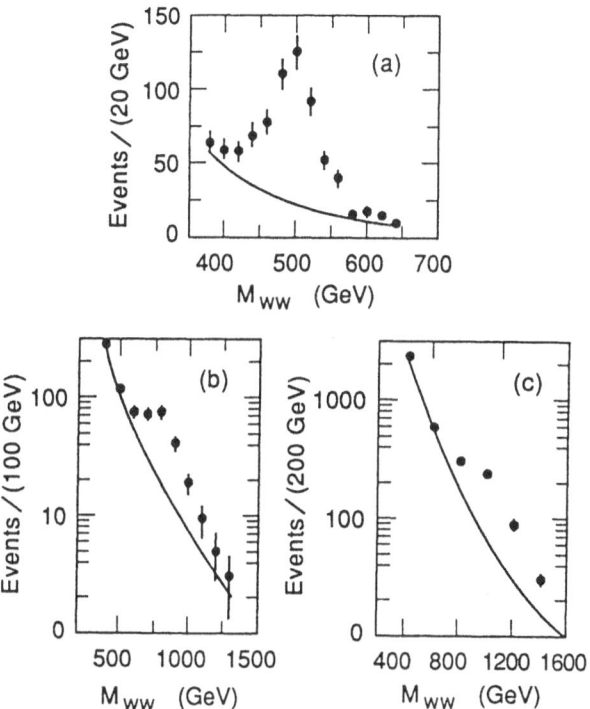

Figure 11. Reconstructed invariant mass distribution for signal $(e^+e^- \to \nu_e \bar{\nu}_e H^0)$ and background $(e^+e^- \to e^+e^- W^+W^-$ and $e^+e^- \to e^+\nu_e W^- Z^0)$ events which satisfy the selection criteria outlined in the text at $\sqrt{s} = 2$ TeV for a Higgs boson mass of (a) 500 GeV, (b) 800 GeV and (c) 1 TeV. The sample size corresponds to 10 fb^{-1} for (a) and (b), and 50 fb^{-1} for (c). (From Ref. 15.) The solid line corresponds to the background alone; the dots to signal plus background.

5 fb^{-1} to close the gap from 80 - 120 GeV. However, no detailed Monte Carlo studies have been done for this case.

- Charged Higgs boson - At e^+e^- colliders with a c.m. energy in the range (0.5-1) TeV, one would be sensitive to a charged Higgs boson with a mass less than about 80% of the beam energy with a data sample of (10-30) fb^{-1}. A charged Higgs boson with a mass close to the $W^\pm$ mass would be detectable.

- Nonminimal neutral Higgs bosons - With a data sample of ≈ 10 fb^{-1} at $\sqrt{s} = 1$ TeV, one could detect $e^+e^- \to H^0 A^0$ for $M_{A^0} \gtrsim 120$ GeV. Search strategies for other mass regions have not been studied in detail but might also be successful.

A minimal neutral Higgs boson with mass less than $2M_{W^\pm}$, a charged Higgs boson and nonminimal neutral Higgs bosons of any mass are all expected to be extremely difficult, if not impossible, to detect at the SSC. High energy e^+e^- colliders provide the necessary environment for exploring these aspects of the Higgs sector. The requirements which are placed on detectors are quite modest and can be satisfied with existing technology. Of course, the greatest challenge is to build the e^+e^- collider itself. The study of the Higgs sector provides great motivation to meet this challenge.

REFERENCES

1. See, for example, *The Higgs Hunter's Guide*, J. F. Gunion, H. E. Haber, G. L. Kane and S. Dawson, Preprint Numbers UCD-89-4, SCIPP-89/13 and BNL-41644 (1988), and references therein.

2. A. Seiden, these proceedings.

3. A. J. Weinstein, proceedings of the XIV International Symposium on Lepton and Photon Interactions, Stanford, California, August 7 - 12, 1989.

4. *Search for Neutral Higgs at LEP 200*, J.Boucrot et al., presented by S.L.Wu at the ECFA Workshop - LEP 200, Aachen, Sept. 29 - Oct. 1, 1986; J.Hilgart et al., A. Phys. C35, 347 (1987).

5. C. Ahn et al., SLAC-Report-329, 1988.

6. P. R. Burchat, D. L. Burke and A. Petersen, Phys. Rev. D 38, 2735 (1988); Erratum, *ibid.* 39, 3515 (1989).

7. S. Komamiya, Phys. Rev. D 38, 2158 (1988).

8. J. Alexander et al., SLAC-PUB-4775, to be published in Proceedings of the DPF Summer Study: Snowmass '88, High Energy Physics in the 1990's, Snowmass, Colorado, June 27 - July 15, 1988.

9. F. Richard, in Proceedings of the Workshop on Physics at Future Accelerators, La Thuile (Italy) and Geneva (Switzerland), Jan. 7-13, 1987.

10. J. Gunion, P. Kalyniak, M. Soldate, P. Galison, Phys. Rev. D **34**, 101 (1986).

11. R. Cahn, Nucl. Phys. B **255**, 341 (1985).

12. K. Yokoya, Nucl. Instrum. Methods A251, 1 (1986); P. Chen, SLAC-PUB-4293, 1987.

13. R. B. Palmer, SLAC-PUB 4295 (1987); W. Schnell, Advanced Accelerator Concepts, Madison, WI, AIP Conf. Proc. 156, 12 (1987), and SLAC/AP-61 (1987).

14. T. Sjostrand and M. Bengtsson, Computer Phys. Comm. 43, 367 (1987).

15. F. Richard, *Proceedings of the Workshop on Physics at Future Accelerators*, La Thuile, Italy and Geneva, Switzerland, 1987, CERN 87-07; B. Mele, ibid.

16. E. Yehudai, private communication.

17. J.F. Gunion *et al.*, Phys. Rev. **D38**, 3444 (1988).

18. W. Bartel *et al.*, Z. Phys., **C33**, 23 (1986).

19. See, for example, P. Grosse-Wiesmann, SLAC-PUB-4616, presented at the Workshop on Intermediate Mass and Non-Minimal Higgs Bosons, U. of California, Davis, California, Jan. 4-6, 1988.

FIRST RESULTS FROM THE SLAC LINEAR COLLIDER AND MARK II DETECTOR*

Patricia R. Burchat

Santa Cruz Institute for Particle Physics

University of California, Santa Cruz, California, 95064

Representing the Mark II Collaboration

ABSTRACT

A first measurement of the Z^0 mass, total width, and partial width into invisible decay modes by the Mark II detector at the SLAC Linear Collider (SLC) is presented. The achieved luminosity of the SLC, beam diagnostic devices near the interaction point, and beam-related backgrounds are all discussed. The first Z^0 mass and width measurement with the SLC is the result of a determinatiion of the cross section for Z^0 production at six center-of-mass energies between 89.2 and 93.0 GeV. The measurements of the three quantities necessary to extract the Z^0 resonance parameters are described: the center-of-mass energy, the luminosity and the number of Z^0 bosons produced. Finally, the implications of the measured parameters and future improvements to the SLC physics program are discussed.

INTRODUCTION

With the Mark II detector at the SLAC Linear Collider (SLC), we have measured[1] the mass of the Z^0 to be (91.11 ± 0.23) GeV/c^2, its total width to be $(1.61^{+0.60}_{-0.43})$ GeV, and, constraining the total width to the standard model value of 2.48 GeV, the partial width to invisible decay modes to be (0.62 ± 0.23) GeV, corresponding to 3.8 ± 1.4 neutrino species. This is the first measurement of the parameters of the Z^0 resonance in e^+e^- collisions. All previous measurements are from experiments at $p\bar{p}$ colliders in which the resonance parameters are extracted from a measurement of the energy of *final state* leptons from the decay $Z^0 \rightarrow e^+e^-$ or $Z^0 \rightarrow \mu^+\mu^-$. In contrast, the resonance shape is measured at SLC, and will be measured at the LEP storage ring at CERN, by measuring the cross section for Z^0 production from *initial state* leptons (e^+e^-) and subsequent decay into all visible decay modes at center-of-mass energies in the range $\pm\Gamma$ around the Z^0 mass.

★ WORK SUPPORTED BY THE U. S. DEPARTMENT OF ENERGY.

Higgs Particle(s)
Edited by A. Ali
Plenum Press, New York, 1990

THE SLAC LINEAR COLLIDER

In 1980 the SLAC Linear Collider was proposed for two purposes:[2] to provide the first test of linear collider technology which will be necessary for e^+e^- physics at center-of-mass energies above about 200 GeV, and to produce large numbers of Z^0's to provide precise tests of the standard model and an environment for searches for new particles such as quarks, leptons and the Higgs boson. Although the SLC has yet to produce a large number of Z^0's, much has been learned about linear collider issues and a sufficient number of Z^0's have been produced to enable a measurement of Z^0 parameters more precise than previous measurements.

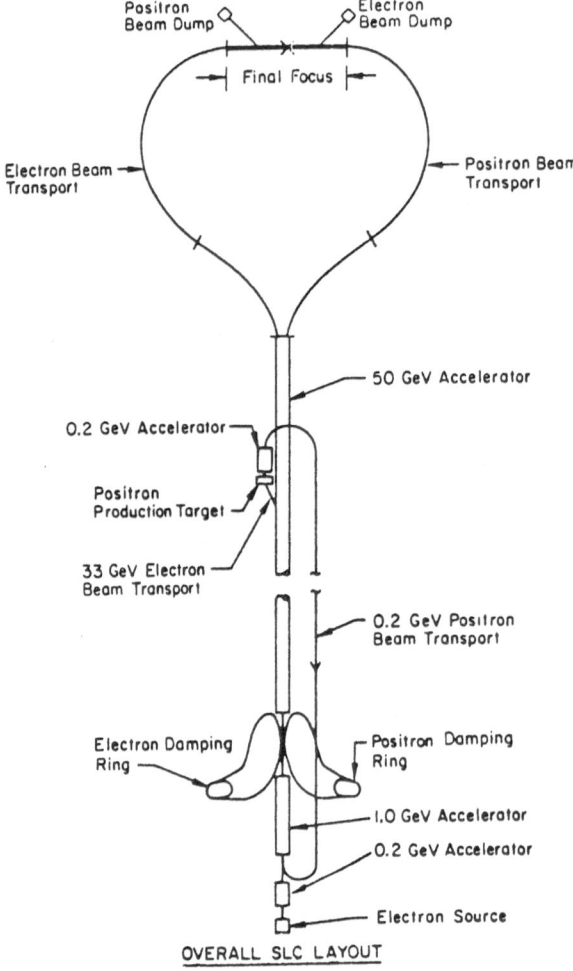

Figure 1. Layout of the SLAC Linear Collider.

The overall layout of the SLC is shown in Figure 1. The same linear accelerator is used to accelerate both the electron and positron beams. Two arcs of bending magnets carry the beams to the final focus systems which focus the beams to a transverse size of a few microns. Two damping rings reduce the emittance of the beams in about five milliseconds. Positrons are produced by bombarding a target with an electron beam about two-thirds of the way down the accelerator. After collision at the interaction point, each beam is deflected out of the final focus system into an extraction line where the beam energy is measured before the beam dump.

416

The luminosity at the interaction point (IP) can be written as follows:

$$\mathcal{L} = 7.0 \times 10^{27} \text{cm}^{-2}\text{s}^{-1} \frac{f}{60\,\text{Hz}} \frac{N_-}{1.3 \times 10^{10}} \frac{N_+}{1.1 \times 10^{10}} \frac{3.5\,\mu\text{m}}{\sigma_x} \frac{2.8\,\mu\text{m}}{\sigma_y}$$

where f is the collision frequency, N_- and N_+ are the number of electrons and positrons per bunch at the IP, respectively, and σ_x and σ_y are the transverse beam sizes at the IP. The rate for Z^0 production at the peak of the Z^0 resonance can be written in terms of the luminosity as follows:

$$Z^0 \text{ rate } \approx \frac{18}{\text{day}} \frac{\mathcal{L}}{7.0 \times 10^{27} \text{cm}^{-2}\text{s}^{-1}} \epsilon$$

where ϵ is the average efficiency for colliding beams.

The achieved values for these parameters are summarized in Table 1 and compared to the design values. The first column lists the best achieved value and the second column lists typical values for recent running, the same values used for reference in the above luminosity formula.

Table 1. SLC parameters at the interaction point.

Parameter	Best Achieved Value	Typical Value for Current Running	Design Value for First Year
f	60 Hz	60 Hz	120 Hz
N_-	1.9×10^{10}	1.3×10^{10}	5.0×10^{10}
N_+	1.4×10^{10}	1.1×10^{10}	5.0×10^{10}
$\Sigma_x \equiv \sqrt{\sigma_{x,e-}^2 + \sigma_{x,e+}^2}$	$4.0\,\mu\text{m}$	$5.0\,\mu\text{m}$	$2.8\,\mu\text{m}$
$\Sigma_y \equiv \sqrt{\sigma_{y,e-}^2 + \sigma_{y,e+}^2}$	$3.3\,\mu\text{m}$	$4.0\,\mu\text{m}$	$2.8\,\mu\text{m}$

The entire machine is currently operating at a repetition rate of at least 60 Hz with parts of the linac running at 120 Hz. The collision frequency is 60 Hz. The design frequency for first year running is 120 Hz.

The design intensity for each beam is 5×10^{10} particles per bunch. Present intensities are limited by beam-related backgrounds (discussed below), wakefield effects in the linac, heating of the positron source, and positron capture and transport efficiency.

The spot sizes listed in Table 1 are the quadratic sum of the electron and positron spot size in each transverse dimension. These are listed rather than the spot size for each beam separately because we monitor the spot sizes using "beam-beam deflection" (described below) which is only sensitive to the quadratic sum Σ_i. The design spot size of 2 μm corresponds to $\Sigma_i = 2.8$ μm to be compared to the best achieved values of $\Sigma_x = 4.0$ μm and $\Sigma_y = 3.3$ μm.

The typical values for current running lead to a luminosity of 7.0×10^{27} cm^{-2}s^{-1} or 18 Z^0/day. The average integrated luminosity per day has been about 4 Z^0 equivalents. Hence, our efficiency is less than 25%. This large inefficiency includes time for tuning the beams, especially to reduce beam-related backgrounds in the Mark II detector, lost time due to hardware failures and Mark II inefficiencies (calibration, tape changes, deadtime, etc.). The most luminosity collected in one 24 hour period has been 15 Z^0 equivalents. The total integrated luminosity for the six scan points used for this first measurement is 5.8 ± 0.5 nb^{-1}.

Beam Diagnostics Near the Interaction Point

Measurements of the transverse beam size at the interaction point (IP) are necessary for tuning optical elements in the final focus system to minimize the spot size. In addition, diagnostics are needed to provide evidence that the two beams are colliding when they are both crossing the IP and that the luminosity has not decreased.

The transverse beam size of individual beams is measured with a fine conductive wire.[3] The beam is moved across a stationary carbon fibre in steps as small as one micron. In each of the two transverse dimensions, there are three wires of different diameters: approximately 4 microns, 7 microns and 28 microns. Two signals are typically recorded for a scan: a secondary emission signal from the wire itself and a bremsstrahlung signal which is detected far from the interaction point. Effects which are not fully understood corrupt the secondary emission signal for electron beam intensities larger than about 6×10^9 particles per bunch. At these intensities we usually use the bremsstrahlung signal to measure the transverse beam profile. This technique will be described in more detail below.

The wire scanner is used in automated procedures for positioning the focal point of each beam at the IP and measuring dispersion and chromaticity at the IP as well as for single beam scans. The wires are also used for initial placement of the two beams at the same point in the transverse plane to within a few microns.

For optimal luminosity, we must ensure that the two beams cross the transverse plane at the IP at the same point to within a fraction of the beam size and that the spot sizes have not increased. In principle, the wire scanner can be used for this purpose. However, scanning the beam across the wire is not compatible with running the Mark II detector with all subsystems at full voltage because of background radiation produced from the wire. In addition, heating of the wire restricts the beam intensities to less than 10^{10} particles per pulse. A method which does not interfere with running the detector has been devised, tested and is now used routinely. This method uses the pattern of angular deflection of one beam as it is scanned across the other.[4] The angular deflection is due to the electromagnetic interaction between the two beams. In terms of the impact parameter $\Delta_{x,y}$ between a round target beam and round probe beam, the deflection of the probe is given by

$$\theta_{x,y} = \frac{-2r_e N \Delta_{x,y}}{\gamma \Delta} \frac{1 - \exp(\frac{-\Delta^2}{2\Sigma^2})}{\Delta}$$

where r_e is the classical radius of the electron, γ is the relativistic Lorentz factor, N is the number of particles in the target bunch, and $\Delta = (\Delta_x^2 + \Delta_y^2)^{1/2}$. Σ^2 is the sum of the squares of the probe and target beam sizes. This function is shown in Figure 2 for a target intensity of 5×10^{10} particles per bunch and for three different target sizes: two, five and ten microns. As the target beam size decreases, the maximum deflection becomes larger and occurs for smaller impact parameter leading to a more rapidly changing deflection angle as the two beams cross each other. This rate of change of deflection angle can be used for tuning on spot size as an alternative to the wire scanner. The symmetry point in the scan gives the point at which the two beams have zero impact parameter. This is used for centering the beams. Once the beams are brought into close proximity (a few to ten microns) with the wire scanner, one beam is scanned across the other in each plane with small air-core dipole magnets which can change the beam position at the IP, between beam pulses, with a resolution of 0.05 μm. The scan range is typically ±40 μm in 2 μm steps. The deflection angle is measured with two beam position monitors (BPM's) on each side of the IP. These monitors of the strip-electrode type are sensitive to each beam separately. For each beam, the BPM readings are used to fit the incoming angle, the deflection angle and the transverse position at the IP, in each plane. The deflection angle is fit to the above equation to determine how much one beam should be moved to center it on the other. The resolution is a small fraction of the beam size.

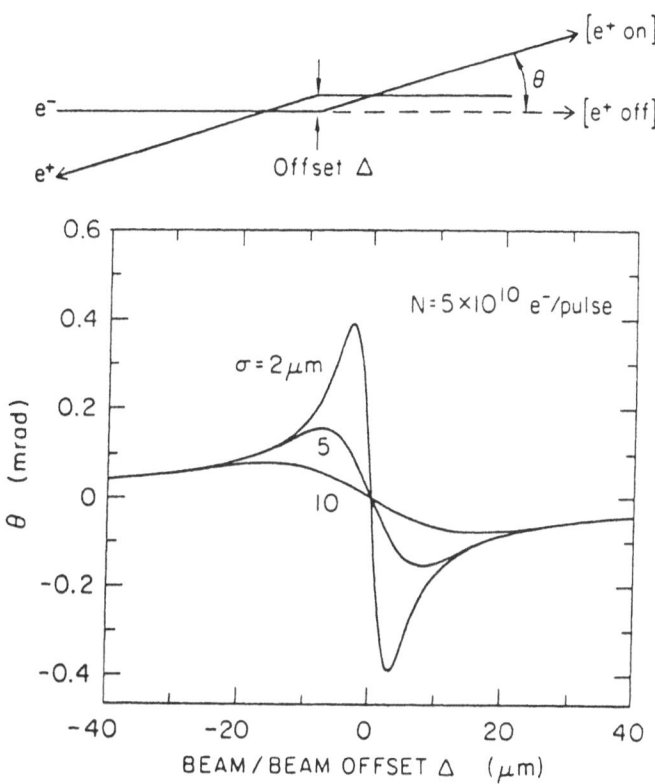

Figure 2. Theoretical deflection angle θ of a point probe as a function of impact parameter Δ between the point size probe beam and a round target beam of three sizes. The intensity of the target beam is assumed to be 5×10^{10}.

This beam deflection method is then used to monitor how well centered the beams remain. The deflection can be measured and the beams recentered in less than 20 seconds. Hence, the beam centering can easily be monitored every 5 minutes or so. It has been found that the beams drift with respect to each other by only a small fraction of the beam width in this length of time. The SLC has operated for weeks at a time with colliding beams (including changes in the center-of-mass energy) without it being necessary to insert the wires to locate the beams. We can recenter the beams through beam-beam deflection if they are smaller than about 10 μm and less than about 100 μm apart.

We can extract the beam sizes and intensities, as well as the offset, from the scan. In this way, we routinely use beam-beam deflection to minimize the spot size at the IP once the beams are in collision. We also use the parameters to estimate the instantaneous luminosity.

Another way to monitor colliding beams is through the observation of beamstrahlung, the radiation emitted by one beam when it passes through the intense electromagnetic field of the other beam. Actually, beamstrahlung is also emitted when the beams are not centered but are deflecting each other. As one beam is scanned across the other, the beamstrahlung intensity is expected to grow as they get closer together and the deflection angle increases, but to actually dip slightly as they become centered. Consequently, the interpretation of the beamstrahlung signal is not obvious.

To monitor the beamstrahlung flux, Čerenkov detectors were built and installed about 40 m downstream on either side of the IP. Stainless steel plates convert gamma rays at about 50 MeV to e^+e^- pairs. These charged particles travel through a Čerenkov detector with a threshold of about 25 MeV, set to reject the background of lower-energy, but far more intense, synchrotron radiation photons generated by the beams as they pass through the final focus magnets. A mirror and a system of light channels are used to focus the Čerenkov light onto an array of photomultiplier tubes. The whole detection system is surrounded by lead shielding for protection from beam-related backgrounds. The beamstrahlung monitor has also been used very effectively to detect the bremsstrahlung photons emitted during a wire scan as mentioned above.

Figure 3 shows data from a beam-beam deflection scan and the corresponding signals from each of the two beamstrahlung detectors. The beam intensities were 1.4×10^{10} e^- per pulse and 1.0×10^{10} e^+ per pulse, and the transverse spot size was approximately 5 μm for both beams. A dip in the beamstrahlung spectrum is observed as expected as the beams pass through the optimum alignment point. The SLAC Linear Collider is the first facility to produce beam-beam deflection signals or beamstrahlung signals. The success of using beam-beam deflection as a diagnostic tool at the IP constitutes an essential step in establishing the viability of the linear collider concept.[5]

BEAM-RELATED BACKGROUNDS

Electrons and positrons in the "tails" of the beam (in position, angle or energy) must be collimated before they reach the region near the IP since one electromagnetic shower from a 50 GeV $e^{\pm}$ can produce unacceptable backgrounds in the detector. Initially, most of the collimation was in the final focus system. Unfortunately, the collimated particles can produce off-energy electrons and positrons, photons and high-energy muons. The electromagnetic debris is mostly stopped with secondary collimators while the muons are very difficult to stop. Two important projects which substantially reduced the beam-related backgrounds were implemented in the last year.

First, large toroidal magnets were installed in the final focus system to bend the muons away from the detector. These toroids provide a reduction of about a factor of ten in the number of muons reaching the Mark II detector.

Secondly, eight collimators were installed at the end of the linac. The positions of the collimators were selected to primarily collimate electrons or positrons, in one of the two transverse directions, at two different points 90° out of phase in betatron phase space. These collimators can be used to remove large beam tails without producing debris which will propagate to the detector. The collimator openings in the linac are less than a mm and consequently require very good and stable linac steering. The beam is also collimated at point in the arcs where the dispersion is large to remove energy tails.

Beam-related backgrounds are now not a serious, but also not a negligible, problem for the Mark II detector. A significant amount of time is required to tune the beams for low backgrounds and the remaining electromagnetic contribution to the background is not fully understood.

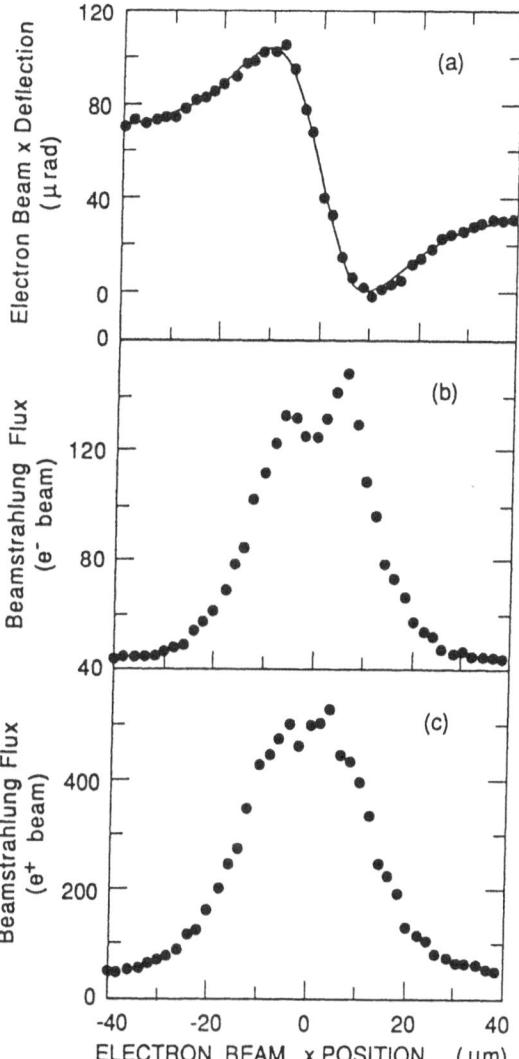

Figure 3. (a) Measured deflection curve of the electron beam as it was scanned across the positron beam. There exists an arbitrary offset in the measured deflection angle due to unknown BPM offsets. (b) The beamstrahlung signals from the electron beam and (c) from the positron beam, as observed during the scan.

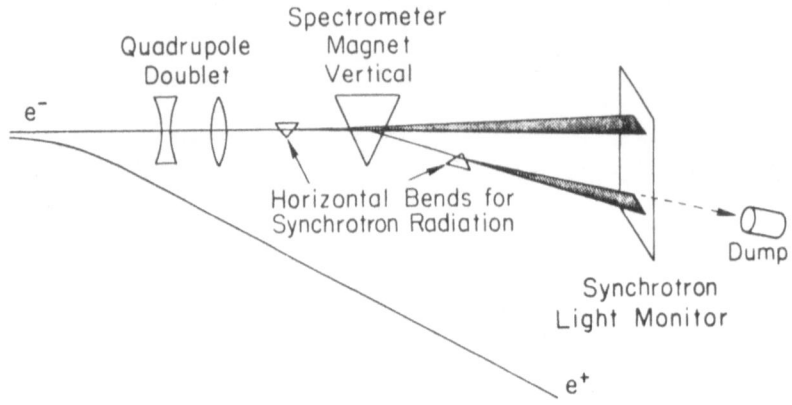

Figure 4. Schematic drawing of the extraction line spectrometer for measuring the beam energy.

EXTRACTION LINE ENERGY SPECTROMETER

The precise measurement of the mass of the Z^0 requires an accurate and precise determination of the electron and positron beam energies. This is provided by a spectrometer system installed in each of the two extraction lines which transport the beams to their dumps.[6] The extraction line spectrometer is shown schematically in Figure 4. The spectrometer magnet is preceded and followed by smaller magnets which bend the beam in the plane perpendicular to the bend plane of the spectrometer magnet. When the beam bends in the small magnets, it produces swaths of synchrotron radiation which are detected about 15 m downstream of the spectrometer magnet.

The synchrotron stripes are intercepted by phosphor screens which are monitored and digitized by a video system.[7] Each digitization channel corresponds to 70 μm on the phosphor screen or 12 MeV in beam energy. An array of fiducial wires is located in front of each screen to provide a calibration for the exact location of the stripes. In addition to providing a measurement of the beam energy, the digitization of the synchrotron stripes enables us to measure the energy spread within each beam by comparing the width of the stripe before and after the spectrometer magnet since we know the dispersion introduced by the magnet. The *rms* center-of-mass energy spread is typically 0.3% and is determined to approximately 30% of itself.

The determination of the beam energy depends on knowledge of the field strength of the spectrometer magnet, the distance between the spectrometer magnet and the synchrotron light monitors, and the distance between the two synchrotron stripes at the plane of the monitors. The field strength of the spectrometer magnet is about 3 T·m and is known and monitored to one part in 10^4.[8] The distance to the light monitor is measured to 0.01%. The distance between the synchrotron stripes at the light monitors is about 26 cm and is known to 0.02%. Rotational errors in the magnet alignment could contribute an error of 16 MeV to the measurement of the beam energy. These uncertainties are summarized in Table 2. They combine to give a total uncertainty on the energy of each beam of 20 MeV. There is an additional uncertainty in the luminosity-weighted center-of-mass energy of less than 30 MeV due to correlations in energy and transverse position for the beam at the IP. Therefore, the total systematic error on the center-of-mass energy is less than 40 MeV.

MARK II DETECTOR AND TRIGGERS

An upgrade of the Mark II detector[9] which involved replacing the drift chamber, endcaps, time-of-flight system and magnet with new systems was completed in the summer of 1985. In the winter of 1986, the upgraded detector was tested and operated at the PEP storage ring. The detector was then moved to the SLC interaction region. In the fall of 1987, the detector was moved onto the beam line. The detector was continuously exercised on cosmic rays until the first colliding beams were provided by the SLC.

Table 2. Summary of the systematic uncertainties in the measurement of the beam energy.

Source of Systematic Uncertainty	Size of Uncertainty
Magnetic measurement	5 MeV
Detector position resolution	10 MeV
Rotational errors in magnet alignment	16 MeV
Survey errors	5 MeV
Total error	20 MeV

A sectional view of one quarter of the detector is shown in Figure 5. The vertex detectors have not yet been installed and will be discussed below. The drift chambers and the calorimeters provide the principle information used in this analysis to identify Z^0 decays. Charged particles are detected and their momentum measured in a 72-layer cylindrical drift chamber in a 4.74 kG solenoidal magnetic field. The drift chamber tracks charged particles with $|\cos\theta| < .092$, where θ is the angle between the track and the incident electron beam. Photons are detected in electromagnetic calorimeters that cover the region $|\cos\theta| < 0.96$. The calorimeters in the central region (barrel calorimeters) are lead-liquid-argon ionization chambers, while the endcap calorimeters are lead-proportional-tube sandwiches.

As well as two small-angle-Bhabha triggers for the luminosity monitors, a cosmic trigger for establishing drift chamber constants, and a random beam-crossing trigger for studying beam-related backgrounds, we have two redundant triggers for Z^0 events: a charged-track trigger and a neutral-energy trigger. With these triggers, our deadtime was typically 10%, occasionally becoming much higher due to beam-related backgrounds, requiring beam-tuning to lower the backgrounds in the detector.

The charged-track trigger requires two charged tracks with $|\cos\theta| < 0.75$ and momentum transverse to the beamline of at least 150 MeV, separated by at least $\approx 11°$ in azimuthal angle. The single track efficiency is greater than 99% for $|\cos\theta| < 0.65$. The charged-track trigger alone is about 97% efficient for hadronic decays of the Z^0. The neutral-energy trigger is a software-based calorimeter trigger which requires a tower of energy of at least 2.2 GeV in the endcap or 3.3 GeV in the barrel calorimeter. This trigger alone is about 92% efficient for hadronic decays of the Z^0.

The combined efficiency of the the charged-track and neutral-energy trigger is 99.8% for hadronic Z^0 decays. Of the 94 hadronic Z^0 decays in our data sample for this scan, 91 satisfied both triggers.

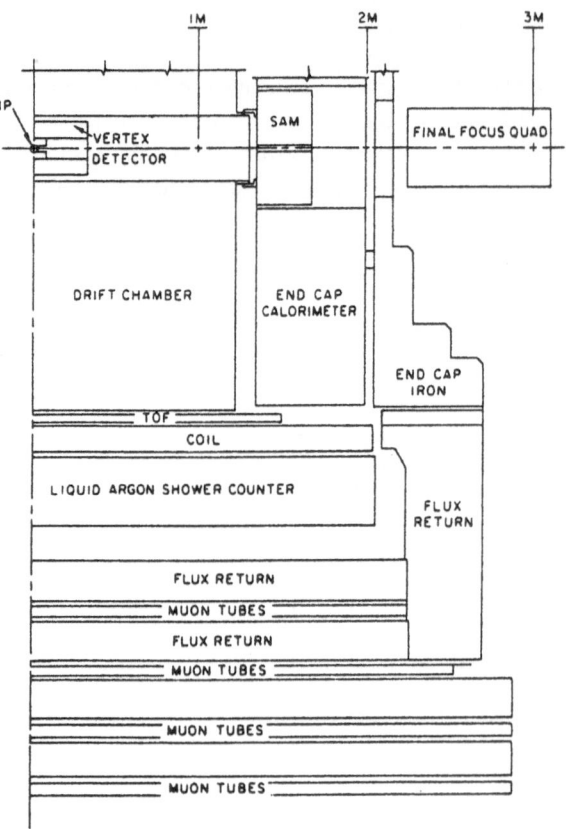

Figure 5. Sectional view of one quarter of the Mark II detector.

SELECTION OF Z^0 EVENTS

In the standard model, the branching fraction of the Z^0 into hadrons is 70%, into charged leptons is 10% and into neutrinos is 20%. For the measurement of the Z^0 mass and width, we use decays to $\tau^+\tau^-$ or $\mu^+\mu^-$ in a restricted volume of the detector and hadronic decays.

We accept $\mu^+\mu^-$ and $\tau^+\tau^-$ events in which the thrust axis satisfies $|\cos\theta| < 0.65$. In this region, the efficiency is well-determined and the identification of the event is unambiguous. Tau events are further required to have visible energy (charged plus neutral) greater than 10% of the center-of-mass energy. The efficiencies are found by Monte Carlo (MC) simulation to be $(99 \pm 1)\%$ for mu pairs and $(96 \pm 1)\%$ for tau pairs. A total of 12 charged lepton pairs satify our criteria.

Candidates for $Z^0 \rightarrow q\bar{q}$ are accepted if they have at least three charged tracks with transverse momentum greater than 110 MeV emerging at $|\cos\theta| < 0.92$ from a cylindrical volume of radius 1 cm and half-length 3 cm parallel to the beam line. To suppress possible contamination from beam gas or two-photon exchange interactions, we also require the energy visible in the forward (E_f) and backward (E_b) hemispheres to each be greater than 5% of the center-of-mass energy E. Figure 6 shows E_f versus E_b for all events satisfying the charged-track requirement. A total of 94 events satisfy our criteria for hadronic decays of the Z^0.

A MC simulation indicates that we expect 0.005 events in our data from two-photon exchange interactions. The number of beam-gas interactions that satisfy the above criteria is expected to be < 0.6 at the 90% confidence level (CL) since no events were found in cylindrical

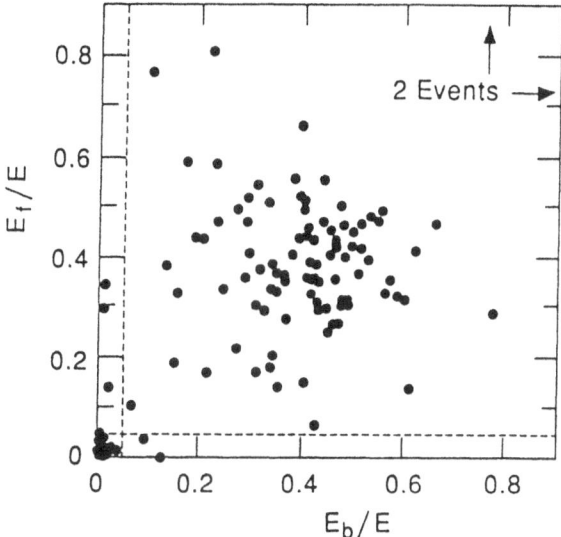

Figure 6. Energy in the forward hemisphere E_f versus energy in the backward hemisphere E_b for all events satisfying the charged-track requirement.

volumes of radius 1 cm and half-length 3 cm displaced ± 10 cm and ± 16 cm from the interaction point.

The efficiency for Z^0 decays to satisfy the selection criteria, including the trigger requirements, is found to be $(93.7 \pm 0.4)\%$ from MC simulation. In estimating the uncertainty on the efficiency, we include such factors as uncertainties in the calorimeter energy scale, trigger inefficiencies, different MC models, and the effects of beam-related backgrounds. We studied the latter by overlaying MC events and real detector data from random beam crossings. The beam-related backgrounds were found to have little effect on this analysis.

MEASUREMENT OF LUMINOSITY

There are two luminosity monitors for measuring the Bhabha rate at small angles with respect to the beam direction: the small angle monitor (SAM) which subtends the region between 50 and 160 mrad and the mini-SAM which subtends the region between 15 and 25 mrad. The Bhabha rate in the SAM is about 40% greater than the peak visible Z^0 rate while the Bhabha rate in the mini-SAM is about seven times greater. Because the mini-SAM is in the region where the Bhabha cross section is changing very rapidly, the absolute cross section is very sensitive to misalignments of the detector. In addition, the efficiency of the detector is sensitive to beam-related backgrounds. We do not use the information from the mini-SAM for this first measurement although the Z^0 parameters determined using the miniSAM for relative normalization between energies, as well as the SAM for absolute normalization, are entirely consistent with those determined using the SAM alone.

The SAM consists of about 14 radiation lengths of lead and proportional tubes (six layers) preceded by nine layers of drift tubes for tracking. Bhabha scattering events in the SAM calorimeter are unmistakable and background free because of the fine segmentation of the calorimeter. The events follow the expected distributions in polar scattering angle, azimuth, and acollinearity. The resolution in θ from reconstructed showers in the SAM calorimeters is 1 mr except near the edges of the calorimeters. To minimize the systematic error of geometric acceptance cuts while retaining as much statistical power as possible, two overlapping classes of Bhabha events are defined. All 234 events in which at least 40% of the beam energy is detected in each SAM are used for calculating the relative luminosity between scan points. The cross section for these events is derived by scaling the 143.5 events which fall into a smaller fiducial volume having an accurately calculable acceptance. These "precise" Bhabha events are those in which $65 < \theta < 160$ mr for both e^- and e^+ showers, plus, with a weight of 0.5, events where one shower is within the precise region and the other shower has $60 < \theta < 65$ mr. This weighting reduces the effects of misalignments and detector resolution. The cross section with this precise angular definition is[10] $24.9 \cdot (91.1\text{GeV}/E)^2$ nb, which includes a +1.5% correction from detector resolution effects, and has an estimated systematic error of 2% from unknown higher order radiative corrections and 2% due to detector resolutions effects. Scaling then to the full sample gives a SAM Bhabha cross section of $\sigma_S = 40.6 \cdot (91.1\text{GeV}/E)^2$ nb, with an additional 5.3% statistical error due to the scaling factor. We also correct σ_S for the effect of $Z^0 - \gamma$ interference which, although small (1%), varies rapidly with E for E near m_Z.

RESULTS OF MEASUREMENT OF THE Z^0 CROSS SECTION FOR SIX CENTER-OF-MASS ENERGIES

For each of the six center-of-mass energies in the scan, Table 3 gives the number of Z^0 hadronic and leptonic decays which satisfied our selection criteria, the number of SAM Bhabha events and the measured cross section for Z^0 production σ_Z. Also listed are the mean energy of the Bhabha events as measured by the energy spectrometer and the rms spread in the center-of-mass energy σ_E generated by the energy spread of the beams and by the pulse-to-pulse jitter and drifts of the beam energies. The cross sections contain a correction for the energy spread, which is largest when the curvature of the resonance peak is largest. The correction is about +3% at the peak and −3% near the tails of the resonance.

Table 3. Average E and σ_E, number of events, and cross sections for the production of hadronic events and muon and tau pairs with $|\cos\theta| < 0.65$ at each energy setting.

$< E >$	$< \sigma_E >$	SAM	Z^0 Decays			σ_Z
(GeV)	(GeV)	e^+e^-	Had.	Lep.	Tot.	(nb)
89.24	0.22	24	3	0	3	$5.5^{+6.1}_{-3.1}$
89.98	0.25	37	8	2	10	$11.8^{+6.0}_{-4.1}$
90.70	0.28	44	27	3	30	$30.4^{+9.1}_{-7.1}$
91.50	0.29	53	32	6	38	$31.9^{+8.4}_{-6.7}$
92.16	0.28	33	11	0	11	$14.2^{+7.0}_{-4.8}$
92.96	0.23	43	13	1	14	$13.5^{+5.6}_{-4.1}$
Totals		234	94	12	106	

In Table 3, the ratio of leptonic to hadronic events is $12/94 = 0.128^{+0.053}_{-0.039}$. The standard model predicts 0.055 for our fiducial region and efficiencies. There are four $\mu^+\mu^-$ and eight $\tau^+\tau^-$ events.

The visible Z cross section (σ_Z) can be represented by a relativistic Breit-Wigner resonance shape:

$$\sigma_Z(E) = \frac{12\pi}{m_Z^2} \frac{s\Gamma_e\Gamma_f}{(s - m_Z^2)^2 + s^2\Gamma^2/m_Z^2}(1 + \delta(E)),$$

where $s \equiv E^2$ and δ is the substantial correction due to initial state radiation calculated using the analytic form given by Cahn.[11] The Z partial widths into electron pairs and into decays in our fiducial volume are Γ_e and Γ_f, respectively. The partial widths into hadrons, muons and taus are related to Γ_f by $\Gamma_f = \Gamma_h + f(\Gamma_\mu + \Gamma_\tau)$, where $f = 0.556$ is the fraction of all muon and tau decays that have $|\cos\theta| < 0.65$. The total Z width is assumed to be $\Gamma = \Gamma_h + \Gamma_e + \Gamma_\mu + \Gamma_\tau + N_\nu\Gamma_\nu$, where N_ν is the number of species of neutrinos and Γ_ν is the Z partial rate to one neutrino species.

We estimate Z resonance parameters by constructing a likelihood function from the probability of observing, at each energy, n_Z Z decays and n_S Bhabha scatters given that we have observed a total of $n_Z + n_S$ events. After eliminating terms that are constant with respect to the fit parameters we obtain for the likelihood L

$$L = \prod \frac{(\epsilon\sigma_Z)^{n_Z}}{(\epsilon\sigma_Z + \sigma_S)^{n_Z + n_S}} .$$

The product is over energy bins and $\epsilon = 0.941$ is the efficiency for detecting Z decays to hadrons, muons and taus in our fiducial volume. We take the limit corresponding to n standard deviations to be the point at which $\ln L$ decreases by $n^2/2$ from its maximum value.

We have performed three fits to the data, which differ in their reliance on the minimal standard model. The first leaves only m_Z as a free parameter. The widths are those expected for Z couplings to the known fermions (5 quarks and 3 lepton doublets), including a QCD correction to the hadronic width of 5%. For $m_Z = 91.11$ GeV/c^2 and $\sin^2\theta_W = 0.2312$, $\Gamma_e = \Gamma_\mu = \Gamma_\tau = 0.083$ GeV, $\Gamma_\nu = 0.165$ GeV and $\Gamma_h = 1.73$ GeV. The second fit leaves both m_Z and N_ν as free parameters but fixes Γ_ν and all other partial widths to their expected values. Finally, the third fit leaves m_Z, N_ν and Γ free but fixes Γ_e, Γ_μ, Γ_τ and Γ_ν to their expected values. For all fits, N_ν depends on the assumed values of Γ_e and ϵ and the absolute normalization of σ_S, while m_Z and Γ are not sensitive to them. The results of these fits are displayed in Fig. 7 and Table 4.

Table 4. Z^0 resonance parameters. The three fits are described in the text.

Fit	m_Z GeV/c^2	N_ν	Γ GeV	χ^2/D. o. F.
1	91.11±0.23	–	–	4.1/5
2	91.11±0.23	3.8 ± 1.4	–	3.7/4
3	91.06±0.17	$3.2^{+1.3}_{-1.1}$	$1.61^{+0.60}_{-0.43}$	1.5/3

The third fit yields $\Gamma = 1.61^{+0.60}_{-0.43}$ GeV, which should be compared to the standard model value of 2.48 GeV. The systematic error on this value is 30 MeV due to uncertainty in the relative energy of scan points. If we assume that Γ can only be larger than this value, then at the 90% CL, $\Gamma < 3.1$ GeV. Previous direct measurements of Γ yielded upper limits at the 90% CL of 5.2 GeV (UA1)[12] and 5.6 GeV (UA2).[13]

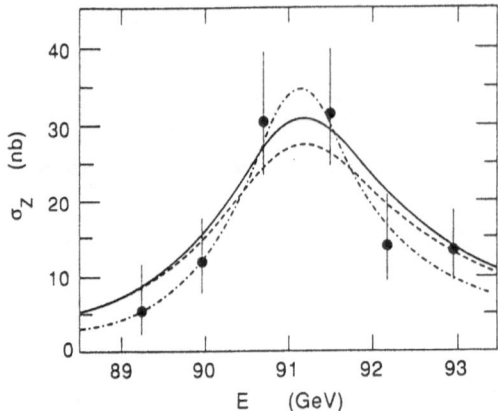

Figure 7. e^+e^- annihilation cross sections to all hadronic events plus μ and τ pairs with $|\cos\theta| < 0.65$. The curves represent the result of different fits: solid – m_Z free; dashed – m_Z and N_ν free, and dot-dashed – m_Z, N_ν, Γ free. The peak σ_Z occurs approximately 100 MeV higher than m_Z due to radiative corrections.

We use the second fit for the number of neutrino generations and obtain $N_\nu = 3.8 \pm 1.4$ corresponding to a partial width to invisible decay modes of $N_\nu \Gamma_\nu = 0.62 \pm 0.23$ GeV. At the 90% CL, $N_\nu < 5.5$. The 6.1% uncertainty in σ_S contributes ± 0.6 to the error on N_ν. Previous measurements of N_ν by detection of single photons in e^+e^- annihilation give $N_\nu < 5.2$.[14] The third fit result for N_ν assumes standard model values of Γ_e and Γ_ν, an assumption that may be incompatible with a measurement of Γ less than the standard model.

Taking the most conservative value from Table 4, we conclude that $m_Z = 91.11 \pm 0.23$ GeV/c^2. The most important systematic error for the mass determination is the ≤ 40 MeV/c^2 uncertainty in the absolute energy scale, which is negligible compared to the statistical error. The error on the mass from the third fit is smaller due to the narrower fit width.

The electroweak mixing angle θ_W, defined by $\sin^2\theta_W \equiv 1 - m_W^2/m_Z^2$,[15] is related to m_Z by

$$\sin 2\theta_W = \left(\frac{4\pi\alpha}{\sqrt{2} G_F m_Z^2 (1 - \Delta r)} \right)^{\frac{1}{2}}.$$

The effects of radiative corrections are represented by Δr.[16] With $m_Z = 91.11 \pm 0.23$ GeV/c^2, $\sin^2\theta_W = 0.2312 \pm 0.0017$ for a top mass (m_t) of 100 GeV/c^2 and a Higgs mass (m_H) of 100 GeV/c^2. The dependence of $\sin^2\theta_W$ on m_t and m_H is shown in Fig. 8.

Previous direct measurements of m_Z at the p$\bar{\text{p}}$ collider at CERN yielded $93.1 \pm 1.0 \pm 3.1$ GeV/c^2 (UA1)[12] and $91.5 \pm 1.2 \pm 1.7$ GeV/c^2 (UA2)[13] where the first error is statistical and the second reflects the uncertainty in the overall mass scale. The CDF experiment at FERMILAB has recently reported[17] a Z^0 mass measurement of $90.9 \pm 0.3 \pm 0.2$ GeV/c^2.

VERTEX DETECTORS FOR THE MARK II

Currently, there are no vertex detectors installed inside the central drift chamber. However, two devices are ready for installation: a three-layer silicon strip detector[18] and a high precision drift chamber.[19] Altogether, the tracking system will provide an impact parameter resolution of about 4 μm for high momentum tracks. The multiple scattering term contributes an extra $(32/p)$ μm where the momentum p is measured in GeV/c. The existing beam pipe and wire scanner are incompatible with the vertex detectors. A new 25 mm radius beampipe (0.6%

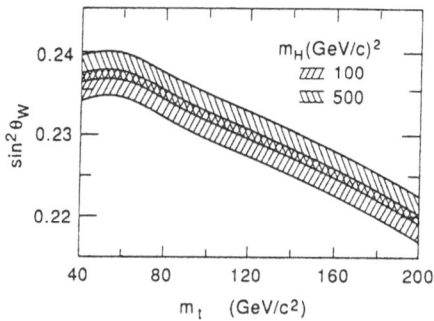

Figure 8. Value of $\sin^2\theta_W$ as a function of the top mass, for Higgs masses of 100 and 500 GeV/c^2. The width of each band represents the uncertainty in $\sin^2\theta_W$ corresponding to the uncertainty in m_Z.

radiation length aluminum) with a compact wire scanner has been designed and will be ready for installation with the vertex detectors October 1, 1989.

The silicon strip detector consists of three layers at radii of 30, 34 and 38 mm. Each layer consists of 12 separate detectors and each detector consists of 512 strips giving a total of 18,000 channels which are read out by custom VLSI chips. Each 6 mm × 6 mm chip reads out 128 channels, integrating and storing the charge deposited on a strip and multiplexing the analog signal onto a serial bus. The strip pitch varies between 25 and 33 μm and the strip length varies between 75 and 94 mm. The vertex detector covers 77% of the solid angle.

Each detector is assembled into a module which consists of two readout chips wire-bonded at each end and a hybrid circuit which provides the control lines, a switchable capacitor bank for power and a differential amplifier and line driver for the analog output signals.

Prototype tests have demonstrated that the signal-to-noise ratio is about 18, the spatial resolution per strip is 5 μm, and the double track resolution is less than 150 μm. To take advantage of the high resolution provided by the strips, their location must be known to better than a few microns. The intermodule alignment procedure uses a collimated x-ray beam. The position of the silicon strip vertex detector with respect to the high precision drift chamber will be monitored with capacitive sensors.

The vertex drift chamber consists of ten jet cells which extend radially between 5 and 17 cm and each provide 38 measurements with a resolution of about 30 to 40 μm per measurement, averaged over 2 cm of drift. The chamber operates at two atmospheres pressure and uses a gas with a slow drift velocity of about 6 μm/ns. The double-track resolution is expected to be less than 1 mm. The vertex drift chamber is complete and has operated outside the Mark II detector with the final gas system, electronics and readout system since December, 1988.

THE POLARIZATION PROJECT

One of the important advantages of a linear collider is the possibility of producing longitudinally polarized electrons at the IP. This allows one to measure the left-right asymmetry A_{LR} which quantifies the difference in coupling between the Z^0 and left- or right-handed electrons. A_{LR} can be related to parameters of the standard model such as $\sin^2\theta_W$ and M_Z. The forward-backward charge asymmetry A_{FB} can also be related to these parameters but has several disadvantages as a measurement compared to the left-right asymmetry. First, the left-right asymmetry is independent of the final state fermion and therefore the total Z^0 cross section contributes to this measurement. The forward-backward charge asymmetry is different

429

for fermions of different electric charge and weak isospin. It will probably only be possible to measure A_{FB} for Z^0 decays to muons and taus (3% of the total Z^0 cross section, each) and perhaps b quarks (about 14% of the cross section). In addition, A_{FB} is changing rapidly across the Z^0 pole resulting in a measurement which is very sensitive to radiative corrections and accurate knowledge of the center-of-mass energy. A_{LR} changes much more slowly than A_{FB} resulting in smaller systematic errors. Finally, A_{LR} is expected to be around 0.13 (for 50% polarization) while A_{FB} is expected to be only about one cent for muon pairs.

To test the standard model, we need two or more measurements of $\sin^2 \theta_W$ of comparable precision. With only a few thousand observed Z^0 decays, we reach the systematic limit on the M_Z measurement corresponding to an uncertainty on $\sin^2 \theta_W$ of less than 0.0005, far better than the precision of previous measurements. For example, Amaldi *et al.*[20] analysed all data pertaining to the weak neutral current and the intermediate vector boson masses in 1987. A global fit to all data yielded an uncertainty on $\sin^2 \theta_W$ of ± 0.0048 including full statistical, systematic and theoretical uncertainties. A measurement of A_{FB} for muon pairs requires more than 10^5 observed Z^0 decays to become competitive with these previous measurements and reaches a systematic uncertainty on $\sin^2 \theta_W$ of about 0.002 for more than 10^7 Z^0's. With a longitudinally polarized electron beam with polarization P_e equal to 45%, one can measure $\sin^2 \theta_W$ more precisely than the earlier measurements with a few thousand observed Z^0 decays. The availability of this polarization is equivalent to about a factor of one hundred in luminosity for precise tests of the standard model. The systematic limit on A_{LR} depends on how well one can measure P_e. A fractional error of $\Delta P_e / P_e = 3\%$ should not be difficult and an eventual precision of 1% should be possible. A 1% precision in the measurement of P_e will result in similar systematic errors on $\sin^2 \theta_W$ for the measurements of A_{LR} and M_Z.

At SLC, longitudinally polarized electrons[21] will be produced by irradiating a GaAs crystal with circularly polarized light. The electron gun has been installed and the laser system will be installed this fall. Three superconducting magnets are needed to rotate the spin before and after the damping ring. These magnets have been delivered to SLAC and tested. They are also awaiting installation which will require about 15 weeks downtime. Møller polarimeters for measuring the polarization at the end of the linac and in the extraction line with a precision of (3-5)% have been installed. The best measurement of the polarization is expected to come from a Compton polarimeter just downstream of the IP. It should provide a measurement with 1% precision in less than a minute. A prototype detector for this system is in place for background studies and the laser has been installed. A vent shaft leading to a counting house above ground is ready for installation of the optical system.

CONCLUSIONS

The feasibility of the e^+e^- collider concept has been demonstrated with the SLAC Linear Collider. Especially important is the implementation of beam-beam deflection as a diagnostic tool for tuning micron-size beams. Beamstrahlung signals have been observed for the first time with colliding beams.

A scan of six center-of-mass energies yields the following measurements of Z^0 resonance parameters: $m_Z = (91.11 \pm 0.23)$ GeV/c^2, $\Gamma = 1.61^{+0.60}_{-0.43}$ GeV or $\Gamma < 3.1$ GeV at the 90% confidence level, and $N_\nu = 3.8 \pm 1.4$ or $N_\nu < 5.5$ at the 90% confidence level. The uncertainty on $\sin^2 \theta_W$ from the Z^0 mass measurement is ± 0.0017 for a fixed top-quark and Higgs-boson mass.

In the near future, two high precision vertex detectors will be installed inside the central drift chamber providing excellent impact parameter resolution. Also nearing completion is a project to provide longitudinally polarized electrons at the interaction point. These last two features represent unique aspects of the SLC program for studying the Z^0 and its decay products.

ADDENDUM

At the time of submission of this paper, the measurements of the Z^0 resonance parameters with the Mark II detector have been improved by tripling the amount of integrated luminosity. A total of 480 Z^0 decays at 10 center-of-mass energies yield the following measurements of the Z^0 resonance parameters:[22] $m_Z = (91.14 \pm 0.12)$ GeV/c^2, $\Gamma = 2.42^{+0.45}_{-0.35}$ GeV, and $N_\nu = 2.8 \pm 0.6$ or $N_\nu < 3.9$ at the 95% confidence level.

REFERENCES

1. G. Abrams *et al.*, Phys. Rev. Lett. 63, 724 (1989).

2. SLAC Linear Collider Conceptual Design Report, SLAC-Report-229 (1980).

3. R. Fulton *et al.*, Nucl. Instr. and Meth. A274, 37 (1989); G. Bowden *et al.*, Nucl. Instr. and Meth. A278, 664 (1989).

4. P. Bambade, SLAC-CN-303 (1985); P. Bambade *et al.*, SLAC-PUB-4767 (1989), submitted to Phys. Rev. Lett.

5. R. Erickson, SLAC-PUB-4974 (1989).

6. J. Kent *et al.*, SLAC-PUB-4922 (1989).

7. M. Levi *et al.*, SLAC-PUB-4921 (1989).

8. M. Levi, J. Nash, and S. Watson, Nucl. Instr. and Meth. A281 (1989).

9. G. Abrams *et al.*, SLAC-PUB-4558 (1989), to be published in Nucl. Instr. and Meth., and references therein.

10. F. A. Berends, R. Kleiss and W. Hollik, Nucl. Phys. B304, 712 (1988); S. Jadach and B. F. L. Ward, UTHEP-88-11-01 (1988), submitted to Phys. Rev. D.

11. R. N. Cahn, Phys. Rev. D 36, 2666 (1987), Eqs. 4.4 and 3.1. See J. Alexander *et al.*, Phys. Rev. D 37, 56 (1988) for a comparison of different radiative correction calculations.

12. C. Albajar *et al.*, CERN-EP/88-168 (1988) (submitted to Z. Phys. C).

13. R. Ansari *et al.*, Phys. Lett. 186B, 440 (1987).

14. The combined ASP, MAC and CELLO result is calculated in C. Hearty *et al.*, Phys. Rev. D 39, 3207 (1989).

15. A. Sirlin, Phys. Rev. D 22, 971 (1980).

16. W. J. Marciano and A. Sirlin, Phys. Rev. D 22, 2695 (1980).

17. K. Ragan, 1989 SLAC Summer Institute, July 11-21, 1989, Stanford, California.

18. C. Adolphsen *et al.*, SLAC-PUB-4452, 1987.

19. J. Jaros, SLAC-PUB-4285, 1987.

20. U. Amaldi *et al.*, Phys. Rev. D 36, 1385 (1987).

21. D. Blockus *et al.*, SLAC Proposal for Polarization at the SLC, 1986; D. Cords *et al.*, Mark II/SLC Note 238, July, 1989.

22. G. Abrams *et al.*, SLAC-PUB-5113, submitted to Phys. Rev. Lett., October, 1989.

RESULTS FROM THE LEP AND SLAC COLLIDERS - AN UPDATE[1]

A. Ali

Sektion Physik, Universität München
and

Deutsches Elektronen Synchrotron, DESY, Hamburg[2]

ABSTRACT

Recent experimental results from the Large Electron Positron Collider LEP and the SLAC Linear Collider SLC are briefly summarized in this
note. The data and analyses shown are selective and only those aspects
having a direct (or indirect) bearing on the searches for the top quark
and/or scalar particles (Higgs bosons and scalar leptons) are included
here.

I. THE STANDARD MODEL Z^0 BOSON PARAMETERS - PRESENT STATUS

The measurements of the mass of the standard model neutral boson,
m_{Z^0} , its total decay width, Γ_{Z^0} , mean leptonic width Γ_{11} , the number

of light neutrino species contributing to the decays $Z^0 \rightarrow \nu\bar{\nu}$, N_ν , and
the hadronic peak cross section, $\sigma(Z)_{peak}$, as published by the MARK II
Collaboration at SLC [1], and the four collaborations ALEPH [2], DELPHI
[3], L3 [4] and OPAL at LEP are summarized in Table 1. The numbers shown
have been corrected for the acceptance, efficiency and radiative effects. In obtaining these numbers a Breit-Wigner Shape has been assumed to
determine m_{Z^0} and Γ_{Z^0} . The typical error in m_{Z^0} due to the uncertainty
in the LEP center of mass energy is ~ 30 MeV. It is interesting to note
that this error is already larger than the statistical and/or other
systematic errors (e.g. 24 MeV and 26 MeV as quoted by the L3 and ALEPH
collaborations, respectively). The experimental resolution on Δm_Z is now
almost comparable to the theoretical precision of the one-loop calculation in the standard model, i.e. $\Delta M \simeq 35$ MeV [6]. The numbers quoted for
the neutrino species N_ν are based on the measurement of the partial width

Γ ($Z^0 \rightarrow$ invisible) and the standard model prediction Γ^{SM} ($Z^0 \rightarrow \nu\bar{\nu}$) $\simeq$

---

[1]Note added by the Editor

[2]Permanent Address

166 MeV. Other methods, such as using the relation $N_\nu = 3 + (\Gamma_Z - \Gamma_Z^{SM})$ / $\Gamma_{\nu\nu}^{SM}$ where Γ_Z is the experimentally measured width , give compatible results.

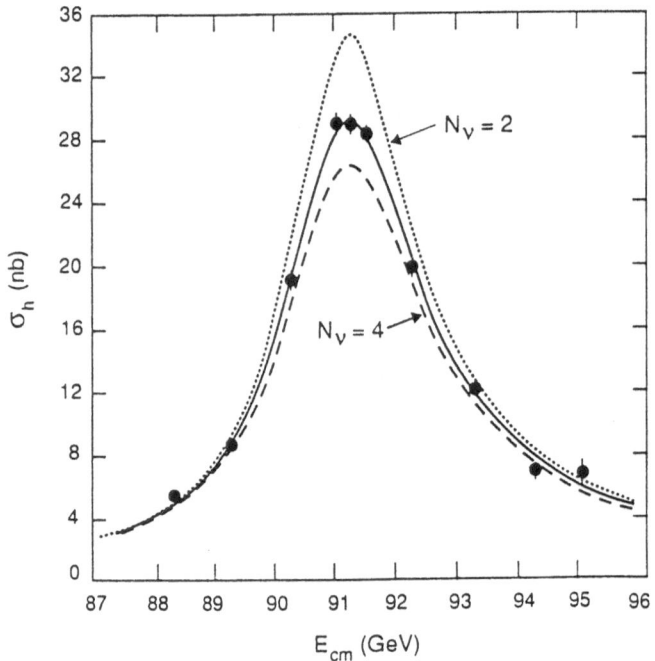

Fig. 1   The measured cross section for $e^+e^- \longrightarrow$ hadrons at LEP as a function of $\sqrt{s}$. The solid curve is a fit to the formula of Borelli et al. [6] in which M_{Z^o} and $\Gamma_{invisible}$ were left free and the partial width Γ_{11} and $\Gamma_{hadrons}$ were the standard model values. The dotted and dashed curves correspond to $N_\nu = 2$ and $N_\nu = 4$ (with the standard model $\Gamma_{\nu\nu}$ width). The curve for $N_\nu = 3$ is indistinguishable from the fitted curve (from L3 Collaboration [4]).

A representative result on N_ν from the Z^o-line shape is shown in Fig. 1. The entries in Table 1 (and Fig. 1) show that the case $N_\nu \geq 4$ can now be ruled out <u>within the standard model</u> at a significance level which is greater than 4σ. In a recent theoretical analysis [7], it has been argued that a fourth light neutrino is also excluded in extended gauge models having an extra U(1) gauge boson. The exclusion of an extra (fourth) light neutrino species is the <u>most</u> significant experimental result from the first phase of the SLC and LEP collider runs.

Table 1. Present experimental measurement from LEP and SLC on the Z boson parameters m_Z, Γ_Z, Γ_{ll}, N_ν and the Z^0 peak hadronic cross-section. The Glashow Salam Weinberg theory predictions are also shown for m_Z = 91.10 ± 0.06 GeV, m_t = 80 – 200 GeV, m_H = 10 – 1000 GeV.

Experimental Collaboration	m_Z(GeV)	Γ_Z(GeV)	Γ_{ll}(MeV)	N_ν	σ_{had}(Z) (nb)
MARK II [1]	91.14 ± 0.12	$2.42 \begin{smallmatrix} + 0.45 \\ - 0.35 \end{smallmatrix}$	—	2.8 ± 0.6	42.0 ± 4
ALEPH [2]	91.182 ± 0.026 (expt.) ± 0.030 (LEP)	2.541 ± 0.056	83.9 ± 2.2	3.01 ± 0.15 (exp.) ± 0.05 (theor.)	41.4 ± 0.8
DELPHI [3]	91.06 ± 0.09 (stat.) ± 0.045 (syst.)	2.42 ± 0.021 (stat.	—	2.4 ± 0.4 (stat.) ± 0.5 (syst.)	—
L3 [4]	91.106 ± 0.024 (expt.) ± 0.030 (LEP)	2.539 ± 0.054	$88 \begin{smallmatrix} + 4 \\ + 3 \end{smallmatrix}$	3.29 ± 0.17	39.8 ± 0.9
OPAL [5]	91.01 ± 0.05 (stat.) ± 0.05 (syst.)	2.60 ± 0.13	88.1 ± 4.6	3.3 ± 0.7	41.7 ± 2.4
GSW [6,7] Theory	91.10 ± 0.06	2.490 ± 0.025	83.5 ± 0.5	3	41.4 ± 0.1

II. DIRECT AND (INDIRECT) BOUNDS ON THE TOP QUARK MASS

The most stringent (lower) bound on m_t in direct searches for the top quark using the standard model decay modes is due to the CDF collaboration which exclude m_t in the mass range $40 < m_t < 77$ GeV at 95 % C.L. (in the Electron + E_T + $\geq$ 2 Jets channel) [8]. The searches in e^+e^- annihilation provide a lower bound, $m_t > 44.5$ GeV (OPAL) [9], and $m_t > 45.8$ GeV (ALEPH) [10] for the standard model decays, and $m_t > 41$ GeV (MARK II) [11] for a t decaying in a charged Higgs by $t \rightarrow bH^+$.

The electroweak data, in particular from the neutral current sector, and recent measurements of the vector gauge boson masses, m_W and m_Z [14,15,16], constrain in principle the top quark and Higgs boson masses through radiative corrections [12]. The consistency of these independent measurements (radiatively corrected at the 1-loop level) has been the subject of a number of recent investigations [13]. It is customary to show this consistency in the parameter $\sin^2 \theta_W \equiv 1 - m_W^2/m_Z^2$. The result of a representative analysis by Langacker [13] is shown in Fig. 2. This analysis is based on the SLC values for m_Z [14]

$$M_Z = 91.17 \pm 0.18 \text{ GeV}$$

the M_W, M_Z values, as obtained by the CDF [15] and UA2 [16] collaborations:

$m_W = 80.0 \pm 0.2 \pm 0.5 \ (\pm 0.3)$ GeV	CDF
$\quad = 80.0 \pm 0.4 \pm 0.4 \ (\pm 1.2)$ GeV	UA2
$m_Z = 90.9 \pm 0.3 \ (\pm 0.2)$ GeV	CDF
$\quad = 90.19 \ ^{+ \ 0.56}_{- \ 0.59} \ (\pm 1.4)$ GeV	UA2

(the first uncertainty is statistical, the second in m_W is systematic, and the uncertainty in parantheses are correlated scale errors)

the weak neutral current data as analysed by Amaldi et al. in ref. [12], and new results on $\nu_\mu e$ scattering and atomic parity violation [17]. It is seen from Fig. 2 that the region 80 GeV $< m_t <$ 210 GeV is favoured (at 90 % C.L.) by the present data. The upper bound on m_t is sensitive to m_H and the value quoted is for an assumed valued $m_H = 1$ TeV. This bound comes down to 190 GeV for $m_H = 100$ GeV. On the other hand, no useful constraints are obtained on m_H from the existing neutral current data, and/ or m_W, m_Z measurements. The present error on m_Z from LEP measurements is much smaller (typically $\Delta m_Z \simeq 40$ MeV) as compared to what has been used in the analyses of Langacker and others in ref. [13]. However, this is going to influence the upper bound on m_t only marginally, as can be judged from Fig. 2.

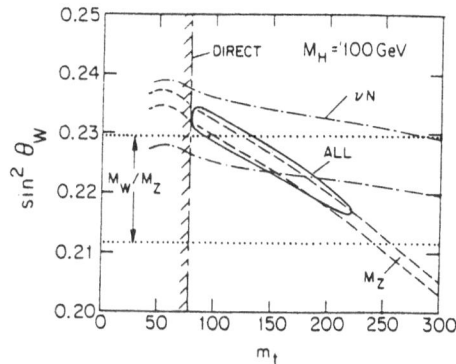

Fig. 2 ± 1σ uncertainties in $\sin^2\theta_W \equiv 1 - M_W^2/M_Z^2$ determined from M_Z (dashed line), M_W/M_Z (dotted line), and νN neutral-current data (dash-dotted line) as a function of m_t for $M_H = 100$ GeV. Also shown is the direct lower limit from the nonobservation of the t in $\overline{p}p \rightarrow \overline{t}t + X$ (long-short line), and the region (solid line) in $\sin^2\theta_W$-m_t allowed by all data at 90 % C.L. ($\Delta\chi^2 = 4.6$) (from Langacker [13]).

III. LIMITS ON THE STANDARD MODEL HIGGS BOSON MASS FROM EXPERIMENTS AT LEP

As discussed at length in these proceedings and elsewhere [18], one hopes to detect a light Higgs boson ($m_H \lesssim m_W$) in e^+e^- collisions at LEP. The search strategies for a standard model Higgs boson at LEP I are based on the Z^0 decays:

$$Z^0 \rightarrow Z^{0*}H \; ; \; Z^* \rightarrow l^+l^-, \; \nu\overline{\nu}, \; q\overline{q} \; ; \; H \rightarrow l^+l^-, \; q\overline{q}$$

and the radiative decay

$$Z^0 \rightarrow H\gamma \; ; \; H \rightarrow l^+l^-, \; q\overline{q}$$

The relative rate $\Gamma(Z^0 \rightarrow H^0 f\overline{f})/\Gamma(Z^0 \rightarrow \mu^+\mu^-)$ in the standard model may have as high a value as 10^{-2} for a very light Higgs boson ($m_H < 1$ GeV) and falls to the level of $\lesssim 10^{-6}$ for $m_H \gtrsim 70$ GeV. The relative rate for the radiative decay process $(Z^0 \rightarrow H\gamma)/\Gamma(Z^0 \rightarrow \mu^+\mu^-)$ has a maximum value of $\sim 6 \times 10^{-4}$ in the standard model even for a very light Higgs boson ($m_H \lesssim 1$ GeV), and hence is not very useful for the Higgs search with the presently available data samples at LEP (typically 20,000 hadronic events/experiment).

The present data at LEP do allow a meaningful search for a Higgs boson with a mass $\lesssim 25$ GeV, and two collaborations (ALEPH and OPAL) have so far published their analyses. The bounds on m_H obtained by the ALEPH and OPAL collaborations are summarized in Table 2. In the first such analysis from the LEP data by the ALEPH collaboration, Higgs boson was searched using the processes

$$Z^0 \rightarrow Hl^+l^- \quad , \quad l = e,\mu,\tau$$
$$H\nu\overline{\nu}$$
$$Hq\overline{q}$$

Table 2. Present limits (95 % C.L.) on the standard model Higgs boson mass from experiments at LEP.

Experimental Collaboration	Number of Hadronic Events	Search Modes	Excluded Higgs Mass Range
ALEPH [19]	~ 11,500	$Z^0 \rightarrow Z^{0*}H$ $\rightarrow l^+l^-$ $(l = e, \mu, \tau)$ $\rightarrow \nu\bar{\nu}$ $\rightarrow q\bar{q}$	$32 \text{ MeV} \leq m_H \leq 15 \text{ GeV}$
ALEPH [20]	~ 25,000	$Z^0 \rightarrow Z^{*0}H$ $\rightarrow e^+e^-$ $\rightarrow \mu^+\mu^-$ $\rightarrow \nu\bar{\nu}$	$11 \text{ GeV} \leq m_H \leq 24 \text{ GeV}$
OPAL [21]	~ 22,700	$Z^0 \rightarrow Z^{*0}H$ $\rightarrow e^+e^-$ $\rightarrow \mu^+\mu^-$ $\rightarrow \nu\bar{\nu}$ $H \rightarrow q\bar{q}$ $\rightarrow \tau^+\tau^-$	$3 \text{ GeV} \leq m_H \leq 19.3 \text{ GeV}$

For the mass range ($\simeq m_H > 2$ GeV), the state $H\nu\bar{\nu}$ with $H \rightarrow q\bar{q}$ provides an efficient way to search for the Higgs boson through its (effective) mono-jet and dijet topologies (+ missing energy), since the inclusive branching ratio for $H \rightarrow q\bar{q}$ is large (~ 70 %). The upper excluded mass in this analysis by ALEPH (15 GeV) is set by the statistics. For the mass range $2m_\mu \leq m_H \leq 2$ GeV, the decays $H \rightarrow \mu^+\mu^-$, $\pi^+\pi^-$, $k\bar{k}$, $\eta\eta'$, $p\bar{p}$, $\omega\omega$, $\omega\rho$,... dominate. For a discussion of theoretical uncertainties in the relative branching ratios in this mass range we refer to the articles of Haber, Wiley and Yu in these proceedings.

For the search of a very light Higgs boson ($2m_e \leq m_H < 2m_\mu$), which incidentally is already ruled out by Kaon experiments [22], the decay $H \rightarrow e^+e^-$ would dominate and the light Higgs boson would be long lived (typically for $m_H = 100$ MeV, the average Higgs decay length in Z^0 decays is 100 cm). For this mass range the Higgs search was performed with the additional requirement on the identification of isolated vertices from the Higgs decay. The lower excluded Higgs boson mass (in this case 32 MeV) is set by the detection efficiency which worsens as m_H decreases. The combined excluded Higgs boson mass range in the first ALEPH analysis is 32 MeV to 15 GeV at 95 % C.L. and 40 MeV to 12 GeV at 99 % C.L.

The second ALEPH analysis [20] for Higgs mass $m_H > 11$ GeV (i.e. m_H above $B\bar{B}$ threshold) is based on the decay process $Z^0 \rightarrow H\nu\bar{\nu}$, followed by the decay $H \rightarrow b\bar{b}(g)$, having the signal of (mostly) dijet events with missing energy and momentum. The excluded Higgs boson mass range from this analysis is 11 GeV to 24 GeV at 95 % C.L. The combined result of the two analyses and the 95 % C.L. bounds on m_H from ALEPH are shown in Fig. 3.

A similar search for the standard model Higgs boson has been performed by the OPAL collaboration [21]. The search modes employed were

$$Z^0 \rightarrow (e^+e^-, \mu^+\mu^-, \nu\bar{\nu})H$$
$$H \rightarrow q\bar{q}, \tau^+\tau^-$$

for $m_H > 3$ GeV. The analysis, based on 825 nb^{-1} of data taken at a center of mass energy between 88.3 and 95.0 GeV, excludes the Higgs boson mass range 3.0 GeV to 19.3 GeV at 95 % C.L. The result of the OPAL collaboration Higgs search and the resulting 95 % C.L. bound are shown in Fig. 4.

It should be remarked here that the searches for the Higgs boson at LEP in the mass range $2m_\mu < m_H < 2$ GeV are probably not overly subject to theoretical uncertainties, since the search technique for H used by ALEPH (2 or more charged tracks) is rather inclusive. Theoretical uncertainties in the decays $H \rightarrow q\bar{q}$, l^+l^- for $m_H > 2$ GeV are expected to be lot less as compared to the ones in the mass range $2m_\mu < m_H < 2$ GeV (due to uncertainties in $H \rightarrow \pi^+\pi^-$, k^+k^- etc.). Hence LEP results provide a significantly larger excluded mass range for m_H than what was known from earlier low energy experiments. We note here that since the decay mode used for the Higgs search by ALEPH and OPAL was $Z \rightarrow Z^*H$, their analyses have no bearing on a composite Higgs (like for example a pseudoscalar pseudo

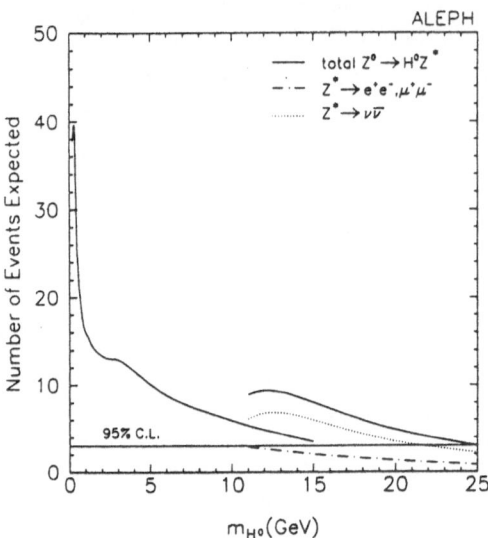

Fig. 3   The total number of events expected for $Z^0 \rightarrow HZ^*$ ($Z^* \rightarrow e^+e^-$, $\mu^+\mu^-$, $\tau^+\tau^-$, $\nu\bar{\nu}$ and $q\bar{q}$ (g)) as a function of the Higgs boson mass. The two solid curves follow from the combined ALEPH analysis I [19] and II [20]. The 95 % C.L. limit at 3 expected events is indicated, corresponding to zero observed candidates (from ALEPH [20]).

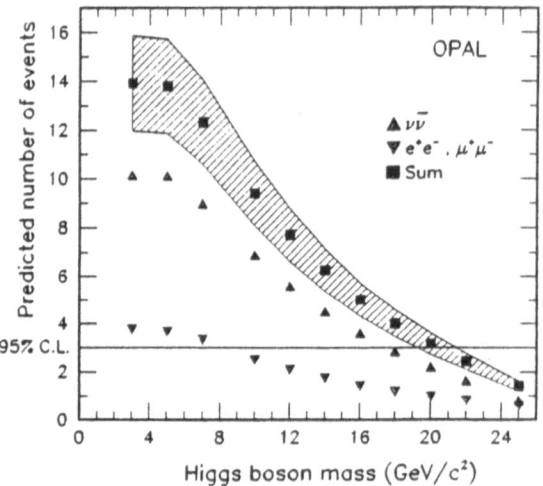

Fig. 4   The expected number of events, as a function of the Higgs boson mass for the indicated search channels. The shaded zone represents the $\pm\ 1\sigma$ systematic level on the predicted number of events. The horizontal line indicates the levels expected for the 95 % C.L. upper limit (from OPAL [21]).

Goldstone boson, π^0, in Technicolour models), since the coupling $Z \rightarrow Z^* \pi^{0\prime}$ in such models is several orders of magnitude smaller than the SM ZZ^*H coupling [24].

IV. LIMITS ON HIGGS BOSON MASSES IN THE SUPERSYMMETRIC EXTENSION OF THE STANDARD MODEL

In the supersymmetric extension of the standard model one expects more than one doublet of Higgs bosons. For example, in the minimal super-symmetric extension one has two Higgs doublets with specific Higgs-fermion couplings. The theory and phenomenology of such models is discussed elsewhere in these proceedings [23]. Recapitulating the salient features very briefly, five physical particles, with masses and couplings constrained by supersymmetry, are expected in this scenario: two charged Higgs bosons, $H^{\pm}$, one neutral Higgs boson, H, (generally) expected to be heavier than the Z, one neutral Higgs, h^0, (generally) expected to be lighter than the Z , and one pseudoscalar A^0. The illustrative relation-ships among m_{h^0} , m_H , m_{A^0} and m_{H^+} are shown in Fig. 2 of Haber's article in these proceedings. A particularly relevant scenario is: $m_{h^0} < m_Z|\cos 2\beta| \lesssim m_Z$ and $m_{A^0} > m_{h^0}$, where the angle β is defined by the ratio of the two vacuum expectation values ϑ_1 and ϑ_2: $\tan \beta = \vartheta_2/\vartheta_1$. It is, therefore, natural in these models to expect the decays $Z \rightarrow hZ^*$ to occur, which is very much like the standard model process discussed above. However, the coupling hZZ is reduced with respect to the standard model coupling ΦZZ (to avoid confusion we denote the SM Higgs boson by Φ) by a factor $\sin(\alpha-\beta)$, giving

$$\frac{\Gamma(Z \rightarrow h^0 Z^*)}{\Gamma(Z \rightarrow \Phi Z^*)} = \sin^2(\alpha-\beta)$$

where α is the mixing angle that arises in the process of diagonalizing the 2×2 neutral Higgs mass matrix (see Haber's article, Eqs. 13 + 14). Similarly, the couplings $hd\bar{d}$, $hu\bar{u}$, $hl\bar{l}$ are also modified with respect to their standard model values (assumed 1 here)

$$h^0 u\bar{u} : \frac{\cos \alpha}{\sin \beta} \quad , \quad h^0 b\bar{b} : \frac{-\sin \alpha}{\cos \beta}$$

In addition to the decay $Z \rightarrow h^0 Z^*$, it is conceivable that for particular choices of $m_{H^{\pm}}$ and β, the pseudoscalar boson mass m_A may approach m_{h^0} and the ZAh coupling may become large. (This happens for $\tan \beta > 1$.) If $m_{A^0} + m_{h^0} < m_Z$, then the following relation holds:

$$\frac{\Gamma(Z^0 \rightarrow A^0 h^0)}{\Gamma(Z^0 \rightarrow \nu\bar{\nu})} = \frac{1}{2}\cos^2(\alpha - \beta)B^3$$

with $B = 2|\vec{p}|/m_Z$, $|\vec{p}|$ being the magnitude of the 3-momenta of one of the final Higgs particles. If <u>kinematically allowed</u>, the decays $\Gamma(Z \rightarrow h^0 Z^*)$ and $\Gamma(Z^0 \rightarrow A^0 h^0)$ are complementary to each other.

The characteristic signatures of the process $Z^0 \rightarrow h^0 A^0$ (if $m_Z > m_A^0 + m_{h^0}$) depend on the values of m_{h^0} and $\tan \beta$ ($= \vartheta_2/\vartheta_1$). For $\vartheta_2/\vartheta_1 > 1$ the following regions and decay channels have been used in the dedicated search for the process $Z^0 \rightarrow h^0 A^0$ by the ALEPH collaboration.

I. $2m_e < m_{h^0} < 2m_\mu$: $h^0 \rightarrow e^+ e^-$, $\gamma\gamma$, with $\tau(h^0) \sim O(10^{-8}$ sec.) giving an h^0 decay length, $d_{h^0} \sim 0$ (1 meter). The signature is an $e^+ e^-$ pair emerging from a vertex well detached from the collision point.

II. $2m_\mu < m_{h^0}$, $m_{A^0} < 2m_\tau(2m_D)$ for $\vartheta_2/\vartheta_1 > 1 (< 1)$, with the decays h^0, $A^0 \rightarrow$ few hadrons, giving rise to two back-back low multiplicity events.

III. $2m_\mu < m_{h^0} < 2m_\tau$; $2m_\tau < m_{A^0} < 2m_B$: h^0 decays to low multiplicity hadron states, A^0 decays to tau pairs.

IV. $2 \, m_\tau < m_{h^0}$, $m_A < 2m_B$: the preferential decays are $(A^0, h^0) \rightarrow \tau^+ \tau^-$. The signature of regions II, III and IV are very similar.

V. $2m_\tau < m_{h^0} < 2m_B$, $2m_B < m_A < m_Z - m_{h^0}$: the preferential decays are $h^0 \rightarrow \tau^+ \tau^-$, $A^0 \rightarrow b\bar{b}$ giving rise to $\tau^+ \tau^-$ jet-Jet events.

VI. $2m_B < m_{h^0} < m_A$, $m_{h^0} < m_{A^0} < m_Z - m_{h^0}$: the preferential decays are: h^0, $A^0 \rightarrow b\bar{b}$ giving 4-jet events but also (at a reduced level) $\tau^+ \tau^-$ jet-jet events.

For $\vartheta_2/\vartheta_1 < 1$, the same classification of final states holds up to the $D\bar{D}$ threshold. From there on, only 4-jet signature remains and hence it is difficult to perform the search for low m_A region in this case.

The results of the ALEPH search based on $Z^0 \rightarrow h^0 A^0$ decay mode are given in Fig. 5a [in the (m_h , ϑ_2/ϑ_1) plane] and Fig. 5b [in the (m_h , m_A) plane for $\vartheta_2/\vartheta_1 > 1$]. The various regions (A, E, F, G on the curves shown) correspond to the follwoing search strategies [25]:

(A):   Same as the search for the standard Higgs, described earlier.
(B):   Low multiplicity final states in $Z^0 \rightarrow h^0 A^0$ (corresponding to the mass region II - IV described above).
(F): $\tau^+ \tau^-$ jet-jet final state (corresponding to the mass regions V + VI above):
(G):   4-jet final states (corresponding to the mass region VI above).

If $\vartheta_2/\vartheta_1 > 1$, then the range $m_h \simeq m_A < 39.2$ GeV is excluded at 95 % C.L. Obviously if $m_{A^0} > m_Z$, then the dedicated search for the process $Z^0 \rightarrow h^0 A^0$ is kinematically forbidden, even for a very light h^0. For a light h^0 the search from the decay $Z^0 \rightarrow h^0 Z^*$ is then the only allowed process leading to the excluded domains in the $(\vartheta_2/\vartheta_1 - m_{h^0})$ plane.

V. LIMITS ON SCALAR LEPTON MASSES FROM EXPERIMENTS AT LEP

Experiments in $e^+ e^-$ annihilation at lower energies and Z^0 decays (at LEP and SLC) have been used to search for the supersymmetric scalar leptons (sleptons) and spin-1/2 charged bosons (higgsino, Wino). The production process assumed in these searches is [26, 27]

$$e^+ e^- \xrightarrow{Z, \gamma} \tilde{e}^+ \tilde{e}^-, \; \tilde{\mu}^+ \tilde{\mu}^-, \; \tilde{\tau}^+ \tilde{\tau}^-, \; \tilde{\chi}^+ \tilde{\chi}^-$$

followed by the decays

$$\tilde{l}^\pm \rightarrow l^\pm + \tilde{\gamma}$$
$$\tilde{\chi}^\pm \rightarrow l^\pm v_1 + \tilde{\gamma}, \; l^\pm \tilde{v}$$

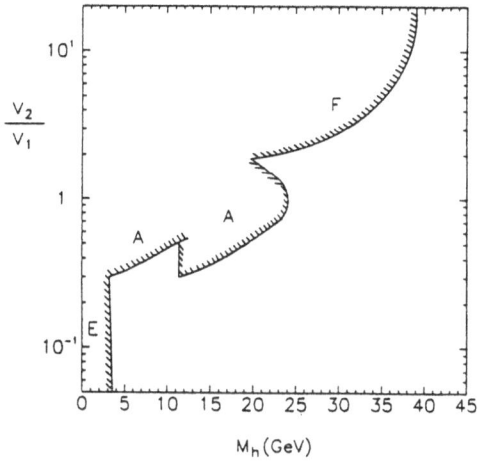

Fig.5 (a)

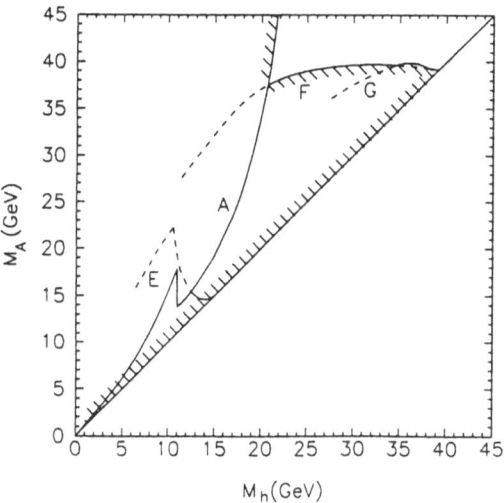

Fig.5 (b)

Fig. 5   Excluded domains (a) in the (M_h , ϑ_2/ϑ_1) plane, (b) in the M_h ,
M_A) plane for $\vartheta_2/\vartheta_1 > 1$. The curves A,E,F,G are defined in the
text (from ALEPH [20]).

where the photino $\tilde{\gamma}$ is assumed to be the lightest SUSY particle and hence stable. The signature of these states is then a pair of charged leptons ($e^{\pm}$, $\mu^{\pm}$) and missing energy and momentum. In the case of scalar taus $\tilde{\tau}^{\pm}$, one replaces the charged lepton $e^{\pm}$, $\mu^{\pm}$ by single charged particles, since BR ($\tau^{\pm} \rightarrow$ 1-charged prong modes) $\simeq$ 85 %. The efficiency in this case is then reduced due to the softer momentum spectrum and the presence of additional photons from π^{0}'s.

There are two sleptons, $\tilde{l}\,^{\pm}_{R}$ and $\tilde{l}\,^{\pm}_{L}$, for each charged lepton, $l^{\pm}$, with their masses, in general, unequal. In most SUSY models the right hand sleptons $\tilde{l}_{R}$ are lighter than the left handed $\tilde{l}_{L}$. A typical analysis at LEP can be seen in Fig. 6, where the 95 % C.L. mass limits in the $m_{\tilde{l}}$ – $m_{\tilde{\gamma}}$ plane are shown from the recent ALEPH slepton searches [28]. The results of the searches by TOPAZ [29] and ASP [30] are also shown. One should remark here that the ASP collaboration exclude $\tilde{e}_{R}$ for $m\tilde{e}_{R} <$ 48 GeV for a small photino mass.

The bounds on $m_{\tilde{e}}$, $m_{\tilde{\mu}}$ and $m_{\tilde{\tau}}$ from the ALEPH [29], L3 [30] and OPAL [31] collaborations (for the assumed photino mass range) are collected in Table 3.

Table 3. Present limits (95 % C.L.) on the scalar-electron mass, $m_{\tilde{e}}$, scalar muon-mass, $m_{\tilde{\mu}}$, and scalar-τau mass, $m_{\tilde{\tau}}$, from experiments at LEP

Scalar Lepton	Assumed Photino Mass (GeV)	Excluded Mass Limit (GeV) ($m_{\tilde{l}_{L}} > m_{\tilde{l}_{R}}$)	Experimental Collaboration
$\tilde{e}$	0 – 35 0 – 20 0 – 30	43.5 41.0 41.5	ALEPH [28] L3 [32] OPAL [31]
$\tilde{\mu}$	0 – 35 0 – 20 0 – 30	42.6 41.0 41.0	ALEPH [28] L3 [32] OPAL [31]
$\tilde{\tau}$	0 – 35 0 – 23	40.4 41.0	ALEPH [28] OPAL [31]

SUMMARY

Recent experiments at LEP and SLC have provided a wealth of new data which have started testing the standard model in a precise manner,

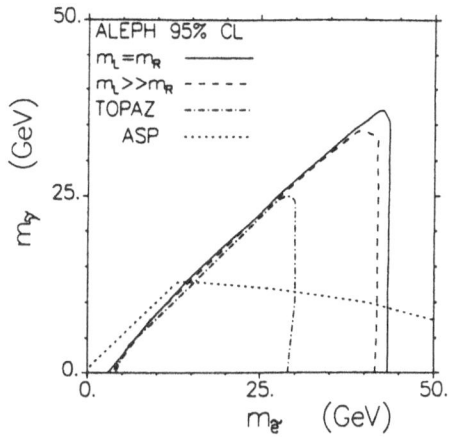

Fig. 6   Excluded region in scalar electron and photino masses excluded
at 95 % C.L. from the LEP data [ALEPH], TOPAZ [29] and ASP [30]
(the result from ASP is at 90 % C.L.) (from ALEPH [28]).

and in mass regions for new particle searches not done before. Particular
applause deserve the precisions reached in m_Z measurements with $\Delta m_Z \simeq 40$
MeV and in the determination of $\Gamma_{(Z^0} \rightarrow$ invisible), which allow only
three generations of light neutrino.

The consistency of electroweak data imposes constraints on m_t ;
however at present the data restrict m_t to lie in the range 80 GeV $< m_t <$
210 GeV. New improved bounds have been obtained on the standard model
Higgs boson mass, with m_H now excluded in the mass range 32 MeV $< m_H < 24$
GeV (at 95 % C.L.) from the LEP data alone. Likewise, new and significant
bounds in the light Higgs boson mass, m_{h^0} , and the pseudoscalar boson
mass, m_{A^0} , in SUSY extensions of the standard model have been obtained
and scalar leptons, $\tilde{l}^{\pm}$, have been excluded in the mass range $m_{\tilde{l}^{\pm}} < 41$
GeV.

REFERENCES

[1]   G.S. Abrams et al. [MARK II Collaboration], Phys. Rev. Lett. 63
(1989) 2173
[2]   D. Decamp et al. [ALEPH Collaboration], Phys. Lett. B231 (1989)
519; CERN Report EP/89 - 169 (1989)
[3]   P. Aninio et al. [DELPHI Collaboration], Phys. Lett. B231 (1989)
539
[4]   B. Adeva et al. [L3 Collaboration], Phys. Lett. B231 (1989) 509; L3
Preprints # 003 and # 004 (1989)
[5]   M.Z. Akrawy et al. [OPAL Collaboration], Phys. Lett. B231 (1989)
530; CERN Report EP/89 - 197 (1989)

[6] D.Yu. Bardin, A. Leike, T. Riemann and M. Sachwitz, Phys. Lett. B206 (1988) 539;
     F.A. Berends, G. Burgers, W. Hollik and W.L. van Neerven, Phys. Lett. B203 (1988) 177;
     A. Borrelli, M. Consoli, L. Maiani and R. Sisto, CERN TH-5441/89 (1989)
[7] G. Altarelli et al., CERN-TH-5591/89 (1989)
[8] S. Geer [CDF Collaboration], these proceedings
[9] M.Z. Akrawy et al. [OPAL Collaboration], CERN-EP/85-154 (1989)
[10] D. Decamp et al. [ALEPH Collaboration], CERN-EP/89-165 (1989)
[11] G.S. Abrams et al. [MARK II Collaboration], SLAC PUB-5106 (1989
[12] M. Veltman, Nucl. Phys. B123 (1977) 89;
     U. Amaldi et al., Phys. Rev. D36 (1987) 1385;
     P. Langacker, W.J. Marciano and A. Sirlin, Phys. Rev. D36 (1987) 2191;
     G. Costa et al., Nucl. Phys. B297 (1988) 244
[13] J. Ellis and G.L. Fogli, CERN-TH.5511/89 1989);
     D. Haidt, DESY 89-073 and contributed paper to the proceedings of the EPS Conference on High Energy Physics, Madrid, Spain, Sept. 1989 (to be published);
     P. Langacker, Phys. Rev. Lett. 63 (1989) 1920
[14] G.S. Abrams et al. (MARK II Collaboration], Phys. Rev. Lett. 63 (1989) 724; ibid 63 (1989) 2173
[15] F. Abe et al. [CDF Collaboration], Phys. Rev. Lett. 63 (1989) 720
[16] A. Weidberg, in Proceedings of the Fourteenth International Symposium on Lepton and Photon Interactions, Stanford, California, August 1989 (to be published)
[17] D. Geiregat et al. [CHARM II Collaboration], CERN Report (to be published;
     J. Dorenbosch et al. [CHARM I Collaboration], Z. Phys. C41 (1989 567;
     K. Abe et al. [E-734 Collaboration], Phys. Rev. Lett. 62 (1989) 1709;
     M.C. Noecker et al., Phys. Rev. Lett. 61 (1988) 310
[18] P. Burchat, these proceedings:
     M. Drees et al., CERN-TH. 5487/89 (1989) and in Z Physics at LEP1 CERN 89/08 (1989) (editors: G. Altarelli, R. Kleiss and C. Verzegnassi) and references quoted therein.
[19] D. Decamp et al. [ALEPH Collaboration], CERN-EP/89-157 (1989)
[20] D. Decamp et al. [ALEPH Collaboration], CERN-EP/90-16 (1990).
[21] M.Z. Akrawy et al. [OPAL Collaboration], CERN-EP/89-174 (1989).
[22] M. Zeller, these proceedings.
[23] H. Haber, these proceedings.
[24] See, for example, A. Ali and M.A.B. Beg, Phys. Lett B103 (1981) 376
[25] D. Decamp et al. [ALEPH Collaboration], CERN-EP/189-168 (1989) and CERN-EP/90-60 (1990).
[26] M. Chen, C. Dionisi, M. Martinez and X. Tata, Phys. Rep. 159 (1988) 201 and references quoted therein.
[27] A. Bartel, H. Fraas and W. Majerotto, Z. Phys. C30 (1986) 441; ibid C41 (1988 475
[28] D. Decamp et al. [ALEPH Collaboration], CERN-EP/89-158 (1989)
[29] I. Adachi et al. [TOPAZ Collaboration], Phys. Lett. B218 (1989) 105
[30] C. Hearty et al. [ASP Collaboration], Phys. Rev. D39 (1989) 3207
[31] M.Z. Akrawy et al. [OPAL Collaboration], CERN-EP/89-176 (1989)
[32] B. Adeva et al. [L3 Collaboration], Phys. Lett. B233 (1989) 530

RECENT RESULTS FROM THE CDF EXPERIMENT AT THE TEVATRON PROTON-ANTIPROTON COLLIDER

The CDF Collaboration
presented by

S. Geer

Harvard University
High Energy Physics Laboratory
42 Oxford Street, Cambridge MA 02138

Abstract

Recent results from the CDF experiment are described. The Standard Model gives a good description of jet production, and W/Z production and decay. There is no evidence yet for the top quark, for fourth generation quarks, or for deviations from the Standard Model ascribable to quark substructure, supersymmetric particles, or heavy additional W-like or Z-like bosons. Limits are given where applicable. A search for a light Higgs Boson is also described.

1. Introduction

The Collider Detector at Fermilab (CDF) had its first physics run in 1987 and recorded 25 nb^{-1} of proton-antiproton collisions at a centre-of-mass energy of 1.8 TeV at the Tevatron collider. In 1988 CDF began a second highly successful physics run which ended in June 1989, and recorded a total integrated luminosity of 4.7 pb^{-1}. In the following sections physics results are reported from both the 1987 and 1988/9 runs. The topics covered include hadronic jet production and fragmentation, W and Z production and decay, the search for supersymmetric particles, the top quark, and the Higgs boson.

Higgs Particle(s)
Edited by A. Ali
Plenum Press, New York, 1990

1.1   The CDF Detector

The CDF detector is a 5000 t magnetic detector. A side view of the detector is shown in fig. 1. Event analysis is based on charged particle tracking, magnetic momentum analysis, and fine-grained calorimetry. The combined electromagnetic and hadron calorimetery has approximately uniform granularity in pseudorapidity (η) - azimuthal angle (ϕ) space, and extends down to 2° from the beam directions. Various tracking chambers cover the calorimeter acceptance and extend charged particle tracking down to 2 mrad from the beam directions. Charged particle momenta are analyzed in a 1.5T solenoidal magnetic field, generated by a superconducting coil which is 3 m in diameter and 5 m in length. The central tracking chamber (CTC) measures particle momenta with a resolution better than $\delta p_T/p_T^2 = 0.002$ (GeV/c)$^{-1}$ in the region $40° < \theta < 140°$ and $\delta p_T/p_T^2 \leq 0.004$ (GeV/c)$^{-1}$ for $21° < \theta < 40°$ and $140 < \theta < 159°$. The calorimeters, which have full azimuthal angle coverage, consist of electromagnetic (EM) shower counters and hadron calorimeters, and are segmented into about 5000 projective towers. Each tower is 0.1 units of η wide by 15° in ϕ in the central region ($|\eta| < 1.1$) and 5° in ϕ elsewhere. Muon coverage is provided by drift chambers in the region $56° < \theta < 124°$, and by large forward toroid systems in the range $3° < \theta < 16°$ and $164° < \theta < 177°$. Isolated high momentum muons can be identified in the intermediate angular range in many cases by a comparison of the tracking and calorimeter information. Custom-built front-end electronics followed by a large Fastbus network provides the readout of approximately 100 000 detector channels. Fast level 1 and level 2 triggers make a detailed pre-analysis of calorimeter and tracking information. A level 3 trigger system uses on-line processors to perform parallel event processing. A more detailed description of the detector can be found in ref. [1].

2.   Jet Physics

Jets are reconstructed in CDF using a cone algorithm which clusters the energy around a central seed calorimeter tower (transverse energy $E_T > 1$ GeV) in a cone in (η,ϕ)-space with radius $\Delta R \equiv \sqrt{\Delta\eta^2 + \Delta\phi^2} = 0.7$. The cluster direction is then calculated and any additional towers within a cone $\Delta R = 0.7$ around the cluster direction added to the cluster. This process is repeated until no new towers are added to the cluster. To obtain the energy of the underlying parton the cluster energies are corrected for non-linear calorimeter response to low energy pions, leakage, uninstrumented regions of the detector, and the contribution from the underlying spectator event. The energy corrections depend on the jet transverse energy, and are 33% for jets with E_T of 20 GeV, decreasing to 17% for jets with E_T of 400 GeV. The uncertainty on these corrections is 12% at 20 GeV and 4% at 400 GeV. The dominant source of the uncertainty comes from the lack of knowledge of the jet

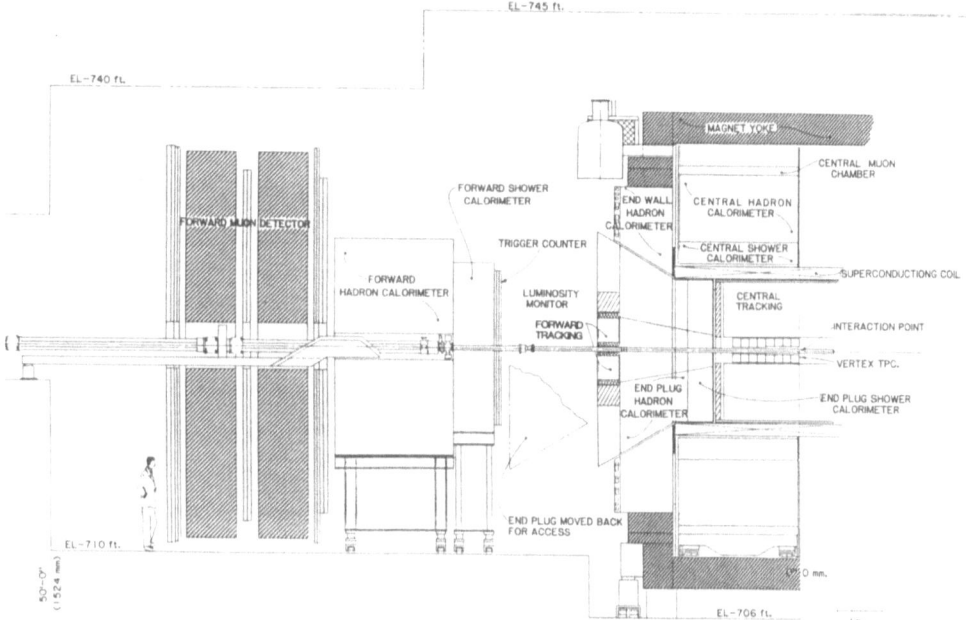

Fig. 1.   Cut-away view through the forward half of the CDF detector. The detector is forward-backward symmetric about the interaction point.

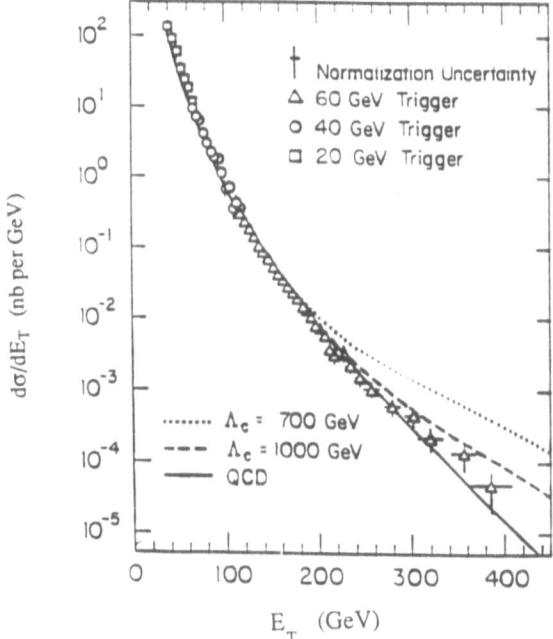

Fig. 2.   Inclusive differential jet cross-section. The curves show the leading order QCD expectation ($\Lambda_C=\infty$) using DO2 structure functions with $Q^2 = E_T^2/2$, and the expected modification which would arise from quark substructure [4] associated with the scale Λ_C.

449

fragmentation function combined with the non-linearity of the calorimeter response. Further details of the jet reconstruction method and jet energy corrections can be found in ref. [2].

2.1 The Inclusive Jet Cross-Section

To ensure good containment in the central calorimeter, jets have been analysed if the jet axis is within the region $0.1 < |\eta| < 0.7$, and the event vertex co-ordinate along the beam direction is within 60 cm of the center of the detector. To eliminate background from cosmic rays and beam halo, events with a significant ammount of energy not in time with the beam crossing were eliminated. To further reject backgrounds, jets with $E_T > 80$ GeV were required to deposit at least 10% and not more than 95% of their energy in the EM calorimeter, and the missing E_T in the event was required to be small ($< 4.8\sqrt{E_T} \sim 6\sigma$). These cuts are estimated to reject > 99% of the background, retaining > 97% of the true jet cross-section.

A preliminary measurement of the inclusive differential jet cross-section per unit of rapidity in the central region is shown in fig. 2 based on 1.4 pb^{-1} of 1988/9 data. The differential cross-section has been corrected for the experimental energy resolution, which can be parameterized by

$$\frac{\sigma(E_{JET})}{E_{JET}} = \frac{110\%}{\sqrt{E_{JET}}} \ .$$

The smearing correction increases the measured cross-section by 70% at the lowest measured jet energies, and 12% at the highest energies. The uncertainties shown on the measurements are a combination of statistical and E_T dependent systematic errors. In addition there is an E_T independent systematic error arising from uncertainties in the integrated luminosity (15%), smearing correction (10%), and jet energy scale (50%). The normalization uncertainty is indicated on the figure. The inclusive differential jet cross-section is well described by the lowest order prediction evaluated at $Q^2 = E_T^2/2$ using the structure functions of Duke and Owens set 2 [3]. There is no hint of a flattening of the jet E_T distribution at high E_T due to quark substructure [4].

2.2 Two-jet Angular- and Mass-Distributions

To obtain the two-jet angular distribution a preliminary analysis has been made of 800 nb^{-1} of 1988/9 data, requiring at least one jet with uncorrected $E_T > 80$ GeV and pseudorapidity $|\eta| < 0.7$ that is coplanar ($\pm 30°$) with a second jet. The jets are treated as

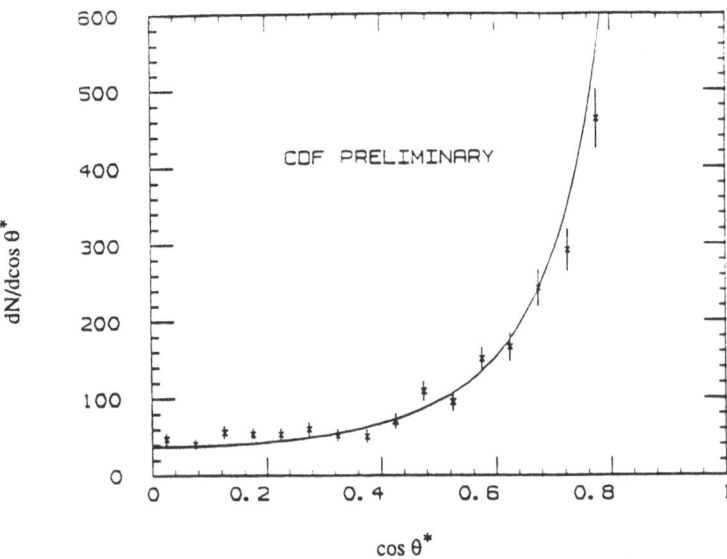

Fig. 3. Uncorrected two-jet angular distribution ($m_{jj} > 300$ GeV/c^2) compared with the leading order QCD prediction (curve).

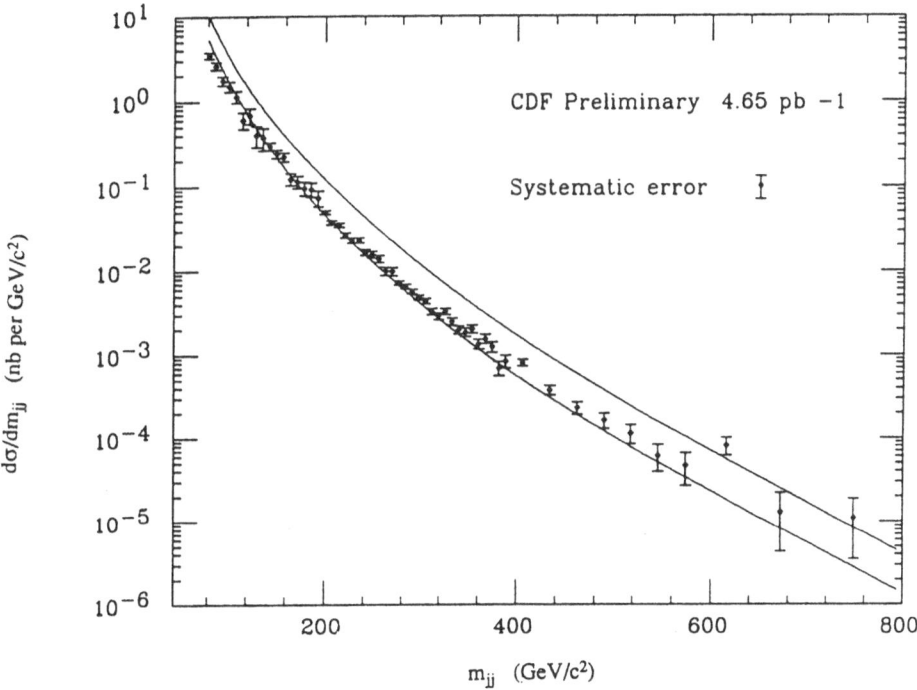

Fig. 4. Two-jet mass distribution. The band shows the envelope of leading order QCD predictions which accommodates the GHR, DO1, DO2, and EHLQ1 structure functions.

massless objects, and their pseudorapidities η_1 and η_2 determined from the centroids of the associated clusters. Defining

$$\eta^* \equiv \frac{(\eta_1 - \eta_2)}{2}$$

we require $|\eta^*| < 1.0$. The center-of-mass momentum

$$p^* \equiv p_T \cosh(\eta^*)$$

is required to be in excess of 150 GeV/c so that the angular distribution is unbiased by the η^* cut. The center-of-mass angular variable $\cos\theta^*$ is given by

$$\cos\theta^* = \tanh(\eta^*) \, .$$

Fig. 3 shows a preliminary measurement of the resulting two-jet angular distribution dN/dcosθ*. The distribution has not been corrected for acceptance, however these corrections are expected to be less than 15%. As expected the measured distribution is similar to the predicted leading order QCD prediction.

A preliminary measurement of the two-jet mass distribution based on the full 1988/9 statistics (4.7 pb^{-1}) is shown in fig. 4. The distribution extends to masses of ~ 800 GeV/c^2, and is well described by leading order QCD expectations.

2.3   Jet Fragmentation

Jet fragmentation has been studied using two-jet data from the 1987 run. Events were selected with two leading jets which were coplanar (±30°), and for which there were no other large clusters ($E_T < 20$ GeV or 0.2 ($E_T^1 + E_T^2$), whichever is smaller) in the event. The jets were required to be contained within the central calorimeter ($0.1 < |\eta| < 0.7$), and the boost variable $\eta_{boost} \equiv 0.5 [\eta_1 + \eta_2]$ was required to be small ($|\eta_{boost}| < 0.6$). Events were boosted along the beam axis by η_{boost} and well measured charged tracks associated to the primary vertex were associated to a jet if they fell within a cone of opening angle 48° around the jet axis and if their projected momentum along the jet axis $p_{//} > 0.6$ GeV/c. The measured charged fragmentation function $D(z) \equiv \frac{1}{N_{jet}} \frac{dN_{charged}}{dz}$

is shown in fig. 5, where the fragmentation variable $z \equiv p_{//} / E_{jet}$. The track reconstruction efficiency has been estimated as a function of z and two-jet mass using a Monte Carlo, and corrected for. The reconstruction efficiency exceeds 85% for moderate two-jet masses (< 200 GeV/c^2). Corrections have also been applied for tracks outside of

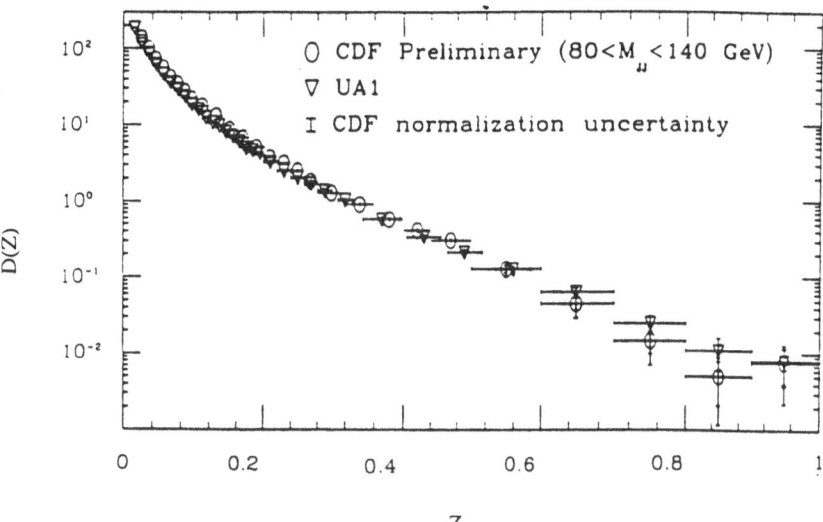

Fig. 5. Jet fragmantation function. The preliminary CDF result is compared with the result from the UA1 experiment [5] for jets with $P_T > 25$ GeV/c.

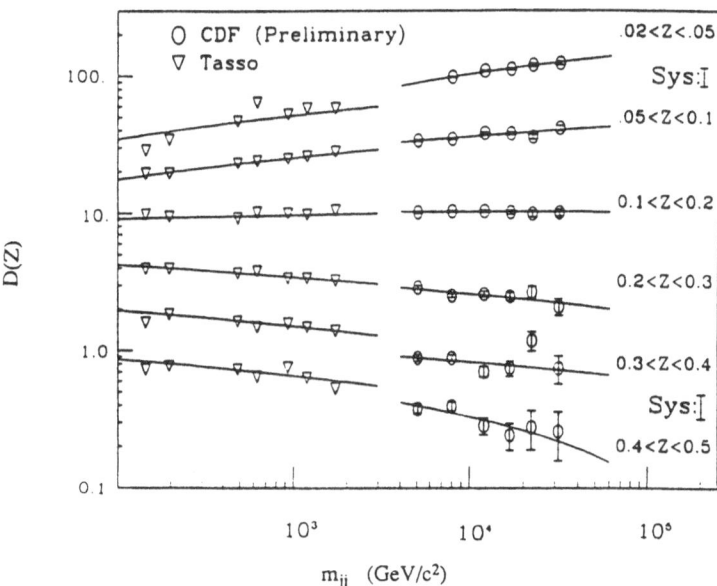

Fig. 6. Dependence of jet fragmentation on the Q^2-scale and the fragmentation variable z. Results from jets produced in e^+e^- annihilation measured by the TASSO experiment [6] are shown for comparison.

the CTC acceptance or the jet reconstruction cone, and for the contribution within the cone from uncorrelated tracks from the underlying spectator event. These corrections are only significant at small z. Finally D(z) has been corrected for the experimental resolution on the measurement of E_{jet} and $p_{//}$. The resulting D(z) is similar to the corresponding measurement of jets produced in proton-antiproton collisions at $\sqrt{s}$ = 630 GeV by the UA1 collaboration [5]. The variation of D(z) with z and two-jet mass is shown in fig. 6, where it is compared with e^+e^- data from the TASSO experiment [6]. Both experiments show the same trend : namely that the fragmentation becomes more peaked at low z as the four-momentum-transfer squared (Q^2) increases. Fits to the two data sets are of the form expected as a result of the Altarelli-Parisi evolution : $D(z,m_{jj}) = \alpha + \beta \log(m_{jj})$. It should be noted that the TASSO jets are typically quark jets, whereas the CDF jets in this m_{jj} range are predominantly gluon jets.

3. W and Z Physics

The production and decay properties of the charged (W) and neutral (Z) Intermediate Vector Bosons provide us with an excellent test of both the electroweak and QCD sectors of the standard model. The CDF experiment has observed the decay modes W $\rightarrow$ ev, W $\rightarrow$ $\mu\nu$, Z $\rightarrow$ e^+e^-, and Z $\rightarrow$ $\mu^+\mu^-$. Work is also in progress on the tau-lepton decay modes of the W and Z.

3.1 W Mass

A preliminary measurement of the W mass has been made using 4.7 pb^{-1} of 1988/9 data. Electron-neutrino decays of the W were selected by requiring an energetic EM cluster (E_T > 25 GeV) produced in association with a large significant missing transverse energy (ΔE_T > 25 GeV, $\Delta E_T/\sqrt{\Sigma E_T}$ > 2.4). The calorimeter cluster was further required to have at least 85% of its energy in the EM calorimeter, and an associated charged track such that E/P < 1.4. To obtain a clean well measured sample of W decays, events were rejected if there were any additional clusters (E_T > 7 GeV) in the event, and the electron was required to be central ($|\eta|$ < 1.0). Since the longitudinal component of the neutrino momentum is not measured in the CDF detector, the invariant mass of the electron-neutrino system cannot be reconstructed directly. Instead the electron neutrino transverse mass

$$m_T^{ev} \equiv \sqrt{2E_T^e E_T^v(1-\cos\phi_{ev})}$$

is used, where E_T^v is identified with the missing transverse energy in the event. The m_T^{ev} distribution for the selected W sample is shown in fig. 7. To extract the W mass m_W this distribution has been fitted using a Monte Carlo calculation to model the expected shape of

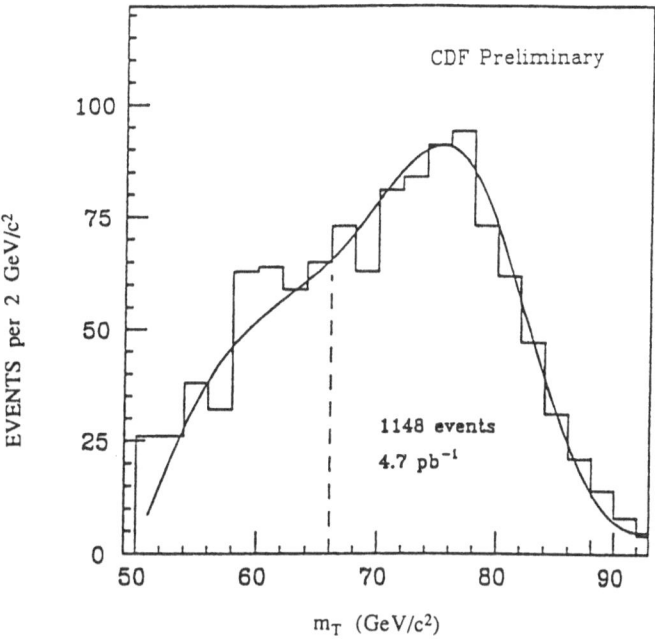

Fig. 7. Electron-neutrino transverse mass distribution for W → eν candidates. The curve shows the best fit .

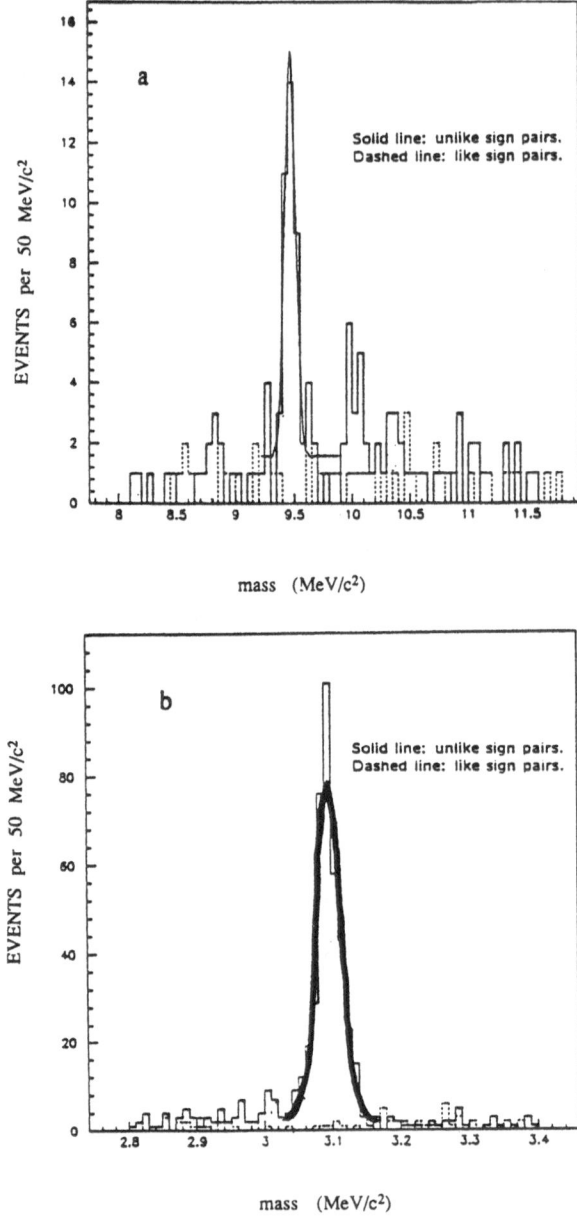

Fig. 8.   The $\mu^+\mu^-$ mass spectrum in the neighbourhood of (a) the Υ, and (b) the J/ψ
resonances. The tracks have been constrained to come from the beam axis.

the distribution as a function of m_W. The Monte Carlo includes the distortion of the m_T^{ev} distribution due to the finite transverse momentum of the W (P_T^W) and due to the calorimeter resolution. Allowing both the W mass and width to vary the resulting best fit (fig. 7) gives a good description of the data, and yields a preliminary result for the W mass of m_W = 80.0 ± 0.2 (stat) ± 0.3 (scale) ± 0.5 (sys) GeV/c². The main systematic uncertainties arise from the uncertainty in modelling the P_T^W distribution, the ΔE_T resolution ($\delta m_W \sim 400$ MeV/c²), and the uncertainty on the structure function ($\delta m_W \sim 300$ MeV/c²).

3.2   Z Mass

CDF has measured the mass of the Z using $Z \rightarrow \mu^+\mu^-$ tracking data and $Z \rightarrow e^+e^-$ calorimeter data. Results are based on an integrated luminosity of 4.7 pb⁻¹. The analyses are restricted to the central region to exploit the optimum track momentum and calorimeter energy resolutions.

Dimuon events were selected by requiring (1) two tracks with $P_T > 20$ GeV/c; (2) at least one match in ϕ between a muon chamber track segment and a CTC track; (3) non-zero hadronic and EM energy deposition; but less than 6 GeV and 2 GeV respectively in a single calorimeter tower associated with each track; (4) no jets with $E_T > 15$ GeV within 10° of these tracks. Events with two muons back to back ($\Delta\eta=\pm0.1$, $\Delta\phi=\pm 1.5°$) were rejected as cosmic rays. Events having muon pairs with invariant masses between 50 and 150 GeV/c² were selected (132 events).

Transverse momenta are calculated from track curvature in the 1.4116 Tesla magnetic field, known to ± 0.05%. The CTC alignment was adjusted using electrons from W decay so that the ratio of track momentum to calorimeter energy was charge independent. The alignment was checked using cosmic ray muons. The CTC alignment and the magnitude of the magnetic field were verified by studying $K^0_s \rightarrow \pi^+\pi^-$, J/$\psi$ (fig. 8b) and Y(1S) $\rightarrow \mu^+\mu^-$ (fig. 8a) decays. The tracks were constrained to come from the beam axis (beam constraint) in the latter two data samples. The reconstructed masses of 0.498 ± 0.002, 3.097 ± 0.001, and 9.469 ± 0.010 GeV/c² agree well with world-average values. Based on these measurements we estimate a mass error < 0.2% due to systematic momentum scale uncertainties. The beam constraint has been applied to the lepton tracks in the Z data samples. The momentum resolution after applying this constraint is $\delta P_T/P_T^2 = 0.0011$ (GeV/c)⁻¹. Radiative corrections (internal and external bremsstrahlung) were studied using a Monte Carlo event generator which used the exact matrix elements to order α^2 [7] and a detailed simulation of the CDF detector.

The mass distribution for $Z \rightarrow \mu^+\mu^-$ (fig. 9a) was fitted using a maximum likelihood fit with a signal modeled by a relativistic Breit-Wigner convoluted with a

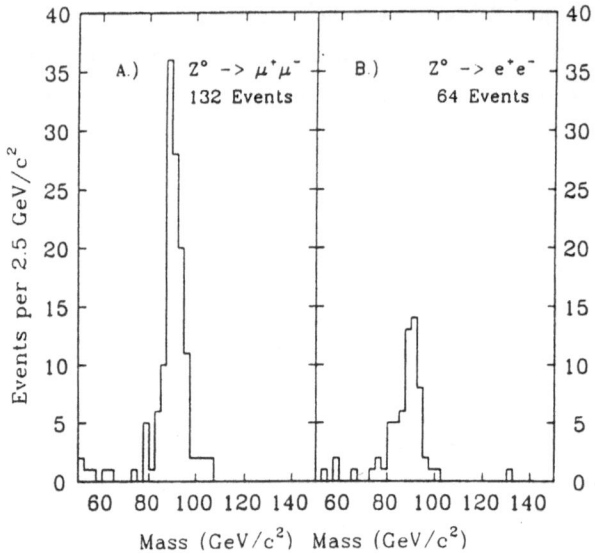

Fig. 9. Mass distribution for (a) $Z \to \mu^+\mu^-$ and (b) $Z \to e^+e^-$ candidates using tracking information.

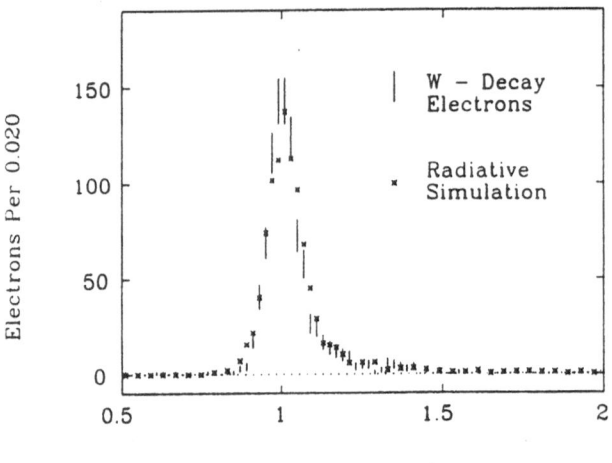

E/P (After Calibrations)

Fig. 10. Ratio of electromagnetic energy to track momentum for electrons in the W sample compared to a Monte Carlo prediction which includes radiative corrections and the CTC resolution.

Gaussian resolution in $1/P_T$. The fitted mass and width are 90.7 ± 0.4 (stat) ± 0.2 (scale) GeV/c^2 and 4.0 ± 1.2 (stat) ± 1.0 (syst) GeV. The fit is insensitive to the non-resonant Drell-Yan contribution. The effects of radiative corrections, different structure functions, and the mass window used are included in the estimate of the uncertainties (Table 1).

An inclusive electron sample was obtained by requiring at least one electron candidate satisfying the following: (1) the electron is away from calorimeter tower edges; (2) a ratio of hadronic to EM calorimeter energy of < 0.1; (3) a ratio of EM energy to track momentum E/P < 1.4; (4) a transverse shower profile in the strip chambers consistent with an electron shower; (5) a match between the strip chamber shower position and the extrapolated track position. A sample of 73 events have electron pairs with both particles satisfying the above criteria and with invariant mass between 50 and 150 GeV/c^2. The previously described mass fitting technique using track information was applied to the Z $\rightarrow$ e$^+$e$^-$ event sample (fig. 9b and Table 1). The radiative effects on the observed mass are appreciably larger than in the muon mode; consequently the best measurement of the Z mass in this mode is obtained using calorimeter information. To determine the Z mass from the calorimeter, the EM calorimeter was calibrated on a tower-by-tower basis using the fitted means of the E/P distributions from a sample of ~17000 inclusive electrons. The measured energy resolution of the calorimeter for EM showers is

$$\left(\frac{\sigma_E}{E}\right)^2 = \left(\frac{13.5\%}{\sqrt{E \sin\theta}}\right)^2 + (1.7\%)^2$$

where the constant term arises from the average uncertainty in the individual tower calibrations. The overall energy scale was established from the momentum scale using the mean E/P from ~ 1000 W decay electrons. The expected shape and mean of the E/P distribution for these W electrons was simulated including external and small angle internal bremsstrahlung (fig. 10). For E/P < 1.4 the mean E/P is 1.026. The systematic uncertainty in E/P is estimated to be $\pm 0.4\%$. The E/P distribution for Z decay electrons is consistent with the predictions. A small correction was applied to the Z mass for internal wide-angle photon emission (Table 1).

The mass and width of the Z peak (fig. 11) were fitted using the maximum likelihood method. The corrected fitted values for the Z mass and width are $91.1 \pm 0.3 \pm 0.4$ GeV/c^2 and $3.6 \pm 1.1 \pm 1.0$ GeV respectively. The quoted systematic uncertainties reflect reasonable variations in the energy resolution, mass window, choice of structure functions, and fitting procedure.

The corrections and uncertainties in each of the mass measurements are summarized in Table 1. Our best value for the Z mass is a weighted mean of the tracking measurement of the $\mu^+\mu^-$ sample and the calorimeter measurement of the e$^+$e$^-$ sample. The resulting Z mass is 90.9 ± 0.3 (stat+sys) ± 0.2 (scale) GeV/c^2 and the width is $3.8 \pm 0.8 \pm 1.0$ GeV. Further details of this analysis can be found in ref. [8].

Table 1 .   Summary of Z mass measurement corrections and results.

	$Z \rightarrow \mu^+\mu^-$ (tracking)	$Z \rightarrow e^+e^-$ (tracking)	$Z \rightarrow e^+e^-$ (Calorimeter)
# events used in fit	123	58	65
Observed Fitted Mass	90.41 ± 0.40	89.27 ± 0.80	90.93 ± 0.34
Radiative Corrections	$+0.22 \pm 0.03$	$+2.19 \pm 0.30$	$+0.11 \pm 0.03$
Structure Functions	$+0.08 \pm 0.03$	$+0.08 \pm 0.03$	$+0.08 \pm 0.03$
E/P Calibration			± 0.20
Mass Scale	± 0.20	± 0.20	± 0.20
Corrected Mass	$90.7 \pm 0.4 \pm 0.2$	$91.5 \pm 0.8 \pm 0.4$	$91.1 \pm 0.3 \pm 0.4$

3.3   Standard Model Parameters

Defining

$$\sin^2 \theta_W \equiv 1 - \left(\frac{m_W}{m_Z}\right)^2$$

and using the measured m_W and m_Z obtained from the electron channels, we obtain $\sin^2 \theta_W$ = 0.229 ± 0.012, which is in excellent agreement with the world average value of 0.230 ± 0.005 (fig. 12a). Due to radiative corrections the W and Z masses also depend upon the top quark mass and, to a lesser extent, the Higgs boson mass. The dependence is shown in fig. 12b.

3.4   Search for Heavy Ws and Zs

No high-mass peaks have been observed by CDF in the ev or e^+e^- mass spectra in excess of those associated with the decays of the standard W and Z bosons. Limits can therefore be deduced on the production and decay of heavy W-like (W') and Z-like (Z') bosons. We obtain preliminary limits (95% C.L.) :

$$\sigma_{W'}.B_{ev} \leq 7.6 \text{ pb}$$

$$\sigma_{Z'}.B_{e^+e^-} \leq 1 \text{ pb}.$$

These limits can be used to obtain mass limits on the W' and Z' bosons provided we specify their couplings. If the heavy bosons have the same weak charge as the standard model W and Z, and the same leptonic branching ratios, then the corresponding mass limits (95% C.L.) are :

$$m_{W'} \geq 380 \text{ GeV/c}^2$$

$$m_{Z'} \geq 400 \text{ GeV/c}^2.$$

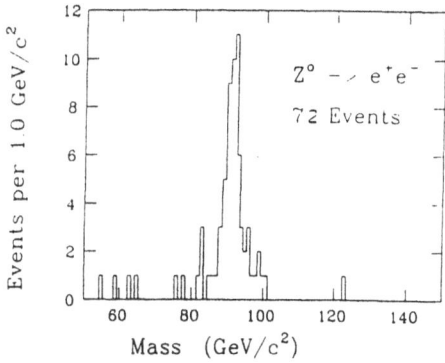

Fig. 11. Mass distribution for $Z \rightarrow e^+e^-$ candidates using calorimeter energies.

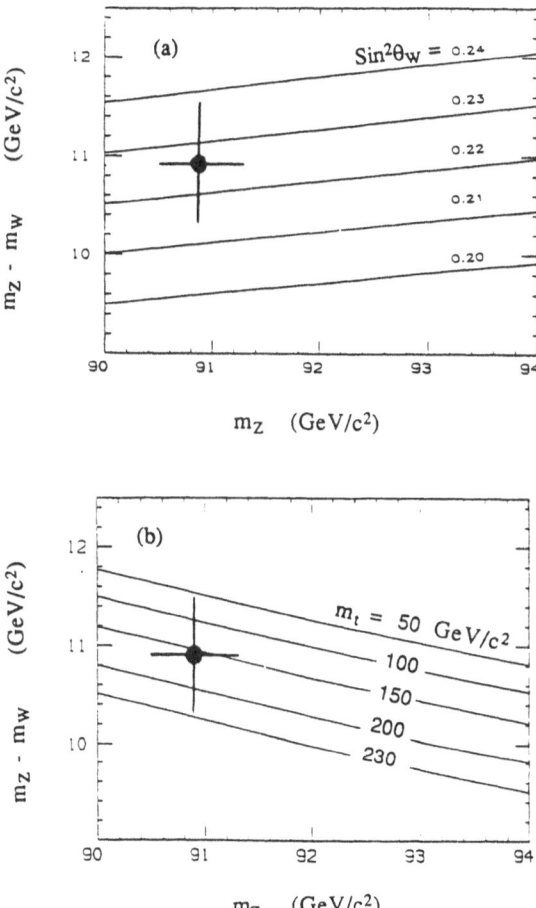

Fig. 12. The measured values of $(m_Z - m_W)$ and m_Z compared with Standard Model expectations as a function of (a) $\sin^2\theta_W$, and (b) top quark mass (assuming $m_{HIGGS} = 100\,\text{GeV}/c^2$).

3.5   The Number of Light Neutrino Types

The ratio $R \equiv (\sigma_W . B_{ev})/(\sigma_Z . B_{e^+e^-})$ has been measured by CDF to be $R = 10.3 \pm 0.8 \pm 0.5$ (preliminary), and $R \leq 11.4$ (90% C.L.). The statistical error on R (± 0.8) arises predominantly from the limited Z statistics. Since the predicted W branching ratio is sensitive to the partial width for the decay $W \rightarrow tb$ and the predicted Z branching ratio is sensitive to the number of $Z \rightarrow \nu\nu$ channels open, the predicted value of R is sensitive to both the top quark mass (m_t) and the number of light neutrino types (N_ν). The measured value of R is compared with the predictions as a function of N_ν and m_t in fig. 13. We conclude that for a heavy top quark ($m_t \geq 80$ GeV/c^2) there are either 3 or 4 light neutrino types.

3.6   W Plus Jets

High transverse momentum W bosons are expected and observed [9] to be produced in association with one or more hadronic jets arising predominantly from gluon bremsstrahlung off the incoming interacting partons. The measured rate of W+1 jet, W+2 jets, and W+3 jets in CDF agrees well with $O(\alpha_s^3)$ tree-level QCD calculations (fig. 14). Note that there is a theoretical uncertainty of $\pm 30\%$ on the W+1 jet prediction and $\pm 50\%$ on the W+2 jet prediction due to uncertainties on α_s and the Q^2-scale. A preliminary analysis of the properties of the W + jet(s) events also shows good agreement with QCD expectations based on the Papageno Monte Carlo and a full simulation of the CDF detector. In the following preliminary jet energy corrections have been applied. There are $\pm 10\%$ uncertainties remaining on the energy scale. Figs. 15a and 15b show that the P_T^W distributions for the W+1 jet and W+2 jet samples are well described by the QCD expectations, and fig. 15c shows good agreement between the measured and expected jet P_T distributions for W+1 jet events. For events with electron $E_T > 20$ GeV, $\Delta E_T > 20$ GeV, and $m_T^{ev} > 40$ GeV/c^2 figs. 15d and 15e show the agreement between the expected and measured m_T^{ev} distributions for W+1 jet and W+2 jet events, where jets with $E_T > 10$ GeV and $|\eta| < 2$ are counted. Finally fig. 15f shows that the two-jet mass distribution for W+2 jet events is also well described by QCD expectations.

4.   Search for Supersymmetry

In supersymmetric models in which the SUSY quantum number is conserved the lightest supersymmetric particle is stable. If this particle is also neutral it will escape the interaction region undetected, giving rise to events in which there is a large ΔE_T. The other supersymmetric particles produced (e.g. squarks and gluinos) will decay to lighter supersymmetric particles, and normal quarks and gluons. Thus the final event will contain high-P_T jets associated with a large ΔE_T.

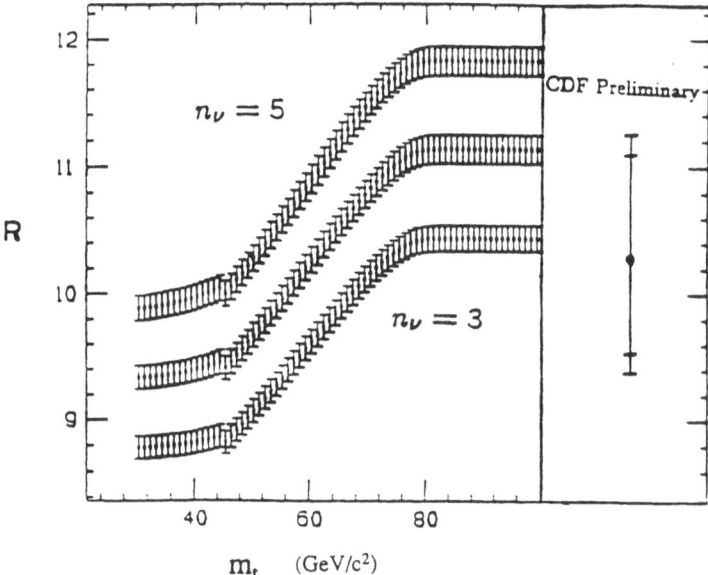

Fig. 13.   The measured value of R $\equiv (\sigma_W.B_{ev})/(\sigma_Z.B_{e^+e^-})$ compared with Standard Model expectations as a function of top quark mass. The bands show the predictions for 3, 4, and 5 neutrino types, and accommodate the MRS structure functions based on BCDMS data (MRSB) and EMC data (MRSE).

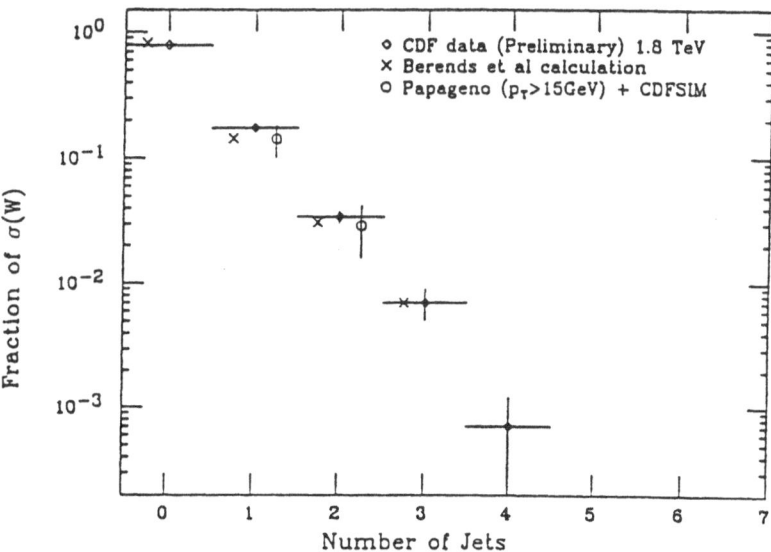

Fig. 14.   Observed fraction of W events in which 0, 1, 2, 3, and 4 jets (uncorrected $E_T > 10$ GeV, $|\eta| < 2.2$) have been produced in association with the W. Measurements are compared with tree-level predictions.

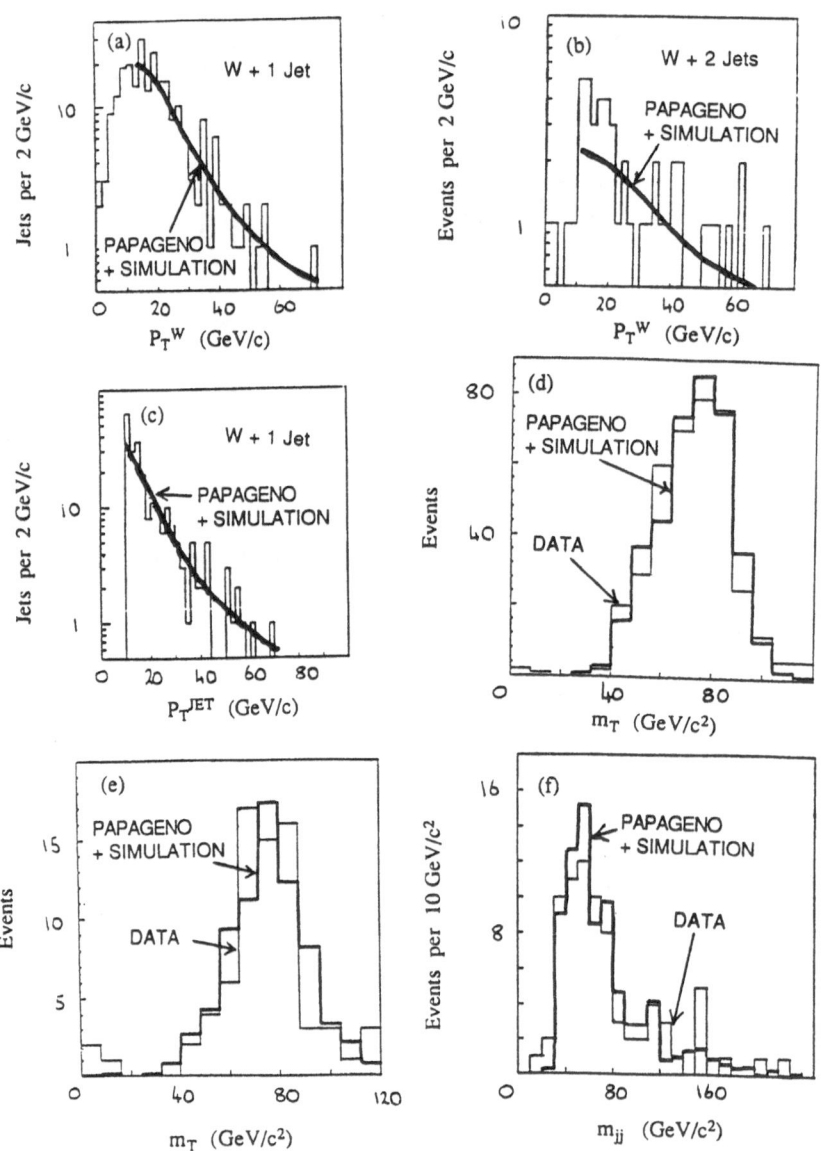

Fig. 15. Properties of W + jet(s) events (see text) compared with expectations based on the Papageno Monte Carlo.

A preliminary search for supersymmetric particles in CDF has been performed on the 1988/9 data (4.6 pb^{-1}). Events have been selected with a large significant ΔE_T ($\Delta E_T > 40$ GeV, $\Delta E_T / \sqrt{E_T} > 2.8$). To remove two-jet fluctuations, events with a jet ($E_T > 5$ GeV) coplanar ($\pm 30°$) to the ΔE_T vector have been rejected. Events with at least two calorimeter clusters ($E_T > 15$ GeV, $|\eta| < 3.5$) were then retained for further analysis if the clusters had an EM fraction between 0.1 amd 0.9, and if at least one cluster was central ($|\eta| < 1$). To remove W and Z decays, events with a cluster ($E_T > 15$ GeV) with EM fraction > 0.9, or with a high-P_T muon candidate ($P_T > 15$ GeV/c) were rejected. Pathological events (noise, cosmics, readout problems, beam-gas interactions) were rejected by scanning.

Table 2 .   Expected number of events in large-missing-E_T data sample from standard model processes compared to the observed number of events.

EXPECT	$\Delta E_T > 40$ GeV	$\Delta E_T > 60$ GeV
W, Z Decays	116 ± 30	24 ± 15
Heavy Quarks	42 ± 42	14 ± 14
TOTAL	158	38
OBSERVE	184	34

The ΔE_T distribution for the 184 events that survive the cuts is shown in fig. 16. The expected event rate from standard model processes is tabulated in table 2. The standard model expectations account for the observed event rate. There is no evidence for an excess of events ascribable to supersymmetric processes. To obtain a limit on the squark mass we assume that the photino is the lightest supersymmetric particle and is massless, and that

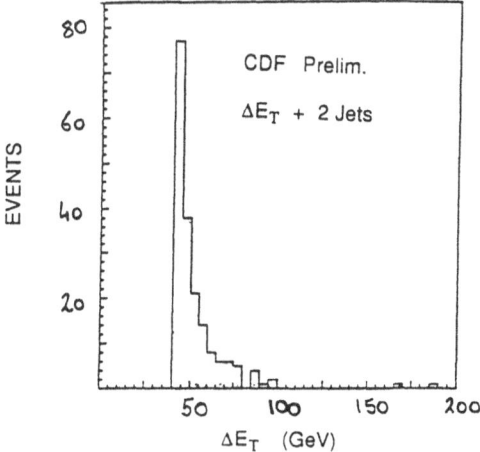

Fig. 16.   Missing E_T distribution for events with two jets passing the selection described in the text.

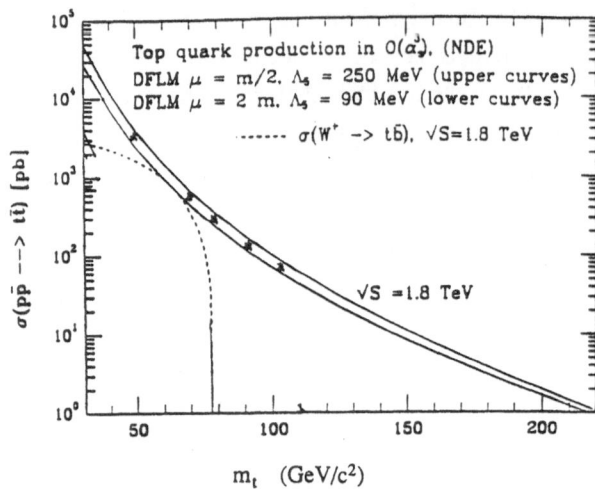

Fig. 17. Expected top quark production cross-section at the Tevatron Collider shown as a function of top quark mass. The contributions from strong and weak production are shown separately.

there are six degenerate squarks. The supersymmetric event rate and characteristics can then be predicted as a function of the squark and gluino masses. We obtain

$$m_q > 140 \, \text{GeV/c}^2 \; (90\% \; \text{C.L.})$$

for all gluino masses. To obtain this preliminary result no standard model background subtraction has been made. Further work is in progress to obtain the corresponding gluino mass limit.

5. Search for the Top Quark

The dominant top quark production mechanism at the Tevatron collider is expected to be via the process gg → tt (fig. 17). The purely hadronic final states for the top quark decay are swamped by QCD light quark multijet backgrounds. In the CDF analysis a search has therefore been made for semileptonic top quark decays. This search is sensitive provided the top quark decays predominantly via the charged weak current as expected in the standard model. Searches in two final states have been made: (i) electron plus ΔE_T plus ≥ 2 jets, and (ii) electron plus muon.

5.1 Electron + ΔE_T + ≥ 2 Jets

The measured inclusive electron P_T distribution is well described [10] in terms of bb production and semileptonic decay (which dominates in the region $P_T < 20$ GeV/c) and W/Z decay (which dominates in the region $P_T > 20$ GeV/c). To search for top decays, events with central electrons have been selected ($|\eta| < 1$, good fiducial region, good transverse shower shape, EM fraction > 0.5, 0.5 < E/P < 1.4, good track-strip chamber match, good isolation, remove photon conversions). Events with ≥ 2 jets ($E_T > 10$ GeV, $|\eta|$

466

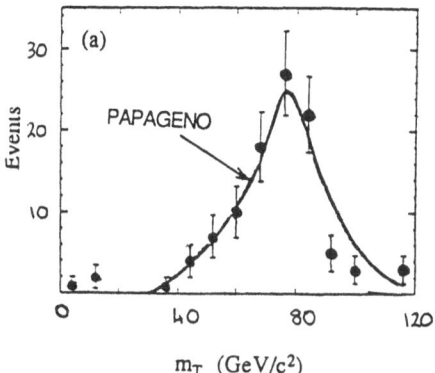

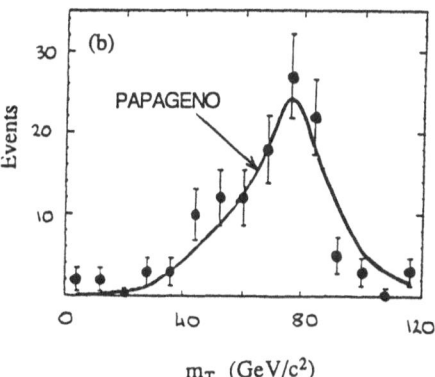

Fig. 18.   Electron-neutrino transverse mass distributions for the (a) high-mass top quark sample, and (b) low-mass top quark sample (see text) compared with expectations (curve) for W + 2 jet production followed by W → eν decay.

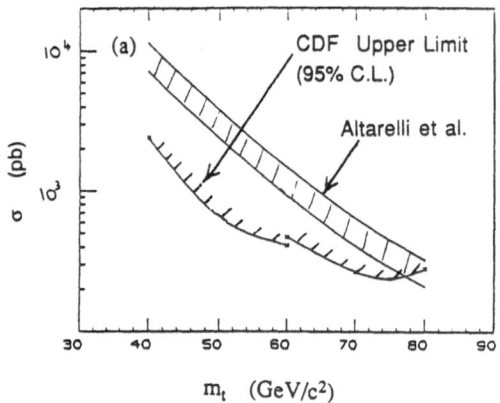

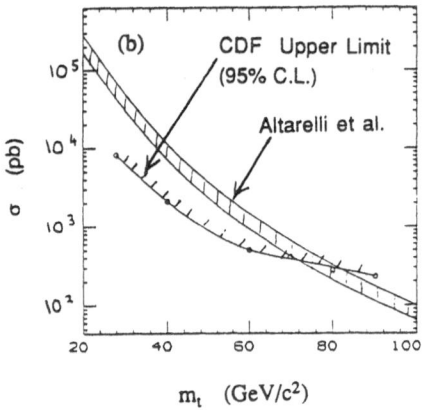

Fig. 19. Upper limits on the top quark production cross-section shown as a function of top quark mass, from (a) the electron + ΔE_T + ≥ 2 jets search, and (b) the electron + muon search. The theoretical expectation is also shown. The band reflects structure function and Q^2-scale uncertainties.

< 2.2) have been retained for further analysis. Two data samples have then been extracted :
(1) High-mass top quark sample; requiring electron E_T > 20 GeV and ΔE_T > 20 GeV, and
(2) Low-mass top quark sample; requiring electron E_T > 15 GeV, ΔE_T > 15 GeV, and
(electron $E_T + \Delta E_T$) > 40 GeV. The electron -"neutrino" transverse mass distributions for
the two data samples are consistent with expectations for W → ev decay (fig. 18), and
show no evidence for a contribution from the decay of a heavy top quark. The resulting top
cross-section limit is shown as a function of m_t in fig. 19a. Comparing this with theoretical
expectations for tt production at the collider we conclude that

$$40 < m_t < 77 \text{ GeV/c}^2 \text{ is excluded (95\% C.L.).}$$

5.2   Electron plus Muon

A search for high-P_T central electrons (E_T > 15 GeV, |η| < 1) produced in
association with a high P_T central muon (P_T > 45 GeV/c, |η| < 1.2) has been made in 4.4
pb^{-1} of 1988/9 data. One event passes these cuts. We expect 0.7 events to satisfy the cuts
from standard model processes (Z → τ$^+$τ$^-$ [0.5 events], W$^+$W$^-$ [0.15 events], WZ [0.05
events]). However, to be conservative, we compute upper limits on the tt production cross-
section (fig. 19b) as a function of m_t based on the observation of one event (no background
subtraction). Comparing this limit with theoretical expectations for tt production we
conclude that

$$30 < m_t < 72 \text{ GeV/c}^2 \text{ is excluded (95\% C.L.)}$$

We note that in the region of lower electron and muon transverse momenta we observe
many events with kinematic characteristics in broad agreement with our expectations for bb
production and semileptonic decay. Further details of the top quark searches in CDF can be
found in ref. [10].

6.   Search for a Light Higgs Boson

A search for a light Higgs Boson (200 MeV/c^2 < m_H < 1.5 GeV/c^2) produced in
association with W or Z Bosons (fig. 20) is in progress in CDF. The method is similar to
that described in ref. [11]. The predicted fraction of W and Z events containing an
associated Higgs Boson is shown as a function of m_H in fig. 21. A light Higgs Boson is
produced in about 1% of all W and Z events. Higgs Bosons with mass above the e$^+$e$^-$
threshold and below the ρ$^+$ρ$^-$ threshold will decay predominantly to a charged track pair
(e$^+$e$^-$, μ$^+$μ$^-$, π$^+$π$^-$ or K$^+$K$^-$). The predicted branching fractions are shown as a function of
m_H in fig. 22. The predicted lifetime of the Higgs Boson increases rapidly below the μ$^+$μ$^-$
threshold. Above the μ$^+$μ$^-$ threshold the lifetime is short and the charged track pair will be
associated to the vertex (fig. 23). The Higgs Boson is expected to be produced at relatively
high transverse momentum (p_T), resulting in a high-p_T charged track pair (fig. 24). The
signature for a light Higgs Boson produced in association with a W or Z Boson is therefore
an isolated high-p_T charged track pair.

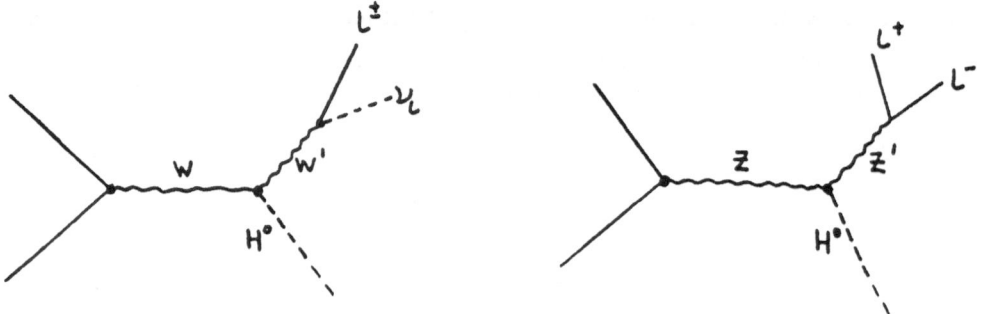

Fig. 20.   Higgs production in association with W and Z bosons.

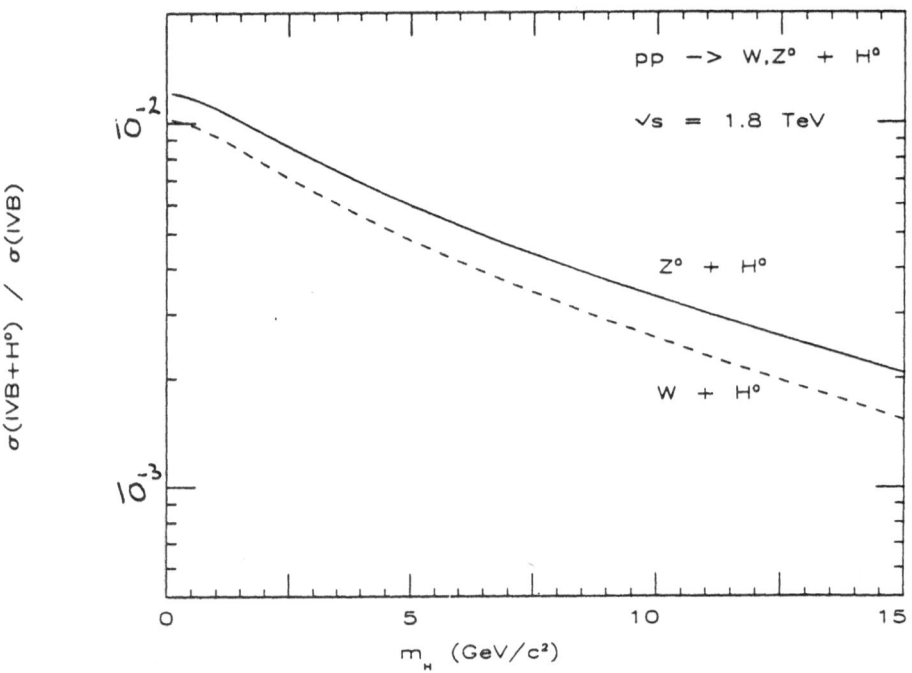

Fig. 21.   Predicted fraction of W and Z bosons produced in association with a Higgs boson at the Tevatron Collider shown as a function of Higgs boson mass.

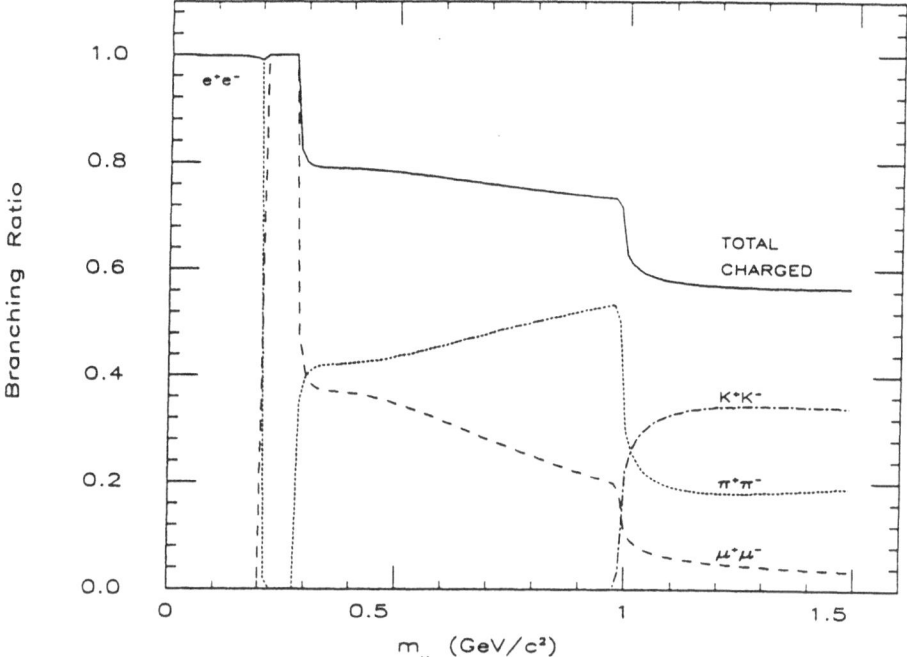

Fig. 22. Higgs boson decay branching ratios shown as a function of Higgs boson mass.

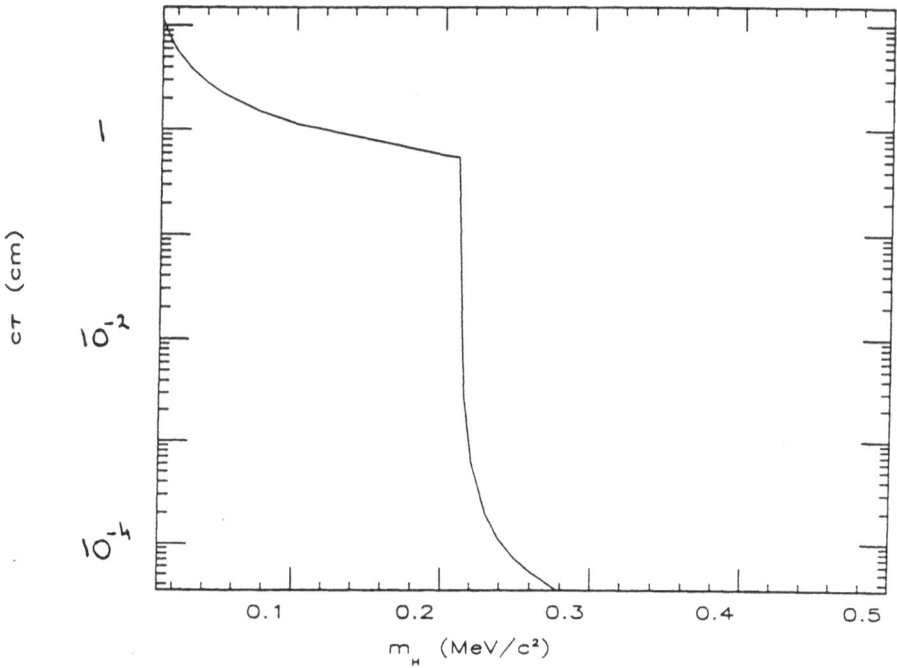

Fig. 23. Higgs boson decay length cτ shown as a function of Higgs boson mass.

Preliminary results from a search for isolated high-P_T track pairs in W and Z events indicate that there is a substantial background from fluctuations of initial state bremsstrahlung jets produced in association with the weak boson. Never-the-less when the analysis is complete it is expected that the search will have sufficient sensitivity to find or exclude a Higgs Boson with mass between the $\mu^+\mu^-$ and K^+K^- thresholds.

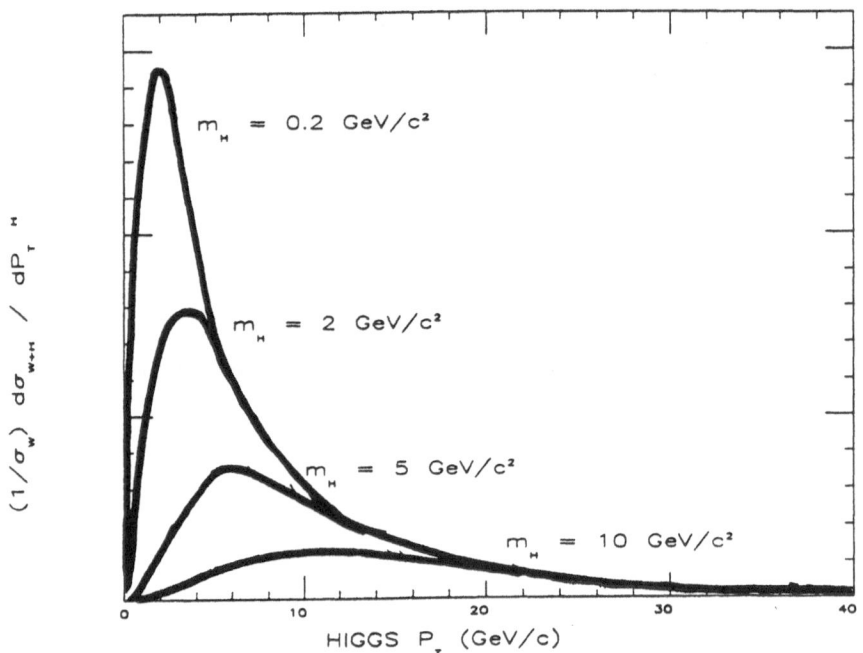

Fig. 24. Expected transverse momentum spectrum for Higgs bosons produced in association with W bosons (fig. 20) at the Tevatron Collider shown for several Higgs boson masses.

7. Conclusions

Results from the 1987 CDF run and preliminary results from the 1988/9 run indicate that the Standard Model gives an excellent description of the observed phenomenology of hard proton-antiproton interactions (the physics of jets, W, Z, heavy flavors, and missing E_T) at $\sqrt{s} = 1.8$ TeV. There is no direct evidence as yet for the top quark, or for fourth generation quarks or leptons, quark substructure, heavy W-like or Z-like bosons, or supersymmetric particles. In the coming months further analysis of the recent data will improve the sensitivity of the searches for deviations from standard model expectations and yield precision measurements of standard model physics.

References

[1]   F. Abe et al. (CDF Collab.) Nucl. Instr. and Meth. **A271**(1988)387.

[2]   J. Huth (CDF Collab.), Proc. of the SSC Workshop on Calorimetery for the SSC, Tuscaloosa, Alabama, March 1989; Fermilab-Conf-89/117-E.

[3]   D. Duke and J. Owens, Phys. Rev. **D27**(1984)508.

[4]   E. Eichten et al., Phys. Rev. Lett. **50**(1983)811.

[5]   G. Arnison et al. (UA1 Collab.) Nucl. Phys. **B276**(1986)253.

[6]   M. Althoff et al. (TASSO Collab.) Z. Phys. **C22**(1984)307.

[7]   F. Berends et al., Z. Phys. **C27**(1985)155; F. Berends and R. Kleiss, Z. Phys. **C27**(1985)365. Also used were calculations by B. Marchesini and B. Webber, Nucl. Phys. **B310**(1988)461; G. Bonvicini and L. Trentadue, UM-HE-88-36.

[8]   F. Abe et al. (CDF Collab.) Phys. Rev. Lett. **63**(1989)720.

[9]   S. Geer and W. J. Stirling, Phys. Lett. **152B**(1985)373.

[10]   F. Abe et al. (CDF Collab.), in preparation.

[11]   S. Geer, UA1 technical note UA1 TN 88-29(1988), unpublished.

TOP SEARCH IN UA1

G. Busetto
Padova (University and I.N.F.N.)

UA1 COLLABORATION

Aachen - Amsterdam (NIKHEF) - Annecy (LAPP) - Birmingham - Boston - CERN - Harvard - Helsinki - Kiel - Imperial College, London - Queen Mary College, London - Madrid (CIEMAT) - MIT - Padua - Paris (Collège de France) - Rome - Rutherford Appleton Lab. - Saclay (CEN) - UCLA - Vienna Collaboration

ABSTRACT

The UA1 Collaboration devotes a large effort in the search for new heavy quarks, in particular for the top quark. The preliminary results presented here are based on the statistics collected during the '88 and '89 $\bar{p}p$ Collider runs which amounts to about 4.6 pb^{-1}. No evidence for the top quark is found and a preliminary lower limit on its mass from the analysis on different semileptonic decay channels is 61 GeV (95% c.l.).

1. Introduction

The last element of the three known quark families, the top, has not yet been observed. The mass region still unexplored is however confined; existing lower limits from $e^+ e^-$ experiments are $m_t > 29.7$ GeV (95% c.l.) from Tristan[1] and from $p\bar{p}$ collider experiments $m_t > 41$ GeV (95% c.l.)[2] while the upper limit lies around 200 GeV and is related to the $\sin^2\vartheta_W$ value from the low energy data and from the intermediate vector boson masses[3].

In this paper a new lower limit on the top mass is obtained using the whole statistics collected by UA1.

2. 1988 - 1989: Detector and Luminosity

The UA1 detector during the '88 and '89 data taking periods was modified respect to the previous configuration; muon detection was improved by installing additional 820 tons of iron shielding in the forward-backward directions.

Muons were detected in the $| \eta | < 2.3$ pseudorapidity region and the single muon trigger operated for $| \eta | < 1.5$. More than 80% of the muons produced in the semileptonic decay of the top-quark are expected to be found in this region for $m_t > 40$ GeV.

The calorimetry was performed only by the already existent hadron calorimeter as the electromagnetic elements (gondolas and bouchons) were removed in order to leave place for the new Uranium - TMP detector, now under construction.

In such a configuration electrons from heavy-quark semileptonic decays cannot be detected.

The hadron calorimeter has been equipped with the new readout electronics and the trigger processor developed for the Uranium TMP calorimeter.

The calorimetry operational during the '88 and '89 runs was however still powerful enough to trigger and measure jet and missing energy topologies.

Calibration of the calorimeter was performed using radioactive sources, cosmics, minimum bias events to monitor the uniformity of the response, laser pulses and test beams of electrons muons and pions.

As a result the systematic error on the jet energy scale was found to be 10% and the missing transverse energy resolution about 15% worse than in the old detector configuration.

The $S p \bar{p} S$ Collider luminosity was increased by almost one order of magnitude giving a peak luminosity of ~2×10^{30} cm^{-2} sec^{-1}; UA1 collected 4.6 pb^{-1} of data to be compared to the 6 pb^{-1} collected before 1988.

3. Heavy quarks production

Top quark in $p \bar{p}$ collisions at $\sqrt{s}$ = 630 GeV can be produced mainly through W decays in t $\bar{b}$ and through direct t t production; while the first channel is known from UA1 measurements of the W production cross section[4], the t $\bar{t}$ production process is evaluated in QCD with an uncertainty of $\pm$ 30%. In the 40 < m_t < 70 GeV the first process is dominant as shown in Fig. 1.

The semileptonic decay channels of the t quark should produce event topologies with one or two leptons and one or more jets. The analysis presented here concerns the μ + 2 jets, 2μ + jet(s) and μ + e + jet(s) topologies where in the last case statistics is limited to ~0.6 pb^{-1} of data taken when the UA1 calorimeter was able to detect the electrons.

The main background to these channels is due to the b$\bar{b}$ (g) and c$\bar{c}$ (g) production where $\overset{(-)}{b}$ and $\overset{(-)}{c}$ decay semileptonically.

Fig. 2 shows the t-quark contribution from t $\bar{t}$ and t $\bar{b}$ channels respect to the contribution coming from known processes in the inclusive μ production p_T spectrum: the non-top μ production exceeds the top one for at least a factor of 10 in the overall spectrum.

The knowledge of b$\bar{b}$ production is then fundamental in order to extract informations on the t-quark contribution from the data. UA1 has performed this study[5] in different data samples:

1.   J/$\Psi \rightarrow \mu^+\mu^-$
2.   high mass dimuons, M ($\mu\mu$) > 6 GeV, $p_T^{\mu\,(1,\,2)}$ > 3 GeV/c;

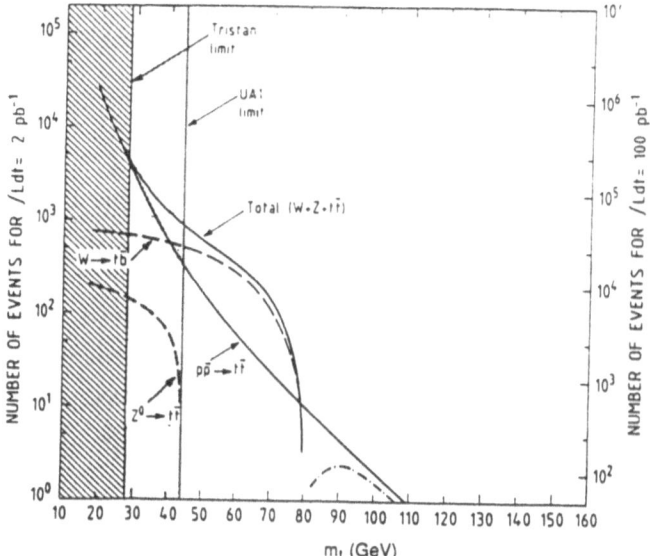

Fig. 1

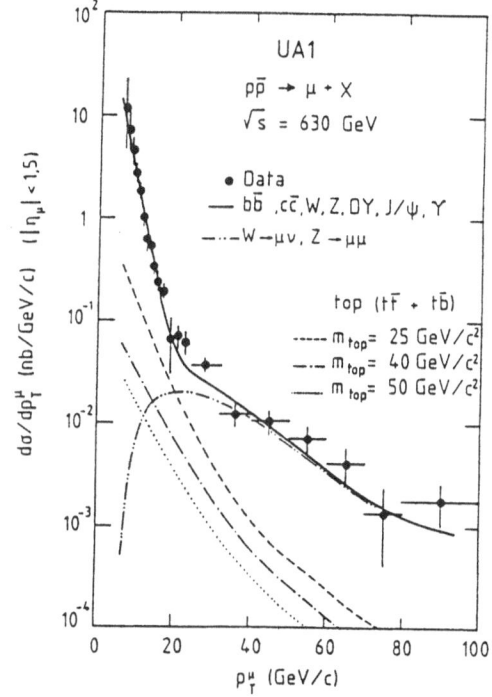

Fig. 2

477

3.   Low mass dimuons M ($\mu\mu$) < 6 GeV, $p_T^{\mu\,(1,\,2)}$ > 3 GeV/c;

4.   Inclusive muon with p_T > 6 GeV/c.

Results are shown in Fig. 3 where they are compared to the QCD predictions of Nason et al. [6] and to the ISAJET Monte Carlo.

The $b\bar{b}$ production cross section extrapolated down to $p_T^{\mu} \geq 0$ GeV/c is $\sigma = 10.2 \pm 3.3$ μb.

The separation of the top-like events from b and c like events in the semileptonic decay channel is based on the isolation variable $I = \left[\left(\Sigma E_T/\,3\right)^2 + \left(\Sigma p_T/\,2\right)^2\right]^{1/2}$ in a cone of semiaperture $\Delta R = (\Delta\eta^2 + \Delta\phi^2)^{1/2} = 0.7$ around the μ; as the lepton-jet angle is expected to be larger in the top case than in the b or c case due to the larger mass, the isolation variable I is expected to be statistically smaller for the t-decays.

4. Muons + jets channels

The $b\bar{b}$ and $c\bar{c}$ production in this analysis are studied selecting a data sample where the b and c decays give the dominant contribution; one muon with 12 < p_T^{μ} < 15 GeV/c in the $|\eta^{\mu}|$ < 1.5 region and one jet with E_T > 12 GeV in the $|\eta^{jet}|$ < 2.5 with the axis separated by ΔR > 1 from the muon direction are required.

All the known source (W, Z, J/Ψ, Υ, D.Y. etc) are included in the Monte Carlo and an equivalent statistics of 50 pb^{-1} has been produced, reconstructed in the simulated UA1 apparatus and selected.

Figs. 4 to 9 show a comparison of the data to the simulated events in the distributions of the most significant kinematical variables; background contribution from K and π decays is shown in the hatched part of the distributions: data are well reproduced by the Monte Carlo samples.

The data sample selected for the top search requires one μ with p_T > 12 GeV, I < 2 GeV and at least two jets with E_T^{j1} > 15 GeV and E_T^{j2} > 7 GeV.

The W contribution is removed requiring an invariant μ-neutrino transverse mass smaller than 60 GeV. These cuts optimize the signal to background ration as can be seen for instance from the I-variable distribution in Fig. 10.

No evidence for a top contribution is found as the data sample can be totally explained by the known sources.

Table 1 resumes the results and Table 2 shows the expected contributions from top-quark of different masses. The data sample amounts to 76 events whilst the expected contribution from standard sources is 77 $\pm$ 10 and from top is 29 $\pm$ 7 if m_t = 50 GeV.

5. Muon pair channel

The aim of this analysis is the isolation of the semileptonic decay of both t and $\bar{b}$ ($\bar{t}$ and b) in the pp $\rightarrow$ W $\rightarrow$ t $\bar{b}$ channel and is based on 5.4 pb^{-1}, the whole statistics collected by UA1 from the beginning of the experiment.

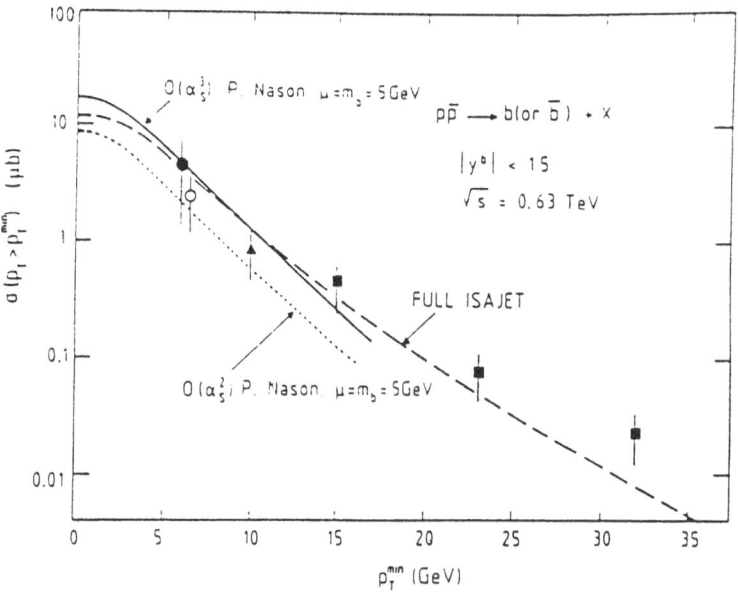

Fig. 3

TABLE 1

	K / π decays	W/Z	Drell-Yan J/Ψ, Υ	bb̄ cc̄	Total MC	Data
μ + ≥ 2 jets	22 ± 5	6 ± 2	6 ± 3	44 ± 8	77 ± 10	76

1988 data and expectations from standard processes.

TABLE 2

m_t (GeV)		40	50	60	70
Isolate μ ± ≥ 2 jets	t b̄	21	20	10	4
	t t̄	18	9	4	1
	total	39 ± 10	29 ± 7	14 ± 4	5 ± 1

Numbers of t-quark events expected for different values of the mass.

479

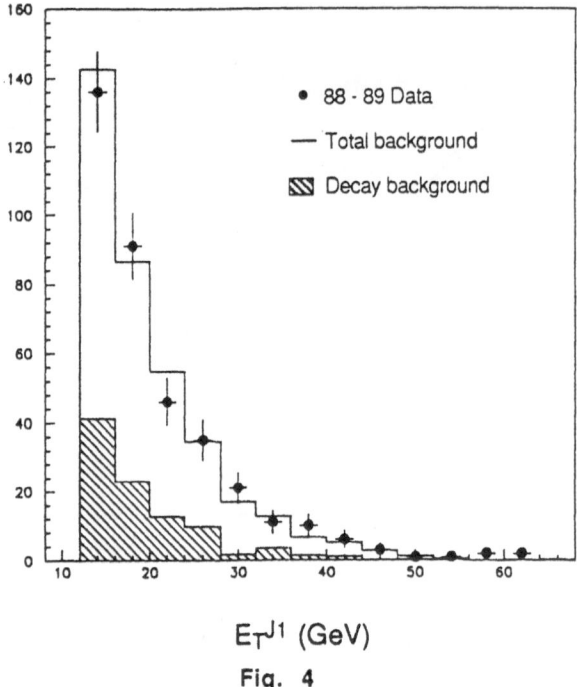

E_T^{J1} (GeV)

Fig. 4

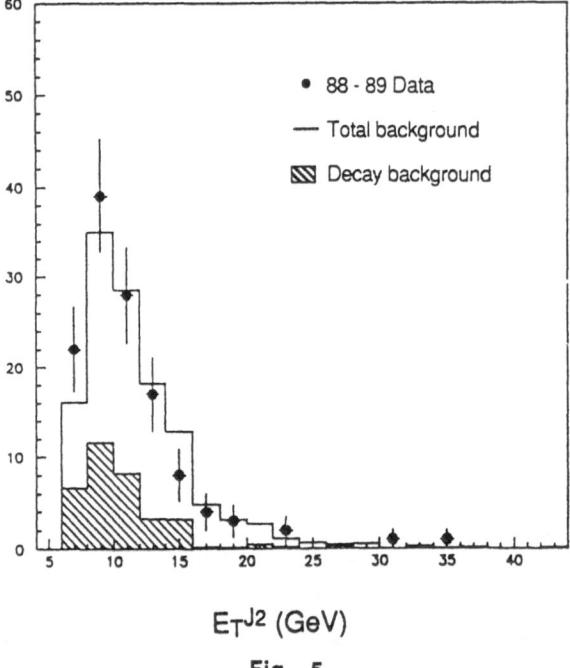

E_T^{J2} (GeV)

Fig. 5

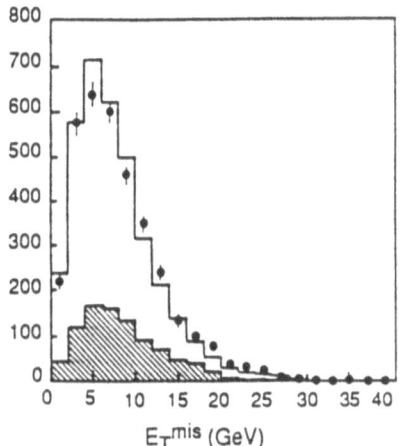

Fig. 6
Missing energy distribution (control sample).

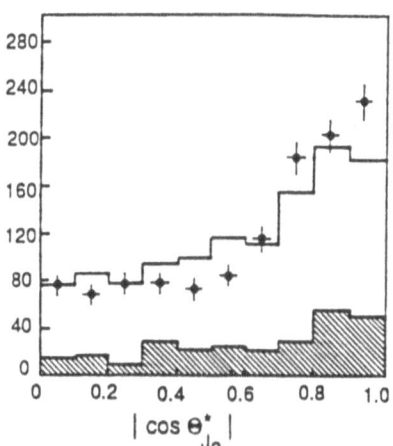

Fig. 7
$|cos\theta^{*}_{j2}|$ *distribution: in the centre-of-mass frame of the $(\mu, jet1, jet2, \nu)$ system θ^{*}_{j2} is the angle between jet2 and the antiproton beam (control sample).*

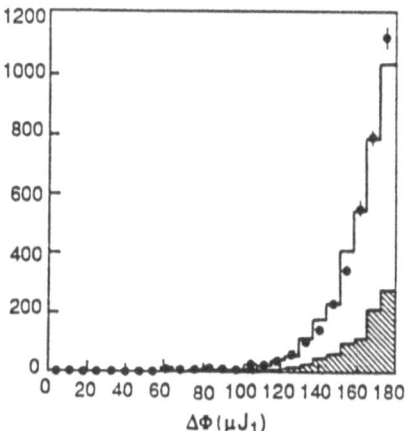

Fig. 8
Distribution of the angle between muon and jet momenta projected in the transverse plane (control sample).

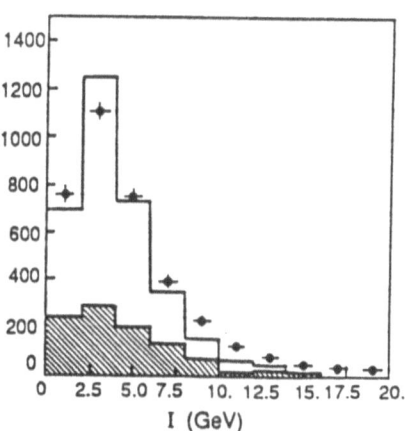

Fig. 9
Muon isolation distribution (control sample).

Two muons are required, the first with $p_T^{\mu 1} > 8$ GeV in the $|\eta_{\mu 1}| < 1.6$ pseudorapidity region and the second with $p_T^{\mu 2} > 3$ GeV; the dimuon invariant mass must be $M_{\mu\mu} > 6$ GeV to suppress Drell-Yan and J/Ψ contribution and $M_{\mu\mu} < 50$ GeV to suppress Z^0 contribution.

The first muon has to be isolated, I (μ_1) < 6 GeV and the second non-isolated, I(μ_2) > 2 GeV. One jet with $E_T > 10$ GeV is required.

These cuts lead to a sample of 102 events where 16 and 93 are the expected contributions from π/K decay and from non-top channels. The top contribution, for instance in the case $m_t = 50$ GeV, is 10 events and a reasonable lower limit to m_t cannot be given at this level. The shape of kinematical distributions in the isolation variable I, in the transverse momentum of the first muon and the azimuthal angle between the two muons, could however help the statistical separation between the $b\bar{b} + c\bar{c}$ events from top events (Fig. 11 a, b, c).

Combining in a likelihood variable the three distributions L = $\prod_i$ P$_{top}$ (x$_i$)/P$_{b+c}$ (x$_i$) [P$_{top}$ and P$_{b+c}$ are the probability density for top and non-top events for the different kinematical variables x$_i$ = I, p$_T^{\mu 1}$, $\Delta\phi$ (μ_1, μ_2)] the ln(L) spectrum shown in Fig. 12 is obtained. Non-top and top ($m_t = 50$ GeV), expectations are compared to the data sample.

No top contribution is found and in the ln(L) > 1, ten events are found while 1.9 ± 0.2 ± 0.4 events, 8.9 ± 1.9 ± 2.3 and 1.0 ± 0.4 ± 0.4 are expected from π/K decays, $b\bar{b} + c\bar{c}$ and J/ψ + Drell-Yan respectively. As the expected number of events from top are 0.9 ± 0.6 ± 1.6, 71 ± 0.5 ± 1.3 and 3.2 ± 0.2 ± 0.6 for $m_t = 40$, 50 and 60 GeV respectively, the lower limit that can be obtained is $m_t > 46$ GeV (95% c.l.).

6. Muon-electron pair channel

This analysis is based on the statistics collected up to 1985, 0.55 pb^{-1}, when the electromagnetic calorimeter was operational. The physical consideration are analogue to the two muon analysis. The data sample, obtained requiring one muon with p$_T^\mu$ > 3 GeV and one electron with $E_T > 8$ GeV in the | η | < 1.5 pseudorapidity region, consists of 10 events.

The expected number of non-top events are: 12 from $b\bar{b}$ and $c\bar{c}$, 0.13 from $Z^0 \rightarrow \tau^+ \tau^-$ and 0.06 from Drell-Yan production of $\tau^+ \tau^-$, while 5.7, 3.8 and 1.9 are expected from top of $m_t = 30$, 40, 50 GeV respectively.

As the neutrino transverse energy is expected to be larger in the top channel then in the b or c ones, a cut on this variable $E_{T\gamma}$ should enhance the top over non-top contribution. A cut $E_{T\gamma} > 10$ GeV leaves no events while 1.6 events from $b\bar{b} + c\bar{c}$ and 1.7 from an $m_t = 40$ GeV top are expected.

The limit obtained on m_t is $m_t > 25$ GeV at 95% c.l.

7. Combined limits

These preliminary results and those obtained in the muon plus jets and electron plus jets, already published by the UA1 collaboration[2], are resumed in table 4. Most of the systematic error are correlated in the different sample (luminosity, jet energy scale,

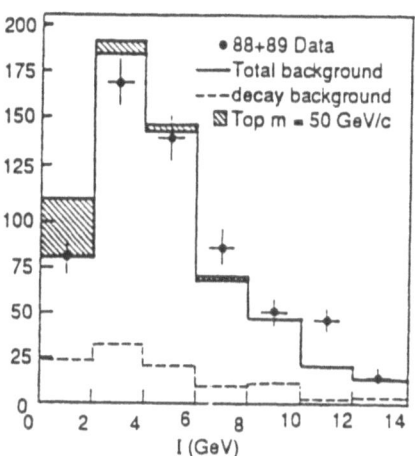

Fig. 10
Muon isolation distribution for the top-sample (see text).

TABLE 3

Channel	Data	Background ± syst. error	Signal m_t = 50 GeV	m_t = 60 GeV	m_t 95% limit (GeV)	∫ L dt (pb⁻¹)
e+jets (≤1985)	26	26.0 ± 2.8	8.5	----	41	0.69
μ+jets (≤1985)	10	11.4 ± 2.8	4.7	----	40	0.55
e+μ (≤1985)	0	1.6 ± 0.1	1.2	----	25	0.55
μ+jets (88-89)	76	77.0 ± 11.0	29.0	13.8	53	4.60
μμ (1982-89)	10	11.8 ± 2.3	7.1	3.2	46	5.40

Summary of present UA1 limits on t-quark mass (95% c.l.)

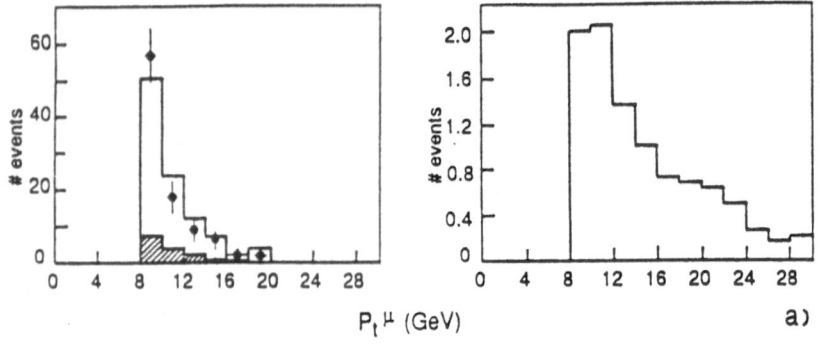

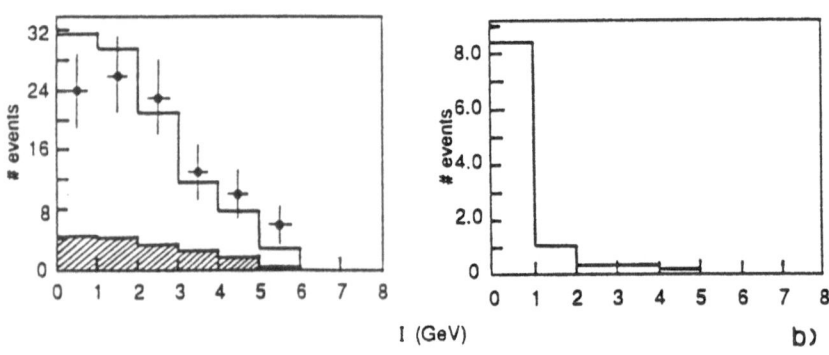

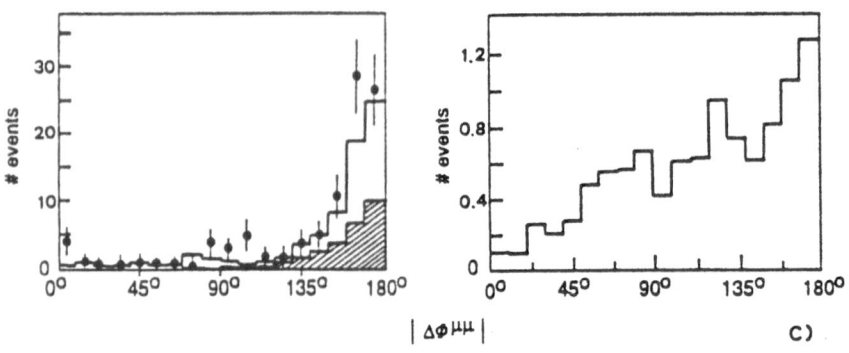

Fig. 11

Left: dimuon data and $b\bar{b} + c\bar{c}$ expectations (continuous histogram) and decay background (hatched)
Right: expectations for top with $m_t = 50$ GeV.

production cross section); the assumption that this correlation is complete seems conservative. Combining in quadrature statistical and systematic errors for the different results, the lower limit obtained on the top mass is $m_t > 61$ GeV at 95% c.l.

A lower limit $m_t > 58$ GeV at 95% c.l. can be obtained if the $W \rightarrow t\bar{b}$ channel only is considered; the systematics in the cross section evaluation is small as the W production cross section has been measured by UA1.

8. Conclusions

The search for top-quark by UA1 collaboration has been performed with the whole statistics collected up to now. No top signal has been found and the best lower limit on the top mass, $m_t > 61$ GeV (95% c.l.) has been obtained combining different results in the lepton(s) + jet(s) samples.

Reference

[1]   Proceeding of the IX Int. Conf. on "Physics in Collision", 1989 Jerusalem.
[2]   C. Albajar et al. (UA1 Collab.), Z. Phys. C 37 (1988) 505.
[3]   U. Amaldi et al., Phys. Rev. D36 (1987) 1385.
[4]   C. Albajar et al. (UA1 Collab.), Z. Phys. C44 (1989) 15.
[5]   C. Albajar et al. (UA1 Collab.), Z. Phys. C37 (1988) 489;
       C. Albajar et al. (UA1 Collab.), Phys. Lett. 213B (1988) 405.
[6]   P. Nason, S. Dawson and R.K. Ellis, Nucl. Phys. B303 (1988) 607;
       G. Altarelli, M. Diemoz, G. Martinelli and P. Nason, Nucl. Phys. B308 (1988) 724.

H. Albrecht
(ARGUS Collaboration)
DESY
Notkestr. 85
D-2000 Hamburg 52, FRG

Ahmed Ali
DESY
Notkestr. 85
D-2000 Hamburg 52, FRG

Farhad Ardalan
Physics Department
Sharif University
Teheran, Iran

Wulfrin Bartel
DESY
Notkestr. 85
D-2000 Hamburg 52, FRG

Jochen Bartels
II Institut f. Theor. Physik
Universität Hamburg
Notkestr. 85
D-2000 Hamburg 52, FRG

Ulrich Baur
TH Division
CERN
CH-1211 Geneva 23, Switzerland

M.A.B. Bég
Physics Department
The Rockefeller University
1230 York Avenue
New York, NY 10021, USA

Stefano Bertolini
DESY
Notkestr. 85
D-2000 Hamburg 52, FRG

Pierre Binétruy
LAPP
IN2P3
B.O. 909
D-74019 Annecy-le-Vieux Cedex
France

Wilfried Buchmüller
Inst. f. Theor. Physik der
Universität Hannover
D-3000 Hannover, FRG

Patricia Burchat
Physics Department
UC Santa Cruz
Santa Cruz, Calif., USA

Giovanni Busetto
(UA1 Collaboration)
Univ. of Padova
Padova, Italy

Michael Chanowitz
Theoretical Physics
Lawrence Berkeley Laboratory
Berkeley, CA 94720, USA

Giacomo D'Ali
Physics Department
Univ. of Palermo
Palermo, Italy

Yu. Dokshitzer
Leningrad Nucl. Res. Inst.
High Energy Theory Dept.
118350 Gatchina (Leningrad), USSR

J.F. Gunion
Department of Physics
University of California
Davis, CA 95616, USA

Juliet Lee-Franzini
Department of Physics
SUNY at Stony Brooke
Stony Brooke, NY 11794, USA

Paolo Franzini
Department of Physics
Pupin Laboratory
Columbia University
New York, NY 10027, USA

Steven Geer
CDF Collaboration
Fermilab
Batavia, Illinois, USA

Graciela Gelmini
ICTP
Strada Costiera 11
Trieste, Italy

Fabiola Gianotti
INFN Milano
Milano, Italy

Michel Gourdin
Université Pierre et Marie Curie
4, Place Jussieu
F-75230 Paris, France

Christoph Greub
Inst. f. Theor. Phys. der
Universtität Bern
Silderstr. 5
CH-3012 Bern, Switzerland

Howard Haber
High Energy Physics
Div. of Natural Sciences
UC Santa Cruz
Santa Cruz, CA 95060, USA

Peter W. Higgs
Physics Department
University of Edinburgh
Kings Building
Edinburg EH9 3JZ, United Kingdom

Pervez Hoodbhoy
Physics Department
Quaid-e-Azam University
Islamabad, Pakistan

J. Jersak
HLRZ (KFA)
Postfach 1913
D-5170 Jülich 1, FRG

P. Kienle
GSI Darmstadt
D-6100 Darmstadt, FRG

V. Khoze
Leningrad Nucl. Res. Inst.
High Energy Theory Dept.
118350 Gatchina (Leningrad), USSR

Karl Koller
Sektion Physik der
Universität München
Theresienstr. 37
D-8000 München 2, FRG

T. Matsuura
Inst. Lorenz
Univ. of Leiden
Nieuwsteeg 18
2311 SR-Leiden, Netherlands

Ruibin Meng
II Inst. f. Theor. Physik der
Universität Hamburg
Notkestr. 85
D-2000 Hamburg 52, FRG

Nello Paver
Inst. di Fisica Teorica
Universita di Trieste
Strada Costiera 11
I-34014 Trieste, Italy

Asghar Qadir
Department of Mathematics
Quaid-e-Azam University
Islamabad, Pakistan

Riazuddin
Physics Department
King Fahd University of Petroleum
& Minerals
Dhahran, Saudi Arabia

Qaiser Shafi
Bartol Research Foundation
Delaware, Conn., USA

Herwig Schopper
World Laboratory
6, Place de la Riponne
CH-1005 Lausanne, Switzerland

A.V. Sarantsev
Leningrad Nucl. Res. Inst.
High Energy Theory Dept.
118350 Gatchina (Leningrad), USSR

Abraham Seiden
High Energy Physics
Div. of Natural Sciences
UC Santa Cruz
Santa Cruz, CA 95060, USA

A.B. Sokornov
Leningrad Nucl. Res. Inst.
High Energy Theory Dept.
118350 Gatchina (Leningrad), USSR

S. Troyan
Leningrad Nucl. Res. Inst.
High Energy Theory Dept.
118350 Gatchina (Leningrad), USSR

Roberto Vega
Department of Physics
University of California
Davis, CA 95616, USA

Lucia Votano
INFN Laboratory (INFN-LNF)
Via E. Fermi 40
C.P. 13
I-00044 Frascati, Italy

Steven Weinberg
Univ. of Texas at Austin
Austin, USA

Christoph Wetterich
DESY
Notkestr. 85
D-2000 Hamburg 52, FRG

Hoi-Lai Yu
Institute of Physics
Academia Sinica
Taipei, Tawain, Republic of China

Michael E. Zeller
Department of Physics
Yale University
P.O. Box 6666
New Haven, Conn. 06511, USA

Antonino Zichichi
CERN, Geneva & CCSEM, Erice
CH-1211 Geneva 23, Switzerland

493